THE LOEB CLASSICAL LIBRARY

FOUNDED BY JAMES LOEB, LL.D.

THE GEOGRAPHY OF STRABO

V

THE GEOGRAPHY OF STRABO

WITH AN ENGLISH TRANSLATION BY

HORACE LEONARD JONES, PH.D., LL.D.
CORNELL UNIVERSITY

IN EIGHT VOLUMES

V

CAMBRIDGE, MASSACHUSETTS
HARVARD UNIVERSITY PRESS
LONDON
WILLIAM HEINEMANN LTD
MCMLXI

First printed 1928
Reprinted 1944, 1954, 1961

Printed in Great Britain

CONTENTS

THE

GEOGRAPHY OF STRABO

BOOK X

ΣΤΡΑΒΩΝΟΣ ΓΕΩΓΡΑΦΙΚΩΝ

Ι′

I

1.[1] Ἐπειδὴ ἡ Εὔβοια παρὰ πᾶσαν τὴν παραλίαν ταύτην παραβέβληται τὴν ἀπὸ Σουνίου μέχρι Θετταλίας, πλὴν τῶν ἄκρων ἑκατέρωθεν, οἰκεῖον ἂν εἴη συνάψαι τοῖς εἰρημένοις τὰ περὶ τὴν νῆσον, εἶθ' οὕτω μεταβῆναι πρός τε τὰ Αἰτωλικὰ καὶ τὰ Ἀκαρνανικά, ἅπερ λοιπά ἐστι τῶν τῆς Εὐρώπης μερῶν.

2. Παραμήκης μὲν τοίνυν ἐστὶν ἡ νῆσος ἐπὶ χιλίους σχεδόν τι καὶ διακοσίους σταδίους ἀπὸ Κηναίου πρὸς Γεραιστόν, τὸ δὲ πλάτος ἀνώμαλος κατὰ δὲ τὸ πλέον ὅσον πεντήκοντα καὶ ἑκατὸν σταδίων. τὸ μὲν οὖν Κήναιόν ἐστι κατὰ Θερμοπύλας καὶ τὰ ἔξω Θερμοπυλῶν ἐπ' ὀλίγον, Γεραιστὸς δὲ καὶ Πεταλία πρὸς Σουνίῳ. γίνεται οὖν ἀντίπορθμος τῇ τε Ἀττικῇ καὶ Βοιωτίᾳ καὶ Λοκρίδι καὶ τοῖς Μαλιεῦσι. διὰ δὲ τὴν στενότητα καὶ τὸ λεχθὲν μῆκος ὑπὸ τῶν παλαιῶν
C 445 Μάκρις ὠνομάσθη. συνάπτει δὲ τῇ ἠπείρῳ κατὰ Χαλκίδα μάλιστα, κυρτὴ προπίπτουσα πρὸς τοὺς κατὰ τὴν Αὐλίδα τόπους τῆς Βοιωτίας καὶ

[1] The Paris MS. No. 1397 (A) ends with Book ix (see Vol. I., p. xxxii).

THE GEOGRAPHY OF STRABO

BOOK X

I

1. Since Euboea lies parallel to the whole of the coast from Sunium to Thessaly, with the exception of the ends on either side,[1] it would be appropriate to connect my description of the island with that of the parts already described before passing on to Aetolia and Acarnania, which are the remaining parts of Europe to be described.

2. In its length, then, the island extends parallel to the coast for a distance of about one thousand two hundred stadia from Cenaeum to Geraestus, but its breadth is irregular and generally only about one hundred and fifty stadia. Now Cenaeum lies opposite to Thermopylae and, to a slight extent, to the region outside Thermopylae, whereas Geraestus and Petalia lie towards Sunium. Accordingly, the island lies across the strait and opposite Attica, Boeotia, Locris, and the Malians. Because of its narrowness and of the above-mentioned length, it was named Macris[2] by the ancients. It approaches closest to the mainland at Chalcis, where it juts out in a convex curve towards the region of Aulis in Boeotia and forms the

[1] *i.e.* the promontories of Thermopylae and Sunium, which lie beyond the corresponding extremities of Euboea—Cenaeum and Geraestus.

[2] *i.e.* "Long" Island (see Map VIII, end of Vol. IV).

ποιοῦσα τὸν Εὔριπον, περὶ οὗ διὰ πλειόνων εἰρήκαμεν, σχεδὸν δέ τι καὶ περὶ τῶν ἀντιπόρθμων ἀλλήλοις τόπων κατά τε τὴν ἤπειρον καὶ κατὰ τὴν νῆσον ἐφ' ἑκάτερα τοῦ Εὐρίπου, τά τε ἐντὸς καὶ τὰ ἐκτός. εἰ δέ τι ἐλλέλειπται, νῦν προσδιασαφήσομεν. καὶ πρῶτον, ὅτι τῆς Εὐβοίας τὰ Κοῖλα λέγουσι τὰ μεταξὺ Αὐλίδος[1] καὶ τῶν περὶ Γεραιστὸν τόπων· κολποῦται[2] γὰρ ἡ παραλία, πλησιάζουσα δὲ τῇ Χαλκίδι κυρτοῦται πάλιν πρὸς τὴν ἤπειρον.

3. Οὐ μόνον δὲ Μάκρις ἐκλήθη ἡ νῆσος, ἀλλὰ καὶ Ἀβαντίς. Εὔβοιαν γοῦν εἰπὼν ὁ ποιητὴς τοὺς ἀπ' αὐτῆς Εὐβοέας οὐδέποτε εἴρηκεν, ἀλλ' Ἄβαντας ἀεί·

> οἳ δ' Εὔβοιαν ἔχον μένεα πνείοντες Ἄβαντες.
> τῷ δ' ἅμ' Ἄβαντες ἕποντο.

φησὶ δ' Ἀριστοτέλης ἐξ Ἄβας τῆς Φωκικῆς Θρᾷκας ὁρμηθέντας ἐποικῆσαι τὴν νῆσον καὶ ἐπονομάσαι Ἄβαντας τοὺς ἔχοντας αὐτήν· οἱ δ' ἀπὸ ἥρωός φασι, καθάπερ καὶ Εὔβοιαν ἀπὸ ἡρωίνης. τάχα δ' ὥσπερ Βοὸς αὐλὴ λέγεταί τι ἄντρον ἐν τῇ πρὸς Αἰγαῖον τετραμμένῃ παραλίᾳ, ὅπου τὴν Ἰώ φασι τεκεῖν Ἔπαφον, καὶ ἡ νῆσος

[1] Αὐλίδος, Du Theil, Corais, and Groskurd would emend to Χαλκίδος.

[2] For κολποῦται, Jones conjectures κοιλοῦται, to correspond with Κοῖλα.

---

[1] 9. 2. 2, 8.

[2] "Inside" means the lower or south-eastern region, "outside" the upper or north-western.

[3] Elephenor.

Euripus. Concerning the Euripus I have already spoken rather at length,[1] as also to a certain extent concerning the places which lie opposite one another across the strait, both on the mainland and on the island, on either side of the Euripus, that is, the regions both inside and outside[2] the Euripus. But if anything has been left out, I shall now explain more fully. And first, let me explain that the parts between Aulis and the region of Geraestus are called the Hollows of Euboea; for the coast bends inwards, but when it approaches Chalcis it forms a convex curve again towards the mainland.

3. The island was called, not only Macris, but also Abantis; at any rate, the poet, although he names Euboea, never names its inhabitants "Euboeans," but always "Abantes": "And those who held Euboea, the courage-breathing Abantes . . . . And with him[3] followed the Abantes."[4] Aristotle[5] says that Thracians, setting out from the Phocian Aba, recolonised the island and renamed those who held it "Abantes." Others derive the name from a hero,[6] just as they derive "Euboea" from a heroine.[7] But it may be, just as a certain cave on the coast which fronts the Aegaean, where Io is said to have given birth to Epaphus, is called Böos Aulê,[8] that the

[4] *Iliad* 2. 536, 542.

[5] Aristotle of Chalcis wrote a work on Euboea, but it is no longer extant. He seems to have flourished in the fourth century B.C.

[6] Abas, founder of Aba, who later conquered Euboea and reigned over it (Stephanus Byzantinus, *s.vv.* Ἄβαι and Ἀβαντίς).

[7] On the heroine "Euboea," see Pauly-Wissowa, *s.v.* "Euboea" (4).

[8] Cow's Stall.

ἀπὸ τῆς αὐτῆς αἰτίας ἔσχε τοῦτο τοὔνομα. καὶ Ὄχη δὲ ἐκαλεῖτο ἡ νῆσος καὶ ἔστιν ὁμώνυμον αὐτῇ τὸ μέγιστον τῶν ἐνταῦθα ὀρῶν. καὶ Ἐλλοπία δ' ὠνομάσθη ἀπὸ Ἔλλοπος τοῦ Ἴωνος· οἱ δὲ Αἴκλου[1] καὶ Κόθου ἀδελφόν φασιν, ὃς καὶ τὴν Ἐλλοπίαν κτίσαι λέγεται, χωρίον ἐν τῇ Ὠρίᾳ καλουμένῃ τῆς Ἱστιαιώτιδος πρὸς τῷ Τελεθρίῳ ὄρει, καὶ τὴν Ἱστίαιαν προσκτήσασθαι καὶ τὴν Περιάδα[2] καὶ Κήρινθον καὶ Αἰδηψὸν[3] καὶ Ὀροβίας, ἐν ᾧ μαντεῖον ἦν ἀψευδέστατον· ἦν δὲ μαντεῖον καὶ τοῦ Σελινουντίου Ἀπόλλωνος· μετῴκησαν δ' εἰς[4] τὴν Ἱστίαιαν οἱ Ἐλλοπιεῖς,[5] καὶ ηὔξησαν τὴν πόλιν Φιλιστίδου τοῦ τυράννου βιασαμένου μετὰ τὰ Λευκτρικά. Δημοσθένης δ' ὑπὸ Φιλίππου κατασταθῆναι τύραννόν φησι καὶ τῶν Ὠρειτῶν τὸν Φιλιστίδην· οὕτω γὰρ ὠνομάσθησαν ὕστερον οἱ Ἱστιαιεῖς, καὶ ἡ πόλις ἀντὶ Ἱστιαίας Ὠρεός· ἔνιοι δ' ὑπ' Ἀθηναίων ἀποικισθῆναί φασι τὴν Ἱστίαιαν ἀπὸ τοῦ δήμου τοῦ Ἱστιαιέων, ὡς καὶ ἀπὸ τοῦ Ἐρετριέων τὴν Ἐρέτριαν. Θεόπομπος δέ φησι, Περικλέους χειρουμένου Εὔβοιαν, τοὺς Ἱστιαιεῖς καθ' ὁμολογίας εἰς Μακεδονίαν μεταστῆναι, δισχιλίους δ' ἐξ Ἀθηναίων ἐλθόντας τὸν Ὠρεὸν οἰκῆσαι, δῆμον ὄντα πρότερον τῶν Ἱστιαιέων.

4. Κεῖται δ' ὑπὸ τῷ Τελεθρίῳ ὄρει ἐν τῷ C 446 Δρυμῷ καλουμένῳ παρὰ τὸν Κάλλαντα ποταμὸν

[1] Αἴκλου BDE*ghlnopu*, Ἀέκλου *y*, Ἀβίκλου *k*.
[2] Meineke emends Περιάδα (otherwise unknown) to πεδίαδα.
[3] Αἰδηψόν, Xylander, for Ἐδηψόν; so the later editors.
[4] δ' εἰς, Corais, for δέ; so the later editors.
[5] Ἐλλοπιεῖς, Tzschucke, for Ἐλλοπεῖς; so the later editors.

island got the name Euboea[1] from the same cause. The island was also called Ochê; and the largest of its mountains bears the same name. And it was also named Ellopia, after Ellops the son of Ion. Some say that he was the brother of Aïclus and Cothus; and he is also said to have founded Ellopia, a place in Oria, as it is called, in Histiaeotis[2] near the mountain Telethrius, and to have added to his dominions Histiaea, Perias, Cerinthus, Aedepsus, and Orobia; in this last place was an oracle most averse to falsehood (it was an oracle of Apollo Selinuntius). The Ellopians migrated to Histiaea and enlarged the city, being forced to do so by Philistides the tyrant, after the battle of Leuctra. Demosthenes says that Philistides was set up by Philip as tyrant of the Oreitae too;[3] for thus in later times the Histiaeans were named, and the city was named Oreus instead of Histiaea. But according to some writers, Histiaea was colonised by Athenians from the deme of the Histiaeans, as Eretria was colonised from that of the Eretrians. Theopompus says that when Pericles overpowered Euboea the Histiaeans by agreement migrated to Macedonia, and that two thousand Athenians who formerly composed the deme of the Histiaeans came and took up their abode in Oreus.

4. Oreus is situated at the foot of the mountain Telethrius in the Drymus,[4] as it is called, on the River Callas, upon a high rock; and hence, perhaps,

[1] *i.e.* from the Greek words "eu" (well) and "bous" cow).

[2] Or Hestiaeotis (see 9. 5. 3 and foot-note 2).

[3] *Third Philippic* 32 (119 Reiske).

[4] "Woodland."

ἐπὶ πέτρας ὑψηλῆς, ὥστε τάχα καὶ διὰ τὸ τοὺς Ἐλλοπιεῖς ὀρείους εἶναι τοὺς προοικήσαντας ἐτέθη τοὔνομα τοῦτο τῇ πόλει· δοκεῖ δὲ καὶ ὁ Ὠρίων ἐνταῦθα τραφεὶς οὕτως ὠνομασθῆναι· ἔνιοι δὲ τοὺς Ὠρείτας, πόλιν ἔχοντας ἰδίαν, φασὶ πολεμουμένους ὑπὸ τῶν Ἐλλοπιέων μεταβῆναι καὶ συνοικῆσαι τοῖς Ἱστιαιεῦσι, μίαν δὲ γενηθεῖσαν πόλιν ἀμφοτέροις χρήσασθαι τοῖς ὀνόμασι, καθάπερ Λακεδαίμων τε καὶ Σπάρτη ἡ αὐτή. εἴρηται δ' ὅτι καὶ ἐν Θετταλίᾳ Ἱστιαιῶτις ἀπὸ τῶν ἀνασπασθέντων ἐνθένδε ὑπὸ Περραιβῶν ὠνόμασται.

5. Ἐπεὶ δ' ἡ Ἐλλοπία τὴν ἀρχὴν ἀπὸ τῆς Ἱστιαίας καὶ τοῦ Ὠρεοῦ προσηγάγετο ἡμᾶς ποιήσασθαι, τὰ συνεχῆ λέγωμεν[1] τοῖς τόποις τούτοις. ἔστι δ' ἐν τῷ Ὠρεῷ τούτῳ τό τε Κήναιον[2] πλησίον,[3] καὶ ἐπ' αὐτῷ τὸ Δῖον καὶ Ἀθῆναι αἱ Διάδες, κτίσμα Ἀθηναίων, ὑπερκείμενον τοῦ ἐπὶ Κῦνον[4] πορθμοῦ· ἐκ δὲ τοῦ[5] Δίου Κάναι τῆς Αἰολίδος ἀπῳκίσθησαν.[6] ταῦτά τε δὴ τὰ χωρία περὶ τὴν Ἱστίαιάν ἐστι καὶ ἔτι Κήρινθος πολείδιον ἐπὶ τῇ θαλάττῃ· ἐγγὺς δὲ Βούδορος ποταμὸς ὁμώνυμος τῷ κατὰ τὴν Σαλαμῖνα ὄρει τῷ πρὸς τῇ Ἀττικῇ.

6. Κάρυστος δέ ἐστιν ὑπὸ τῷ ὄρει τῇ Ὄχῃ.[7] πλησίον δὲ τὰ Στύρα καὶ τὸ Μαρμάριον, ἐν ᾧ τὸ λατόμιον τῶν Καρυστίων κιόνων, ἱερὸν ἔχον

[1] λέγωμεν, Corais, for λέγομεν; so the later editors.
[2] Κήναιον, Hopper, for Κλειναῖον and Κλιναῖον; so the later editors.
[3] πλησίον, E omits; so Kramer and Müller-Dübner.
[4] Κῦνον, Tzschucke, for Καῦνον; so the later editors.

it was because the Ellopians who formerly inhabited it were mountaineers that the name Oreus[1] was assigned to the city. It is also thought that Orion was so named because he was reared there. Some writers say that the Oreitae had a city of their own, but because the Ellopians were making war on them they migrated and took up their abode with the Histiaeans; and that, although they became one city, they used both names, just as the same city is called both Lacedaemon and Sparta. As I have already said,[2] Histiaeotis in Thessaly was also named after the Histiaeans who were carried off from here into the mainland by the Perrhaebians.

5. Since Ellopia induced me to begin my description with Histiaea and Oreus, let me speak of the parts which border on these places. In the territory of this Oreus lies, not only Cenaeum, near Oreus, but also, near Cenaeum, Dium[3] and Athenae Diades, the latter founded by the Athenians and lying above that part of the strait where passage is taken across to Cynus; and Canae in Aeolis was colonised from Dium. Now these places are in the neighbourhood of Histiaea; and so is Cerinthus, a small city by the sea; and near it is the Budorus River, which bears the same name as the mountain in Salamis which is close to Attica.

6. Carystus is at the foot of the mountain Ochê; and near it are Styra and Marmarium, in which latter are the quarry of the Carystian columns[4] and a

[1] *i.e.* from "oreius" (mountaineer). [2] 9. 5. 17.
[3] Mentioned in *Iliad* 2. 538. [4] See 9. 5. 16.

---

[5] τῆς B(τοῦ in *sec. man.* above τῆς)CD*ghiuv*.
[6] ἀπῳκίσθησαν D, ἐπῳκίσθησαν other MSS. [7] ὄχθῃ C*glnoy*.

Ἀπόλλωνος Μαρμαρίνου, ὅθεν διάπλους εἰς Ἁλὰς τὰς Ἀραφηνίδας·[1] ἐν δὲ τῇ Καρύστῳ καὶ ἡ λίθος φύεται ἡ ξαινομένη[2] καὶ ὑφαινομένη, ὥστε τὰ ὕφη[3] χειρόμακτρα γίνεσθαι, ῥυπωθέντα δ' εἰς φλόγα βάλλεσθαι καὶ ἀποκαθαίρεσθαι τῇ πλύσει τῶν λίνων[4] παραπλησίως· ᾠκίσθαι δὲ τὰ χωρία ταῦτά φασιν ὑπὸ τῶν ἐκ Τετραπόλεως τῆς περὶ Μαραθῶνα καὶ Στειριέων·[5] κατεστράφη δὲ τὰ Στύρα ἐν τῷ Μαλιακῷ[6] πολέμῳ ὑπὸ Φαίδρου τοῦ Ἀθηναίων στρατηγοῦ· τὴν δὲ χώραν ἔχουσιν Ἐρετριεῖς. Κάρυστος δέ ἐστι καὶ ἐν τῇ Λακωνικῇ τόπος τῆς Αἴγυος πρὸς Ἀρκαδίαν, ἀφ' οὗ Καρύστιον οἶνον Ἀλκμὰν εἴρηκε.

7. Γεραιστὸς δ' ἐν μὲν τῷ Καταλόγῳ τῶν νεῶν οὐκ εἴρηται, μέμνηται δ' ὁ ποιητὴς ὅμως αὐτοῦ·

> ἐς δὲ Γεραιστὸν
> ἐννύχιοι κατάγοντο·

καὶ δηλοῖ, διότι τοῖς διαίρουσιν ἐκ τῆς Ἀσίας εἰς τὴν Ἀττικὴν ἐπικαιρίως κεῖται τῷ Σουνίῳ πλησιάζον τὸ χωρίον· ἔχει δ' ἱερὸν Ποσειδῶνος ἐπισημότατον τῶν ταύτῃ καὶ κατοικίαν ἀξιόλογον.

8. Μετὰ δὲ τὸν Γεραιστὸν Ἐρέτρια, πόλις μεγίστη τῆς Εὐβοίας μετὰ Χαλκίδα, ἔπειθ' ἡ Χαλκὶς μητρόπολις τῆς νήσου τρόπον τινά, ἐπ' αὐτῷ τῷ Εὐρίπῳ ἱδρυμένη· ἀμφότεραι δὲ πρὸ

[1] Ἀραφηνίδας, Xylander, following D *pr. man.*, for Ἀραφηνίας; so the later editors.
[2] On an interpolation after ξαινομένη in the Ald. Ed., see Müller's *Ind. Var. Lect.* p. 1007.
[3] ὑφάσματα *kno* Ald.

temple of Apollo Marmarinus; and from here there is a passage across the strait to Halae Araphenides. In Carystus is produced also the stone which is combed and woven,[1] so that the woven material is made into towels, and, when these are soiled, they are thrown into fire and cleansed, just as linens are cleansed by washing. These places are said to have been settled by colonists from the Marathonian Tetrapolis[2] and by Steirians. Styra was destroyed in the Malian war by Phaedrus, the general of the Athenians; but the country is held by the Eretrians. There is also a Carystus in the Laconian country, a place belonging to Aegys, towards Arcadia; whence the Carystian wine of which Alcman speaks.

7. Geraestus is not named in the *Catalogue of Ships*, but still the poet mentions it elsewhere: "and at night they landed at Geraestus."[3] And he plainly indicates that the place is conveniently situated for those who are sailing across from Asia to Attica, since it comes near to Sunium. It has a temple of Poseidon, the most notable of those in that part of the world, and also a noteworthy settlement.

8. After Geraestus one comes to Eretria, the greatest city in Euboea except Chalcis; and then to Chalcis, which in a way is the metropolis of the island, being situated on the Euripus itself. Both

[1] *i.e.* asbestos. [2] See 8. 7. 1.
[3] *Od.* 3. 177.

---

[4] τῶν λίνων *Epit.*, for τὸν πίνον (filth); and so the editors in general.

[5] Στειριέων, Palmer, for Στυρίεων *Dhi*, Στυριαίων *BCklnox*; so the later editors.

[6] Μαλιακῷ, Meineke, following conj. of Casaubon, emends to Λαμιακῷ. Perhaps rightly, but evidence is lacking.

C 447 τῶν Τρωικῶν ὑπ' Ἀθηναίων ἐκτίσθαι λέγονται, καὶ μετὰ τὰ Τρωικὰ Ἄϊκλος καὶ Κόθος, ἐξ Ἀθηνῶν ὁρμηθέντες, ὁ μὲν τὴν Ἐρέτριαν ᾤκισε, Κόθος δὲ τὴν Χαλκίδα· καὶ τῶν Αἰολέων δέ τινες ἀπὸ τῆς Πενθίλου στρατιᾶς κατέμειναν ἐν τῇ νήσῳ, τὸ δὲ παλαιὸν καὶ Ἄραβες οἱ Κάδμῳ συνδιαβάντες. αἱ δ' οὖν πόλεις αὗται διαφερόντως αὐξηθεῖσαι καὶ ἀποικίας ἔστειλαν ἀξιολόγους εἰς Μακεδονίαν· Ἐρέτρια μὲν γὰρ συνῴκισε τὰς περὶ Παλλήνην καὶ τὸν Ἄθω πόλεις, ἡ δὲ Χαλκὶς τὰς ὑπὸ Ὀλύνθῳ, ἃς Φίλιππος διελυμήνατο. καὶ τῆς Ἰταλίας δὲ καὶ Σικελίας πολλὰ χωρία Χαλκιδέων ἐστίν· ἐστάλησαν δε αἱ ἀποικίαι αὗται, καθάπερ εἴρηκεν Ἀριστοτέλης, ἡνίκα ἡ τῶν Ἱπποβοτῶν καλουμένη ἐπεκράτει πολιτεία· προέστησαν γὰρ αὐτῆς ἀπὸ τιμημάτων ἄνδρες ἀριστοκρατικῶς ἄρχοντες. κατὰ δὲ τὴν Ἀλεξάνδρου διάβασιν καὶ τὸν περίβολον τῆς πόλεως ηὔξησαν, ἐντὸς τείχους λαβόντες τόν τε Κάνηθον και τὸν Εὔριπον, ἐπιστήσαντες τῇ γεφύρᾳ πύργους καὶ πύλας καὶ τεῖχος.

9. Ὑπέρκειται δὲ τῆς τῶν Χαλκιδέων πόλεως τὸ Λήλαντον καλούμενον πεδίον. ἐν δὲ τούτῳ θερμῶν τε ὑδάτων εἰσὶν ἐκβολαὶ πρὸς θεραπείαν νόσων εὐφυεῖς, οἷς ἐχρήσατο καὶ Σύλλας Κορνήλιος, ὁ τῶν Ῥωμαίων ἡγεμών, καὶ μέταλλον δ' ὑπῆρχε θαυμαστὸν χαλκοῦ καὶ σιδήρου κοινόν, ὅπερ οὐχ ἱστοροῦσιν ἀλλαχοῦ συμβαῖνον· νυνὶ μέντοι ἀμφότερα ἐκλέλοιπεν, ὥσπερ καὶ Ἀθήνησι

[1] Son of Orestes (13. 1. 3).
[2] See note on Aristotle, 10. 1. 3. [3] "Knights."

are said to have been founded by the Athenians before the Trojan War. And after the Trojan War, Aïclus and Cothus, setting out from Athens, settled inhabitants in them, the former in Eretria and the latter in Chalcis. There were also some Aeolians from the army of Penthilus[1] who remained in the island, and, in ancient times, some Arabians who had crossed over with Cadmus. Be this as it may, these cities grew exceptionally strong and even sent forth noteworthy colonies into Macedonia; for Eretria colonised the cities situated round Pallenê and Athos, and Chalcis colonised the cities that were subject to Olynthus, which later were treated outrageously by Philip. And many places in Italy and Sicily are also Chalcidian. These colonies were sent out, as Aristotle[2] states, when the government of the Hippobotae,[3] as it is called, was in power; for at the head of it were men chosen according to the value of their property, who ruled in an aristocratic manner. At the time of Alexander's passage across,[4] the Chalcidians enlarged the circuit of the walls of their city, taking inside them both Canethus and the Euripus, and fortifying the bridge with towers and gates and a wall.[5]

9. Above the city of the Chalcidians lies the so-called Lelantine Plain. In this plain are fountains of hot water suited to the cure of diseases, which were used by Cornelius Sulla, the Roman commander. And in this plain was also a remarkable mine which contained copper and iron together, a thing which is not reported as occurring elsewhere; now, however, both metals have given out, as in the case of the

[4] Across the Hellespont to Asia, 334 B.C.
[5] Cf. 9. 2. 8 and foot-notes.

τἀργυρεῖα.[1] ἔστι δὲ καὶ ἅπασα μὲν ἡ Εὔβοια εὔσειστος, μάλιστα δ' ἡ περὶ τὸν πορθμόν, καὶ δεχομένη πνευμάτων ὑποφοράς, καθάπερ καὶ ἡ Βοιωτία καὶ ἄλλοι τόποι, περὶ ὧν ἐμνήσθημεν διὰ πλειόνων πρότερον. ὑπὸ τοιοῦδε πάθους καὶ ἡ ὁμώνυμος τῇ νήσῳ πόλις καταποθῆναι λέγεται, ἧς μέμνηται καὶ Αἰσχύλος ἐν τῷ Ποντίῳ Γλαύκῳ·

> Εὐβοΐδα καμπτὴν[2] ἀμφὶ Κηναίου Διὸς
> ἀκτήν, κατ' αὐτὸν τύμβον ἀθλίου Λίχα.

Χαλκὶς δ' ὁμωνύμως λέγεται καὶ ἐν Αἰτωλίᾳ·

> Χαλκίδα τ' ἀγχίαλον, Καλυδῶνά τε πετρήεσσαν·

καὶ ἐν τῇ νῦν Ἠλείᾳ·

> βὰν δὲ παρὰ Κρουνοὺς καὶ Χαλκίδα πετρήεσσαν

οἱ περὶ Τηλέμαχον ἀπιόντες παρὰ Νέστορος εἰς τὴν οἰκείαν.

10. Ἐρέτριαν[3] δ' οἱ μὲν ἀπὸ Μακίστου τῆς Τριφυλίας ἀποικισθῆναί φασιν ὑπ' Ἐρετριέως, οἱ δ' ἀπὸ τῆς Ἀθήνησιν Ἐρετρίας, ἣ νῦν ἐστὶν C 448 ἀγορά· ἔστι δὲ καὶ περὶ Φάρσαλον Ἐρέτρια. ἐν δὲ τῇ Ἐρετρικῇ πόλις ἦν Ταμύναι, ἱερὰ τοῦ Ἀπόλλωνος· Ἀδμήτου δ' ἵδρυμα λέγεται τὸ ἱερόν, παρ' ᾧ θητεῦσαι λέγουσι τὸν θεὸν ἐνιαυτόν,[4] πλησίον τοῦ πορθμοῦ· Μελανηὶς δ' ἐκαλεῖτο πρότερον ἡ Ἐρέτρια καὶ Ἀρότρια· ταύτης δ' ἐστὶ κώμη ἡ Ἀμάρυνθος ἀφ' ἑπτὰ σταδίων τοῦ

[1] ὥσπερ . . . τἀργυρεῖα, preserved only in the *Epit.*, and inserted by Groskurd and Meineke.

[2] καμπτὴν *Bkl* Ald., instead of καμπήν; so Meineke.

silver mines at Athens. The whole of Euboea is much subject to earthquakes, but particularly the part near the strait, which is also subject to blasts through subterranean passages, as are Boeotia and other places which I have already described rather at length.[1] And it is said that the city which bore the same name as the island was swallowed up by reason of a disturbance of this kind. This city is also mentioned by Aeschylus in his *Glaucus Pontius*:[2] "Euboïs, about the bending shore of Zeus Cenaeus, near the very tomb of wretched Lichas." In Aetolia, also, there is a place called by the same name Chalcis: "and Chalcis near the sea, and rocky Calydon,"[3] and in the present Eleian country: "and they went past Cruni and rocky Chalcis,"[4] that is, Telemachus and his companions, when they were on their way back from Nestor's to their homeland.

10. As for Eretria, some say that it was colonised from Triphylian Macistus by Eretrieus, but others say from the Eretria at Athens, which now is a market-place. There is also an Eretria near Pharsalus. In the Eretrian territory there was a city Tamynae, sacred to Apollo; and the temple, which is near the strait, is said to have been founded by Admetus, at whose house the god served as an hireling for a year. In earlier times Eretria was called Melaneïs and Arotria. The village Amarynthus, which is seven stadia distant from the walls,

[1] 1. 3. 16. [2] *Frag.* 30 (Nauck).
[3] *Iliad* 2. 640. [4] *Od.* 15. 295.

---

[3] Ἐρετρίας BCD*hiklno*; Ἐρετριέας *x* (?) and the editors before Kramer.

[4] ἐνιαυτόν, Müller-Dübner, from conj. of Meineke, for αὐτόν.

τείχους. τὴν μὲν οὖν ἀρχαίαν πόλιν κατέσκαψαν Πέρσαι, σαγηνεύσαντες, ὥς φησιν Ἡρόδοτος, τοὺς ἀνθρώπους τῷ πλήθει, περιχυθέντων τῶν βαρβάρων τῷ τείχει (καὶ δεικνύουσιν ἔτι τοὺς θεμελίους, καλοῦσι δὲ παλαιὰν Ἐρέτριαν), ἡ δὲ νῦν ἐπέκτισται. τὴν δὲ δύναμιν τὴν Ἐρετριέων, ἣν ἔσχον ποτέ, μαρτυρεῖ ἡ στήλη, ἥν ἀνέθεσάν ποτε ἐν τῷ ἱερῷ τῆς Ἀμαρυνθίας Ἀρτέμιδος· γέγραπται δ' ἐν αὐτῇ, τρισχιλίοις μὲν ὁπλίταις, ἑξακοσίοις δ' ἱππεῦσιν, ἑξήκοντα δ' ἅρμασι ποιεῖν τὴν πομπήν· ἐπῆρχον δὲ καὶ Ἀνδρίων καὶ Τηνίων καὶ Κείων καὶ ἄλλων νήσων. ἐποίκους δ' ἔσχον ἀπ' Ἤλιδος, ἀφ' οὗ καὶ τῷ γράμματι τῷ ῥῷ πολλῷ χρησάμενοι, οὐκ ἐπὶ τέλει μόνον τῶν ῥημάτων ἀλλὰ καὶ ἐν μέσῳ, κεκωμῴδηνται. ἔστι δὲ καὶ Οἰχαλία κώμη τῆς Ἐρετρικῆς, λείψανον τῆς ἀναιρεθείσης πόλεως ὑπὸ Ἡρακλέους, ὁμώνυμος τῇ Τραχινίᾳ καὶ τῇ [1] περὶ Τρίκκην καὶ τῇ Ἀρκαδικῇ, ἣν Ἀνδανίαν οἱ ὕστερον ἐκάλεσαν, καὶ τῇ ἐν Αἰτωλίᾳ περὶ τοὺς Εὐρυτᾶνας.

11. Νυνὶ μὲν οὖν ὁμολογουμένως ἡ Χαλκὶς φέρεται τὰ πρωτεῖα καὶ μητρόπολις αὕτη λέγεται τῶν Εὐβοέων, δευτερεύει δ' ἡ Ἐρέτρια. ἀλλὰ καὶ πρότερον αὗται μέγα εἶχον ἀξίωμα καὶ πρὸς

[1] ἡ BCD*hklnox* ; οἱ Ald.

---

[1] "Whenever they took one of the islands, the barbarians, as though capturing each severally, would net the people.

belongs to this city. Now the old city was rased to the ground by the Persians, who "netted" the people, as Herodotus[1] says, by means of their great numbers, the barbarians being spread about the walls (the foundations are still to be seen, and the place is called Old Eretria); but the Eretria of to-day was founded on it.[2] As for the power the Eretrians once had, this is evidenced by the pillar which they once set up in the temple of Artemis Amarynthia. It was inscribed thereon that they made their festal procession with three thousand heavy-armed soldiers, six hundred horsemen, and sixty chariots. And they ruled over the peoples of Andros, Teos, Ceos, and other islands. They received new settlers from Elis; hence, since they frequently used the letter *r*,[3] not only at the end of words, but also in the middle, they have been ridiculed by comic writers. There is also a village Oechalia in the Eretrian territory, the remains of the city which was destroyed by Heracles; it bears the same name as the Trachinian Oechalia and that near Triccê, and the Arcadian Oechalia, which the people of later times called Andania, and that in Aetolia in the neighbourhood of the Eurytanians.

11. Now at the present time Chalcis by common consent holds the leading position and is called the metropolis of the Euboeans; and Eretria is second. Yet even in earlier times these cities were held in

They net them in this way: the men link hands and form a line extending from the northern sea to the southern, and then advance through the whole island hunting out the people" (6. 31).

[2] *i.e.* on a part of the old site.

[3] *i.e.* like the Eleians, who regularly rhotacised final *s* (see Buck, *Greek Dialects*, § 60).

πόλεμον καὶ πρὸς εἰρήνην, ὥστε καὶ φιλοσόφοις ἀνδράσι παρασχεῖν διαγωγὴν ἡδεῖαν καὶ ἀθόρυβον. μαρτυρεῖ δ' ἥ τε τῶν Ἐρετρικῶν φιλοσόφων σχολὴ τῶν περὶ Μενέδημον ἐν τῇ Ἐρετρίᾳ γενομένη, καὶ ἔτι πρότερον ἡ Ἀριστοτέλους ἐν τῇ Χαλκίδι διατριβή, ὅς γε κἀκεῖ [1] κατέλυσε τὸν βίον.

12. Τὸ μὲν οὖν πλέον ὡμολόγουν ἀλλήλαις αἱ πόλεις αὗται, περὶ δὲ Ληλάντου διενεχθεῖσαι οὐδ' οὕτω τελέως ἐπαύσαντο, ὥστε τῷ πολέμῳ κατὰ αὐθάδειαν δρᾶν ἕκαστα, ἀλλὰ συνέθεντο, ἐφ' οἷς συστήσονται τὸν ἀγῶνα. δηλοῖ δὲ καὶ τοῦτο ἐν τῷ Ἀμαρυνθίῳ στήλη τις, φράζουσα μὴ χρῆσθαι τηλεβόλοις. [2] καὶ γὰρ δὴ καὶ τῶν πολεμικῶν ἐθῶν καὶ τῶν ὁπλισμῶν οὐχ ἓν [3] οὔτ' ἐστὶν οὔτ' ἦν [4] ἔθος· ἀλλ' οἱ μὲν τηλεβόλοις χρῶνται, καθάπερ οἱ τοξόται καὶ οἱ σφενδονῆται καὶ οἱ ἀκοντισταί, οἱ δ' ἀγχεμάχοις, καθάπερ οἱ ξίφει καὶ δόρατι τῷ ὀρεκτῷ χρώμενοι· διττὴ γὰρ ἡ τῶν δοράτων χρῆσις, ἡ μὲν ἐκ χειρός, ἡ δ' ὡς παλτοῖς, καθάπερ καὶ ὁ κοντὸς ἀμφοτέρας τὰς χρείας ἀποδίδωσι· καὶ γὰρ συστάδην καὶ κοντοβολούντων, ὅπερ καὶ ἡ σάρισσα δύναται καὶ ὁ ὑσσός.

13. Οἱ δ' Εὐβοεῖς ἀγαθοὶ πρὸς μάχην ὑπῆρξαν τὴν σταδίαν, ἣ καὶ συστάδην λέγεται καὶ ἐκ

[1] ὅς γε κἀκεῖ Meineke, for ὥς γε καί CDghi; ὥστε καί s; οὗ γε καί kx; ὅς γε B (?); ὅς γε καὶ ἐκεῖ Casaubon.

[2] καὶ γὰρ . . . ὁ ὑσσός Meineke, following conj. of Kramer, rejects as an interpolation.

[3] οὐχ ἕν, Meineke, for οὐθέν CDEkx, Ald., οὔθ' ἕν lnos, Casaubon.

[4] ἦν is omitted by all MSS. except E.

great esteem, not only in war, but also in peace; indeed, they afforded philosophers a pleasant and undisturbed place of abode. This is evidenced by the school of the Eretrian philosophers, Menedemus and his disciples, which was established in Eretria, and also, still earlier, by the sojourn of Aristotle in Chalcis, where he also ended his days.[1]

12. Now in general these cities were in accord with one another, and when differences arose concerning the Lelantine Plain they did not so completely break off relations as to wage their wars in all respects according to the will of each, but they came to an agreement as to the conditions under which they were to conduct the fight. This fact, among others, is disclosed by a certain pillar in the Amarynthium, which forbids the use of long-distance missiles. [2] In fact among all the customs of warfare and of the use of arms there neither is, nor has been, any single custom; for some use long-distance missiles, as, for example, bowmen and slingers and javelin-throwers, whereas others use close-fighting arms, as, for example, those who use sword, or outstretched spear; for the spear is used in two ways, one in hand-to-hand combat and the other for hurling like a javelin; just as the pike serves both purposes, for it can be used both in close combat and as a missile for hurling, which is also true of the sarissa[3] and the hyssus.[4]

13. The Euboeans excelled in "standing" combat, which is also called "close" and "hand-to-hand"

[1] 322 B.C.

[2] The rest of the paragraph is probably an interpolation; see critical note.

[3] Used by the Macedonian phalanx.

[4] The Roman "pilum."

χειρός. δόρασι δ' ἐχρῶντο τοῖς ὀρεκτοῖς, ὥς φησιν ὁ ποιητής,

C 449 αἰχμηταὶ μεμαῶτες ὀρεκτῇσι μελίῃσι
θώρηκας ῥήσσειν.

ἀλλοίων ἴσως ὄντων τῶν παλτῶν, οἵαν εἰκὸς εἶναι τὴν Πηλιάδα μελίην, ἥν, ὥς φησιν ὁ ποιητής,

οἶος ἐπίστατο[1] πῆλαι Ἀχιλλεύς

καὶ ὁ εἰπών·

δουρὶ δ' ἀκοντίζω, ὅσον οὐκ ἄλλος τις ὀϊστῷ,

τῷ παλτῷ λέγει δόρατι. καὶ οἱ μονομαχοῦντες τοῖς παλτοῖς χρώμενοι δόρασιν εἰσάγονται πρότερον, εἶτα ἐπὶ τὰ ξίφη βαδίζοντες· ἀγχέμαχοι δ' εἰσὶν οὐχ οἱ ξίφει χρώμενοι μόνον, ἀλλὰ καὶ δόρατι ἐκ χειρός, ὥς φησιν·

οὔτησε ξυστῷ χαλκήρει, λῦσε δὲ γυῖα.

τοὺς μὲν οὖν Εὐβοέας τούτῳ τῷ τρόπῳ χρωμένους εἰσάγει, περὶ δὲ Λοκρῶν τἀναντία λέγει, ὡς

οὔ σφιν σταδίης ὑσμίνης ἔργα μέμηλεν,
ἀλλ' ἄρα τόξοισιν καὶ ἐϋστρόφῳ οἰὸς ἀώτῳ
Ἴλιον εἰς ἅμ' ἕποντο.

περιφέρεται[2] δὲ καὶ χρησμὸς ἐκδοθεὶς Αἰγιεῦσιν,

ἵππον Θεσσαλικόν,[3] Λακεδαιμονίαν δὲ γυναῖκα,
ἄνδρας θ', οἳ πίνουσιν ὕδωρ ἱερῆς Ἀρεθούσης,

τοὺς Χαλκιδέας λέγων ὡς ἀρίστους· ἐκεῖ γὰρ ἡ Ἀρέθουσα.

14. Εἰσὶ δὲ νῦν Εὐβοῖται ποταμοὶ Κηρεὺς καὶ Νηλεύς, ὧν ἀφ' οὗ μὲν πίνοντα τὰ πρόβατα

combat; and they used their spears outstretched, as the poet says: "spearmen eager with outstretched ashen spears to shatter corselets."[1] Perhaps the javelins were of a different kind, such as probably was the "Pelian ashen spear," which, as the poet says, "Achilles alone knew how to hurl";[2] and he[3] who said, "And the spear I hurl farther than any other man can shoot an arrow,"[4] means the javelin-spear. And those who fight in single combat are first introduced as using javelin-spears, and then as resorting to swords. And close-fighters are not those who use the sword alone, but also the spear hand-to-hand, as the poet says: "he pierced him with bronze-tipped polished spear, and loosed his limbs."[5] Now he introduces the Euboeans as using this mode of fighting, but he says the contrary of the Locrians, that "they cared not for the toils of close combat, . . . but relying on bows and well-twisted slings of sheep's wool they followed with him to Ilium."[6] There is current, also, an oracle which was given out to the people of Aegium, "Thessalian horse, Lacedemonian woman, and men who drink the water of sacred Arethusa," meaning that the Chalcidians are best of all, for Arethusa is in their territory.

14. There are now two rivers in Euboea, the Cereus and the Neleus; and the sheep which drink

[1] *Iliad* 2. 543.
[2] *Iliad* 19. 389.
[3] Odysseus.
[4] *Od.* 8. 229.
[5] *Iliad* 4. 469.
[6] *Iliad* 13. 713, 716.

---

[1] ἐπίστατο *no*; other MSS ἐπίσταται.
[2] περιφέρεται, Corais and later editors, for παραφέρεται.
[3] Θεσσαλικήν *k* by correction.

λευκὰ γίνεται, ἀφ' οὗ δὲ μέλανα· καὶ περὶ τὸν Κρᾶθιν δὲ εἴρηται τοιοῦτόν τι συμβαῖνον.

15. Τῶν δ' ἐκ Τροίας ἐπανιόντων Εὐβοέων τινὲς εἰς Ἰλλυριοὺς ἐκπεσόντες, ἄραντες[1] οἴκαδε διὰ τῆς Μακεδονίας περὶ Ἔδεσσαν ἔμειναν, συμπολεμήσαντες τοῖς ὑποδεξαμένοις, καὶ ἔκτισαν πόλιν Εὔβοιαν· ἦν δὲ καὶ ἐν Σικελίᾳ Εὔβοια, Χαλκιδέων τῶν ἐκεῖ κτίσμα, ἣν Γέλων ἐξανέστησε, καὶ ἐγένετο φρούριον Συρακουσίων· καὶ ἐν Κερκύρᾳ δὲ καὶ ἐν Λήμνῳ τόπος ἦν Εὔβοια καὶ ἐν τῇ Ἀργείᾳ λόφος τις.

16. Ἐπεὶ δὲ τοῖς Θετταλοῖς καὶ Οἰταίοις τὰ πρὸς ἑσπέραν Αἰτωλοὶ καὶ Ἀκαρνᾶνές εἰσι καὶ Ἀθαμᾶνες, εἰ χρὴ καὶ τούτους Ἕλληνας εἰπεῖν, λοιπὸν ἐξηγήσασθαι περὶ τούτων, ἵν' ἔχωμεν τὴν περίοδον ἅπασαν τὴν τῆς Ἑλλάδος· προσθεῖναι δὲ καὶ τὰς νήσους τὰς προσχώρους μάλιστα τῇ Ἑλλάδι καὶ οἰκουμένας ὑπὸ τῶν Ἑλλήνων, ὅσας μὴ περιωδεύκαμεν.

II

1. Αἰτωλοὶ μὲν τοίνυν καὶ Ἀκαρνᾶνες ὁμοροῦσιν ἀλλήλοις, μέσον ἔχοντες τὸν Ἀχελῷον ποταμόν, ῥέοντα ἀπὸ τῶν ἄρκτων καὶ τῆς Πίνδου πρὸς C 450 νότον διά τε Ἀγραίων, Αἰτωλικοῦ ἔθνους, καὶ Ἀμφιλόχων· Ἀκαρνᾶνες μὲν τὸ πρὸς ἑσπέραν

[1] ἄραντες, T. G. Tucker, for Ἄβαντες; ἀναβάντες, Xylander; μεταβαίνοντες, Corais; ἀποβάντες, Kramer; ἀποβαίνοντες, Meineke.

from one of them turn white, and from the other black. A similar thing takes place in connection with the Crathis River, as I have said before.[1]

15. When the Euboeans were returning from Troy, some of them, after being driven out of their course to Illyria, set out for home through Macedonia, but remained in the neighbourhood of Edessa, after aiding in war those who had received them hospitably; and they founded a city Euboea. There was also a Euboea in Sicily, which was founded by the Chalcidians of Sicily, but they were driven out of it by Gelon; and it became a stronghold of the Syracusans. In Corcyra, also, and in Lemnos, there were places called Euboea; and in the Argive country a hill of that name.

16. Since the Aetolians, Acarnanians, and Athamanians (if these too are to be called Greeks) live to the west of the Thessalians and the Oetaeans, it remains for me to describe these three, in order that I may complete the circuit of Greece; I must also add the islands which lie nearest to Greece and are inhabited by the Greeks, so far as I have not already included them in my description.

II

1. Now the Aetolians and the Acarnanians border on one another, having between them the Acheloüs River, which flows from the north and from Pindus on the south through the country of the Agraeans, an Aetolian tribe, and through that of the Amphilochians, the Acarnanians holding the western side of the river

[1] 6. 1. 13.

μέρος ἔχοντες τοῦ ποταμοῦ μέχρι τοῦ Ἀμβρακικοῦ κόλπου τοῦ κατὰ Ἀμφιλόχους καὶ τὸ ἱερὸν τοῦ Ἀκτίου Ἀπόλλωνος, Αἰτωλοὶ δὲ τὸ πρὸς ἕω μέχρι τῶν Ὀζολῶν Λοκρῶν καὶ τοῦ Παρνασσοῦ καὶ τῶν Οἰταίων. ὑπέρκεινται δ' ἐν τῇ μεσογαίᾳ καὶ τοῖς προσβορείοις μέρεσι τῶν μὲν Ἀκαρνάνων Ἀμφίλοχοι, τούτων δὲ Δόλοπες καὶ ἡ Πίνδος, τῶν δ' Αἰτωλῶν Περραιβοί τε καὶ Ἀθαμᾶνες καὶ Αἰνιάνων τι μέρος τῶν τὴν Οἴτην ἐχόντων· τὸ δὲ νότιον πλευρόν, τό τε Ἀκαρνανικὸν ὁμοίως καὶ τὸ Αἰτωλικόν, κλύζεται τῇ ποιούσῃ θαλάττῃ τὸν Κορινθιακὸν κόλπον, εἰς ὃν καὶ ὁ Ἀχελῷος ποταμὸς ἐξίησιν, ὁρίζων τὴν τῶν Αἰτωλῶν παραλίαν καὶ τὴν Ἀκαρνανικήν· ἐκαλεῖτο δὲ Θόας ὁ Ἀχελῷος πρότερον. ἔστι δὲ καὶ ὁ παρὰ Δύμην ὁμώνυμος τούτῳ, καθάπερ εἴρηται, καὶ ὁ περὶ Λαμίαν. εἴρηται δὲ καί, ὅτι ἀρχὴν τοῦ Κορινθιακοῦ κόλπου τὸ στόμα τοῦδε τοῦ ποταμοῦ φασί.

2. Πόλεις δ' εἰσὶν ἐν μὲν τοῖς Ἀκαρνᾶσιν Ἀνακτόριόν τε ἐπὶ χερρονήσου ἱδρυμένον Ἀκτίου πλησίον, ἐμπόριον τῆς νῦν ἐκτισμένης ἐφ' ἡμῶν Νικοπόλεως, καὶ Στράτος, ἀνάπλουν ἔχουσα τῷ Ἀχελῴῳ πλειόνων ἢ διακοσίων σταδίων, καὶ Οἰνειάδαι,[1] καὶ αὐτὴ ἐπὶ τῷ ποταμῷ, ἡ μὲν παλαιὰ οὐ κατοικουμένη, ἴσον ἀπέχουσα τῆς τε θαλάττης καὶ τοῦ[2] Στράτου, ἡ δὲ νῦν ὅσον ἑβδομήκοντα σταδίους ὑπὲρ τῆς ἐκβολῆς διέχουσα. καὶ ἄλλαι δ' εἰσί, Παλαιρός τε καὶ Ἀλυζία καὶ

[1] Οἰνειάδαι, Meineke from conj. of Kramer, for Ἠναία δέ *Bk*, Αἰνεία δέ *l* (?), Ald

[2] But τῆς is the reading of *noxy* (cp. Stephanus: Στράτος . . . θηλυκῶς καὶ ἀρσενικῶς).

as far as that part of the Ambracian Gulf which is near Amphilochi and the temple of the Actian Apollo, but the Aetolians the eastern side as far as the Ozalian Locrians and Parnassus and the Oetaeans. Above the Acarnanians, in the interior and the parts towards the north, are situated the Amphilochians, and above these the Dolopians and Pindus, and above the Aetolians are the Perrhaebians and Athamanians and a part of the Aenianians who hold Oeta. The southern side, of Acarnania and Aetolia alike, is washed by the sea which forms the Corinthian Gulf, into which empties the Acheloüs River, which forms the boundary between the coast of the Aetolians and that of Acarnania. In earlier times the Acheloüs was called Thoas. The river which flows past Dymê bears the same name as this, as I have already said,[1] and also the river near Lamia.[2] I have already stated, also, that the Corinthian Gulf is said to begin at the mouth of this river.[3]

2. As for cities, those of the Acarnanians are Anactorium, which is situated on a peninsula near Actium and is a trading-centre of the Nicopolis of to-day, which was founded in our times;[4] Stratus, where one may sail up the Acheloüs River more than two hundred stadia; and Oeneiadae, which is also on the river—the old city, which is equidistant from the sea and from Stratus, being uninhabited, whereas that of to-day lies at a distance of about seventy stadia above the outlet of the river. There are also other cities, Palaerus, Alyzia, Leucas,[5] Argos

[1] 8. 3. 11. [2] 9. 5. 10. [3] 8. 2. 3.

[4] This Nicopolis ("Victory City") was founded by Augustus Caesar in commemoration of his victory over Antony and Cleopatra at Actium in 31 B.C. See 7. 7. 5.

[5] Amaxiki, now in ruins.

Λευκὰς καὶ Ἄργος τὸ Ἀμφιλοχικὸν καὶ Ἀμβρακία, ὧν αἱ πλεῖσται περιοικίδες γεγόνασιν ἢ καὶ πᾶσαι τῆς Νικοπόλεως· κεῖται δ' ὁ[1] Στράτος κατὰ μέσην τὴν ἐξ Ἀλυζίας ὁδὸν εἰς Ἀνακτόριον.

3. Αἰτωλῶν δ' εἰσὶ Καλυδών τε καὶ Πλευρών, νῦν μὲν τεταπεινωμέναι, τὸ δὲ παλαιὸν πρόσχημα τῆς Ἑλλάδος ἦν ταῦτα τὰ κτίσματα. καὶ δὴ καὶ διῃρῆσθαι συνέβαινε δίχα τὴν Αἰτωλίαν, καὶ τὴν μὲν ἀρχαίαν λέγεσθαι, τὴν δ' ἐπίκτητον· ἀρχαίαν μὲν τὴν ἀπὸ τοῦ Ἀχελώου μέχρι Καλυδῶνος παραλίαν, ἐπὶ πολὺ καὶ τῆς μεσογαίας ἀνήκουσαν, εὐκάρπου τε καὶ πεδιάδος, ᾗ ἐστὶ καὶ Στράτος καὶ τὸ Τριχώνιον,[2] ἀρίστην ἔχον γῆν· ἐπίκτητον δὲ τὴν τοῖς Λοκροῖς συνάπτουσαν, ὡς ἐπὶ Ναύπακτόν τε καὶ Εὐπάλιον, τραχυτέραν τε οὖσαν καὶ λυπροτέραν, μέχρι τῆς Οἰταίας καὶ τῆς Ἀθαμάνων καὶ τῶν ἐφεξῆς ἐπὶ τὴν ἄρκτον ἤδη περιισταμένων ὀρῶν τε καὶ ἐθνῶν.

4. Ἔχει δὲ καὶ ἡ Αἰτωλία ὄρος μέγιστον μὲν τὸν Κόρακα, συνάπτοντα τῇ Οἴτῃ, τῶν δ' ἄλλων C 451 ἐν μέσῳ μὲν μᾶλλον[3] τὸν Ἀράκυνθον, περὶ ὃν τὴν νεωτέραν Πλευρῶνα συνῴκισαν ἀφέντες τὴν παλαιάν, ἐγγὺς κειμένην Καλυδῶνος, οἱ οἰκήτορες, εὔκαρπον οὖσαν καὶ πεδιάδα, πορθοῦντος τὴν χώραν Δημητρίου τοῦ ἐπικληθέντος Αἰτωλικοῦ· ὑπὲρ δὲ τῆς Μολυκρείας[4] Ταφιασσὸν καὶ Χαλκίδα,

[1] ἡ *nox*, instead of ὁ, other MSS.

[2] Τριχώνιον, Palmer, for Τραχήνιον οs, Τραχίνιον, other MSS. So the later editors.

[3] μᾶλλον, Casaubon, for μαλαόν BC*ghilnosxy*, μάλα ὄντων marg. *h*, μάλα ὂν D*k*, omitted in E; so the later editors.

[4] Μολυκρείας, Tzschucke, for Μολυκρίας; so the later editors.

Amphilochicum, and Ambracia, most of which, or rather all, have become dependencies of Nicopolis. Stratus is situated about midway of the road between Alyzia and Anactorium.[1]

3. The cities of the Aetolians are Calydon and Pleuron, which are now indeed reduced, though in early times these settlements were an ornament to Greece. Further, Aetolia has come to be divided into two parts, one part being called Old Aetolia and the other Aetolia Epictetus.[2] The Old Aetolia was the seacoast extending from the Acheloüs to Calydon, reaching for a considerable distance into the interior, which is fertile and level; here in the interior lie Stratus and Trichonium, the latter having excellent soil. Aetolia Epictetus is the part which borders on the country of the Locrians in the direction of Naupactus and Eupalium, being a rather rugged and sterile country, and extends to the Oetaean country and to that of the Athamanians and to the mountains and tribes which are situated next beyond these towards the north.

4. Aetolia also has a very large mountain, Corax, which borders on Oeta; and it has among the rest of its mountains, and more in the middle of the country than Corax, Aracynthus, near which New Pleuron was founded by the inhabitants of the Old, who abandoned their city, which had been situated near Calydon in a district both fertile and level, at the time when Demetrius, surnamed Aetolicus,[3] laid waste the country; above Molycreia are Taphiassus

[1] An error either of Strabo or of the MSS. "Stratus" and "Alyzia" should exchange places in the sentence.

[2] *i.e.* the Acquired.

[3] Son of Antigonus Gonatas; reigned over Macedonia 239–229 B.C.

ὄρη ἱκανῶς ὑψηλά, ἐφ' οἷς πολίχνια ἵδρυτο[1] Μακυνία τε καὶ Χαλκίς, ὁμώνυμος τῷ ὄρει, ἣν καὶ Ὑποχαλκίδα καλοῦσι· Κούριον δὲ πλησίον τῆς παλαιᾶς Πλευρῶνος, ἀφ' οὗ τοὺς Πλευρωνίους Κουρῆτας ὀνομασθῆναί τινες ὑπέλαβον.

5. Ὁ δ' Εὔηνος [2] ποταμὸς ἄρχεται μὲν ἐκ Βωμιέων [3] τῶν ἐν Ὀφιεῦσιν, Αἰτωλικῷ ἔθνει (καθάπερ καὶ οἱ Εὐρυτᾶνες καὶ Ἀγραῖοι καὶ Κουρῆτες καὶ ἄλλοι), ῥεῖ δ' οὐ διὰ τῆς Κουρητικῆς κατ' ἀρχάς, ἥτις ἐστὶν ἡ αὐτὴ τῇ Πλευρωνίᾳ, ἀλλὰ διὰ τῆς προσεῴας μᾶλλον παρὰ τὴν Χαλκίδα καὶ Καλυδῶνα· εἶτ' ἀνακάμψας ἐπὶ τὰ τῆς Πλευρῶνος πεδία τῆς παλαιᾶς καὶ παραλλάξας εἰς δύσιν ἐπιστρέφει πρὸς τὰς ἐκβολὰς καὶ τὴν μεσημβρίαν· ἐκαλεῖτο δὲ Λυκόρμας[4] πρότερον, καὶ ὁ Νέσσος ἐνταῦθα λέγεται πορθμεὺς ἀποδεδειγμένος ὑφ' Ἡρακλέους ἀποθανεῖν, ἐπειδὴ πορθμεύων τὴν Δηιάνειραν ἐπεχείρει βιάσασθαι.

6. Καὶ Ὤλενον δὲ καὶ Πυλήνην ὀνομάζει πόλεις ὁ ποιητὴς Αἰτωλικάς, ὧν τὴν μὲν Ὤλενον ὁμωνύμως τῇ Ἀχαϊκῇ λεγομένην Αἰολεῖς κατέσκαψαν, πλησίον οὖσαν τῆς νεωτέρας Πλευρῶνος, τῆς δὲ χώρας ἠμφισβήτουν Ἀκαρνᾶνες· τὴν δὲ Πυλήνην μετενέγκαντες εἰς τοὺς ἀνώτερον τόπους ἤλλαξαν αὐτῆς καὶ τοὔνομα, Πρόσχιον καλέσαντες. Ἑλλάνικος δ' οὐδὲ τὴν περὶ ταύτας ἱστο-

[1] ἵδρυται *Bkno*.
[2] Εὔηνος *no*, ὁ δὲ Τῆνος BC*Dhilsx*.

and Chalcis, rather high mountains, on which were situated the small cities Macynia and Chalcis, the latter bearing the same name as the mountain, though it is also called Hypochalcis. Near Old Pleuron is the mountain Curium, after which, as some have supposed, the Pleuronian Curetes were named.

5. The Evenus River begins in the territory of those Bomians who live in the country of the Ophians, the Ophians being an Aetolian tribe (like the Eurytanians and Agraeans and Curetes and others), and flows at first, not through the Curetan country, which is the same as the Pleuronian, but through the more easterly country, past Chalcis and Calydon; and then, bending back towards the plains of Old Pleuron and changing its course to the west, it turns towards its outlets and the south. In earlier times it was called Lycormas. And there Nessus, it is said, who had been appointed ferryman, was killed by Heracles because he tried to violate Deïaneira when he was ferrying her across the river.

6. The poet also names Olenus and Pylenê as Aetolian cities.[1] Of these, the former, which bears the same name as the Achaean city, was rased to the ground by the Aeolians; it was near New Pleuron, but the Acarnanians claimed possession of the territory. The other, Pylenê, the Aeolians moved to higher ground, and also changed its name, calling it Proschium. Hellanicus does not know the

[1] *Iliad* 2. 639.

---

[3] βωμιαίων DC*ghinox*, Βωιαίων B*kl*; emended by Tzschucke and so by the later editors.

[4] Λυκόρμας E, Λυκέρνας CD*ghilxy* and by corr. in B*k*, and Λυκάρνας *no* but corr. to Λυκόρμος.

ρίαν οἶδεν, ἀλλ' ὡς ἔτι καὶ αὐτῶν οὐσῶν ἐν τῇ ἀρχαίᾳ καταστάσει μέμνηται, τὰς δ' ὕστερον καὶ τῆς τῶν Ἡρακλειδῶν καθόδου κτισθείσας, Μακυνίαν[1] καὶ Μολύκρειαν,[2] ἐν ταῖς ἀρχαίαις καταλέγει, πλείστην εὐχέρειαν ἐπιδεικνύμενος ἐν πάσῃ σχεδόν τι τῇ γραφῇ.

7. Καθόλου μὲν οὖν ταῦτα περὶ τῆς χώρας ἐστὶ τῆς τῶν Ἀκαρνάνων καὶ τῶν Αἰτωλῶν, περὶ δὲ τῆς παραλίας καὶ τῶν προκειμένων νήσων ἔτι καὶ ταῦτα προσληπτέον· ἀπὸ γὰρ τοῦ στόματος ἀρξαμένοις[3] τοῦ Ἀμβρακικοῦ κόλπου πρῶτόν ἐστιν Ἀκαρνάνων χωρίον τὸ Ἄκτιον. ὁμωνύμως δὲ λέγεται τό τε ἱερὸν τοῦ Ἀκτίου Ἀπόλλωνος καὶ ἡ ἄκρα ἡ ποιοῦσα τὸ στόμα τοῦ κόλπου, ἔχουσα καὶ λιμένα ἐκτός. τοῦ δ' ἱεροῦ τετταράκοντα μὲν σταδίους ἀπέχει τὸ Ἀνακτόριον ἐν τῷ κόλπῳ ἱδρυμένον, διακοσίους δὲ καὶ τετταράκοντα ἡ Λευκάς.

8. Αὕτη δ' ἦν τὸ παλαιὸν μὲν χερρόνησος τῆς Ἀκαρνάνων γῆς, καλεῖ δ' ὁ ποιητὴς αὐτὴν ἀκτὴν ἠπείροιο, τὴν περαίαν τῆς Ἰθάκης καὶ τῆς Κεφαλ-C 452 ληνίας ἤπειρον καλῶν· αὕτη δ' ἐστὶν ἡ Ἀκαρνανία· ὥστε, ὅταν φῇ ἀκτὴν ἠπείροιο, τῆς Ἀκαρνανίας ἀκτὴν δέχεσθαι δεῖ. τῆς δὲ Λευκάδος ἥ τε Νήρικος,[4] ἥν φησιν ἑλεῖν ὁ Λαέρτης,

ἦ μὲν[5] Νήρικον[6] εἷλον ἐΰκτίμενον πτολίεθρον,
ἀκτὴν ἠπείροιο, Κεφαλλήνεσσιν ἀνάσσων·

[1] Μακυνίαν, the editors, for Μακίνιον.
[2] Μολύκρειαν, the editors, for Μολύκριαν.
[3] The MSS., except *k*, have καί after ἀρξαμένοις.
[4] Νήρικος, Jones restores, following BED (though in D the Νήρικος is written above Νήριτος in first hand), instead of Νήριτος (Kramer and later editors).

history of these cities either, but mentions them as though they too were still in their early status; and among the early cities he names Macynia and Molycreia, which were founded even later than the return of the Heracleidae, almost everywhere in his writings displaying a most convenient carelessness.

7. Upon the whole, then, this is what I have to say concerning the country of the Acarnanians and the Aetolians, but the following is also to be added concerning the seacoast and the islands which lie off it: Beginning at the mouth of the Ambracian Gulf, the first place which belongs to the Acarnanians is Actium. The temple of the Actian Apollo bears the same name, as also the cape which forms the mouth of the Gulf and has a harbour on the outer side. Anactorium, which is situated on the gulf, is forty stadia distant from the temple, whereas Leucas is two hundred and forty.

8. In early times Leucas was a peninsula of Acarnania, but the poet calls it "shore of the mainland,"[1] using the term "mainland" for the country which is situated across from Ithaca and Cephallenia; and this country is Acarnania. And therefore, when he says, "shore of the mainland," one should take it to mean "shore of Acarnania." And to Leucas also belonged, not only Nericus, which Laertes says he took ("verily I took Nericus, well-built citadel, shore of the mainland, when I was lord over the

[1] Homer specifically mentions Leucas only once, as the "rock Leucas" (*Od.* 24. 11). On the Ithaca-Leucas problem, see *Appendix* in this volume.

---

[5] Instead of ἦ μέν, Homer (*Od.* 24. 376) has οἷος; B reads both, ἦ μὲν οἷος.

[6] Νήρικον, Jones restores, following MSS., except B, which reads Νήριτον.

καὶ ἃς ἐν Καταλόγῳ φησί·

καὶ Κροκύλει'[1] ἐνέμοντο καὶ Αἰγίλιπα τρηχεῖαν.

Κορίνθιοι δὲ πεμφθέντες ὑπὸ Κυψέλου καὶ Γόργου[2] ταύτην τε κατέσχον τὴν ἀκτήν, καὶ μέχρι τοῦ Ἀμβρακικοῦ κόλπου προῆλθον, καὶ ἥ τε Ἀμβρακία συνῳκίσθη καὶ Ἀνακτόριον, καὶ τῆς χερρονήσου διορύξαντες τὸν ἰσθμὸν ἐποίησαν νῆσον τὴν Λευκάδα, καὶ μετενέγκαντες τὴν Νήρικον[3] ἐπὶ τὸν τόπον, ὃς ἦν ποτὲ μὲν ἰσθμός, νῦν δὲ πορθμὸς γεφύρᾳ ζευκτός, μετωνόμασαν Λευκάδα ἐπώνυμον, δοκῶ μοι, τοῦ Λευκάτα· πέτρα γάρ ἐστι λευκὴ τὴν χρόαν, προκειμένη τῆς Λευκάδος εἰς τὸ πέλαγος καὶ τὴν Κεφαλληνίαν, ὥς ἐντεῦθεν τοὔνομα λαβεῖν.

9. Ἔχει δὲ τὸ τοῦ Λευκάτα Ἀπόλλωνος ἱερὸν καὶ τὸ ἅλμα, τὸ τοὺς ἔρωτας παύειν πεπιστευμένον·

> οὗ δὴ λέγεται πρώτη Σαπφώ,

(ὥς φησιν ὁ Μένανδρος)

> τὸν ὑπέρκομπον θηρῶσα Φάων',
> οἰστρῶντι πόθῳ ῥῖψαι πέτρας
> ἀπὸ τηλεφανοῦς ἅλμα[4] κατ' εὐχὴν
> σήν, δέσποτ' ἄναξ.

ὁ μὲν οὖν Μένανδρος πρώτην ἁλέσθαι λέγει τὴν Σαπφώ, οἱ δ' ἔτι ἀρχαιολογικώτεροι Κέφαλόν φασιν ἐρασθέντα Πτερέλα,[5] τὸν[6] Δηιονέως. ἦν

[1] Κροκύλει' E, Κροκύλην other MSS.
[2] Γόργου, Runke, for Γαργάσουσος CD*hil*, Γαργάσου other MSS.; so Meineke.
[3] Νήρικον, the reading of the MSS. (except B where Νήριτον is corrected), Jones restores.

Cephallenians"),[1] but also the cities which Homer names in the *Catalogue* ("and dwelt in Crocyleia and rugged Aegilips").[2] But the Corinthians sent by Cypselus[3] and Gorgus took possession of this shore and also advanced as far as the Ambracian Gulf; and both Ambracia and Anactorium were colonised at this time; and the Corinthians dug a canal through the isthmus of the peninsula and made Leucas an island; and they transferred Nericus to the place which, though once an isthmus, is now a strait spanned by a bridge, and they changed its name to Leucas, which was named, as I think, after Leucatas; for Leucatas is a rock of white[4] colour jutting out from Leucas into the sea and towards Cephallenia, and therefore it took its name from its colour.

9. It contains the temple of Apollo Leucatas, and also the "Leap," which was believed to put an end to the longings of love. "Where Sappho is said to have been the first," as Menander says, "when through frantic longing she was chasing the haughty Phaon, to fling herself with a leap from the far-seen rock, calling upon thee in prayer, O lord and master." Now although Menander says that Sappho was the first to take the leap, yet those who are better versed than he in antiquities say that it was Cephalus, who was in love with Pterelas the son of

[1] *Od.* 24. 377.
[2] *Iliad* 2. 633.
[3] See *Dictionary* in Vol. IV.
[4] "leuca."

---

[4] ἅλμα, Wordsworth (note on Theocritus 3. 25), for ἀλλά; so Meineke.

[5] Πτερέλα, Tzschucke, for Περόλα *Dh*, but Πτερόλα in margin of *h* and *Ci*, Πταρόλα *Bglmno*, Πταροχα *x*, Παρόλα *k*; so the later editors.

[6] τόν, Kramer, for τοῦ, from corr. in B.

δὲ καὶ πάτριον τοῖς Λευκαδίοις κατ᾽ ἐνιαυτὸν ἐν τῇ θυσίᾳ τοῦ Ἀπόλλωνος ἀπὸ τῆς σκοπῆς ῥιπτεῖσθαί τινα τῶν ἐν αἰτίαις ὄντων ἀποτροπῆς χάριν, ἐξαπτομένων ἐξ αὐτοῦ παντοδαπῶν πτερῶν καὶ ὀρνέων ἀνακουφίζειν δυναμένων τῇ πτήσει τὸ ἅλμα, ὑποδέχεσθαι δὲ κάτω μικραῖς ἁλιάσι κύκλῳ περιεστῶτας πολλοὺς καὶ περισώζειν εἰς δύναμιν τῶν ὅρων ἔξω τὸν ἀναληφθέντα. ὁ δὲ τὴν Ἀλκμαιωνίδα γράψας· Ἰκαρίου, τοῦ Πηνελόπης πατρός, υἱεῖς γενέσθαι δύο, Ἀλυζέα καὶ Λευκάδιον, δυναστεῦσαι δ᾽ ἐν τῇ Ἀκαρνανίᾳ τούτους μετὰ τοῦ πατρός· τούτων οὖν ἐπωνύμους τὰς πόλεις Ἔφορος λέγεσθαι δοκεῖ.

10. Κεφαλλῆνας δὲ νῦν μὲν τοὺς ἐκ τῆς νήσου τῆς Κεφαλληνίας λέγουσιν, Ὅμηρος δὲ πάντας τοὺς ὑπὸ τῷ Ὀδυσσεῖ, ὧν εἰσὶ καὶ οἱ Ἀκαρνᾶνες· εἰπὼν γάρ·

αὐτὰρ Ὀδυσσεὺς ἦγε Κεφαλλῆνας,
οἵ ῥ᾽ Ἰθάκην εἶχον καὶ Νήριτον εἰνοσίφυλλον,

(τὸ ἐν ταύτῃ ὄρος ἐπιφανές· ὡς καί

οἳ δ᾽ ἐκ Δουλιχίοιο Ἐχινάων θ᾽ ἱεράων,

καὶ αὐτοῦ τοῦ Δουλιχίου τῶν Ἐχινάδων ὄντος· καί

C 453 οἳ δ᾽ ἄρα Βουπράσιόν τε καὶ Ἦλιδα,

καὶ τοῦ Βουπρασίου ἐν Ἤλιδι ὄντος·

οἳ δ᾽ Εὔβοιαν ἔχον καὶ Χαλκίδα τ᾽ Εἰρέτριάν τε,

ὡς [1] τούτων ἐν Εὐβοίᾳ οὐσῶν· καί

[1] ὡς, all MSS., except E and the editors (καί), Jones restores.

Deïoneus. It was an ancestral custom among the Leucadians, every year at the sacrifice performed in honour of Apollo, for some criminal to be flung from this rocky look-out for the sake of averting evil, wings and birds of all kinds being fastened to him, since by their fluttering they could lighten the leap, and also for a number of men, stationed all round below the rock in small fishing-boats, to take the victim in, and, when he had been taken on board,[1] to do all in their power to get him safely outside their borders. The author of the *Alcmaeonis* [2] says that Icarius, the father of Penelope, had two sons, Alyzeus and Leucadius, and that these two reigned over Acarnania with their father; accordingly, Ephorus thinks that the cities were named after these.

10. But though at the present time only the people of the island Cephallenia are called Cephallenians, Homer so calls all who were subject to Odysseus, among whom are also the Acarnanians. For after saying, "but Odysseus led the Cephallenians, who held Ithaca and Neritum with quivering foliage" [3] (Neritum being the famous mountain on this island, as also when he says, "and those from Dulichium and the sacred Echinades," [4] Dulichium itself being one of the Echinades; and "those who dwelt in Buprasium and Elis," [5] Buprasium being in Elis; and "those who held Euboea and Chalcis and Eiretria," [6] meaning that these cities

[1] Or perhaps "resuscitated."

[2] The author of this epic poem on the deeds of Alcmaeon is unknown.

[3] *Iliad* 2. 631.

[4] *Iliad* 2. 625.

[5] *Iliad* 2. 615.

[6] *Iliad* 2. 536.

Τρῶες καὶ Λύκιοι καὶ Δάρδανοι,

ὡς καὶ ἐκείνων Τρώων ὄντων)· πλὴν μετά γε Νήριτόν φησι·

καὶ Κροκύλει'[1] ἐνέμοντο καὶ Αἰγίλιπα τρηχεῖαν,
οἵ τε Ζάκυνθον ἔχον ἠδ' οἳ Σάμον ἀμφενέμοντο,
οἵ τ' ἤπειρον ἔχον ἠδ' ἀντιπέραι' ἐνέμοντο.

ἤπειρον μὲν οὖν[2] τὰ ἀντιπέρα τῶν νήσων βούλεται λέγειν, ἅμα τῇ Λευκάδι καὶ τὴν ἄλλην Ἀκαρνανίαν συμπεριλαβεῖν βουλόμενος, περὶ ἧς καὶ οὕτω λέγει·

δώδεκ' ἐν ἠπείρῳ ἀγέλαι, τόσα πώεα μήλων·[3]

τάχα τῆς Ἠπειρώτιδος τὸ παλαιὸν μέχρι δεῦρο διατεινούσης καὶ ὀνόματι κοινῷ ἠπείρου λεγομένης· Σάμον δὲ τὴν νῦν Κεφαλληνίαν, ὡς καὶ ὅταν φῇ·

ἐν πορθμῷ Ἰθάκης τε Σάμοιό τε παιπαλοέσσης.

τῷ γὰρ ἐπιθέτῳ τὴν ὁμωνυμίαν διέσταλται, ὡς οὐκ ἐπὶ τῆς πόλεως, ἀλλ' ἐπὶ τῆς νήσου τιθεὶς τοὔνομα. τετραπόλεως γὰρ οὔσης τῆς νήσου, μία τῶν τεττάρων ἐστὶν ἡ καὶ Σάμος καὶ Σάμη καλουμένη καθ' ἑκάτερον τοὔνομα, ὁμωνυμοῦσα τῇ νήσῳ. ὅταν δ' εἴπῃ·

ὅσσοι γὰρ νήσοισιν ἐπικρατέουσιν ἄριστοι,
Δουλιχίῳ τε Σάμῃ τε καὶ ὑλήεντι Ζακύνθῳ,

τῶν νήσων ἀριθμὸν ποιῶν[4] δῆλός ἐστι, καὶ Σάμην καλῶν τὴν νῆσον, ἣν πρότερον Σάμον ἐκάλεσεν.

[1] Κροκύλην *nox*.
[2] καί, after οὖν, marked out in B and omitted by *kno*.
[3] οἰῶν, not μήλων, is Homer's word (*Od.* 14. 100).
[4] ποιῶν *hi* and D *man. pr.*, instead of ποιεῖσθαι; so Meineke.

were in Euboea; and "Trojans and Lycians and Dardanians,"[1] meaning that the Lycians and Dardanians were Trojans)—however, after mentioning "Neritum,"[2] he says, "and dwelt in Crocyleia and rugged Aegilips, and those who held Zacynthos and those who dwelt about Samos, and those who held the mainland and dwelt in the parts over against the islands." By "mainland,"[3] therefore, he means the parts over against the islands, wishing to include, along with Leucas, the rest of Acarnania as well,[4] concerning which he also speaks in this way, "twelve herd on the mainland, and as many flocks of sheep,"[5] perhaps because Epeirotis extended thus far in early times and was called by the general name "mainland." But by "Samos" he means the Cephallenia of to-day, as, when he says, "in the strait between Ithaca and rugged Samos";[6] for by the epithet he differentiates between the objects bearing the same name, thus making the name apply, not to the city, but to the island. For the island was a Tetrapolis,[7] and one of its four cities was the city called indifferently either Samos or Samê, bearing the same name as the island. And when the poet says, "for all the nobles who hold sway over the islands, Dulichium and Samê and woody Zacynthos,"[8] he is evidently making an enumeration of the islands and calling "Samê" that island which he had formerly[9] called Samos. But

[1] *Iliad* 8. 173. [2] *Iliad* 2. 632.

[3] "epeirus" (cp. "Epeirus").

[4] On Homer's use of this "poetic figure," in which he specifies the part with the whole, cp. 8. 3. 8 and 1. 2. 23.

[5] *Od.* 14. 100. [6] *Od.* 4. 671.

[7] *i.e.* politically it was composed of four cities.

[8] *Od.* 1. 245. [9] *Iliad* 2. 634.

Ἀπολλόδωρος δέ, τοτὲ μὲν[1] τῷ ἐπιθέτῳ λέγων διεστάλθαι τὴν ἀμφιβολίαν, εἰπόντα

Σάμοιό τε παιπαλοέσσης,

ὡς τὴν νῆσον λέγοντα· τοτὲ δὲ ἀντιγράφεσθαι[2] δεῖν

Δουλιχίῳ τε Σάμῳ τε,

ἀλλὰ μή

Σάμῃ τε,

δῆλός ἐστι τὴν μὲν πόλιν Σάμην καὶ Σάμον συνωνύμως ὑπολαμβάνων ἐκφέρεσθαι, τὴν δὲ νῆσον Σάμον μόνον· ὅτι γὰρ Σάμη λέγεται ἡ πόλις, δῆλον εἶναι ἔκ τε τοῦ διαριθμούμενον τοὺς ἐξ ἑκάστης πόλεως μνηστῆρας φάναι,

ἐκ δὲ Σάμης πίσυρές τε καὶ εἴκοσι φῶτες ἔασι,

καὶ ἐκ τοῦ περὶ τῆς Κτιμένης λόγου·

τὴν μὲν ἔπειτα Σάμηνδ' ἔδοσαν.

C 454 ἔχει δὲ ταῦτα λόγον, οὐ γὰρ εὐκρινῶς ἀποδίδωσιν ὁ ποιητὴς οὔτε περὶ τῆς Κεφαλληνίας, οὔτε περὶ τῆς Ἰθάκης καὶ τῶν ἄλλων πλησίον[3] τόπων, ὥστε καὶ οἱ ἐξηγούμενοι διαφέρονται καὶ οἱ ἱστοροῦντες.

11. Αὐτίκα γὰρ ἐπὶ τῆς Ἰθάκης, ὅταν φῇ·

οἵ ῥ' Ἰθάκην εἶχον καὶ Νήριτον εἰνοσίφυλλον,

ὅτι μὲν τὸ Νήριτον ὄρος λέγει, τῷ ἐπιθέτῳ δηλοῖ. ἐν ἄλλοις δὲ καὶ ῥητῶς ὄρος·

ναιετάω δ' Ἰθάκην εὐδείελον· ἐν δ' ὄρος αὐτῇ,
Νήριτον εἰνοσίφυλλον ἀριπρεπές.

Apollodorus,[1] when he says in one passage that ambiguity is removed by the epithet when the poet says "and *rugged* Samos,"[2] showing that he meant the island, and then, in another passage, says that one should copy the reading, "Dulichium and Samos,"[3] instead of "Samê," plainly takes the position that the city was called "Samê" or "Samos" indiscriminately, but the island "Samos" only; for that the city was called Samê is clear, according to Apollodorus, from the fact that, in enumerating the wooers from the several cities, the poet[4] said, "from Samê came four and twenty men,"[5] and also from the statement concerning Ktimenê, "they then sent her to Samê to wed."[6] But this is open to argument, for the poet does not express himself distinctly concerning either Cephallenia or Ithaca and the other places near by; and consequently both the commentators and the historians are at variance with one another.

11. For instance, when Homer says in regard to Ithaca, "those who held Ithaca and Neritum with quivering foliage,"[7] he clearly indicates by the epithet that he means the mountain Neritum; and in other passages he expressly calls it a mountain; "but I dwell in sunny Ithaca, wherein is a mountain, Neritum, with quivering leaves and conspicuous from afar."[8] But whether by Ithaca he means the

[1] See *Dictionary* in Vol. I. [2] *Od.* 4. 671. [3] *Od.* 1. 246.
[4] In the words of Telemachus. [5] *Od.* 16. 249.
[6] *Od.* 15. 367. [7] *Iliad* 2. 632. [8] *Od.* 9. 21.

---

[1] ἐν, after μέν, Corais omits.
[2] ἀντιγράφεσθαι, Tzschucke and Corais, following *ox*, for γράφεσθαι E, ἂν γράφεσθαι BCD*hikln*.
[3] πλησίον, *h* and the editors, instead of πλησίων.

Ἰθακην δ' εἴτε τὴν πόλιν, εἴτε τὴν νῆσον λέγει, οὐ δῆλον ἐν τούτῳ γε τῷ ἔπει·

οἵ ῥ'[1] Ἰθάκην εἶχον καὶ Νήριτον.

κυρίως μὲν γὰρ ἀκούων τις τὴν πόλιν δέξαιτ' ἄν, ὡς καὶ Ἀθήνας καὶ Λυκαβηττὸν εἴ τις λέγοι, καὶ Ῥόδον καὶ Ἀτάβυριν, καὶ ἔτι Λακεδαίμονα καὶ Ταΰγετον· ποιητικῶς δὲ τοὐναντίον. ἐν μέντοι τῷ

ναιετάω δ' Ἰθάκην εὐδείελον· ἐν δ' ὄρος αὐτῇ Νήριτον

δῆλον·[2] ἐν γὰρ τῇ νήσῳ, οὐκ ἐν τῇ πόλει τὸ ὄρος. ὅταν δὲ[3] οὕτω φῇ·

ἡμεῖς ἐξ Ἰθάκης ὑπὸ Νηίου εἰλήλουθμεν,

ἄδηλον,[4] εἴτε τὸ αὐτὸ τῷ Νηρίτῳ λέγει τὸ Νήιον, εἴτε ἕτερον, ἢ ὄρος ἢ χωρίον. [5]ὁ μέντοι ἀντὶ Νηρίτου γράφων Νήρικον, ἢ ἀνάπαλιν, παραπαίει τελέως· τὸ μὲν γὰρ εἰνοσίφυλλον καλεῖ ὁ ποιητής, τὸ δ' ἐϋκτίμενον πτολίεθρον, καὶ τὸ μὲν ἐν Ἰθάκῃ, τὸ δ' ἀκτὴν ἠπείροιο.

12. Καὶ τοῦτο δὲ δοκεῖ ὑπεναντιότητά τινα δηλοῦν·

αὐτὴ δὲ χθαμαλὴ πανυπερτάτη εἰν ἁλὶ κεῖται·

χθαμαλὴ μὲν γὰρ ἡ ταπεινὴ καὶ χαμηλή, πανυπερτάτη δὲ ἡ ὑψηλή, οἵαν διὰ πλειόνων σημαίνει, κραναὴν καλῶν· καὶ τὴν ὁδὸν τὴν ἐκ τοῦ λιμένος

[1] οἵ ῥ', *nosx* and the editors, instead of οἵ τ'.
[2] δῆλον, after Νήριτον, Corais inserts; so the later editors.
[3] δέ, after ὅταν, *o* and the editors, instead of τε.
[4] ἄδηλον, Xylander and later editors, instead of οὐ ἄδηλον B by corr. and *x*, δῆλον other MSS.
[5] ὁ μέντοι . . . ἠπείροιο, Kramer suspects and Meineke rejects.

city or the island, is not clear, at least in the following verse, "those who held Ithaca and Neritum";[1] for if one takes the word in its proper sense, one would interpret it as meaning the city, just as though one should say "Athens and Lycabettus," or "Rhodes and Atabyris," or "Lacedaemon and Taÿgetus"; but if he takes it in a poetical sense the opposite is true. However, in the words, "but I dwell in sunny Ithaca, wherein is a mountain Neritum,"[2] his meaning is clear, for the mountain is in the island, not in the city. But when he says as follows, "we have come from Ithaca below Neïum,"[3] it is not clear whether he means that Neïum is the same as Neritum or different, or whether it is a mountain or place. However, the critic who writes Nericum[4] instead of Neritum, or the reverse, is utterly mistaken; for the poet refers to the latter as "quivering with foliage,"[5] but to the former as "well-built citadel,"[6] and to the latter as "in Ithaca,"[7] but to the former as "shore of the mainland."[8]

12. The following verse also is thought to disclose a sort of contradiction: "Now Ithaca itself lies *chthamalê, panypertatê* on the sea";[9] for *chthamalê* means "low," or "on the ground," whereas *panypertatê* means "high up," as Homer indicates in several places when he calls Ithaca "rugged."[10] And so when he refers to the road that leads from

[1] *Iliad* 2. 632. [2] *Od.* 9. 21. [3] *Od.* 3. 81.
[4] Accusative of "Nericus." [5] *Iliad* 2. 632.
[6] *Od.* 24. 377. [7] *Od.* 9. 21. [8] *Od.* 24. 378.
[9] *Od.* 9. 25 (see 1. 2. 20 and foot-note).
[10] *Iliad* 3. 201; *Od.* 1. 247; 9. 27; 10. 417, 463; 15. 510; 16. 124; 21. 346

τρηχεῖαν ἀταρπὸν
χῶρον ἀν' ὑλήεντα·

καὶ

οὐ γάρ τις νήσων εὐδείελος,[1] οὐδ' εὐλείμων,
αἵ θ' ἁλὶ κεκλίαται· Ἰθάκη δέ τε καὶ περὶ πασέων.

ἔχει μὲν οὖν ἀπεμφάσεις τοιαύτας ἡ φράσις, ἐξηγοῦνται δὲ οὐ κακῶς· οὔτε γὰρ χθαμαλὴν δέχονται ταπεινὴν ἐνταῦθα, ἀλλὰ πρόσχωρον τῇ ἠπείρῳ, ἐγγυτάτω οὖσαν αὐτῆς· οὔτε πανυπερτάτην ὑψηλοτάτην, ἀλλὰ πανυπερτάτην πρὸς ζόφον, οἷον ὑπὲρ πάσας ἐσχάτην[2] τετραμμένην πρὸς ἄρκτον· τοῦτο γὰρ βούλεται λέγειν τὸ πρὸς ζόφον, τὸ δ' ἐναντίον πρὸς νότον·

C 455 αἱ δέ τ' ἄνευθε πρὸς ἠῶ τ' ἠέλιόν τε·

τὸ γὰρ ἄνευθε πόρρω καὶ χωρίς ἐστιν, ὡς τῶν μὲν ἄλλων πρὸς νότον κεκλιμένων καὶ ἀπωτέρω τῆς ἠπείρου, τῆς δ' Ἰθάκης ἐγγύθεν καὶ[3] πρὸς ἄρκτον. ὅτι δ' οὕτω λέγει τὸ νότιον μέρος, καὶ ἐν τοῖσδε φανερόν·

εἴτ' ἐπὶ δεξί' ἴωσι, πρὸς ἠῶ τ' ἠέλιόν τε,
εἴτ' ἐπ' ἀριστερὰ τοίγε, ποτὶ ζόφον ἠερόεντα·

καὶ ἔτι μᾶλλον ἐν τοῖσδε·

ὦ φίλοι, οὐ γάρ τ' ἴδμεν, ὅπη ζόφος, οὐδ' ὅπη ἠώς,
οὐδ' ὅπη ἠέλιος φαεσίμβροτος εἶσ' ὑπὸ γαῖαν,
οὐδ' ὅπη ἀννεῖται·

[1] Instead of εὐδείελος the margin of B has ἱππήλατος, the Homeric reading.

[2] ἐσχάτην E, πρὸς ἐσχάτην BCklno, ὡς ἐσχάτην x; ἐσχάτην omitted by Dhi.

[3] καί, after ἐγγύθεν, omitted by MSS. except E.

the harbour as "rugged path up through the wooded place,"[1] and when he says "for not one of the islands which lean upon the sea is *eudeielos*[2] or rich in meadows, and Ithaca surpasses them all."[3] Now although Homer's phraseology presents incongruities of this kind, yet they are not poorly explained; for, in the first place, writers do not interpret *chthamalê* as meaning "low-lying" here, but "lying near the mainland," since it is very close to it, and, secondly, they do not interpret *panypertatê* as meaning "highest," but "highest towards the darkness," that is, farthest removed towards the north beyond all the others; for this is what he means by "towards the darkness," but the opposite by "towards the south," as in "but the other islands lie *aneuthe* towards the dawn and the sun,"[4] for the word *aneuthe* is "at a distance," or "apart," implying that the other islands lie towards the south and farther away from the mainland, whereas Ithaca lies near the mainland and towards the north. That Homer refers in this way to the southerly region is clear also from these words, "whether they go to the right, towards the dawn and the sun, or yet to the left towards the misty darkness,"[5] and still more clear from these words, "my friends, lo, now we know not where is the place of darkness, nor of dawn, nor where the sun, that gives light to men, goes beneath the earth, nor where he rises."[6] For

[1] *Od.* 14. 1.

[2] On *eudeielos*, see 9. 2. 41 and foot-note.

[3] *Od.* 4. 607; but in this particular passage the Homeric text has *hippêlatos* ("fit for driving horses") instead of *eudeielos*, although in *Od.* 9. 21, and elsewhere, Homer does apply the latter epithet to Ithaca.

[4] *Od.* 9. 26.

[5] *Iliad* 12. 239.

[6] *Od.* 10. 190

ἔστι μὲν γὰρ δέξασθαι τὰ τέτταρα κλίματα, τὴν ἠῶ δεχομένους τὸ νότιον μέρος, ἔχει τέ[1] τινα τοῦτ' ἔμφασιν, ἀλλὰ βέλτιον τὸ κατὰ τὴν πάροδον τοῦ ἡλίου νοεῖν ἀντιτιθέμενον τῷ ἀρκτικῷ μέρει· ἐξάλλαξιν γάρ τινα τῶν οὐρανίων πολλὴν βούλεται σημαίνειν ὁ λόγος, οὐχὶ ψιλὴν ἐπίκρυψιν τῶν κλιμάτων, δεῖ γὰρ κατὰ πάντα συννεφῆ[2] καιρόν, ἄν θ' ἡμέρας, ἄν τε νύκτωρ συμβῇ, παρακολουθεῖν· τὰ δ' οὐράνια ἐξαλλάττει ἐπὶ πλέον τῷ πρὸς μεσημβρίαν μᾶλλον ἢ ἧττον προχωρεῖν[3] ἡμᾶς ἢ εἰς τοὐναντίον. τοῦτο δὲ οὐ δύσεως καὶ ἀνατολῆς ἐγκαλύψεις ποιεῖ, ἀλλὰ μεσημβρίας καὶ ἄρκτου, καὶ γὰρ αἰθρίας οὔσης συμβαίνει.[4] μάλιστα γὰρ ἀρκτικός ἐστιν ὁ πόλος· τούτου δὲ κινουμένου καὶ ποτὲ μὲν κατὰ κορυφὴν ἡμῖν γινομένου, ποτὲ δὲ ὑπὸ γῆς ὄντος, καὶ οἱ ἀρκτικοὶ συμμεταβάλλουσι, ποτὲ δὲ συνεκλείπουσι κατὰ τὰς τοιαύτας προχωρήσεις,[5] ὥστε οὐκ ἂν εἰδείης ὅπου ἐστὶ τὸ ἀρκτικὸν κλίμα, οὐδὲ ἀρχή.[6] εἰ δὲ τοῦτο, οὐδὲ τοὐναντίον ἂν

[1] τέ, Kramer, for δέ; so the later editors.

[2] συννεφῆ, Casaubon, for συναφῆ BCD*hikl*, συναφῆς *nox*; so the later editors.

[3] προχωρεῖν, Jones, for παραχωρεῖν (cp. similar emendation below).

[4] καὶ γὰρ . . . συμβαίνει, Jones transfers from position after ποιεῖ to position after ἄρκτου.

[5] προχωρήσεις, Jones, for παραχωρήσεις.

[6] ἐστιν, after ἀρχή, Jones deletes. Corais and Meineke, following conj. of Tyrwhitt, read οὐδ' εἰ ἀρχὴν ἐστίν ("or whether there is a northern clima at all"); Groskurd, following Tzschucke, reads οὐδ' ὅπου ἀρχή ἐστιν.

---

[1] But in this passage "climata" is used in a different sense from that in 1. 1. 10 (see also foot-note 2 *ad loc.*, Vol. I,

it is indeed possible to interpret this as meaning the four "climata,"[1] if we interpret "the dawn" as meaning the southerly region (and this has some plausibility), but it is better to conceive of the region which is along the path of the sun as set opposite to the northerly region, for the poetic words are intended to signify a considerable change in the celestial phenomena,[2] not merely a temporary concealment of the "climata," for necessarily concealment ensues every time the sky is clouded, whether by day or by night; but the celestial phenomena change to a greater extent as we travel farther and farther towards the south or in the opposite direction. Yet this travel causes a hiding, not of the western or eastern sky, but only of the southern or northern, and in fact this hiding takes place when the sky is clear; for the pole is the most northerly point of the sky, but since the pole moves and is sometimes at our zenith and sometimes below the earth, the arctic circles also change with it and in the course of such travels sometimes vanish with it,[3] so that you cannot know where the northern "clima" is, or even where it begins.[4] And if this is true,

p. 22). It means here the (four) quarters of the sky, (1) where the sun sets, (2) where it rises, (3) the region of the celestial north pole, and (4) the region opposite thereto south of the equator.

[2] Odysseus was at the isle of Circe when he uttered the words in question, and hence, relatively, the celestial phenomena had changed (see 1. 1. 21).

[3] *i.e.* the infinite number of possible northern arctic circles vanish when the traveller (going south) crosses the equator, and, in the same way, the corresponding quarter of the southern sky vanishes when the traveller, going north, crosses the equator (see Vol. I, p. 364, note 2).

[4] See critical note.

γνοίης. κύκλος δὲ τῆς Ἰθάκης ἐστὶν ὡς ὀγδοήκοντα[1] σταδίων. περὶ μὲν Ἰθάκης ταῦτα.

13. Τὴν δὲ Κεφαλληνίαν, τετράπολιν οὖσαν, οὔτ' αὐτὴν εἴρηκε τῷ νῦν ὀνόματι, οὔτε τῶν πόλεων οὐδεμίαν, πλὴν μιᾶς, εἴτε Σάμης εἴτε Σάμου, ἣ νῦν μὲν οὐκέτ' ἐστίν, ἴχνη δ' αὐτῆς δείκνυται κατὰ μέσον τὸν πρὸς Ἰθάκῃ πορθμόν· οἱ δ' ἀπ' αὐτῆς Σαμαῖοι καλοῦνται· αἱ δ' ἄλλαι καὶ νῦν εἰσὶν ἔτι, μικραὶ πόλεις τινές, Παλεῖς,[2] Πρώνησος καὶ Κράνιοι. ἐφ' ἡμῶν δὲ καὶ ἄλλην προσέκτισε Γάϊος Ἀντώνιος, ὁ θεῖος Μάρκου Ἀντωνίου, ἡνίκα φυγὰς γενόμενος μετὰ τὴν ὑπατείαν, ἣν συνῆρξε Κικέρωνι τῷ ῥήτορι, ἐν τῇ Κεφαλληνίᾳ διέτριψε καὶ τὴν ὅλην νῆσον ὑπήκοον ἔσχεν, ὡς ἴδιον κτῆμα· οὐκ ἔφθη μέντοι συνοικίσας, ἀλλὰ καθόδου τυχών, πρὸς ἄλλοις μείζοσιν ὢν κατέλυσε τὸν βίον.

14. Οὐκ ὤκνησαν δέ τινες τὴν Κεφαλληνίαν C 456 τὴν αὐτὴν τῷ Δουλιχίῳ φάναι, οἱ δὲ τῇ Τάφῳ, καὶ Ταφίους τοὺς Κεφαλληνίους, τοὺς δ' αὐτοὺς καὶ Τηλεβόας, καὶ τὸν Ἀμφιτρύωνα δεῦρο στρατεῦσαι μετὰ Κεφάλου τοῦ Δηιονέως, ἐξ Ἀθηνῶν φυγάδος, παραληφθέντος, κατασχόντα δὲ τὴν νῆσον παραδοῦναι τῷ Κεφάλῳ, καὶ ταύτην μὲν ἐπώνυμον ἐκείνου γενέσθαι, τὰς δὲ πόλεις τῶν παίδων αὐτοῦ. ταῦτα δ' οὐχ Ὁμηρικά· οἱ μὲν γὰρ Κεφαλλῆνες ὑπὸ Ὀδυσσεῖ καὶ Λαέρτῃ, ἡ δὲ Τάφος ὑπὸ τῷ Μέντῃ·

[1] But the Ithaca of to-day is nearer 300 stadia in circuit. Pliny says 25 Roman miles (*Nat. Hist.* 4. 12). Strabo must have written 180 (σ' π') or 280 (τ' π') instead of 80 (π'). And if he meant Leucas, the error would be far greater.

[2] Παλεῖς, Casaubon inserts; so the later editors.

neither can you know the opposite "clima." The circuit of Ithaca is about eighty stadia.[1] So much for Ithaca.

13. As for Cephallenia, which is a Tetrapolis, the poet mentions by its present name neither it nor any of its cities except one, Samê or Samos, which now no longer exists, though traces of it are to be seen midway of the passage to Ithaca; and its people are called Samaeans. The other three, however, survive even to this day in the little cities Paleis, Pronesus, and Cranii. And in our time Gaius Antonius, the uncle of Marcus Antonius, founded still another city, when, after his consulship, which he held with Cicero the orator, he went into exile,[2] sojourned in Cephallenia, and held the whole island in subjection as though it were his private estate. However, before he could complete the settlement he obtained permission to return home,[3] and ended his days amid other affairs of greater importance.

14. Some, however, have not hesitated to identify Cephallenia with Dulichium, and others with Taphos, calling the Cephallenians Taphians, and likewise Teleboans, and to say that Amphitryon made an expedition thither with Cephalus, the son of Deïoneus, whom, an exile from Athens, he had taken along with him, and that when Amphitryon seized the island he gave it over to Cephalus, and that the island was named after Cephalus and the cities after his children. But this is not in accordance with Homer; for the Cephallenians were subject to Odysseus and Laertes, whereas Taphos was subject

[1] See critical note. [2] 59 B.C.
[3] Probably from Caesar. He was back in Rome in 44 B.C.

Μέντης Ἀγχιάλοιο δαΐφρονος εὔχομαι εἶναι
υἱός, ἀτὰρ Ταφίοισι φιληρέτμοισιν ἀνάσσω.

καλεῖται δὲ νῦν Ταφιοῦς[1] ἡ Τάφος. οὐδ' Ἑλλάνικος Ὁμηρικός, Δουλίχιον τὴν Κεφαλληνίαν λέγων. τὸ μὲν γὰρ ὑπὸ Μέγητι εἴρηται καὶ αἱ λοιπαὶ Ἐχινάδες, οἵ τε ἐνοικοῦντες Ἐπειοὶ ἐξ Ἤλιδος ἀφιγμένοι· διόπερ καὶ τὸν Ὦτον τὸν Κυλλήνιον

Φυλείδεω[2] ἕταρον μεγαθύμων ἀρχὸν Ἐπειῶν

καλεῖ·

αὐτὰρ Ὀδυσσεὺς ἦγε Κεφαλλῆνας μεγαθύμους.

οὔτ' οὖν Δουλίχιον ἡ Κεφαλληνία καθ' Ὅμηρον, οὔτε τῆς Κεφαλληνίας τὸ Δουλίχιον, ὡς Ἄνδρων φησί· τὸ μὲν[3] γὰρ Ἐπειοὶ κατεῖχον, τὴν δὲ Κεφαλληνίαν ὅλην Κεφαλλῆνες, καὶ οἱ μὲν[4] ὑπὸ Ὀδυσσεῖ, οἱ δ' ὑπὸ Μέγητι. οὐδὲ[5] Παλεῖς Δουλίχιον ὑφ' Ὁμήρου λέγονται, ὡς γράφει Φερεκύδης. μάλιστα δ' ἐναντιοῦται Ὁμήρῳ ὁ τὴν Κεφαλληνίαν τὴν αὐτὴν τῷ Δουλιχίῳ λέγων, εἴπερ τῶν μνηστήρων ἐκ μὲν Δουλιχίοιο δύω καὶ πεντήκοντα ἦσαν, ἐκ δὲ Σάμης πίσυρές τε καὶ εἴκοσι. οὐ γὰρ τοῦτ' ἂν εἴη λέγων, ἐξ ὅλης μὲν τόσους, ἐκ δὲ μιᾶς τῶν τεττάρων παρὰ δύο[6] τοὺς ἡμίσεις; εἰ δ' ἄρα τοῦτο δώσει τις, ἐρησόμεθα, τίς ἂν εἴη ἡ Σάμη, ὅταν οὕτω φῇ·

Δουλίχιόν τε Σάμην τ' ἠδ' ὑλήεντα Ζάκυνθον.

[1] Ταφιοῦς, Meineke, following Pliny, emends to Ταφιάς; but see Ταφιοῦς in § 20 below.

[2] Φυλειδέω, Casaubon, for Φυλιέως *CDhiksx*, Φυλλιέως *Bl*, Φυλιδέω *Epit.*

[3] τὸ μέν, Tzschucke, for τὴν μέν; so the later editors.

[4] οἱ μέν, *k* inserts; Meineke omits the καί instead.

to Mentes: "I declare that I am Mentes the son of wise Anchialus, and I am lord over the oar-loving Taphians."[1] Taphos is now called Taphius. Neither is Hellanicus[2] in accord with Homer when he identifies Cephallenia with Dulichium, for Homer[3] makes Dulichium and the remainder of the Echinades subject to Meges; and their inhabitants were Epeians, who had come there from Elis; and it is on this account that he calls Otus the Cyllenian "comrade of Phyleides[4] and ruler of the high-hearted Epeians";[5] "but Odysseus led the high-hearted Cephallenians."[6] According to Homer, therefore, neither is Cephallenia Dulichium nor is Dulichium a part of Cephallenia, as Andron[7] says; for the Epeians held possession of Dulichium, whereas the Cephallenians held possession of the whole of Cephallenia and were subject to Odysseus, whereas the Epeians were subject to Meges. Neither is Paleis called Dulichium by the poet, as Pherecydes writes. But that writer is most in opposition to Homer who identifies Cephallenia with Dulichium, if it be true that "fifty-two" of the suitors were "from Dulichium" and "twenty-four from Samê";[8] for in that case would not Homer say that fifty-two came from the island as a whole and a half of that number less two from a single one of its four cities? However, if one grants this, I shall ask what Homer can mean by "Samê" in the passage, "Dulichium and Samê and woody Zacynthos."[9]

[1] *Od.* 1. 180. [2] See *Dictionary* in Vol. I. [3] *Iliad* 2. 625.
[4] Son of Phyleus (Meges). [5] *Iliad* 15. 519.
[6] *Iliad* 2. 631. [7] See foot-note on Andron, 10. 4. 6.
[8] *Od.* 16. 247, 249. [9] *Od.* 1. 246.

---

[5] *οὐδέ*, Groskurd, for *οἱ δέ*; so the later editors.
[6] *παρὰ δύο x*, *παρ' ἕνα* other MSS.

15. Κεῖται δ' ἡ Κεφαλληνία κατὰ Ἀκαρνανίαν, διέχουσα τοῦ Λευκάτα περὶ πεντήκοντα (οἱ δὲ τετταράκοντά φασι) σταδίους, τοῦ δὲ Χελωνάτα περὶ ἑκατὸν[1] ὀγδοήκοντα. αὐτὴ δ' ἐστὶν ὡς τριακοσίων[2] τὴν περίμετρον, μακρὰ δ' ἀνήκουσα πρὸς Εὖρον, ὀρεινή· μέγιστον δ' ὄρος ἐν αὐτῇ Αἶνος,[3] ἐν ᾧ τὸ τοῦ Διὸς Αἰνησίου ἱερόν· καθ' ὃ δὲ στενωτάτη ἐστὶν ἡ νῆσος, ταπεινὸν ἰσθμὸν ποιεῖ, ὥσθ' ὑπερκλύζεσθαι πολλάκις ἐκ θαλάττης εἰς θάλατταν· πλησίον δ' εἰσὶ τῶν στενῶν ἐν τῷ κόλπῳ Κράνιοί τε καὶ Παλεῖς.

16. Μεταξὺ δὲ τῆς Ἰθάκης καὶ τῆς Κεφαλληνίας ἡ Ἀστερία νησίον· Ἀστερὶς δ' ὑπὸ τοῦ ποιητοῦ λέγεται· ἣν ὁ μὲν Σκήψιος μὴ μένειν τοιαύτην, οἵαν φησὶν ὁ ποιητής,

> λιμένες δ' ἔνι ναύλοχοι αὐτῇ
> ἀμφίδυμοι,

C 457 ὁ δὲ Ἀπολλόδωρος μένειν καὶ νῦν, καὶ πολίχνιον λέγει ἐν αὐτῇ Ἀλαλκομενάς, τὸ ἐπ' αὐτῷ τῷ ἰσθμῷ κείμενον.

17. Καλεῖ δ' ὁ ποιητὴς Σάμον καὶ τὴν Θρᾳκίαν, ἣν νῦν Σαμοθρᾴκην καλοῦμεν. τὴν δ' Ἰωνικὴν οἶδε[4] μέν, ὡς εἰκός· καὶ γὰρ τὴν Ἰωνικὴν ἀποικίαν εἰδέναι φαίνεται· οὐκ ἂν[5] ἀντιδιέστειλε δὲ τὴν ὁμωνυμίαν, περὶ τῆς Σαμοθρᾴκης λέγων, τοτὲ μὲν τῷ ἐπιθέτῳ·

[1] ἑκατὸν (ρ′), Jones inserts, following conj. of C. Müller.

[2] Instead of τριακοσίων (τ′ = 300), Strabo probably wrote ἑπτακοσίων (ψ′ = 700), which, not counting the sinuosities of the gulfs, is about correct. Pliny (4. 19) says "93 miles" (744 stadia).

15. Cephallenia lies opposite Acarnania, at a distance of about fifty stadia from Leucatas (some say forty), and about one hundred and eighty from Chelonatas. It has a perimeter of about three hundred [1] stadia, is long, extending towards Eurus,[2] and is mountainous. The largest mountain upon it is Aenus, whereon is the temple of Zeus Aenesius; and where the island is narrowest it forms an isthmus so low-lying that it is often submerged from sea to sea. Both Paleis and Crannii are on the gulf near the narrows.

16. Between Ithaca and Cephallenia is the small island Asteria (the poet calls it Asteris), which the Scepsian [3] says no longer remains such as the poet describes it, "but in it are harbours safe for anchorage with entrances on either side";[4] Apollodorus, however, says that it still remains so to this day, and mentions a town Alalcomenae upon it, situated on the isthmus itself.

17. The poet also uses the name "Samos" for that Thrace which we now call Samothrace. And it is reasonable to suppose that he knows the Ionian Samos, for he also appears to know of the Ionian migration; otherwise he would not have differentiated between the places of the same name when referring to Samothrace, which he designates at one time by the

[1] See critical note.
[2] *i.e.* towards the direction of winter sunrise (rather south-east) as explained by Poseidonius (see discussion in 1. 2. 21).
[3] Demetrius of Scepsis. [4] *Od.* 4. 846.

---

[3] Αἶνος, Xylander inserts; so the later editors.
[4] εἶδε *Bkl.*
[5] ἄν, Corais inserts; so the later editors.

ὑψοῦ ἐπ' ἀκροτάτης κορυφῆς Σάμου ὑλήεσσης,
Θρηικίης·

τοτὲ δὲ τῇ συζυγίᾳ τῶν πλησίον νήσων·

ἐς Σάμον ἔς τ' Ἴμβρον καὶ Λῆμνον ἀμιχθαλόεσσαν·

καὶ πάλιν·

μεσσηγύς τε Σάμοιο καὶ Ἴμβρου παιπαλοέσσης.

ᾔδει μὲν οὖν, οὐκ ὠνόμακε δ' αὐτήν· οὐδ' ἐκαλεῖτο τῷ αὐτῷ ὀνόματι πρότερον, ἀλλὰ Μελάμφυλος, εἶτ' Ἀνθεμίς, εἶτα Παρθενία ἀπὸ τοῦ ποταμοῦ τοῦ Παρθενίου, ὃς Ἴμβρασος μετωνομάσθη. ἐπεὶ οὖν κατὰ τὰ Τρωικὰ Σάμος μὲν καὶ ἡ Κεφαλληνία ἐκαλεῖτο καὶ ἡ Σαμοθρᾴκη (οὐ γὰρ ἂν Ἑκάβη εἰσήγετο λέγουσα, ὅτι τοὺς παῖδας αὐτῆς πέρνασχ', ὅν κε λάβοι, ἐς Σάμον ἔς τ' Ἴμβρον),[1] Ἰωνικὴ δ'[2] οὐκ ἀπῴκιστό πω, δῆλον δ'[3] ὅτι ἀπὸ τῶν προτέρων τινὸς τὴν ὁμωνυμίαν ἔσχεν· ἐξ ὧν κἀκεῖνο δῆλον, ὅτι παρὰ τὴν ἀρχαίαν ἱστορίαν ὃ λέγουσιν οἱ φήσαντες, μετὰ τὴν Ἰωνικὴν ἀποικίαν καὶ τὴν Τεμβρίωνος παρουσίαν ἀποίκους ἐλθεῖν ἐκ Σάμου καὶ ὀνομάσαι Σάμον τὴν Σαμοθρᾴκην, ὡς οἱ Σάμιοι τοῦτ' ἐπλάσαντο δόξης χάριν. πιθανώτεροι δ' εἰσὶν οἱ[4] ἀπὸ τοῦ σάμους[5] καλεῖσθαι τὰ ὕψη φήσαντες εὑρῆσθαι τοῦτο τοὔνομα τὴν νῆσον· ἐντεῦθεν γάρ

ἐφαίνετο πᾶσα μὲν Ἴδη,
φαίνετο δὲ Πριάμοιο πόλις καὶ νῆες Ἀχαιῶν.

[1] Before Ἰωνική *hi* have ἡ, *x* ὥστ', *y* ὥστε ἡ, Corais ἡ δ'.
[2] Kramer inserts δ' before οὐκ; so the later editors.
[3] Kramer inserts δ' before ὅτι; so the later editors.

epithet, "high on the topmost summit of woody Samos, the Thracian,"[1] and at another time by connecting it with the islands near it, "unto Samos and Imbros and inhospitable[2] Lemnos." And again, "between Samos and rugged Imbros." He therefore knew the Ionian island, although he did not name it; in fact it was not called by the same name in earlier times, but Melamphylus, then Anthemis, then Parthenia, from the River Parthenius, the name of which was changed to Imbrasus. Since, then, both Cephallenia and Samothrace were called Samos at the time of the Trojan War (for otherwise Hecabe would not be introduced as saying that he[3] was for selling her children whom he might take captive "unto Samos and unto Imbros"),[4] and since the Ionian Samos had not yet been colonised, it plainly got its name from one of the islands which earlier bore the same name. Whence that other fact is also clear, that those writers contradict ancient history who say that colonists came from Samos after the Ionian migration and the arrival of Tembrion[5] and named Samothrace Samos, since this story was fabricated by the Samians to enhance the glory of their island. Those writers are more plausible who say that the island came upon this name from the fact that lofty places are called "samoi,"[6] "for thence all Ida was plain to see, and plain to see were the city of Priam and the ships of the Achaeans."[7] But some say that the island was

[1] *Iliad* 13. 12.
[2] Or "smoky"; the meaning of the Greek word is doubtful.
[3] Achilles. [4] *Iliad* 24. 752. [5] See 14. 1. 3.
[6] See 8. 3. 19. [7] *Iliad* 13. 13.

---

[4] *οἱ*, before *ἀπό*, CD*hil* omit.
[5] *σάμους* E, *σαμαίους* other MSS.

τινὲς δὲ Σάμον καλεῖσθαί φασιν ἀπὸ Σαΐων, τῶν οἰκούντων Θρᾳκῶν πρότερον, οἳ καὶ τὴν ἤπειρον ἔσχον τὴν προσεχῆ, εἴτε οἱ αὐτοὶ τοῖς Σαπαίοις ὄντες ἢ τοῖς Σιντοῖς, οὓς Σίντιας καλεῖ ὁ ποιητής, εἴθ' ἕτεροι. μέμνηται δὲ τῶν Σαΐων Ἀρχίλοχος·

ἀσπίδα μὲν Σαΐων τις ἀνείλετο,[1] τὴν παρὰ θάμνῳ
ἔντος ἀμώμητον κάλλιπον οὐκ ἐθέλων.

18. Λοιπὴ δ' ἐστὶ τῶν ὑπὸ τῷ Ὀδυσσεῖ τεταγμένων νήσων ἡ Ζάκυνθος, μικρῷ πρὸς C 458 ἑσπέραν μᾶλλον τῆς Κεφαλληνίας κεκλιμένη[2] τῆς Πελοποννήσου, συνάπτουσα δ' αὐτῇ[3] πλέον. ἔστιν ὁ κύκλος τῆς Ζακύνθου σταδίων ἑκατὸν[4] ἑξήκοντα· διέχει δὲ καὶ τῆς Κεφαλληνίας ὅσον ἑξήκοντα σταδίους, ὑλώδης μέν, εὔκαρπος δέ· καὶ ἡ πόλις ἀξιόλογος ὁμώνυμος. ἐντεῦθεν εἰς Ἑσπερίδας τῆς Λιβύης στάδιοι τρισχίλιοι τριακόσιοι.[5]

19. Καὶ ταύτης δὲ καὶ τῆς Κεφαλληνίας πρὸς ἕω τὰς Ἐχινάδας ἱδρῦσθαι νήσους συμβέβηκεν· ὧν τό τε Δουλίχιόν ἐστι (καλοῦσι δὲ νῦν Δολίχαν) καὶ αἱ Ὀξεῖαι καλούμεναι, ἃς Θοὰς ὁ ποιητὴς εἶπε· καὶ ἡ μὲν Δολίχα κεῖται κατὰ Οἰνειάδας καὶ τὴν ἐκβολὴν τοῦ Ἀχελῴου, διέ-

[1] ἀνείλετο *Epit.* and corr. in B, ἀνείλατο B*gy*, ἀφείλατο *s*, ἀγείλατο *i*, ἀγάλλεται editors before Kramer (cp. readings of same passage in 12. 3. 20).

[2] Palmer omits καί before τῆς; so Tzschucke, Groskurd, and Meineke.

[3] αὐτῇ, Kramer, for αὐτή (*gxy*); συνάπτων δ' αὐτὴν (πλέον ἐστὶν ὁ κτλ.), other MSS.; so the later editors.

[4] Instead of ἑκατὸν (ρ' = 100) Strabo almost certainly

called Samos after the Saïi, the Thracians who inhabited it in earlier times, who also held the adjacent mainland, whether these Saïi were the same people as the Sapaeï or Sinti (the poet calls them Sinties) or a different tribe. The Saïi are mentioned by Archilochus: "One of the Saïi robbed me of my shield, which, a blameless weapon, I left behind me beside a bush, against my will."[1]

18. Of the islands classified as subject to Odysseus, Zacynthos remains to be described. It leans slightly more to the west of the Peloponnesus than Cephallenia and lies closer to the latter. The circuit of Zacynthos is one hundred and sixty stadia.[2] It is about sixty stadia distant from Cephallenia. It is indeed a woody island, but it is fertile; and its city, which bears the same name, is worthy of note. The distance thence to the Libyan Hesperides is three thousand three hundred stadia.

19. To the east of Zacynthos and Cephallenia are situated the Echinades Islands, among which is Dulichium, now called Dolicha, and also what are called the Oxeiae, which the poet called Thoae.[3] Dolicha lies opposite Oeniadae and the outlet of the Achelous, at a distance of one hundred stadia from

[1] Bergk, *Frag.* 6 (51). Two more lines are preserved: "but I myself escaped the doom of death. Farewell to that shield! I shall get another one as good."

[2] See critical note.

[3] In Greek "Oxeiai" and "Thoai," both words meaning "sharp" or "pointed" (see 8. 3. 26 and foot-note, and *Od.* 15. 299).

wrote πεντακόσιοι (φ' = 500). 560 stadia is about correct for the circuit. Pliny's text has 36 miles (4. 12).

[5] Meineke emends τριακόσιοι (τ') to ἑξακόσιοι (χ' = 600), as in 17. 3. 20, but this is doubtful.

χουσα Ἀράξου, τῆς τῶν Ἠλείων ἄκρας, ἑκατόν. καὶ αἱ λοιπαὶ δ' Ἐχινάδες (πλείους εἰσί, πᾶσαι λυπραὶ καὶ τραχεῖαι)[1] πρὸ τῆς ἐκβολῆς τοῦ Ἀχελῴου, πεντεκαίδεκα σταδίους ἀφεστῶσα ἡ ἀπωτάτω, ἡ δ' ἐγγυτάτω πέντε, πελαγίζουσαι πρότερον· ἀλλ' ἡ χοῦς τὰς μὲν ἐξηπείρωκεν αὐτῶν ἤδη, τὰς δὲ μέλλει, πολλὴ καταφερομένη· ἥπερ καὶ τὴν Παραχελωῖτιν[2] καλουμένην χώραν,[3] ἣν ὁ ποταμὸς ἐπικλύζει, περιμάχητον[4] ἐποίει τὸ παλαιόν, τοὺς ὅρους συγχέουσα ἀεὶ τοὺς ἀποδεικνυμένους τοῖς Ἀκαρνᾶσι καὶ τοῖς Αἰτωλοῖς· ἐκρίνοντο γὰρ τοῖς ὅπλοις, οὐκ ἔχοντες διαιτητάς, ἐνίκων δ' οἱ πλέον δυνάμενοι· ἀφ' ἧς αἰτίας καὶ μῦθος ἐπλάσθη τις, ὡς Ἡρακλέους καταπολεμήσαντος τὸν Ἀχελῷον καὶ ἐνεγκαμένου τῆς νίκης ἆθλον τὸν Δηιανείρας γάμον, τῆς Οἰνέως θυγατρός, ἣν πεποίηκε Σοφοκλῆς τοιαῦτα λέγουσαν·

> μνηστὴρ γὰρ ἦν μοι ποταμός, Ἀχελῷον λέγω,
> ὅς μ' ἐν τρισὶν μορφαῖσιν ἐξῄτει πατρός,
> φοιτῶν ἐναργὴς ταῦρος, ἄλλοτ' αἰόλος
> δράκων ἑλικτός, ἄλλοτ' ἀνδρείῳ κύτει[5]
> βούπρωρος.

προστιθέασι δ' ἔνιοι καὶ τὸ τῆς Ἀμαλθείας τοῦτ' εἶναι λέγοντες κέρας, ὃ ἀπέκλασεν ὁ Ἡρακλῆς τοῦ Ἀχελῴου καὶ ἔδωκεν Οἰνεῖ τῶν γάμων ἕδνον·

[1] Corais omits καί before πρό; so Meineke.
[2] Παραχελῴιν *Bkl*, Παραχελῴην *nosx*, Παραχελῷτιν D.
[3] After χώραν *x* adds ἐστὶ προσχοῦσα; so Corais.
[4] Xylander omits δέ before ἐποίει; so Meineke.
[5] τύπῳ *Dhil.*

Araxus, the promontory of the Eleians; the rest of the Echinades (they are several in number, all poor-soiled and rugged) lie off the outlet of the Acheloüs, the farthermost being fifteen stadia distant and the nearest five. In earlier times they lay out in the high sea, but the silt brought down by the Acheloüs has already joined some of them to the mainland and will do the same to others. It was this silt which in early times caused the country called Paracheloïtis,[1] which the river overflows, to be a subject of dispute, since it was always confusing the designated boundaries between the Acarnanians and the Aetolians; for they would decide the dispute by arms, since they had no arbitrators, and the more powerful of the two would win the victory; and this is the cause of the fabrication of a certain myth, telling how Heracles defeated Acheloüs and, as the prize of his victory, won the hand of Deïaneira, the daughter of Oeneus, whom Sophocles represents as speaking as follows: "For my suitor was a river-god, I mean Acheloüs, who would demand me of my father in three shapes, coming now as a bull in bodily form, now as a gleaming serpent in coils, now with trunk of man and front of ox."[2] Some writers add to the myth, saying that this was the horn of Amaltheia,[3] which Heracles broke off from Acheloüs and gave to Oeneus as a wedding gift. Others, conjecturing the

[1] *i.e.* "Along the Acheloüs."

[2] *Trachiniae* 7–11. One vase-painting shows Acheloüs fighting with Achilles as a serpent with the head and arms of a man, and with ox-horns, and another as a human figure, except that he had the forehead, horns, and ears of an ox (Jebb, note *ad loc.*).

[3] Cf. 3. 2. 14 and foot-note.

οἱ δ', εἰκάζοντες ἐξ αὐτῶν τἀληθές, ταύρῳ μὲν ἐοικότα λέγεσθαι τὸν Ἀχελῷόν φασι, καθάπερ καὶ τοὺς ἄλλους ποταμούς, ἀπό τε τῶν ἤχων καὶ τῶν κατὰ τὰ ῥεῖθρα καμπῶν, ἃς καλοῦσι κέρατα, δράκοντι δὲ διὰ τὸ μῆκος καὶ τὴν σκολιότητα, βούπρωρον δὲ διὰ τὴν αὐτὴν αἰτίαν, δι' ἣν καὶ ταυρωπόν· τὸν Ἡρακλέα δέ, καὶ ἄλλως εὐεργετικὸν ὄντα καὶ τῷ Οἰνεῖ κηδεύσοντα, παραχώμασί τε καὶ διοχετείαις βιάσασθαι τὸν ποταC 459 μὸν πλημμελῶς ῥέοντα καὶ πολλὴν τῆς Παραχελωίτιδος [1] ἀναψῦξαι [2] χαριζόμενον τῷ Οἰνεῖ· καὶ τοῦτ' εἶναι τὸ τῆς Ἀμαλθείας κέρας. τῶν μὲν οὖν Ἐχινάδων καὶ τῶν Ὀξειῶν κατὰ τὰ Τρωικὰ Μέγητα ἄρχειν φησὶν Ὅμηρος,

ὃν τίκτε Διὶ φίλος ἱππότα Φυλεύς,
ὅς ποτε Δουλιχίονδ' ἀπενάσσατο, πατρὶ χολωθείς.

πατὴρ δ' ἦν Αὐγέας, ὁ τῆς Ἠλείας καὶ τῶν Ἐπειῶν ἄρχων· ὥστ' Ἐπειοὶ τὰς νήσους ταύτας εἶχον οἱ συνεξάραντες εἰς τὸ Δουλίχιον τῷ Φυλεῖ.

20. Αἱ δὲ τῶν Ταφίων νῆσοι, πρότερον δὲ Τηλεβοῶν, ὧν ἦν καὶ ἡ Τάφος, νῦν δὲ Ταφιοῦς [3] καλουμένη, χωρὶς ἦσαν τούτων, οὐ τοῖς διαστήμασιν [4] (ἐγγὺς γὰρ κεῖνται), ἀλλὰ ὑφ' ἑτέροις ἡγεμόσι ταττόμεναι, Ταφίοις καὶ Τηλεβόαις· πρότερον μὲν οὖν Ἀμφιτρύων, ἐπιστρατεύσας

[1] After Παραχελωίτιδος, *Bnox* add φθείροντα.
[2] ἀναψῦξαι, Villebrun, for ἀναψύξιν; so the later editors.
[3] Ταφιοῦς, Meineke, following Pliny, emends to Ταφιάς; but see Ταφίους in § 14 above.

truth from the myths, say that the Acheloüs, like the other rivers, was called "like a bull" from the roaring of its waters, and also from the the bendings of its streams, which were called Horns, and "like a serpent" because of its length and windings, and "with front of ox"[1] for the same reason that he was called "bull-faced"; and that Heracles, who in general was inclined to deeds of kindness, but especially for Oeneus, since he was to ally himself with him by marriage, regulated the irregular flow of the river by means of embankments and channels, and thus rendered a considerable part of Paracheloïtis dry, all to please Oeneus; and that this was the horn of Amaltheia.[2] Now, as for the Echinades, or the Oxeiae, Homer says that they were ruled over in the time of the Trojan War by Meges, "who was begotten by the knightly Phyleus, dear to Zeus, who once changed his abode to Dulichium because he was wroth with his father."[3] His father was Augeas, the ruler of the Eleian country and the Epeians; and therefore the Epeians who set out for Dulichium with Phyleus held these islands.

20. The islands of the Taphians, or, in earlier times, of the Teleboans, among which was Taphos, now called Taphius, were distinct from the Echinades; not in the matter of distances (for they lie near them), but in that they are classified as under different commanders, Taphians and Teleboans.[4] Now in earlier times Amphitryon made an expedition

[1] Literally, "ox-prowed" (see Jebb, *loc. cit.*).
[2] Cp. 3. 2. 14.
[3] *Iliad* 2. 628.
[4] The latter name is not found in the *Iliad* or *Odyssey*.

[4] διαστήμασιν, Xylander, for διαιτήμασιν BDE*klnox*.

truth from the myths, say that the Acheloüs, like the other rivers, was called "like a bull" from the roaring of its waters, and also from the the bendings of its streams, which were called Horns, and "like a serpent" because of its length and windings, and "with front of ox"[1] for the same reason that he was called "bull-faced"; and that Heracles, who in general was inclined to deeds of kindness, but especially for Oeneus, since he was to ally himself with him by marriage, regulated the irregular flow of the river by means of embankments and channels, and thus rendered a considerable part of Paracheloïtis dry, all to please Oeneus; and that this was the horn of Amaltheia.[2] Now, as for the Echinades, or the Oxeiae, Homer says that they were ruled over in the time of the Trojan War by Meges, "who was begotten by the knightly Phyleus, dear to Zeus, who once changed his abode to Dulichium because he was wroth with his father."[3] His father was Augeas, the ruler of the Eleian country and the Epeians; and therefore the Epeians who set out for Dulichium with Phyleus held these islands.

20. The islands of the Taphians, or, in earlier times, of the Teleboans, among which was Taphos, now called Taphius, were distinct from the Echinades; not in the matter of distances (for they lie near them), but in that they are classified as under different commanders, Taphians and Teleboans.[4] Now in earlier times Amphitryon made an expedition

[1] Literally, "ox-prowed" (see Jebb, *loc. cit.*).
[2] Cp. 3. 2. 14.
[3] *Iliad* 2. 628.
[4] The latter name is not found in the *Iliad* or *Odyssey*.

---

[4] διαστήμασιν, Xylander, for διαιτήμασιν BDE*klnox*.

αὐτοῖς μετὰ Κεφάλου τοῦ Δηιονέως[1] ἐξ Ἀθηνῶν φυγάδος, ἐκείνῳ τὴν ἀρχὴν παρέδωκεν αὐτῶν· ὁ δὲ ποιητὴς ὑπὸ Μέντῃ τετάχθαι φησί, λῃστὰς καλῶν αὐτούς, καθάπερ καὶ τοὺς Τηλεβόας ἅπαντάς φασι.[2] τὰ μὲν περὶ τὰς νήσους τὰς πρὸ τῆς Ἀκαρνανίας ταῦτα.

21. Μεταξὺ δὲ Λευκάδος καὶ τοῦ Ἀμβρακικοῦ κόλπου λιμνοθάλαττά ἐστι, Μυρτούντιον λεγομένη. ἀπὸ δὲ Λευκάδος ἑξῆς Πάλαιρος καὶ Ἀλυζία τῆς Ἀκαρνανίας εἰσὶ[3] πόλεις,[4] ὧν ἡ Ἀλυζία πεντεκαίδεκα ἀπὸ θαλάττης διέχει σταδίους, καθ' ἥν ἐστι λιμὴν Ἡρακλέους ἱερὸς καὶ τέμενος, ἐξ οὗ[5] τοὺς Ἡρακλέους ἄθλους, ἔργα Λυσίππου, μετήνεγκεν εἰς Ῥώμην τῶν ἡγεμόνων τις, παρὰ τόπον[6] κειμένους διὰ τὴν ἐρημίαν. εἶτα ἄκρα Κριθωτὴ[7] καὶ αἱ[8] Ἐχινάδες καὶ πόλις Ἀστακός, ὁμώνυμος τῇ περὶ Νικομήδειαν καὶ τὸν Ἀστακηνὸν κόλπον, θηλυκῶς[9] λεγομένη. καὶ ἡ Κριθωτὴ δ' ὁμώνυμος πολίχνῃ[10] τῶν ἐν τῇ Θρᾳκίᾳ Χερρονήσῳ. πάντα δ' εὐλίμενα τὰ μεταξύ· εἶτ' Οἰνιάδαι καὶ ὁ Ἀχελῷος· εἶτα λίμνη τῶν Οἰνιαδῶν, Μελίτη καλουμένη, μῆκος μὲν ἔχουσα τριάκοντα σταδίων, πλάτος δὲ εἴκοσι, καὶ ἄλλη Κυνία, διπλασία ταύτης καὶ

[1] Δηιονέως E and Eustathius (note on *Od.* 1. 105), Δηίονος CDB*hlnsx*, Δηιόνεος B*o* by corr., Δηίωνος *k*.

[2] φασι, Corais, for φησι; so the later editors.

[3] εἰσί, Palmer, for ἐστί (all MSS. except *nox*, which omit the word).

[4] πόλεις *x*, πόλις other MSS.

[5] οὗ, Casaubon, for αὐτοῦ; so the later editors.

[6] παρατόπων *g*, παρατόπως Corais.

[7] Κριθωτή, *h* and by corr. in D, Κορινθώτη BC*klnosx* and *man. pr.* in D and in margin of *h*.

which is twice the size of Melitê, both in length and in breadth; and to a third, Uria, which is much smaller than those. Now Cynia empties into the sea, but the others lie about half a stadium above it. Then one comes to the Evenus, to which the distance from Actium is six hundred and seventy stadia. After the Evenus one comes to the mountain Chalcis, which Artemidorus has called Chalcia; then to Pleuron; then to the village Halicyrna, above which, thirty stadia in the interior, lies Calydon; and near Calydon is the temple of the Laphrian Apollo. Then one comes to the mountain Taphiassus; then to the city Macynia; then to Molycreia and, near by, to Antirrhium, the boundary between Aetolia and Locris, to which the distance from the Evenus is about one hundred and twenty stadia. Artemidorus, indeed, does not give this account of the mountain, whether we call it Chalcis or Chalcia, since he places it between the Acheloüs and Pleuron, but Apollodorus, as I have said before,[1] places both Chalcis and Taphiassus above Molycreia, and he also says that Calydon is situated between Pleuron and Chalcis. Perhaps, however, we should postulate two mountains, one near Pleuron called Chalcis, and the other near Molycreia called Chalcis. Near Calydon, also, is a lake, which is large and

[1] 10. 2. 4.

---

[7] Ταφιασσόν, the editors, for Ταφίασσον B, Ταφίασος other MSS.

[8] δέ, Kramer, from conj. of Tzschucke, for τε (BCD*hk*); other MSS. omit the word.

[9] φησι, the editors, for φασι.

[10] For τις καί Palmer conj. Ὄνθις; so Kiepert in *Tab. Graec.*

μεγάλη καὶ εὔοψος,[1] ἣν ἔχουσιν οἱ ἐν Πάτραις Ῥωμαῖοι.

22. Τῆς δὲ μεσογαίας κατὰ μὲν τὴν Ἀκαρνανίαν Ἐρυσιχαίους τινάς φησιν Ἀπολλόδωρος λέγεσθαι, ὧν Ἀλκμὰν μέμνηται·

> οὐδ' Ἐρυσιχαῖος οὐδὲ[2] ποιμήν,
> ἀλλὰ Σαρδίων ἀπ' ἀκρᾶν.

κατὰ δὲ τὴν Αἰτωλίαν ἦν Ὤλενος, ἧς[3] ἐν τῷ Αἰτωλικῷ καταλόγῳ μέμνηται Ὅμηρος, ἴχνη δ' αὐτῆς λείπεται μόνον ἐγγὺς τῆς Πλευρῶνος ὑπὸ τῷ Ἀρακύνθῳ·[4] ἦν δὲ καὶ Λυσιμαχία πλησίον, ἠφανισμένη καὶ αὐτή, κειμένη πρὸς τῇ λίμνῃ, τῇ νῦν μὲν Λυσιμαχίᾳ, πρότερον δ' Ὕδρᾳ, μεταξὺ Πλευρῶνος καὶ Ἀρσινόης πόλεως, ἣ κώμη μὲν ἦν πρότερον, καλουμένη Κωνώπα,[5] κτίσμα δ' ὑπῆρξεν Ἀρσινόης, τῆς Πτολεμαίου τοῦ δευτέρου γυναικὸς ἅμα καὶ ἀδελφῆς, εὐφυῶς ἐπικειμένη πως τῇ τοῦ Ἀχελῴου διαβάσει· παραπλήσιον δέ τι καὶ ἡ Πυλήνη τῷ Ὠλένῳ πέπονθεν. ὅταν δὲ φῇ τὴν Καλυδῶνα αἰπεινήν τε καὶ πετρήεσσαν, ἀπὸ τῆς χώρας δεκτέον· εἴρηται γάρ, ὅτι τὴν χώραν δίχα διελόντες τὴν μὲν ὀρεινὴν καὶ ἐπίκτητον τῇ Καλυδῶνι προσένειμαν, τὴν πεδιάδα δὲ τῇ Πλευρῶνι.

23. Νυνὶ μὲν οὖν ἐκπεπόνηται καὶ ἀπηγόρευκεν ὑπὸ τῶν συνεχῶν πολέμων ἥ τ' Ἀκαρνανία καὶ Αἰτωλοί, καθάπερ καὶ πολλὰ τῶν ἄλλων ἐθνῶν·

[1] εὔυψος BCD*ghlnox* ; εὔψυχος *k*.

[2] Before ποιμήν Bergk (note to *Frag.* 24) reads merely οὐδέ instead of Καλυδωναίου δέ DH*isn*, Καλυδωνέου δέ B*k*, Κλυδωναίου δέ C ; Καλυδώνιος οὐδέ, Corais from conj. of Casaubon.

well supplied with fish; it is held by the Romans who live in Patrae.

22. Apollodorus says that in the interior of Acarnania there is a people called Erysichaeans, who are mentioned by Alcman: "nor yet an Erysichaean nor shepherd, but from the heights of Sardeis."[1] But Olenus, which Homer mentions in the Aetolian catalogue, was in Aetolia, though only traces of it are left, near Pleuron at the foot of Aracynthus. Near it, also, was Lysimachia; this, too, has disappeared; it was situated by the lake now called Lysimachia, in earlier times Hydra, between Pleuron and the city Arsinoê. In earlier times Arsinoê was only a village, and was called Conopa, but it was first founded as a city by Arsinoê, who was both wife and sister of Ptolemy the second;[2] it was rather happily situated at the ford across the Acheloüs. Pylenê[3] has also suffered a fate similar to that of Olenus. When the poet calls Calydon both "steep"[4] and "rocky,"[5] one should interpret him as referring to the country; for, as I have said,[6] they divided the country into two parts and assigned the mountainous part, or Epictetus,[7] to Calydon and the level country to Pleuron.

23. At the present time both the Acarnanians and the Aetolians, like many of the other tribes, have been exhausted and reduced to impotence by their

[1] *Frag.* 24 (Bergk). [2] She married him in 279 B.C. [3] Cf. 10. 2. 6. [4] *Iliad* 13. 217. [5] *Iliad* 2. 640. [6] 10. 2. 3. [7] *i.e.* Aetolia the "Acquired" (10. 2. 3).

---

[3] ᾗς, Corais, for ὡς; so the later editors.
[4] Ἀρακύνθῳ, the editors, for Ἀρακίνθῳ.
[5] Κωνώπα, Tzschucke, for Κονώπα; so the later editors.

πλεῖστον μέντοι χρόνον συνέμειναν Αἰτωλοὶ μετὰ τῶν Ἀκαρνάνων πρός τε τοὺς Μακεδόνας καὶ τοὺς ἄλλους Ἕλληνας, ὕστατα δὲ καὶ πρὸς Ῥωμαίους περὶ τῆς αὐτονομίας ἀγωνιζόμενοι. ἐπεὶ δὲ καὶ Ὅμηρος αὐτῶν ἐπὶ πολὺ μέμνηται καὶ οἱ ἄλλοι ποιηταί τε καὶ συγγραφεῖς, τὰ μὲν εὐσήμως τε καὶ ὁμολογουμένως, τὰ δ' ἧττον γνωρίμως (καθάπερ τοῦτο [1] καὶ ἐν τοῖς ἤδη λεχθεῖσι περὶ αὐτῶν ἀποδέδεικται), προσληπτέον καὶ τῶν παλαιοτέρων τινὰ τῶν ἀρχῆς ἐχόντων τάξιν ἢ διαπορουμένων.

C 461 24. Εὐθὺς ἐπὶ τῆς Ἀκαρνανίας, ὅτι μὲν αὐτὴν ὁ Λαέρτης καὶ οἱ Κεφαλλῆνες κατεκτήσαντο, εἴρηται ἡμῖν, τίνων δὲ κατεχόντων πρότερον, πολλοὶ μὲν εἰρήκασιν, οὐχ ὁμολογούμενα δ' εἰπόντων, ἐπιφανῆ δέ, ἀπολείπεταί τις λόγος ἡμῖν διαιτητικὸς περὶ αὐτῶν. φασὶ γὰρ τοὺς Ταφίους τε καὶ Τηλεβόας λεγομένους οἰκεῖν τὴν Ἀκαρνανίαν πρότερον, καὶ τὸν ἡγεμόνα αὐτῶν Κέφαλον τὸν κατασταθέντα ὑπὸ Ἀμφιτρύωνος κύριον τῶν περὶ τὴν Τάφον νήσων κυριεῦσαι καὶ ταύτης τῆς χώρας· ἐντεῦθεν δὲ καὶ τὸ ἀπὸ τοῦ Λευκάτα νομιζόμενον ἅλμα τούτῳ πρώτῳ προσμυθεύουσιν, ὡς προείρηται. ὁ δὲ ποιητής, ὅτι μὲν ἦρχον οἱ Τάφιοι τῶν Ἀκαρνάνων, πρὶν ἢ τοὺς Κεφαλλῆνας καὶ τὸν Λαέρτην ἐπελθεῖν, οὐ λέγει, διότι δ' ἦσαν φίλοι τοῖς Ἰθακησίοις λέγει, ὥστ' ἢ οὐδ' ὅλως ἐπῆρξαν

[1] τοῦτο *no*, τούτου BCD*hkl*.

---

[1] 10. 2. 8, 10. [2] Cf. 10. 2. 9.

continual wars. However, for a very long time the Aetolians, together with the Acarnanians, stood firm, not only against the Macedonians and the other Greeks, but also finally against the Romans, when fighting for autonomy. But since they are often mentioned by Homer, as also both by the other poets and by historians, sometimes in words that are easy to interpret and about which there is no disagreement, and sometimes in words that are less intelligible (this has been shown in what I have already said about them), I should also add some of those older accounts which afford us a basis of fact to begin with, or are matters of doubt.

24. For instance, in the case of Acarnania, Laertes and the Cephallenians acquired possession of it, as I have said;[1] but as to what people held it before that time, many writers have indeed given an opinion, but since they do not agree in their statements, which have, however, a wide currency, there is left for me a word of arbitration concerning them. They say that the people who were called both Taphians and Teleboans lived in Acarnania in earlier times, and that their leader Cephalus, who had been set up by Amphitryon as master over the islands about Taphos, gained the mastery over this country too. And from this fact they go on to add the myth that Cephalus was the first to take the leap from Leucatas which became the custom, as I have said before.[2] But the poet does not say that the Taphians were ruling the Acarnanians before the Cephallenians and Laertes came over, but only that they were friends to the Ithacans, and therefore, according to the poet, they either had not ruled over the region at all, or had yielded Acarnania to the

τῶν τόπων κατ' αὐτόν, ἢ ἑκόντες παρεχώρησαν ἢ καὶ σύνοικοι ἐγένοντο. φαίνονται δὲ καὶ ἐκ Λακεδαίμονός τινες ἐποικῆσαι τὴν Ἀκαρνανίαν, οἱ μετ' Ἰκαρίου τοῦ Πηνελόπης πατρός· καὶ γὰρ τοῦτον καὶ τοὺς ἀδελφοὺς αὐτῆς ζῶντας παραδίδωσιν ὁ ποιητὴς κατὰ τὴν Ὀδύσσειαν·

> οἳ πατρὸς μὲν ἐς οἶκον ἀπερρίγασι νέεσθαι
> Ἰκαρίου, ὥς κ' αὐτὸς ἐεδνώσαιτο θύγατρα·

καὶ περὶ τῶν ἀδελφῶν·

> ἤδη γάρ ῥα πατήρ τε κασίγνητοί τε κέλονται
> Εὐρυμάχῳ γήμασθαι.

οὔτε γὰρ ἐν Λακεδαίμονι πιθανὸν αὐτοὺς οἰκεῖν· οὐ γὰρ ἂν ὁ Τηλέμαχος παρὰ Μενελάῳ κατήγετο ἀφιγμένος ἐκεῖσε· οὔτ' ἄλλην οἴκησιν παρειλήφαμεν αὐτῶν. φασὶ δὲ Τυνδάρεων καὶ τὸν ἀδελφὸν αὐτοῦ τὸν Ἰκάριον,[1] ἐκπεσόντας ὑπὸ Ἱπποκόωντος τῆς οἰκείας, ἐλθεῖν παρὰ Θέστιον, τὸν τῶν Πλευρωνίων ἄρχοντα, καὶ συγκατακτήσασθαι τὴν πέραν[2] τοῦ Ἀχελῴου πολλὴν[3] ἐπὶ μέρει· τὸν μὲν οὖν Τυνδάρεων ἐπανελθεῖν οἴκαδε, γήμαντα Λήδαν, τὴν τοῦ Θεστίου θυγατέρα, τὸν δ' Ἰκάριον[4] ἐπιμεῖναι,[5] τῆς Ἀκαρνανίας ἔχοντα μέρος, καὶ τεκνοποιήσασθαι τήν τε Πηνελόπην ἐκ Πολυκάστης τῆς Λυγαίου θυγατρὸς καὶ τοὺς ἀδελφοὺς αὐτῆς. ἡμεῖς μὲν οὖν ἀπεδείξαμεν ἐν τῷ καταλόγῳ τῶν νεῶν καὶ τοὺς Ἀκαρνᾶνας καταριθμουμένους καὶ

[1] Ἰκάριον, Xylander, for Ἴκαρον.
[2] For τὴν πέραν (τὴν περαίαν BE*kno*) Tzschucke and Corais, from conj. of Casaubon, read τῆς περαίας.
[3] πόλιν CDE*ghislx*, πολλά *k*.

Ithacans voluntarily, or had become joint-occupants with them. It appears that also a colony from Lacedaemon settled in Acarnania, I mean Icarius, father of Penelope, and his followers; for in the *Odyssey* the poet represents both Icarius and the brothers of Penelope as living: "who[1] shrink from going to the house of her father, Icarius, that he himself may exact the bride-gifts for his daughter,"[2] and, concerning her brothers, "for already her father and her brothers bid her marry Eurymachus";[3] for, in the first place, it is improbable that they were living in Lacedaemon, since in that case Telemachus would not have lodged at the home of Menelaüs when he went to Lacedaemon, and, secondly, we have no tradition of their having lived elsewhere. But they say that Tyndareus and his brother Icarius, after being banished by Hippocoön from their home-land, went to Thestius, the ruler of the Pleuronians, and helped him to acquire possession of much of the country on the far side of the Acheloüs on condition that they should receive a share of it; that Tyndareus, however, went back home, having married Leda, the daughter of Thestius, whereas Icarius stayed on, keeping a portion of Acarnania, and by Polycastê, the daughter of Lygaeus, begot both Penelope and her brothers. Now I have already set forth that the Acarnanians were enumerated in the *Catalogue of Ships*,[4] that they took part in the

[1] The suitors. [2] *Od.* 2. 52. [3] *Od.* 15. 16.

[4] 10. 2. 25; but Homer nowhere specifically mentions the "Acarnanians."

---

[4] Ἴκαρον MSS. except E.

[5] ἐπιμεῖναι, Meineke emends to ὑπομεῖναι.

μετασχόντας τῆς ἐπὶ Ἴλιον στρατείας, ἐν οἷς κατωνομάζοντο οἵ τε τὴν ἀκτὴν οἰκοῦντες καὶ ἔτι

οἵ τ' ἤπειρον ἔχον ἠδ' ἀντιπέραι' ἐνέμοντο.

οὔτε δ' ἡ ἤπειρος Ἀκαρνανία ὠνομάζετό πω, οὔθ' ἡ ἀκτὴ Λευκάς.

C 462 25. Ἔφορος δ' οὔ φησι συστρατεῦσαι· Ἀλκμαίωνα[1] γὰρ τὸν Ἀμφιάρεω, στρατεύσαντα[2] μετὰ Διομήδους καὶ τῶν ἄλλων Ἐπιγόνων καὶ κατορθώσαντα τὸν πρὸς Θηβαίους πόλεμον, συνελθεῖν Διομήδει καὶ τιμωρήσασθαι μετ' αὐτοῦ τοὺς Οἰνέως ἐχθρούς, παραδόντα δ' ἐκείνοις[3] τὴν Αἰτωλίαν, αὐτὸν εἰς τὴν Ἀκαρνανίαν παρελθεῖν καὶ ταύτην καταστρέφεσθαι. Ἀγαμέμνονα δ', ἐν τούτῳ τοῖς Ἀργείοις ἐπιθέμενον, κρατῆσαι ῥᾳδίως τῶν πλείστων τοῖς περὶ Διομήδη συνακολουθησάντων. μικρὸν δ' ὕστερον ἐπιπεσούσης τῆς ἐπ' Ἴλιον ἐξόδου, δείσαντα, μὴ ἀπόντος αὐτοῦ κατὰ τὴν στρατείαν ἐπανελθόντες οἴκαδε οἱ περὶ τὸν Διομήδη (καὶ γὰρ ἀκούεσθαι μεγάλην περὶ αὐτὸν συνεστραμμένην δύναμιν) κατάσχοιεν τὴν μάλιστα προσήκουσαν αὐτοῖς ἀρχήν, τὸν μὲν γὰρ Ἀδράστου, τὸν δὲ τοῦ πατρὸς εἶναι κληρονόμον, ταῦτα δὴ διανοηθέντα καλεῖν αὐτοὺς ἐπί τε τὴν τοῦ Ἄργους ἀπόληψιν καὶ τὴν κοινωνίαν τοῦ πολέμου· τὸν μὲν οὖν Διομήδη πεισθέντα μετασχεῖν τῆς στρατείας, τὸν δὲ Ἀλκμαίωνα ἀγανακτοῦντα μὴ φροντίσαι· διὰ δὲ τοῦτο μηδὲ κοινωνῆσαι τῆς στρατείας μόνους τοὺς Ἀκαρνᾶνας τοῖς Ἕλλησι·

[1] Ἀλκμαίωνα, Meineke emends to Ἀλκμέωνα.
[2] συστρατεύσαντα *Clo*.
[3] ἐκείνῳ C (?) and editors before Kramer.

expedition to Ilium, and that among these were named "those who lived on the 'shore,'"[1] and also "those who held the mainland and dwelt in parts opposite."[2] But as yet neither had the mainland been named "Acarnania" nor the shore "Leucas."

25. Ephorus denies that they joined the Trojan expedition, for he says that Alcmaeon, the son of Amphiaraüs, made an expedition with Diomedes and the other Epigoni, and had brought to a successful issue the war against the Thebans, and then joined Diomedes and with him took vengeance upon the enemies of Oeneus, after which he himself, first giving over Aetolia to them,[3] passed into Acarnania and subdued it; and meanwhile Agamemnon attacked the Argives and easily prevailed over them, since [illegible] most of them had accompanied the army of D[illegible]medes; but a little later, when the expedition against Ilium confronted him, he conceived the fear that, when he was absent on the expedition, Diomedes and his army might come back home (and in fact it was reported that a great army had gathered round him) and seize the empire to which they had the best right, for one[4] was the heir of Adrastus and the other[5] of his father;[6] and accordingly, after thinking this all over, Agamemnon invited them both to resume possession of Argos and to take part in the war; and although Diomedes was persuaded to take part in the expedition, Alcmaeon was vexed and refused to heed the invitation; and for this reason the Acarnanians alone refused to share in the ex-

[1] "Shore of the mainland," *Od.* 24. 378.
[2] See 10. 2. 8.
[3] Diomedes and Oeneus.
[4] Diomedes.
[5] Alcmaeon.
[6] Amphiaraüs.

τούτοις δ', ὡς εἰκός, τοῖς λόγοις ἐπακολουθήσαντες οἱ Ἀκαρνᾶνες σοφίσασθαι Ῥωμαίους καὶ τὴν αὐτονομίαν παρ' αὐτῶν ἐξανύσασθαι, λέγοντες, ὡς οὐ μετάσχοιεν μόνοι τῆς ἐπὶ τοὺς προγόνους τοὺς ἐκείνων στρατείας· οὔτε γὰρ ἐν τῷ Αἰτωλικῷ καταλόγῳ φράζοιντο, οὔτε ἰδίᾳ· οὐδὲ γὰρ ὅλως τοὔνομα τοῦτ' ἐμφέροιτο ἐν τοῖς ἔπεσιν.

26. Ὁ μὲν οὖν Ἔφορος, πρὸ τῶν Τρωικῶν ἤδη τὴν Ἀκαρνανίαν ὑπὸ τῷ Ἀλκμαίωνι ποιήσας, τό τε Ἄργος τὸ Ἀμφιλοχικὸν ἐκείνου κτίσμα ἀποφαίνει καὶ τὴν Ἀκαρνανίαν ὠνομάσθαι φησὶν ἀπὸ τοῦ παιδὸς αὐτοῦ Ἀκαρνᾶνος, Ἀμφιλόχους δὲ ἀπὸ τοῦ ἀδελφοῦ Ἀμφιλόχου· ὥστε ἐκπίπτει εἰς τὰ παρὰ τὴν Ὁμηρικὴν ἱστορίαν λεγόμενα. Θουκυδίδης δὲ καὶ ἄλλοι τὸν Ἀμφίλοχον, ἀπ[ὸ] τῆς στρατείας τῆς Τρωικῆς ἐπανιόντα, οὐκ ἀρεσκό[μενον] τοῖς ἐν Ἄργει, ταύτην οἰκῆσαί φασι[1] τὴν χώραν, οἱ μὲν κατὰ διαδοχὴν ἥκοντα τῆς τοῦ ἀδελφοῦ δυναστείας, οἱ δ' ἄλλως. καὶ ἰδίᾳ μὲν περὶ Ἀκαρνάνων ταῦτα λέγοιτ' ἄν, κοινῇ δ' ὅσα καὶ τοῖς Αἰτωλικοῖς ἐπιπλέκεται νῦν ἐροῦμεν, τὰ Αἰτωλικὰ λέγοντες ἐφεξῆς, ὅσα προσλαβεῖν τοῖς εἰρημένοις ἔγνωμεν.

[1] φησι BC*hino*.

---

[1] *Iliad* 2. 638 ff. [2] 2. 68.

pedition with the Greeks. And it was probably by following this account that the Acarnanians tricked the Romans, as they are said to have done, and obtained from them their autonomy, urging that they alone had had no part in the expedition against the ancestors of the Romans, for they were named neither in the Aetolian catalogue[1] nor separately, and in fact their name was not mentioned in the Epic poems at all.

26. Ephorus, then, makes Acarnania subject to Alcmaeon even before the Trojan War; and he not only declares that the Amphilochian Argos was founded by him, but also says that Acarnania was named after Alcmaeon's son Acarnan, and the Amphilochians after Alcmaeon's brother Amphilochus; therefore his account is to be cast out amongst those contrary to Homeric history. But Thucydides[2] and others say that Amphilochus, on his return from the Trojan expedition, was displeased with the state of affairs at Argos, and took up his abode in this country, some saying that he came by right of succession to the domain of his brother, others giving a different account. So much may be said of the Acarnanians specifically; I shall now speak of their history in a general way, in so far as their history is interwoven with that of the Aetolians, relating next in order the history of the Aetolians, in so far as I have thought best to add to my previous narrative.

III

1. Τοὺς δὲ Κουρῆτας τῶν μὲν Ἀκαρνᾶσι, τῶν δ' Αἰτωλοῖς προσνεμόντων, καὶ τῶν μὲν ἐκ Κρήτης, τῶν δ' ἐξ Εὐβοίας τὸ γένος εἶναι φασκόντων, C 163 ἐπειδὴ καὶ Ὅμηρος αὐτῶν μέμνηται, τὰ παρ' ἐκείνου πρῶτον ἐπισκεπτέον. οἴονται δ' αὐτὸν λέγειν Αἰτωλοὺς μᾶλλον ἢ Ἀκαρνᾶνας, εἴπερ οἱ Πορθαονίδαι ἦσαν

> Ἄγριος ἠδὲ Μέλας, τρίτατος δ' ἦν ἱππότα Οἰνεύς·
> ᾤκεον δ' ἐν Πλευρῶνι καὶ αἰπεινῇ Καλυδῶνι.[1]

αὗται δ' εἰσὶν Αἰτωλικαὶ πόλεις ἀμφότεραι καὶ φέρονται ἐν Αἰτωλικῷ καταλόγῳ, ὥστε, ἐπεὶ τὴν Πλευρῶνα οἰκοῦντες φαίνονται καὶ κατ' αὐτὸν οἱ Κουρῆτες, Αἰτωλοὶ ἂν εἶεν. οἱ δ' ἀντιλέγοντες τῷ τρόπῳ τῆς φράσεως παράγονται, ὅταν φῇ,

> Κουρῆτές τ' ἐμάχοντο καὶ Αἰτωλοὶ μενεχάρμαι ἀμφὶ πόλιν Καλυδῶνα.[2]

οὐδὲ γὰρ ἂν κυρίως εἶπεν οὕτως· ἐμάχοντο Βοιωτοὶ καὶ Θηβαῖοι πρὸς ἀλλήλους, οὐδ' Ἀργεῖοι καὶ Πελοποννήσιοι. ἐδείχθη δ' ἐν τοῖς ἔμπροσθεν, ὅτι ἐστὶ καὶ Ὁμηρικὸν τὸ ἔθος τοῦτο τῆς φράσεως καὶ ὑπὸ τῶν ἄλλων ποιητῶν τετριμμένον· τοῦτο μὲν οὖν εὐαπολόγητον. ἐκεῖνοι δὲ λεγέτωσαν πῶς ἂν μὴ ὁμοεθνεῖς ὄντας μηδ' Αἰτωλοὺς τοὺς Πλευρωνίους ἐν τοῖς Αἰτωλοῖς κατέλεγεν.

2. Ἔφορος δὲ τοὺς Αἰτωλοὺς εἰπὼν ἔθνος εἶναι μηδεπώποτε γεγενημένον ὑφ' ἑτέροις, ἀλλὰ πάντα

---

[1] *Iliad* 14. 117. [2] *Iliad* 14. 116.

III

1. As for the Curetes, some assign them to the Acarnanians, others to the Aetolians; and some assert that the tribe originated in Crete, but others in Euboea; but since Homer mentions them, I should first investigate his account. It is thought that he means that they were Aetolians rather than Acarnanians, if indeed the sons of Porthaon were "Agrius and Melas, and, the third, Oeneus the knight";[1] "and they lived in Pleuron and steep Calydon."[2] These are both Aetolian cities, and are referred to in the Aetolian catalogue; and therefore, since, even according to the poet, the Curetes obviously lived in Pleuron, they would be Aetolians. Those writers who oppose this view are misled by Homer's mode of expression when he says, "the Curetes were fighting, and the Aetolians steadfast in battle, about the city of Calydon";[3] for, they add, neither would he have spoken appropriately if he had said, "the Boeotians and the Thebans were fighting against one another"; or "the Argives and the Peloponnesians." But, as I have shown heretofore,[4] this habit of expression not only is Homeric, but is much used by the other poets also. This interpretation, then, is easy to defend; but let those writers explain how the poet could catalogue the Pleuronians among the Aetolians if they were not Aetolians or at least of the same race.

2. Ephorus,[5] after saying that the Aetolians were a race which had never become subject to any other

[3] *Iliad* 9. 529. [4] 8. 3. 8, 10. 2. 10.
[5] See *Dictionary* in Vol. I.

τὸν μνημονευόμενον χρόνον μεμενηκὸς ἀπόρθητον διά τε[1] τὰς δυσχωρίας τῶν τόπων καὶ διὰ τὴν περὶ τὸν πόλεμον ἄσκησιν, ἐξ ἀρχῆς μέν φησιν[2] ἅπασαν τὴν χώραν Κουρῆτας κατασχεῖν, ἀφικομένου δ' ἐξ Ἤλιδος Αἰτωλοῦ τοῦ Ἐνδυμίωνος καὶ τοῖς πολέμοις κρατοῦντος αὐτῶν, τοὺς μὲν Κουρῆτας εἰς τὴν νῦν καλουμένην Ἀκαρνανίαν ὑποχωρῆσαι, τοὺς δ' Αἰτωλοὺς συγκατελθόντας Ἐπειοῖς τὰς ἀρχαιοτάτας κτίσαι τῶν ἐν Αἰτωλίᾳ πόλεων, δεκάτῃ δ'[3] ὕστερον γενεᾷ τὴν Ἦλιν ὑπὸ Ὀξύλου τοῦ Αἵμονος συνοικισθῆναι, περαιωθέντος ἐκ τῆς Αἰτωλίας. παρατίθησι δὲ τούτων μαρτύρια τὰ ἐπιγράμματα, τὸ μὲν ἐν Θέρμοις τῆς Αἰτωλίας, ὅπου τὰς ἀρχαιρεσίας ποιεῖσθαι πάτριον αὐτοῖς ἐστίν, ἐγκεχαραγμένον τῇ βάσει τῆς Αἰτωλοῦ εἰκόνος·

χώρης οἰκιστῆρα, παρ' Ἀλφειοῦ ποτὲ δίναις
  θρεφθέντα,[4] σταδίων γείτον' Ὀλυμπιάδος,
Ἐνδυμίωνος παῖδ' Αἰτωλοὶ τόνδ' ἀνέθηκαν
  Αἰτωλόν, σφετέρας μνῆμ' ἀρετῆς ἐσορᾶν.

τὸ δ' ἐν τῇ ἀγορᾷ τῶν Ἠλείων[5] ἐπὶ τῷ Ὀξύλου ἀνδριάντι·

Αἰτωλός ποτε τόνδε λιπὼν αὐτόχθονα δῆμον
  κτήσατο Κουρῆτιν γῆν, δορὶ πολλὰ καμών·
C 464 τῆς δ' αὐτῆς γενεᾶς δεκατόσπορος Αἵμονος υἱός
  Ὄξυλος ἀρχαίην ἔκτισε τήνδε πόλιν.

[1] τε, Tzschucke, for δέ; so the later editors.
[2] φησιν, Tzschucke, for φασι; so the later editors.
[3] δεκάτῃ δ', Corais, for δέκα, τῇ δ'; so the later editors.

people, but throughout all time of which there is any record had remained undevastated, both because of the ruggedness of their country and because of their training in warfare, says at the outset that the Curetes held possession of the whole country, but when Aetolus,[1] the son of Endymion, arrived from Elis and overpowered them in war, the Curetes withdrew to what is now called Acarnania, whereas the Aetolians came back with Epeians and founded the earliest of the cities of Aetolia, and in the tenth generation after that Elis was settled by Oxylus[2] the son of Haemon, who had crossed over from Aetolia. And he cites as evidence of all this two inscriptions, the one at T[illegible]erma in Aetolia (where it is their ancestral custom [illegible] hold their elections of magistrates), engrav[illegible]n the base of the statue of Aetolus: "Fou[illegible]of the country, once reared beside the eddie[illegible] the Alpheius, neighbour of the race-courses of Olympia, son of Endymion, this Aetolus has been set up by the Aetolians as a memorial of his valour to behold"; and the other inscription in the market-place of the Eleians on the statue of Oxylus: "Aetolus once left this autochthonous people, and through many a toil with the spear took possession of the land of Curetis; but the tenth scion of the same stock, Oxylus, the son of Haemon, founded this city in early times."

[1] Cp. 8. 3. 33. [2] Cf. 8. 3. 33.

---

[4] θρεφθέντα, Jacobs, Corais, and later editors, for τραφέντα *nx*, τρεφθέντα other MSS.
[5] Ἠλείων, correction in *n*, and Pletho, for Αἰτωλῶν; so the editors.

3. Τὴν μὲν οὖν συγγένειαν τὴν πρὸς ἀλλήλους τῶν τε Ἠλείων καὶ τῶν Αἰτωλῶν ὀρθῶς ἐπισημαίνεται διὰ τῶν ἐπιγραμμάτων, ἐξομολογουμένων ἀμφοῖν οὐ τὴν συγγένειαν μόνον, ἀλλὰ καὶ τὸ ἀρχηγέτας ἀλλήλων εἶναι· δι' οὗ καλῶς ἐξελέγχει ψευδομένους τοὺς φάσκοντας τῶν μὲν Αἰτωλῶν ἀποίκους εἶναι τοὺς Ἠλείους, μὴ μέντοι τῶν Ἠλείων τοὺς Αἰτωλούς. τὴν δ' ἀνομολογίαν τῆς γραφῆς καὶ τῆς ἀποφάσεως φαίνεται τὴν αὐτὴν ἐπιδεδειγμένος κἀνταῦθα, ἥνπερ ἐπὶ τοῦ μαντείου τοῦ ἐν Δελφοῖς παρεστήσαμεν. εἰπὼν γὰρ ἀπόρθητον ἐκ τοῦ μνημονευομένου χρόνου παντὸς τὴν Αἰτωλίαν, εἰπὼν δὲ καὶ ἐξ ἀρχῆς τὴν χώραν ταύτην τοὺς Κουρῆτας κατασχεῖν, ὤφειλε μὲν [1] τοῖς εἰρημένοις ἀκόλουθον τοῦτο ἐπιφέρειν,[2] ὅτι οἱ Κουρῆτες διέμειναν ἕως εἰς αὐτὸν κατέχοντες τὴν Αἰτωλίαν γῆν, οὕτω γὰρ ἔμελλεν ἀπόρθητός τε καὶ οὐδέποτε ἐπ' [3] ἄλλοις γεγονυῖα ὀρθῶς λεχθήσεσθαι· ὁ δ' ἐκλαθόμενος τῆς ὑποσχέσεως οὐ τοῦτ' ἐπιφέρει, ἀλλὰ τοὐναντίον, ὡς ἀφικομένου ἐξ Ἤλιδος Αἰτωλοῦ καὶ τοῖς πολέμοις κρατοῦντος αὐτῶν, οἱ Κουρῆτες ἀπῆλθον εἰς τὴν Ἀκαρνανίαν· τί οὖν ἄλλο πορθήσεως ἴδιον ἢ τῷ πολέμῳ κρατηθῆναι καὶ τὴν χώραν ἐκλιπεῖν; τοῦτο δὲ καὶ τὸ ἐπίγραμμα μαρτυρεῖ τὸ παρὰ τοῖς Ἠλείοις, ὁ γὰρ Αἰτωλός, φησί,

κτήσατο Κουρῆτιν γῆν, δορὶ πολλὰ καμών.

[1] Corais and Meineke delete τοίγε, before τοῖς.

[2] ἐπιφέρειν, Meineke, following conj. of Casaubon, for φέρειν.

[3] ὑπ' *x*, Corais, and Meineke.

3. Now through these inscriptions Ephorus correctly signifies the kinship of the Eleians and Aetolians with one another, since both inscriptions agree, not merely as to the kinship of the two peoples, but also that each people was the founder of the other, through which he successfully convicts of falsehood those who assert that, while the Eleians were indeed colonists of the Aetolians, the Aetolians were not colonists of the Eleians. But here, too, Ephorus manifestly displays the same inconsistency in his writing and his pronouncements as in the case of the oracle at Delphi, which I have already set forth;[1] for, after saying that Aetolia has been undevastated throughout all times of which there is any record, and after saying also that in the beginning the Curetes held possession of this country, he should have added as a corollary to what he had already said that the Curetes continued to hold possession of the Aetolian land down to his own time, for only thus could it have been rightly said that the land had been undevastated and that it had never come under the power of others; and yet, utterly forgetting his promise,[2] he does not add this, but the contrary, that when Aetolus arrived from Elis and overpowered the Curetes in war, they withdrew into Acarnania. What else, pray, is specifically characteristic of a devastation than being overpowered in war and abandoning the country? And this is evidenced also by the inscription among the Eleians, for Aetolus, it says, "through many a toil with the spear took possession of the land of Curetis."

[1] 9. 3. 11.

[2] See 9. 3. 11.

4. Ἴσως δή τις ἂν φαίη, λέγειν αὐτὸν ἀπόρθητον τὴν Αἰτωλίαν, ἀφ' οὗ τοὔνομα τοῦτ' ἔσχε μετὰ τὴν Αἰτωλοῦ παρουσίαν· ἀλλ' ἀφῄρηται καὶ τούτου [1] τοῦ νοήματος τὸν λόγον, φήσας ἐν τοῖς ἐφεξῆς τὸ μὲν πλεῖστον τοῦ λαοῦ τοῦ διαμένοντος ἐν τοῖς Αἰτωλοῖς τοῦτο εἶναι, τὸ τῶν Ἐπειῶν λέγων,[2] συμμιχθέντων δ' αὐτοῖς ὕστερον Αἰολέων, τῶν ἅμα Βοιωτοῖς ἐκ Θετταλίας ἀναστάντων, κοινῇ μετὰ τούτων τὴν χώραν κατασχεῖν. ἆρ' οὖν πιστόν [3] ἐστι χωρὶς πολέμου τὴν ἀλλοτρίαν ἐπελθόντας συγκατανείμασθαι τοῖς ἔχουσι, μηδὲν δεομένοις κοινωνίας τοιαύτης; ἢ τοῦτο μὲν οὐ πιστόν, τὸ δὲ κρατουμένοις τοῖς ὅπλοις ἐπ' ἴσοις [4] συμβῆναι πιστόν; τί οὖν ἄλλο πόρθησις ἢ τὸ κρατεῖσθαι τοῖς ὅπλοις; καὶ Ἀπολλόδωρος δ' εἴρηκεν ἐκ τῆς Βοιωτίας ἀπελθόντας [5] Ὕαντας ἱστορεῖσθαι καὶ ἐποίκους τοῖς Αἰτωλοῖς γενομένους· ὁ δ' ὥσπερ κατωρθωκὼς ἐπιλέγει, διότι [6] ταῦτα καὶ τὰ τοιαῦτα διακριβοῦν εἰώθαμεν, ὅταν ᾖ τι τῶν πραγμάτων ἢ παντελῶς ἀπορούμενον ἢ ψευδῆ δόξαν ἔχον.

C 465 5. Τοιοῦτος δ' ὢν Ἔφορος ἑτέρων ὅμως κρείττων ἐστί· καὶ αὐτὸς ὁ ἐσπουδασμένως οὕτως ἐπαινέσας αὐτὸν Πολύβιος καὶ φήσας περὶ τῶν Ἑλληνικῶν καλῶς μὲν Εὔδοξον, κάλλιστα δ' Ἔφορον ἐξη-

[1] τούτου, Corais inserts; so the later editors.

[2] λέγων, Jones restores to the text. Corais emends to ἢ Ἠλείων; Meineke deletes.

[3] πιστόν, Groskurd inserts; so the later editors.

[4] ἴσης Bklnox.

[5] ἀπελθόντας, Corais and Meineke emend to ἐπελθόντος; a tempting emendation.

[6] ὅτι Bklnox.

4. Perhaps, however, one might say that Ephorus means that Aetolia was undevastated from the time when it got this name, that is, after Aetolus arrived there; but Ephorus has deprived himself of the argument in support of this idea by saying in his next words that this, meaning the tribe of the Epeians, constituted the greatest part of the people who stayed on among the Aetolians, but that later, when Aeolians, who at the same time with Boeotians had been compelled to migrate from Thessaly, were intermingled with them, they in common with these held possession of the country. Is it credible, pray, that without war they invaded the country of a different people and divided it up with its possessors, when the latter had no need of such a partnership? Or, since this is not credible, is it credible that those who were overpowered by arms came out on an equality with the victors? What else, pray, is devastation than being overpowered by arms? Apollodorus, also, says that, according to history, the Hyantes left Boeotia and settled among the Aetolians. But Ephorus, as though he had achieved success in his argument, adds: "It is my wont to examine such matters as these with precision, whenever any matter is either altogether doubtful or falsely interpreted."

5. But though Ephorus is such, still he is better than others. And Polybius[1] himself, who praises him so earnestly, and says concerning the Greek histories that Eudoxus[2] indeed gave a good account, but Ephorus gave the best account of the foundings of

[1] Book 34, Frag. 1.
[2] Eudoxus of Cnidus (fl. about 350 B.C.).

γεῖσθαι περὶ κτίσεων, συγγενειῶν, μεταναστάσεων, ἀρχηγετῶν, ἡμεῖς δέ, φησί, τὰ νῦν ὄντα δηλώσομεν καὶ περὶ θέσεως τόπων καὶ διαστημάτων· τοῦτο γάρ ἐστιν οἰκειότατον χωρογραφίᾳ. ἀλλὰ μὴν σύ γε, ὦ Πολύβιε, ὁ τὰς λαοδογματικὰς[1] ἀποφάσεις περὶ τῶν διαστημάτων εἰσάγων οὐκ ἐν τοῖς ἔξω τῆς Ἑλλάδος μόνον, ἀλλὰ καὶ ἐν τοῖς Ἑλληνικοῖς, καὶ διδοῖς[2] εὐθύνας τὰς μὲν Ποσειδωνίῳ, τὰς δ' Ἀρτεμιδώρῳ, τὰς δ' ἄλλοις πλείοσι· καὶ ἡμῖν οὖν συγγνώμην[3] ἔχειν[4] καὶ οὐ δυσχεραίνειν δεῖ, παρὰ τῶν τοιούτων μεταφέρουσι τὴν πολλὴν ἱστορίαν, ἐάν τι πταίωμεν, ἀλλ' ἀγαπᾷν, ἐὰν τὰ πλείω τῶν εἰρημένων ἑτέροις ἄμεινον λέγωμεν, ἢ τὰ παραλειφθέντα κατ' ἄγνοιαν προστιθῶμεν.

6. Περὶ δὲ Κουρήτων ἔτι καὶ τοιαῦτα λέγεται, τὰ μὲν ἐγγυτέρω ὄντα τῆς περὶ Αἰτωλῶν καὶ Ἀκαρνάνων ἱστορίας, τὰ δ' ἀπωτέρω· ἐγγυτέρω μὲν τὰ τοιαῦτα, οἷα προείρηται, ὅτι τὴν χώραν, ἣ νῦν Αἰτωλία καλεῖται, Κουρῆτες ᾤκουν, ἐλθόντες δ' οἱ Αἰτωλοὶ μετὰ Αἰτωλοῦ τούτους ἐξέβαλον εἰς τὴν Ἀκαρνανίαν· καὶ ἔτι τὰ τοιαῦτα, ὅτι τὴν Πλευρωνίαν ὑπὸ Κουρήτων οἰκουμένην καὶ Κουρῆτιν προσαγορευομένην Αἰολεῖς ἐπελθόντες ἀφείλοντο, τοὺς δὲ κατέχοντας ἐξέβαλον. Ἀρχέ-

[1] τὰς λαοδογματικάς, Tzschucke, from conj. of Tyrwhitt, for τάλας ὁ δογματικάς CDghilnosx, τὰς τῶν ἄλλων δογματικάς Bk; so the later editors.

[2] καὶ διδοῖς, Casaubon, for καὶ διαδούς BCDghikx, καὶ διαδιδούς lno, νὴ Δία, δίδως Corais; so the editors after Corais.

[3] συγγνώμη Bk; so Müller-Dübner.

cities, kinships, migrations, and original founders, "but I," he says, "shall show the facts as they now are, as regards both the position of places and the distances between them; for this is the most appropriate function of Chorography." But assuredly you, Polybius, who introduce "popular notions"[1] concerning distances, not only in dealing with places outside of Greece, but also when treating Greece itself, must also submit to an accounting, not only to Poseidonius,[2] and to Apollodorus, but to several others as well. One should therefore pardon me as well, and not be vexed, if I make any mistakes when I borrow from such writers most of my historical material, but should rather be content if in the majority of cases I improve upon the accounts given by others, or if I add such facts as have elsewhere, owing to lack of knowledge, been left untold.

6. Concerning the Curetes still further accounts, to the following effect, are given, some of them being more closely related to the history of the Aetolians and the Acarnanians, others more remotely. More closely related are such accounts as I have given before—that the Curetes were living in the country which is now called Aetolia, and that the Aetolians came with Aetolus and drove them into Acarnania; and also accounts of this kind, that, when Pleuronia was inhabited by the Curetes and was called Curetis, Aeolians made an invasion and took it away from them, and drove out its occupants.

[1] See 2. 4. 2 and 7. 5. 9. [2] Cf. 2. 3. 1 ff. and 2. 4. 3 ff.

[4] ἔχειν, Jones inserts, following a correction in *n*; Meineke merely indicates a lacuna; Kramer conj. συγγνῶναι.

μαχος δ' ὁ Εὐβοεύς φησι τοὺς Κουρῆτας ἐν Χαλκίδι συνοικῆσαι, συνεχῶς δὲ περὶ τοῦ Ληλάντου πεδίου πολεμοῦντας, ἐπειδὴ οἱ πολέμιοι τῆς κόμης ἐδράττοντο τῆς ἔμπροσθεν καὶ κατέσπων αὐτούς, ὄπισθεν κομῶντας γενέσθαι, τὰ δ' ἔμπροσθεν κείρεσθαι· διὸ καὶ Κουρῆτας ἀπὸ τῆς κουρᾶς κληθῆναι· μετοικῆσαι δ' εἰς τὴν Αἰτωλίαν,[1] καὶ κατασχόντας τὰ περὶ Πλευρῶνα χωρία τοὺς πέραν οἰκοῦντας τοῦ Ἀχελῴου διὰ τὸ ἀκούρους φυλάττειν τὰς κεφαλὰς Ἀκαρνᾶνας καλέσαι.[2] ἔνιοι δ' ἀπὸ ἥρωος τοὔνομα σχεῖν ἑκάτερον τὸ φῦλον· οἱ δ' ἀπὸ τοῦ ὄρους τοῦ Κουρίου τοὺς Κουρῆτας ὀνομασθῆναι τοῦ ὑπερκειμένου τῆς Πλευρῶνος, εἶναί τε φῦλόν τι Αἰτωλικὸν τοῦτο, ὡς Ὀφιεῖς καὶ Ἀγραίους καὶ Εὐρυτᾶνας καὶ ἄλλα πλείω. ὡς δ' εἴρηται, τῆς Αἰτωλίας δίχα διῃρημένης, τὰ μὲν περὶ Καλυδῶνα τὸν Οἰνέα ἔχειν φασί, τῆς δὲ Πλευρωνίας μέρος μέν τι καὶ τοὺς Πορθαονίδας ἔχειν τοὺς περὶ τὸν Ἄγριον, εἴπερ[3]

C 466 ᾤκεον ἐν Πλευρῶνι καὶ αἰπεινῇ Καλυδῶνι·

ἐπικρατεῖν μέντοι Θέστιον τῆς Πλευρωνίας, τὸν πενθερὸν τοῦ Οἰνέως, Ἀλθαίας δὲ πατέρα, ἡγούμενον τῶν Κουρήτων· πολέμου δ' ἐμπεσόντος

[1] Πλευρωνίαν *no*.
[2] καλέσαι, Meineke, from conj. of Kramer, for καλεῖσθαι.
[3] οἵπερ *Bkno*.

---

[1] Archemachus (fl. not later than the third century B.C.) wrote works (now lost) on the *History of Euboea* and *Metonymies* (Change of Names).

[2] "Cura." From this passage one might identify the "Curetes" with the "Abantes" (see 10. 1. 3), whom Homer

Archemachus the Euboean[1] says that the Curetes settled at Chalcis, but since they were continually at war for the Lelantine Plain and the enemy would catch them by the front hair and drag them down, he says, they let their hair grow long behind but cut short the part in front, and because of this they were called "Curetes," from the cut of their hair,[2] and they then migrated to Aetolia, and, after taking possession of the region round Pleuron, called the people who lived on the far side of the Acheloüs "Acarnanians," because they kept their heads "unshorn."[3] But some say that each of the two tribes got its name from a hero; others, that the Curetes were named after the mountain Curium, which is situated about Pleuron, and also that this is an Aetolian tribe, like the Ophians and the Agraeans and the Eurytanians and several others. But, as I have already stated,[4] when Aetolia was divided into two parts, the region round Calydon, they say, was in the possession of Oeneus, whereas a certain part of Pleuronia was in the possession of the sons of Porthaon, that is, Agrius and his followers, if it be true that "they lived in Pleuron and steep Calydon";[5] the mastery over Pleuronia, however, was held by Thestius (the father-in-law of Oeneus and father of Althaea), who was leader of the Curetes; but when war broke out between the

speaks of as "letting their hair grow long behind" (*Iliad* 2. 542). According to a scholium (on *Iliad l. c.*), the Euboeans wore their hair long behind "for the sake of manly strength." The Greeks in general, however, let their hair grow long all over the head in Trojan times, being often referred to by Homer as the "long-haired Achaeans."

[3] The Greek adjective used is ἀκούρους ("acurus").

[4] 10. 2. 3, 22.

[5] *Iliad* 14. 116.

τοῖς Θεστιάδαις πρὸς Οἰνέα καὶ Μελέαγρον, ὡς[1] μὲν ὁ ποιητὴς ἀμφὶ συὸς κεφαλῇ καὶ δέρματι, κατὰ τὴν περὶ τοῦ κάπρου μυθολογίαν, ὡς δὲ τὸ εἰκός, περὶ μέρος τῆς χώρας, οὕτω δὴ λέγεται·[2]

Κουρῆτές τ' ἐμάχοντο καὶ Αἰτωλοὶ μενεχάρμαι.

ταῦτα μὲν τὰ ἐγγυτέρω.

7. Τὰ δ' ἀπωτέρω τῆς ὑποθέσεως ταύτης, ἄλλως δὲ διὰ τὴν ὁμωνυμίαν εἰς ταὐτὸν ὑπὸ τῶν ἱστορικῶν ἀγόμενα, ἅπερ Κουρητικὰ μὲν καὶ περὶ Κουρήτων λέγεται, ὁμοίως ὥσπερ καὶ τὰ περὶ τῶν τὴν Αἰτωλίαν καὶ τὴν Ἀκαρνανίαν οἰκησάντων, ἐκείνων μὲν διαφέρει, ἔοικε δὲ μᾶλλον τῷ περὶ Σατύρων καὶ Σειληνῶν καὶ Βακχῶν καὶ Τιτύρων λόγῳ· τοιούτους γάρ τινας δαίμονας ἢ προπόλους θεῶν τοὺς Κουρῆτάς φασιν οἱ παραδόντες τὰ Κρητικὰ καὶ τὰ Φρύγια, ἱερουργίαις τισὶν ἐμπεπλεγμένα ταῖς μὲν μυστικαῖς, ταῖς δ' ἄλλαις[3] περί τε τὴν τοῦ Διὸς παιδοτροφίαν τὴν ἐν Κρήτῃ καὶ τοὺς τῆς μητρὸς τῶν θεῶν ὀργιασμοὺς ἐν τῇ Φρυγίᾳ καὶ τοῖς περὶ τὴν Ἴδην τὴν Τρωικὴν τόποις. τοσαύτη δ' ἐστὶν ἐν τοῖς λόγοις τούτοις ποικιλία, τῶν μὲν τοὺς αὐτοὺς τοῖς Κουρῆσι τοὺς Κορύβαντας καὶ Καβείρους καὶ Ἰδαίους Δακτύλους καὶ Τελχῖνας ἀποφαινόντων, τῶν δὲ συγγενεῖς ἀλλήλων, καὶ

[1] ὡς is omitted in all MSS. except E.
[2] *Dhi* read διαλέγεται instead of δὴ λέγεται.
[3] ἄλλαις *x*, instead of ἄλλως.

sons of Thestius, on the one hand, and Oeneus and Meleager, on the other ("about the hog's head and skin,"[1] as the poet says, following the mythical story of the boar,[2] but in all probability about the possession of a part of the territory), according to the words of the poet, "the Curetes were fighting, as also the Aetolians steadfast in battle."[3] So much for the accounts which are more closely related.

7. The accounts which are more remotely related, however, to the present subject, but are wrongly, on account of the identity of the names, brought into the same connection by the historians—I mean those accounts which, although they are called "Curetan History" and "History of the Curetes," just as if they were the history of those Curetes who lived in Aetolia and Acarnania, not only are different from that history, but are more like the accounts of the Satyri, Sileni, Bacchae, and Tityri; for the Curetes, like these, are called genii or ministers of gods by those who have handed down to us the Cretan and the Phrygian traditions, which are interwoven with certain sacred rites, some mystical, the others connected in part with the rearing of the child Zeus[4] in Crete and in part with the orgies in honour of the mother of the gods which are celebrated in Phrygia and in the region of the Trojan Ida. But the variation in these accounts is so small that, whereas some represent the Corybantes, the Cabeiri, the Idaean Dactyli, and the Telchines as identical with the Curetes, others

[1] *Iliad* 9. 548.
[2] Known in mythology as "the Calydonian boar."
[3] *Iliad* 9. 529.
[4] 10. 3. 11.

μικράς τινας αὐτῶν πρὸς ἀλλήλους διαφορὰς διαστελλομένων, ὡς δὲ τύπῳ εἰπεῖν καὶ κατὰ τὸ πλέον, ἅπαντας ἐνθουσιαστικούς τινας καὶ Βακχικοὺς καὶ ἐνοπλίῳ κινήσει μετὰ θορύβου καὶ ψόφου καὶ κυμβάλων καὶ τυμπάνων καὶ ὅπλων, ἔτι δ᾿ αὐλοῦ καὶ βοῆς ἐκπλήττοντας κατὰ τὰς ἱερουργίας ἐν σχήματι διακόνων, ὥστε[1] καὶ τὰ ἱερὰ τρόπον τινὰ κοινοποιεῖσθαι ταῦτά τε καὶ τῶν Σαμοθρᾴκων καὶ τὰ ἐν Λήμνῳ καὶ ἄλλα πλείω διὰ τὸ τοὺς προπόλους λέγεσθαι τοὺς αὐτούς. ἔστι μὲν οὖν θεολογικὸς πᾶς ὁ τοιοῦτος τρόπος τῆς ἐπισκέψεως καὶ οὐκ ἀλλότριος τῆς τοῦ φιλοσόφου θεωρίας.

8. Ἐπεὶ δὲ δι᾿ ὁμωνυμίαν[2] τῶν Κουρήτων καὶ οἱ ἱστορικοὶ συνήγαγον εἰς ἓν τὰ ἀνόμοια, οὐδ᾿ ἂν[3] αὐτὸς ὀκνήσαιμ᾿ ἂν εἰπεῖν περὶ αὐτῶν ἐπὶ πλέον ἐν παραβάσει, προσθεὶς τὸν οἰκεῖον τῇ ἱστορίᾳ φυσικὸν λόγον. καίτοι τινὲς καὶ συνοικειοῦν βούλονται ταῦτ᾿ ἐκείνοις, καὶ τυχὸν ἴσως ἔχονταί τινος πιθανοῦ· θηλυστολοῦντας γάρ, ὡς αἱ κόραι, τοὔνομα σχεῖν τοῦτο τοὺς[4] περὶ τὴν Αἰτωλίαν φασίν· εἶναι γὰρ καί τινα τοιοῦτον ζῆλον ἐν τοῖς Ἕλλησι, καὶ Ἰάονας ἑλκεχίτωνας εἰρῆσθαι,[5] καὶ τοὺς περὶ Λεωνίδαν κτενιζομένους, ὅτ᾿ ἐξῄεσαν εἰς τὴν μάχην, καταφρονηθῆναι

C 467

[1] ὥστε, Corais, for τε; so the later editors.

[2] ἐπεὶ δὲ δι᾿ ὁμωνυμίαν, Corais, for ἐπειδὴ δὲ ὁμωνυμία (ἐπεὶ δέ *no*, ἐπεὶ δ᾿ ἡ *x*); so the later editors.

[3] ἄν is omitted by *nox*.

[4] τούς, the editors, for τοῖς.

[5] After εἰρῆσθαι Meineke (from Stephanus, *s.v.* Ἀκαρνανία) inserts the words καὶ κρώβυλον καὶ τέττιγα ἐμπλέκεσθαι.

represent them as all kinsmen of one another and differentiate only certain small matters in which they differ in respect to one another; but, roughly speaking and in general, they represent them, one and all, as a kind of inspired people and as subject to Bacchic frenzy, and, in the guise of ministers, as inspiring terror at the celebration of the sacred rites by means of war-dances, accompanied by uproar and noise and cymbals and drums and arms, and also by flute and outcry; and consequently these rites are in a way regarded as having a common relationship, I mean these and those of the Samothracians and those in Lemnos and in several other places, because the divine ministers are called the same. However, every investigation of this kind pertains to theology, and is not foreign to the speculation of the philosopher.

8. But since also the historians, because of the identity of name of the Curetes, have classed together things that are unlike, neither should I myself shrink from discussing them at greater length, by way of digression, adding such account of their physical habits as is appropriate to history. And yet some historians even wish to assimilate their physical habits with those others, and perhaps there is something plausible in their undertaking. For instance, they say that the Curetes of Aetolia got this name because, like "girls,"[1] they wore women's clothes, for, they add, there was a fashion of this kind among the Greeks, and the Ionians were called "tunic-trailing,"[2] and the soldiers of Leonidas were "dressing their hair"[3] when they were to go forth

[1] "Corai" (see foot-note on "girls" and "youths," p. 91).
[2] *e.g. Iliad* 13. 685.
[3] Herodotus 7. 208, 209.

λέγουσιν ὑπὸ τῶν Περσῶν, ἐν δὲ τῇ μάχῃ θαυμασθῆναι. ἁπλῶς δ' ἡ περὶ τὰς κόμας φιλοτεχνία συνέστηκε περί τε θρέψιν καὶ κουρὰν τριχός, ἄμφω δὲ κόραις καὶ κόροις ἐστὶν οἰκεῖα· ὥστε πλεοναχῶς τὸ ἐτυμολογεῖν τοὺς Κουρῆτας[1] ἐν εὐπόρῳ κεῖται. εἰκὸς δὲ καὶ τὴν ἐνόπλιον ὄρχησιν ὑπὸ τῶν ἠσκημένων οὕτω περὶ κόμην καὶ στολὴν πρῶτον εἰσαχθεῖσαν, ἐκείνων Κουρήτων καλουμένων, παρασχεῖν πρόφασιν καὶ τοῖς στρατιωτικωτέροις ἑτέρων καὶ τὸν βίον ἐνόπλιον ἔχουσιν, ὥσθ' ὁμωνύμως καὶ αὐτοὺς Κουρῆτας λεχθῆναι, τοὺς ἐν Εὐβοίᾳ λέγω καὶ Αἰτωλίᾳ καὶ Ἀκαρνανίᾳ. καὶ Ὅμηρος δὲ τοὺς νέους στρατιώτας οὕτω προσηγόρευσε·[2]

> κρινάμενος κούρητας ἀριστῆας Παναχαιῶν,
> δῶρα θοῆς[3] παρὰ νηὸς ἐνεγκεῖν, ὅσσ' Ἀχιλῆι
> χθιζοὶ ὑπέστημεν·

καὶ πάλιν,

> δῶρα φέρον κούρητες Ἀχαιοί.[4]

περὶ μὲν οὖν τῆς τῶν Κουρήτων ἐτυμολογίας ταῦτα. ἡ δὲ[5] ἐνόπλιος ὄρχησις στρατιωτική, καὶ ἡ πυρρίχη δηλοῖ καὶ ὁ Πύρριχος, ὅν φασιν

[1] τοῖς Κουρῆσι CD*hilsx*.
[2] The editors omit καί, after προσηγόρευσε.
[3] The *Iliad* (19. 193) has ἐμῆς instead of θοῆς.
[4] The *Iliad* (19. 248) has Ἀχαιῶν instead of Ἀχαιοί.
[5] The words ἡ δὲ ἐνόπλιος . . . στρατιωτικά are suspected by Kramer, and relegated to foot of page by Meineke.

---

[1] "Corai" and "Coroi." But the corresponding Homeric forms (κοῦροι, κοῦραι) yield in English "Curae" and "Curoe";

to battle, so that the Persians, it is said, conceived a contempt for them, though in the battle they marvelled at them. Speaking generally, the art of caring for the hair consists both in its nurture and in the way it is cut, and both are given special attention by "girls" and "youths";[1] so that there are several ways in which it is easy to derive an etymology of the word "Curetes." It is reasonable to suppose, also, that the war-dance was first introduced by persons who were trained in this particular way in the matter of hair and dress, these being called Curetes, and that this dance afforded a pretext to those also who were more warlike than the rest and spent their life under arms, so that they too came to be called by the same name, "Curetes"—I mean the Curetes in Euboea, Aetolia, and Acarnania. And indeed Homer applied this name to young soldiers, "choose thou the noblest young men[2] from all the Achaeans, and bring the gifts from the swift ship, all that we promised yesterday to Achilles";[3] and again, "the young men of the Achaeans brought the gifts."[4] So much for the etymology of the word "Curetes." The war-dance was a soldiers' dance; and this is plainly indicated both by the "Pyrrhic dance," and by "Pyrrichus," who is said to be the founder of this

and Strabo evidently had those forms in mind (see note on 10. 3. 11).

[2] "Curetes."   [3] *Iliad* 19. 193.

[4] "The Pyrrhic dance of our time seems to be a sort of Dionysiac dance, being more respectable than that of early times, for the dancers have thyrsi instead of spears, and hurl them at one another, and carry fennel-stalks and torches" (Athenaeus 14. 631 B).

εὑρετὴν εἶναι τῆς τοιαύτης ἀσκήσεως τῶν νέων καὶ [1] τὰ στρατιωτικά.[2]

9. Τὸ δ' εἰς ἓν συμφέρεσθαι τὰ τοσαῦτα ὀνόματα καὶ τὴν ἐνοῦσαν θεολογίαν ἐν τῇ περὶ αὐτῶν ἱστορίᾳ νῦν ἐπισκεπτέον. κοινὸν δὴ τοῦτο καὶ τῶν Ἑλλήνων καὶ τῶν βαρβάρων ἐστὶ τὸ τὰς ἱεροποιίας μετὰ ἀνέσεως ἑορταστικῆς ποιεῖσθαι, τὰς μὲν σὺν ἐνθουσιασμῷ, τὰς δὲ χωρίς· καὶ τὰς μὲν μετὰ μουσικῆς, τὰς δὲ μή· καὶ τὰς μὲν μυστικῶς, τὰς δὲ ἐν φανερῷ· καὶ τοῦθ' ἡ φύσις οὕτως ὑπαγορεύει. ἥ τε γὰρ ἄνεσις τὸν νοῦν ἀπάγει ἀπὸ τῶν ἀνθρωπικῶν ἀσχολημάτων, τὸν δὲ ὄντως νοῦν τρέπει πρὸς τὸ θεῖον· ὅ τε ἐνθουσιασμὸς ἐπίπνευσίν τινα θείαν ἔχειν δοκεῖ καὶ τῷ μαντικῷ γένει πλησιάζειν· ἥ τε κρύψις ἡ μυστικὴ τῶν ἱερῶν σεμνοποιεῖ τὸ θεῖον, μιμουμένη τὴν φύσιν αὐτοῦ φεύγουσαν ἡμῶν τὴν αἴσθησιν· ἥ τε μουσική, περί τε ὄρχησιν οὖσα καὶ ῥυθμὸν καὶ μέλος, ἡδονῇ τε ἅμα καὶ καλλιτεχνίᾳ πρὸς τὸ θεῖον ἡμᾶς συνάπτει κατὰ τοιαύτην αἰτίαν. εὖ μὲν γὰρ εἴρηται καὶ τοῦτο, τοὺς ἀνθρώπους τότε μάλιστα μιμεῖσθαι τοὺς θεούς, ὅταν εὐεργετῶσιν· ἄμεινον δ' ἂν λέγοι τις, ὅταν εὐδαιμονῶσι· τοιοῦτον δὲ τὸ χαίρειν καὶ τὸ ἑορτάζειν καὶ τὸ φιλοσοφεῖν καὶ μουσικῆς ἅπτεσθαι· μὴ γάρ, εἴ τις ἔκπτωσις πρὸς τὸ χεῖρον γεγένηται,[3] τῶν

[1] καί, Xylander, Casaubon, and Corais emend to ἐπί; Kramer conj. κατά.

[2] ἡ στρατιωτική C. [3] γεγένηται, Meineke, for γένηται.

[1] Or, following the conjecture of Kramer (see critical note), we should have, instead of "but . . . affairs," simply "in the work of the soldier."

kind of training for young men, as also by the treatises on military affairs.[1]

9. But I must now investigate how it comes about that so many names have been used of one and the same thing, and the theological element contained in their history. Now this is common both to the Greeks and to the barbarians, to perform their sacred rites in connection with the relaxation of a festival, these rites being performed sometimes with religious frenzy, sometimes without it; sometimes with music, sometimes not; and sometimes in secret, sometimes openly. And it is in accordance with the dictates of nature that this should be so, for, in the first place, the relaxation draws the mind away from human occupations and turns the real mind towards that which is divine; and, secondly, the religious frenzy seems to afford a kind of divine inspiration and to be very like that of the soothsayer; and, thirdly, the secrecy with which the sacred rites are concealed induces reverence for the divine, since it imitates the nature of the divine, which is to avoid being perceived by our human senses; and, fourthly, music, which includes dancing as well as rhythm and melody, at the same time, by the delight it affords and by its artistic beauty, brings us in touch with the divine, and this for the following reason; for although it has been well said that human beings then act most like the gods when they are doing good to others, yet one might better say, when they are happy; and such happiness consists of rejoicing, celebrating festivals, pursuing philosophy, and engaging in music; for, if music is perverted when musicians turn their art to sensual delights

μουσικῶν εἰς ἡδυπαθείας τρεπόντων τὰς τέχνας
C 468 ἐν τοῖς συμποσίοις καὶ θυμέλαις καὶ σκηναῖς καὶ ἄλλοις τοιούτοις, διαβαλλέσθω τὸ πρᾶγμα, ἀλλ' ἡ φύσις ἡ τῶν παιδευμάτων ἐξεταζέσθω τὴν ἀρχὴν ἐνθένδε ἔχουσα.

10. Καὶ διὰ τοῦτο μουσικὴν ἐκάλεσε Πλάτων καὶ ἔτι πρότερον οἱ Πυθαγόρειοι τὴν φιλοσοφίαν, καὶ καθ' ἁρμονίαν τὸν κόσμον συνεστάναι φασί, πᾶν τὸ μουσικὸν εἶδος θεῶν ἔργον ὑπολαμβάνοντες. οὕτω δὲ καὶ αἱ Μοῦσαι θεαὶ καὶ Ἀπόλλων Μουσηγέτης καὶ ἡ ποιητικὴ πᾶσα ὑμνητική.[1] ὡσαύτως δὲ καὶ τὴν τῶν ἠθῶν κατασκευὴν τῇ μουσικῇ προσνέμουσιν, ὡς πᾶν τὸ ἐπανορθωτικὸν τοῦ νοῦ τοῖς θεοῖς ἐγγὺς ὄν. οἱ μὲν οὖν Ἕλληνες οἱ πλεῖστοι τῷ Διονύσῳ προσέθεσαν καὶ τῷ Ἀπόλλωνι καὶ τῇ Ἑκάτῃ καὶ ταῖς Μούσαις καὶ Δήμητρι, νὴ Δία,[2] τὸ ὀργιαστικὸν πᾶν καὶ τὸ βακχικὸν καὶ τὸ χορικὸν καὶ τὸ περὶ τὰς τελετὰς μυστικόν, Ἴακχόν τε καὶ τὸν Διόνυσον καλοῦσι καὶ τὸν ἀρχηγέτην τῶν μυστηρίων, τῆς Δήμητρος δαίμονα· δενδροφορίαι τε καὶ χορεῖαι καὶ τελεταὶ κοιναὶ τῶν θεῶν εἰσὶ τούτων· αἱ δὲ Μοῦσαι καὶ ὁ Ἀπόλλων, αἱ μὲν τῶν χορῶν προεστᾶσιν, ὁ δὲ καὶ τούτων καὶ τῶν κατὰ μαντικήν· πρόπολοι δὲ τῶν Μουσῶν οἱ πεπαιδευμένοι πάντες, καὶ ἰδίως οἱ μουσικοί, τοῦ δ' Ἀπόλλωνος οὗτοί τε καὶ οἱ

[1] οὖσα, after ὑμνητική, Kramer omits; so the later editors.
[2] *x*, Tzschucke, and Corais write καὶ Διί instead of νὴ Δία.

---

[1] Plato, *Phaedo* 61.
[2] Philolaüs, *Frag.* 4 (Stobaeus 1. 458–460). See also

at symposiums and in orchestric and scenic performances and the like, we should not lay the blame upon music itself, but should rather examine the nature of our system of education, since this is based on music.

10. And on this account Plato, and even before his time the Pythagoreians, called philosophy music;[1] and they say that the universe is constituted in accordance with harmony,[2] assuming that every form of music is the work of the gods. And in this sense, also, the Muses are goddesses, and Apollo is leader of the Muses, and poetry as a whole is laudatory of the gods. And by the same course of reasoning they also attribute to music the upbuilding of morals, believing that everything which tends to correct the mind is close to the gods. Now most of the Greeks assigned to Dionysus, Apollo, Hecatê, the Muses, and above all to Demeter, everything of an orgiastic or Bacchic or choral nature, as well as the mystic element in initiations; and they give the name "Iacchus" not only to Dionysus but also to the leader-in-chief of the mysteries, who is the genius of Demeter. And branch-bearing, choral dancing, and initiations are common elements in the worship of these gods. As for the Muses and Apollo, the Muses preside over the choruses, whereas Apollo presides both over these and the rites of divination. But all educated men, and especially the musicians, are ministers of the Muses; and both these and those who have to do with divination are ministers of Apollo;

Athenaeus 14. 632 B-C Aristotle, *Metaphysics* 1. 5, Sextus Empiricus, *Adv. Math.* 4. 6. Cp. Plato, *Timaeus* 32 C, 36 D, 37 A, 41 B, *Republic* 617 B, *Epinomis* 991 E.

περὶ μαντικήν, Δήμητρος δὲ οἵ τε μύσται καὶ δᾳδοῦχοι καὶ ἱεροφάνται, Διονύσου δὲ Σειληνοί τε καὶ Σάτυροι καὶ Βάκχαι, Λῆναί τε καὶ Θυῖαι καὶ Μιμαλλόνες καὶ Ναΐδες καὶ Νύμφαι καὶ Τίτυροι προσαγορευόμενοι.[1]

11. Ἐν δὲ τῇ Κρήτῃ καὶ ταῦτα καὶ τὰ τοῦ Διὸς ἱερὰ ἰδίως ἐπετελεῖτο μετ᾽ ὀργιασμοῦ καὶ τοιούτων προπόλων, οἷοι[2] περὶ τὸν Διόνυσόν εἰσιν οἱ Σάτυροι· τούτους δ᾽ ὠνόμαζον Κουρῆτας, νέους τινὰς ἐνόπλιον κίνησιν μετ᾽ ὀρχήσεως ἀποδιδόντας, προστησάμενοι μῦθον τὸν περὶ τῆς τοῦ Διὸς γενέσεως, ἐν ᾧ τὸν μὲν Κρόνον εἰσάγουσιν εἰθισμένον καταπίνειν τὰ τέκνα ἀπὸ τῆς γενέσεως εὐθύς, τὴν δὲ Ῥέαν πειρωμένην ἐπικρύπτεσθαι τὰς ὠδῖνας καὶ τὸ γεννηθὲν βρέφος ἐκποδὼν ποιεῖν καὶ περισώζειν εἰς δύναμιν· πρὸς δὲ τοῦτο συνεργοὺς λαβεῖν τοὺς Κουρῆτάς φασιν,[3] οἳ μετὰ τυμπάνων καὶ τοιούτων ἄλλων ψόφων καὶ ἐνοπλίου χορείας καὶ θορύβου περιέποντες τὴν θεὸν ἐκπλήξειν ἔμελλον τὸν Κρόνον καὶ λήσειν ὑποσπάσαντες αὐτοῦ τὸν παῖδα, τῇ δ᾽ αὐτῇ ἐπιμελείᾳ καὶ τρεφόμενον ὑπ᾽ αὐτῶν παραδίδοσθαι· ὥσθ᾽ οἱ Κουρῆτες ἤτοι διὰ τὸ νέοι[4] καὶ κόροι ὄντες ὑπουργεῖν ἢ διὰ τὸ κουροτροφεῖν τὸν Δία (λέγεται γὰρ ἀμφοτέρως) ταύτης ἠξιώθησαν τῆς προσηγορίας, οἱονεὶ Σάτυροί τινες ὄντες περὶ τὸν Δία. C 469 οἱ μὲν οὖν Ἕλληνες τοιοῦτοι περὶ τοὺς ὀργιασμούς.

[1] καὶ Τίτυροι προσαγορευόμενοι *no*, for καὶ Σάτυροι προσαγορευόμεναι (other MSS.). Cp. 10. 3. 7.

[2] οἷοι *x*, οἱ other MSS.

[3] φασιν, Jones inserts.

[4] νέοι E, νέον other MSS.

and the initiated and torch-bearers and hierophants, of Demeter; and the Sileni and Satyri and Bacchae, and also the Lenae and Thyiae and Mimallones and Naïdes and Nymphae and the beings called Tityri, of Dionysus.

11. In Crete, not only these rites, but in particular those sacred to Zeus, were performed along with orgiastic worship and with the kind of ministers who were in the service of Dionysus, I mean the Satyri. These ministers they called "Curetes," young men who executed movements in armour, accompanied by dancing, as they set forth the mythical story of the birth of Zeus; in this they introduced Cronus as accustomed to swallow his children immediately after their birth, and Rhea as trying to keep her travail secret and, when the child was born, to get it out of the way and save its life by every means in her power; and to accomplish this it is said that she took as helpers the Curetes, who, by surrounding the goddess with tambourines and similar noisy instruments and with war-dance and uproar, were supposed to strike terror into Cronus and without his knowledge to steal his child away; and that, according to tradition, Zeus was actually reared by them with the same diligence; consequently the Curetes, either because, being young, that is "youths,"[1] they performed this service, or because they "reared" Zeus "in his youth"[2] (for both explanations are given), were accorded this appellation, as if they were Satyrs, so to speak, in the service of Zeus. Such, then, were the Greeks in the matter of orgiastic worship.

[1] "Coroi" (see note on "youths," 10. 3. 8).
[2] "Curo-tropheir," to "rear youth."

12. Οἱ δὲ Βερέκυντες, Φρυγῶν τι φῦλον, καὶ ἁπλῶς οἱ Φρύγες καὶ τῶν Τρώων οἱ περὶ τὴν Ἴδην κατοικοῦντες, Ῥέαν μὲν καὶ αὐτοὶ τιμῶσι καὶ ὀργιάζουσι ταύτῃ, μητέρα καλοῦντες θεῶν καὶ Ἄγδιστιν[1] καὶ Φρυγίαν θεὸν μεγάλην, ἀπὸ δὲ τῶν τόπων Ἰδαίαν καὶ Δινδυμήνην καὶ Σιπυλήνην[2] καὶ Πεσσινουντίδα[3] καὶ Κυβέλην καὶ Κυβήβην.[4] οἱ δ' Ἕλληνες τοὺς προπόλους αὐτῆς ὁμωνύμως Κουρῆτας λέγουσιν, οὐ μήν γε ἀπὸ τῆς αὐτῆς μυθοποιίας, ἀλλ' ἑτέρους, ὡς ἂν ὑπουργούς τινας, τοῖς Σατύροις ἀνὰ λόγον· τοὺς δ' αὐτοὺς καὶ Κορύβαντας καλοῦσι.

13. Μάρτυρες δ' οἱ ποιηταὶ τῶν τοιούτων ὑπονοιῶν· ὅ τε γὰρ Πίνδαρος ἐν τῷ διθυράμβῳ, οὗ ἡ ἀρχή

> Πρὶν μὲν εἷρπε σχοινοτένειά[5] τ' ἀοιδά[6]
> διθυράμβων,[7]

μνησθεὶς[8] τῶν περὶ τὸν Διόνυσον ὕμνων τῶν τε παλαιῶν καὶ τῶν ὕστερον, μεταβὰς ἀπὸ τούτων φησί

> σοὶ μὲν κατάρχειν,[9]
> μᾶτερ μεγάλα, πάρα[10] ῥόμβοι κυμβάλων,

[1] Ἄγδιστιν (word omitted by *x*), Casaubon, for Αἴεστιν; so the later editors.

[2] Σιπυλήνην, Tzschucke, for Πυλήνην; so the later editors.

[3] Πεσσινουντίδα, the editors, for Περισσινοῦντα B, Πισινοῦντα *x*, Πισσινοῦντα other MSS.

[4] καὶ Κυβήβην, omitted by MSS. except *Eno*.

[5] σχοινοτένεια Bergk, for σχοῖνος τονίας *k*, σχοινοχονίας *hi*, σχοινοτονίας other MSS.

[6] ἀοιδαί *Bklnox*.

[7] διθυράμβων *x* and Dionys. (*de Comp. Verb.* 14); διθυράμβῳ, other MSS.

[8] δέ, after μνησθείς, Corais and Meineke eject.

12. But as for the Berecyntes,[1] a tribe of Phrygians, and the Phrygians in general, and those of the Trojans who live round Ida, they too hold Rhea in honour and worship her with orgies, calling her Mother of the gods and Agdistis and Phrygia the Great Goddess, and also, from the places where she is worshipped, Idaea and Dindymenê and Sipylenê and Pessinuntis and Cybelê and Cybebê.[2] The Greeks use the same name "Curetes" for the ministers of the goddess, not taking the name, however, from the same mythical story,[3] but regarding them as a different set of "Curetes," helpers as it were, analogous to the Satyri; and the same they also call Corybantes.

13. The poets bear witness to such views as I have suggested. For instance, when Pindar, in the dithyramb which begins with these words, "In earlier times there marched[4] the lay of the dithyrambs long drawn out," mentions the hymns sung in honour of Dionysus, both the ancient and the later ones, and then, passing on from these, says, "To perform the prelude in thy honour, great Mother, the whirling

[1] See 12. 8. 21.

[2] *i.e.* from Mt. Ida, Mt. Dindymum (12. 5. 3), Mt. Sipylus, Pessinus (*l.c.*), and Mt. Cybela (*l.c.*), and Cybeba. Cf. Diodorus Siculus (3. 58), who spells the next to last name "Cybelum."

[3] The story of the Cretan Curetes.

[4] Or perhaps "was drawled" (*sc.* from the lips of men; see Bergk, or Sandys in *Loeb Classical Library*, *Frag.* 79). Roberts (Dionysius of Halicarnassus, *On Literary Composition* 14) translates the verb "crept in" and Sandys (*l.c.*) "flowed."

---

[9] κατάρχειν, Bergk, following *kx*, instead of κατάρχει other MSS.; so Kramer, Müller-Dübner, and Meineke.

[10] μεγάλα, πάρα Bergk, for πάρα μεγάλαι corr. in B, πάρα μεγάλοι other MSS.

ἐν δὲ καχλάδων[1] κρόταλ', αἰθομένα τε
δᾳ̃ς ὑπὸ ξανθαῖσι πεύκαις,

τὴν κοινωνίαν τῶν περὶ τὸν Διόνυσον ἀποδειχθέντων νομίμων παρὰ τοῖς Ἕλλησι καὶ τῶν παρὰ τοῖς Φρυξὶ περὶ τὴν μητέρα τῶν θεῶν συνοικειῶν ἀλλήλοις.[2] Εὐριπίδης τε ἐν ταῖς Βάκχαις τὰ παραπλήσια ποιεῖ, τοῖς Φρυγίοις ἅμα καὶ τὰ Λύδια συμφέρων διὰ τὸ ὅμοιον·[3]

ἀλλ' ὦ λιποῦσαι Τμῶλον, ἔρυμα Λυδίας,
θίασος ἐμός, γυναῖκες, ἃς ἐκ βαρβάρων
ἐκόμισα παρέδρους καὶ ξυνεμπόρους ἐμοί,
αἴρεσθε τἀπιχώρι' ἐν πόλει Φρυγῶν
τύμπανα, Ῥέας τε μητρὸς ἐμά θ' εὑρήματα

καὶ πάλιν·

ὦ μάκαρ, ὅστις εὐδαίμων τελετὰς θεῶν
εἰδώς, βιοτὰν ἁγιστεύει·
τά τε ματρὸς μεγάλας ὄργια Κυβέλας θεμιτεύων[4]
ἀνὰ θύρσον τε τινάσσων, κισσῷ τε στεφανωθείς,
Διόνυσον θεραπεύει.
ἴτε Βάκχαι, ἴτε Βάκχαι, Βρόμιον παῖδα θεὸν θεοῦ
Διόνυσον κατάγουσαι Φρυγίων ἐξ ὀρέων
Ἑλλάδος εἰς εὐρυχόρους ἀγυιάς.

πάλιν δ' ἐν τοῖς ἑξῆς καὶ τὰ Κρητικὰ συμπλέκει τούτοις·

[1] καχλάδων (= *sistrorum*), Wilamowitz restores the reading of all MSS. For other emendations, see C. Müller, *Ind. Var. Lect.* p. 1010.

[2] ἀλλήλαις BCD*hiklx*.

of cymbals is at hand, and among them, also, the clanging of castanets, and the torch that blazeth beneath the tawny pine-trees," he bears witness to the common relationship between the rites exhibited in the worship of Dionysus among the Greeks and those in the worship of the Mother of the gods among the Phrygians, for he makes these rites closely akin to one another. And Euripides does likewise, in his *Bacchae*, citing the Lydian usages at the same time with those of Phrygia, because of their similarity: "But ye who left Mt. Tmolus, fortress of Lydia, revel-band of mine, women whom I brought from the land of barbarians as my assistants and travelling companions, uplift the tambourines native to Phrygian cities, inventions of mine and mother Rhea."[1] And again, "happy he who, blest man, initiated in the mystic rites, is pure in his life, . . . who, preserving the righteous orgies of the great mother Cybelê, and brandishing the thyrsus on high, and wreathed with ivy, doth worship Dionysus. Come, ye Bacchae, come, ye Bacchae, bringing down[2] Bromius,[3] god the child of god, Dionysus, out of the Phrygian mountains into the broad highways of Greece."[4] And again, in the following verses he connects the Cretan usages also with the Phrygian:

[1] *Bacchae* 55.

[2] The verb is also used in the sense of "*bringing back home*," and in the above case might be construed as a *double entente*.

[3] *i.e.* "*Boisterous*" one.

[4] *Bacchae* 72.

---

[3] διὰ τὸ ὅμοιον, Professor Capps, for διά τε Ὅμηρον (κατὰ τὸν Ὅμηρον B*kno*); οὐ κατὰ τὸν Ὅμηρον, Corais, διὰ τὸ ὅμορον, Meineke.

[4] θεμιτεύων, Musgrave, for θεμιστεύων, on account of metre.

ὦ θαλάμευμα Κουρήτων, ζάθεοί τε Κρήτας
διογενέτορες ἔναυλοι,
ἔνθα τρικόρυθες ἄντροις
βυρσότονον κύκλωμα τόδε
C 470 μοι Κορύβαντες εὗρον,
ἀνὰ δὲ Βακχεῖα συντόνῳ
κέρασαν ἁδυβόᾳ Φρυγίων
αὐλῶν πνεύματι, ματρός τε Ῥέας
εἰς χέρα θῆκαν κτύπον εὐάσμασι Βακχᾶν
παρὰ δὲ μαινόμενοι Σάτυροι
ματέρος ἐξανύσαντο Ῥέας,
εἰς δὲ χορεύματα
προσῆψαν Τριετηρίδων,
αἷς χαίρει Διόνυσος.

καὶ ἐν Παλαμήδει φησὶν ὁ χορός·

Θύσαν Διονύσου
κόραν, ὃς ἀν' Ἴδαν
τέρπεται σὺν ματρὶ φίλᾳ
τυμπάνων ἐπ' ἰαχαῖς.[1]

14. Καὶ Σειληνὸν καὶ Μαρσύαν καὶ Ὄλυμπον συνάγοντες εἰς ἓν καὶ εὑρετὰς αὐλῶν ἱστοροῦντες πάλιν καὶ οὕτως τὰ Διονυσιακὰ καὶ Φρύγια εἰς ἓν συμφέρουσι· τήν τε Ἴδην καὶ τὸν Ὄλυμπον συγκεχυμένως πολλάκις ὡς τὸ αὐτὸ ὄρος κτυποῦσιν. εἰσὶ μὲν οὖν λόφοι τέτταρες Ὄλυμποι καλούμενοι τῆς Ἴδης κατὰ τὴν Ἀντανδρίαν, ἔστι δὲ καὶ ὁ Μυσὸς Ὄλυμπος, ὅμορος μέν, οὐχ ὁ αὐτὸς δὲ τῇ Ἴδῃ. ὁ δ' οὖν Σοφοκλῆς ποιήσας τὸν

[1] The reading and metrical arrangement of this corrupt passage is that of Nauck, *Frag.* 586 (*q.v.*).

"O thou hiding-bower[1] of the Curetes, and sacred haunts of Crete that gave birth to Zeus, where for me[2] the triple-crested[3] Corybantes[4] in their caverns invented this hide-stretched circlet,[5] and blent its Bacchic revelry with the high-pitched, sweet-sounding breath of Phrygian flutes, and in Rhea's hands placed its resounding noise, to accompany the shouts of the Bacchae,[6] and from Mother Rhea frenzied Satyrs obtained it and joined it to the choral dances of the Trieterides,[7] in whom Dionysus takes delight." And in the *Palamedes* the Chorus says, "Thysa, daughter of Dionysus, who on Ida rejoices with his dear mother in the Iacchic revels of tambourines."[8]

14. And when they bring Seilenus and Marsyas and Olympus into one and the same connection, and make them the historical inventors of flutes, they again, a second time, connect the Dionysiac and the Phrygian rites; and they often in a confused manner drum on[9] Ida and Olympus as the same mountain. Now there are four peaks of Ida called Olympus, near Antandria; and there is also the Mysian Olympus, which indeed borders on Ida, but is not the same. At any rate, Sophocles, in his *Polyxena*,

[1] Where Zeus was hid.

[2] The leader of the Chorus in *Bacchae* 120 ff. is spokesman of the chorus, and hence of all the Greeks.

[3] Referring to the triple rim of their helmets (cp. the triple crown of the Pope).

[4] Name of the *Phrygian* priests of Cybelê.

[5] *i.e.* the tambourine.

[6] They shouted "ev-ah!" (εὖα; cf. Lat. *ovatio*), as the Greek word shows.

[7] "Triennial Festivals."

[8] See critical note.

[9] "Drum on" is an effort to reproduce in English Strabo's word-play.

Μενέλαον ἐκ τῆς Τροίας ἀπαίρειν σπεύδοντα ἐν τῇ Πολυξένῃ, τὸν δ' Ἀγαμέμνονα μικρὸν ὑπολειφθῆναι βουλόμενον τοῦ ἐξιλάσασθαι τὴν Ἀθηνᾶν χάριν, εἰσάγει λέγοντα τὸν Μενέλαον·

> σὺ δ' αὖθι μίμνων που [1] κατ' Ἰδαίαν χθόνα
> ποίμνας Ὀλύμπου συναγαγὼν θυηπόλει.

15. Τῷ δ' αὐλῷ καὶ κτύπῳ κροτάλων τε καὶ κυμβάλων καὶ τυμπάνων καὶ ταῖς ἐπιβοήσεσι καὶ εὐασμοῖς καὶ ποδοκρουστίαις οἰκεῖα ἐξεύροντο καί τινα τῶν ὀνομάτων, ἃ τοὺς προπόλους καὶ χορευτὰς καὶ θεραπευτὰς τῶν ἱερῶν ἐκάλουν, Καβείρους καὶ Κορύβαντας καὶ Πᾶνας καὶ Σατύρους καὶ Τιτύρους, καὶ τὸν θεὸν Βάκχον καὶ τὴν Ῥέαν Κυβέλην καὶ Κυβήβην [2] καὶ Δινδυμήνην κατὰ τοὺς τόπους αὐτούς. καὶ ὁ Σαβάζιος δὲ τῶν Φρυγιακῶν ἐστὶ καὶ τρόπον τινὰ τῆς Μητρὸς τὸ παιδίον παραδοὺς τὰ [3] τοῦ Διονύσου καὶ αὐτός.

16. Τούτοις δ' ἔοικε καὶ τὰ παρὰ τοῖς Θρᾳξὶ τά τε Κοτύτια [4] καὶ τὰ Βενδίδεια, [5] παρ' οἷς καὶ τὰ Ὀρφικὰ τὴν καταρχὴν ἔσχε. τῆς μὲν οὖν Κότυος [6] τῆς ἐν τοῖς Ἠδωνοῖς Αἰσχύλος μέμνηται καὶ τῶν περὶ αὐτὴν ὀργάνων. εἰπὼν γάρ·

> σεμνὰ Κότυς ἐν τοῖς Ἠδωνοῖς,
> ὄρεια [7] δ' ὄργαν' ἔχοντες,

[1] που, Corais, from conj. of Xylander, for τοῦ CDhl, τήν Bkno.

[2] Κυβήβην, Tzschucke, for Κύβην; so the later editors.

[3] παραδοὺς τά, Meineke from conj. of Kramer, for παράδων τά x, παραδίδοντα s, παραδιδόμενος τοῖς Bkno.

[4] Κότυα Dh, Κόττυα i, Κοττύτια *Epit.*

[5] Βενδίδια *nox*, Μενδίδια Ckl, Βενθείδια E.

representing Menelaüs as in haste to set sail from Troy, but Agamemnon as wishing to remain behind for a short time for the sake of propitiating Athena, introduces Menelaüs as saying, "But do thou, here remaining, somewhere in the Idaean land collect flocks of Olympus and offer them in sacrifice."[1]

15. They invented names appropriate to the flute, and to the noises made by castanets, cymbals, and drums, and to their acclamations and shouts of "ev-ah," and stampings of the feet;[2] and they also invented some of the names by which to designate the ministers, choral dancers, and attendants upon the sacred rites, I mean "Cabeiri" and "Corybantes" and "Pans" and "Satyri" and "Tityri," and they called the god "Bacchus," and Rhea "Cybelê" or "Cybebê" or "Dindymenê" according to the places where she was worshipped. Sabazius also belongs to the Phrygian group and in a way is the child of the Mother, since he too transmitted the rites of Dionysus.[3]

16. Also resembling these rites are the Cotytian and the Bendideian rites practised among the Thracians, among whom the Orphic rites had their beginning. Now the Cotys who is worshipped among the Edonians, and also the instruments used in her rites, are mentioned by Aeschylus; for he says, "O adorable Cotys among the Edonians, and ye who hold mountain-ranging[4] instruments"; and

[1] *Frag.* 47. 9 (Nauck). [2] Cp. end of § 17 following.
[3] Cp. end of § 18 following.
[4] The instruments, like those who play them (cp. §§ 19 and 23 following), are boldly referred to as "mountain-ranging."

[6] Κόττυος *ino*. [7] ὄρεια D*h*, ὅρια other MSS.

τοὺς περὶ τὸν Διόνυσον εὐθέως ἐπιφέρει·

ὁ μὲν ἐν χερσὶν
βόμβυκας ἔχων, τόρνου κάματον,
δακτυλόδεικτον[1] πίμπλησι μέλος,
μανίας ἐπαγωγὸν ὁμοκλάν,
ὁ δὲ χαλκοδέτοις[2] κοτύλαις ὀτοβεῖ

καὶ πάλιν·

C 471 ψαλμὸς δ' ἀλαλάζει·
ταυρόφθογγοι δ' ὑπομυκῶνται[3]
ποθὲν ἐξ ἀφανοῦς φοβεροὶ μῖμοι,
τυμπάνου δ' εἰκὼν[4] ὥσθ' ὑπογαίου
βροντῆς, φέρεται βαρυταρβής.

ταῦτα γὰρ ἔοικε τοῖς Φρυγίοις· καὶ οὐκ ἀπεικός γε, ὥσπερ αὐτοὶ οἱ Φρύγες Θρᾳκῶν ἄποικοί εἰσιν, οὕτω καὶ τὰ ἱερὰ ἐκεῖθεν μετενηνέχθαι. καὶ τὸν Διόνυσον δὲ καὶ τὸν Ἠδωνὸν Λυκοῦργον συνάγοντες εἰς ἓν τὴν ὁμοιοτροπίαν τῶν ἱερῶν αἰνίττονται.

17. Ἀπὸ δὲ τοῦ μέλους καὶ τοῦ ῥυθμοῦ καὶ τῶν ὀργάνων καὶ ἡ μουσικὴ πᾶσα Θρᾳκία καὶ Ἀσιᾶτις νενόμισται. δῆλον δ' ἔκ τε τῶν τόπων, ἐν οἷς αἱ Μοῦσαι τετίμηνται· Πιερία γὰρ καὶ Ὄλυμπος καὶ Πίμπλα καὶ Λείβηθρον τὸ παλαιὸν ἦν Θρᾴκια χωρία καὶ ὄρη, νῦν δὲ ἔχουσι Μακεδόνες· τόν τε Ἑλικῶνα καθιέρωσαν ταῖς Μούσαις Θρᾷκες οἱ τὴν Βοιωτίαν ἐποικήσαντες, οἵπερ καὶ

[1] δακτυλόδεικτον MSS., but Corais, from conj. of Jacobs, reads δακτυλόθικτον. Perhaps δακτυλόδικτον is right; so Nauck reads, *Frag.* 57, but the interpretation of the word in L. and S. ("of the humming of a top") is wrong.
[2] χαλκοδέτοις, Casaubon, for χαλκοθέοις MSS., χαλκοθέτοις *Epit.*; so the later editors.
[3] ὑπομηκῶνται *Bklno.*

he mentions immediately afterwards the attendants of Dionysus: "one, holding in his hands the bombyces,[1] toilsome work of the turner's chisel, fills full the fingered melody, the call that brings on frenzy, while another causes to resound the bronze-bound cotylae";[2] and again, "stringed instruments raise their shrill cry, and frightful mimickers from some place unseen bellow like bulls, and the semblance[3] of drums, as of subterranean thunder, rolls along, a terrifying sound"; for these rites resemble the Phrygian rites, and it is at least not unlikely that, just as the Phrygians themselves were colonists from Thrace, so also their sacred rites were borrowed from there. Also when they identify Dionysus and the Edonian Lycurgus, they hint at the homogeneity of their sacred rites.

17. From its melody and rhythm and instruments, all Thracian music has been considered to be Asiatic. And this is clear, first, from the places where the Muses have been worshipped, for Pieria and Olympus and Pimpla and Leibethrum were in ancient times Thracian places and mountains, though they are now held by the Macedonians; and again, Helicon was consecrated to the Muses by the Thracians who settled in Boeotia, the same who

[1] A kind of reed-flute.

[2] Literally "cups"; hence, a kind of cymbal.

[3] In connection with this bold use of "semblance" (εἰκών) by Aeschylus, note Strabo's studied use of "resembles" (ἔοικε, twice in this paragraph) and "unlikely" (ἀπεικός). Others either translate εἰκών "echo," or omit the thought.

[4] εἰκών, Kramer restores, instead of ἠχώ *kno* and earlier editors; εἰχών B(by corr.)*lx*.

τὸ τῶν Λειβηθριάδων Νυμφῶν ἄντρον καθιέρωσαν. οἵ τ' ἐπιμεληθέντες τῆς ἀρχαίας μουσικῆς Θρᾷκες λέγονται, Ὀρφεύς τε καὶ Μουσαῖος καὶ Θάμυρις καὶ τῷ Εὐμόλπῳ δὲ τοὔνομα ἐνθένδε, καὶ οἱ τῷ Διονύσῳ τὴν Ἀσίαν ὅλην καθιερώσαντες μέχρι τῆς Ἰνδικῆς ἐκεῖθεν καὶ τὴν πολλὴν μουσικὴν μεταφέρουσι· καὶ ὁ μέν τίς φησιν· κιθάραν Ἀσιᾶτιν ῥάσσων,[1] ὁ δὲ τοὺς αὐλοὺς Βερεκυντίους καλεῖ καὶ Φρυγίους· καὶ τῶν ὀργάνων ἔνια βαρβάρως ὠνόμασται νάβλας[2] καὶ σαμβύκη καὶ βάρβιτος καὶ μαγάδις καὶ ἄλλα πλείω.

18. Ἀθηναῖοι δ' ὥσπερ περὶ τὰ ἄλλα φιλοξενοῦντες διατελοῦσιν, οὕτω καὶ περὶ τοὺς θεούς. πολλὰ γὰρ τῶν ξενικῶν ἱερῶν παρεδέξαντο, ὥστε καὶ ἐκωμῳδήθησαν· καὶ δὴ καὶ τὰ Θρᾴκια καὶ τὰ Φρύγια. τῶν μὲν γὰρ Βενδιδείων[3] Πλάτων μέμνηται, τῶν δὲ Φρυγίων Δημοσθένης, διαβάλλων τὴν Αἰσχίνου μητέρα καὶ αὐτόν, ὡς τελούσῃ τῇ μητρὶ συνόντα καὶ συνθιασεύοντα καὶ ἐπιφθεγγόμενον εὐοῖ σαβοῖ πολλάκις καὶ ὕης ἄττης, ἄττης[4] ὕης· ταῦτα γάρ ἐστι Σαβάζια καὶ Μητρῷα.

19. Ἔτι δ' ἄν τις καὶ ταῦτα εὕροι[5] περὶ τῶν δαιμόνων τούτων καὶ τῆς τῶν ὀνομάτων ποικιλίας καὶ ὅτι οὐ πρόπολοι θεῶν μόνον, ἀλλὰ καὶ αὐτοὶ θεοὶ προσηγορεύθησαν. Ἡσίοδος μὲν γὰρ Ἑκα-

[1] ἀράσσων *nox*.
[2] νάμβλας C*Dilnosx*, νάμβλα E*k* and corr. in B.
[3] Βεδιδείων D*hi*, Βενδιδίων other MSS.
[4] The second ἄττης Kramer restores (for the variant readings see his edition).
[5] εὕροι omitted except in B*kno*.

consecrated the cave of the nymphs called Leibethrides. And again, those who devoted their attention to the music of early times are called Thracians, I mean Orpheus, Musaeus, and Thamyris; and Eumolpus,[1] too, got his name from there. And those writers who have consecrated the whole of Asia, as far as India, to Dionysus, derive the greater part of music from there. And one writer says, "striking the Asiatic cithara"; another calls flutes "Berecyntian" and "Phrygian"; and some of the instruments have been called by barbarian names, "nablas," "sambycê," "barbitos," "magadis," and several others.

18. Just as in all other respects the Athenians continue to be hospitable to things foreign, so also in their worship of the gods; for they welcomed so many of the foreign rites that they were ridiculed therefor by comic writers; and among these were the Thracian and Phrygian rites. For instance, the Bendideian rites are mentioned by Plato,[2] and the Phrygian by Demosthenes,[3] when he casts the reproach upon Aeschines' mother and Aeschines himself that he was with her when she conducted initiations, that he joined her in leading the Dionysiac march, and that many a time he cried out "êvoe saboe" and "hyês attês, attês hyês"; for these words are in the ritual of Sabazius and the Mother.

19. Further, one might also find, in addition to these facts concerning these genii and their various names, that they were called, not only ministers of gods, but also gods themselves. For instance, Hesiod

[1] "Sweet-singer." [2] *Republic* I. 327, II. 354.
[3] *On the Crown* 313.

τέρου[1] καὶ τῆς Φορωνέως θυγατρὸς πέντε γενέσθαι θυγατέρας φησίν,

> ἐξ ὧν οὔρειαι Νύμφαι θεαὶ ἐξεγένοντο,[2]
> καὶ γένος οὐτιδανῶν Σατύρων καὶ ἀμηχανοεργῶν
> Κουρῆτές τε θεοὶ φιλοπαίγμονες, ὀρχηστῆρες.

C 472 ὁ δὲ τὴν Φορωνίδα γράψας[3] αὐλητὰς καὶ Φρύγας τοὺς Κουρῆτας λέγει, ἄλλοι δὲ γηγενεῖς καὶ χαλκάσπιδας· οἱ δ' οὐ τοὺς Κουρῆτας, ἀλλὰ τοὺς Κορύβαντας Φρύγας, ἐκείνους δὲ Κρῆτας, περιθέσθαι δ' ὅπλα χαλκᾶ πρώτους ἐν Εὐβοίᾳ· διὸ καὶ Χαλκιδέας αὐτοὺς κληθῆναι· οἱ δ' ὑπὸ Τιτάνων Ῥέᾳ δοθῆναι προπόλους ἐνόπλους τοὺς Κορύβαντας ἐκ τῆς Βακτριανῆς ἀφιγμένους, οἱ δ' ἐκ Κόλχων φασίν. ἐν δὲ τοῖς Κρητικοῖς λόγοις οἱ Κουρῆτες Διὸς τροφεῖς λέγονται καὶ φύλακες, εἰς Κρήτην ἐκ Φρυγίας μεταπεμφθέντες ὑπὸ τῆς Ῥέας· οἱ δὲ Τελχίνων ἐν Ῥόδῳ ἐννέα ὄντων, τοὺς Ῥέᾳ συνακολουθήσαντας εἰς Κρήτην καὶ τὸν Δία κουροτροφήσαντας Κουρῆτας ὀνομασθῆναι· Κύρβαντα δέ, τούτων ἑταῖρον, Ἱεραπύτνης[4] ὄντα κτίστην, παρὰ τοῖς Ῥοδίοις παρασχεῖν πρόφασιν τοῖς Πρασίοις ὥστε λέγειν ὡς εἶεν Κορύβαντες δαίμονές τινες Ἀθηνᾶς καὶ Ἡλίου παῖδες. ἔτι δὲ

[1] Ἑκατέρου Nauck, following *n* (*man. sec.*) and Göttling; Ἑκατέου B, Ἑκαταίου *k* and editors before Kramer; Ἑκατέρω other MSS. But Hecaterus is otherwise unknown. At any rate, the person mentioned was probably a son or descendant of Hecatê, unless one should read Ἑκήτορος or Ἑκητόρου (see Diod. Sic. 5. 50) or Ἑκάτου (Apollo).

[2] ἐξεγένοντο, Corais, for ἐγένοντο; so the later editors.

says that five daughters were born to Hecaterus and the daughter of Phoroneus, "from whom sprang the mountain-ranging nymphs, goddesses, and the breed of Satyrs, creatures worthless and unfit for work, and also the Curetes, sportive gods, dancers."[1] And the author of *Phoronis*[2] speaks of the Curetes as "flute-players" and "Phrygians"; and others as "earth-born" and "wearing brazen shields." Some call the Corybantes, and not the Curetes, "Phrygians," but the Curetes "Cretes,"[3] and say that the Cretes were the first people to don brazen armour in Euboea, and that on this account they were also called "Chalcidians";[4] still others say that the Corybantes, who came from Bactriana (some say from among the Colchians), were given as armed ministers to Rhea by the Titans. But in the Cretan accounts the Curetes are called "rearers of Zeus," and "protectors of Zeus," having been summoned from Phrygia to Crete by Rhea. Some say that, of the nine Telchines[5] who lived in Rhodes, those who accompanied Rhea to Crete and "reared" Zeus "in his youth"[6] were named "Curetes"; and that Cyrbas, a comrade of these, who was the founder of Hierapytna, afforded a pretext to the Prasians[7] for saying among the Rhodians that the Corybantes were certain genii, sons of Athena and Helius.

[1] *Frag.* 198 (Rzach).
[2] Hellanicus of Lesbos (fl. about 430 B.C.).
[3] "Cretans." [4] "Chalc-" means "brazen."
[5] See 14. 2. 7. [6] See 10. 3. 11. [7] See 10. 4. 12.

---

[3] γράψας, Xylander, following *x*, instead of στέψας, other MSS.; so the later editors.
[4] Ἱεραπύτνης, Casaubon, for Ἱερέα Πύδνης; so the later editors.

Κρόνου τινὲς τοὺς Κορύβαντας,[1] ἄλλοι δὲ Διὸς καὶ Καλλιόπης φασὶ τοὺς Κορύβαντας, τοὺς αὐτοὺς τοῖς Καβείροις ὄντας· ἀπελθεῖν δὲ τούτους εἰς Σαμοθρᾴκην, καλουμένην πρότερον Μελίτην, τὰς δὲ πράξεις αὐτῶν μυστικὰς εἶναι.

20. Ταῦτα δ' οὐκ ἀποδεξάμενος ὁ Σκήψιος ὁ τοὺς μύθους συναγαγὼν τούτους, ὡς μηδενὸς ἐν Σαμοθρᾴκῃ μυστικοῦ λόγου περὶ Καβείρων λεγομένου, παρατίθησιν ὅμως[2] καὶ Στησιμβρότου τοῦ Θασίου δόξαν, ὡς τὰ ἐν Σαμοθρᾴκῃ ἱερὰ τοῖς Καβείροις ἐπιτελοῖτο· καλεῖσθαι δέ φησιν αὐτοὺς ἐκεῖνος ἀπὸ τοῦ ὄρους τοῦ ἐν τῇ Βερεκυντίᾳ Καβείρου. οἱ δ' Ἑκάτης προπόλους νομίζουσι τοὺς Κουρῆτας, τοὺς αὐτοὺς τοῖς Κορύβασιν ὄντας. φησὶ δὲ πάλιν ὁ Σκήψιος ἐν τῇ Κρήτῃ τὰς τῆς Ῥέας τιμὰς μὴ νομίζεσθαι μηδὲ ἐπιχωριάζειν, ὑπεναντιούμενος τῷ τοῦ Εὐριπίδου λόγῳ, ἀλλ' ἐν τῇ Φρυγίᾳ μόνον καὶ τῇ Τρῳάδι, τοὺς δὲ λέγοντας μυθολογεῖν μᾶλλον ἢ ἱστορεῖν, πρὸς τοῦτο δὲ καὶ τὴν τῶν τόπων ὁμωνυμίαν συμπρᾶξαι τυχὸν ἴσως αὐτοῖς· Ἴδη γὰρ τὸ ὄρος τό τε Τρωικὸν καὶ τὸ Κρητικόν, καὶ Δίκτη τόπος ἐν τῇ Σκηψίᾳ καὶ ὄρος ἐν Κρήτῃ· τῆς δὲ Ἴδης λόφος Πύτνα, ἀφ' οὗ Ἱεράπυτνα ἡ πόλις, Ἱπποκορωνά τε τῆς Ἀδραμυττηνῆς καὶ Ἱπποκορώνιον ἐν Κρήτῃ, Σαμώνιόν τε τὸ ἑωθινὸν ἀκρωτήριον τῆς νήσου καὶ πεδίον ἐν τῇ Νεανδρίδι καὶ τῇ Ἀλεξανδρέων.

[1] τοὺς Κορύβαντας, Meineke omits; perhaps rightly.
[2] ὅμως, Corais, from conj. of Xylander, for ὁμοίως.

---

[1] Demetrius of Scepsis.

Further, some call the Corybantes sons of Cronus, but others say that the Corybantes were sons of Zeus and Calliopê and were identical with the Cabeiri, and that these went off to Samothrace, which in earlier times was called Melitê, and that their rites were mystical.

20. But though the Scepsian,[1] who compiled these myths, does not accept the last statement, on the ground that no mystic story of the Cabeiri is told in Samothrace, still he cites also the opinion of Stesimbrotus the Thasian[2] that the sacred rites in Samothrace were performed in honour of the Cabeiri: and the Scepsian says that they were called Cabeiri after the mountain Cabeirus in Berecyntia. Some, however, believe that the Curetes were the same as the Corybantes and were ministers of Hecatê. But the Scepsian again states, in opposition to the words of Euripides,[3] that the rites of Rhea were not sanctioned or in vogue in Crete, but only in Phrygia and the Troad, and that those who say otherwise are dealing in myths rather than in history, though perhaps the identity of the place-names contributed to their making this mistake. For instance, Ida is not only a Trojan, but also a Cretan, mountain; and Dictê is a place in Scepsia[4] and also a mountain in Crete; and Pytna, after which the city Hierapytna[5] was named, is a peak of Ida. And there is a Hippocorona in the territory of Adramyttium and a Hippocoronium in Crete. And Samonium is the eastern promontory of the island and a plain in the territory of Neandria and in that of the Alexandreians.[6]

[2] Fl. about 460 B.C.; only fragments of his works are extant.
[3] Quoted in 10. 3. 13.
[4] 13. 1. 51.
[5] In Crete.
[6] See 13. 1. 47.

21. Ἀκουσίλαος δ' ὁ Ἀργεῖος ἐκ Καβειροῦς[1] καὶ Ἡφαίστου Καδμῖλον[2] λέγει, τοῦ δὲ τρεῖς Καβείρους, ὧν[3] Νύμφας Καβειρίδας· Φερεκύδης δ' ἐξ Ἀπόλλωνος καὶ Ῥητίας[4] Κύρβαντας ἐννέα, οἰκῆσαι δ' αὐτοὺς ἐν Σαμοθρᾴκῃ· ἐκ δὲ Καβειροῦς τῆς Πρωτέως καὶ Ἡφαίστου Καβείρους[5] τρεῖς καὶ Νύμφας τρεῖς Καβειρίδας, ἑκατέροις δ' ἱερὰ
C 473 γίνεσθαι· μάλιστα μὲν οὖν ἐν Ἴμβρῳ καὶ Λήμνῳ τοὺς Καβείρους τιμᾶσθαι συμβέβηκεν, ἀλλὰ καὶ ἐν Τροίᾳ κατὰ πόλεις· τὰ δ' ὀνόματα αὐτῶν ἐστὶ μυστικά. Ἡρόδοτος δὲ καὶ ἐν Μέμφει λέγει τῶν Καβείρων ἱερά, καθάπερ καὶ τοῦ Ἡφαίστου, διαφθεῖραι δ' αὐτὰ Καμβύσην. ἔστι δ' ἀοίκητα τὰ χωρία τῆς τῶν δαιμόνων τούτων τιμῆς, τό τε Κορυβαντεῖον[6] τὸ ἐν τῇ Ἀμαξιτίᾳ τῆς νῦν Ἀλεξανδρέων χώρας ἐγγὺς τοῦ Σμινθίου, καὶ ἡ Κορύβισσα ἐν τῇ Σκηψίᾳ περὶ ποταμὸν Εὐρήεντα καὶ κώμην ὁμώνυμον καὶ ἔτι χείμαρρον Αἰθαλόεντα. πιθανὸν δέ φησιν ὁ Σκήψιος, Κουρῆτας μὲν καὶ Κορύβαντας εἶναι τοὺς αὐτούς, οἳ περὶ τὰς τῆς μητρὸς τῶν θεῶν ἁγιστείας πρὸς ἐνόπλιον ὄρχησιν ἤίθεοι καὶ κόροι τυγχάνουσι παρειλημμένοι, καὶ

[1] Καβείρους *gs*, Καβείρου CD*hi* (ους added above in D), Καβείρης B*klno*.
[2] Καδμῖλον, Jones, for Κάμιλον B*klo*, Κάμιλλον other MSS. and the editors.
[3] ὧν *kno*, οἷς other MSS. and editors.
[4] Ῥυτίας *n*, perhaps rightly, as suggested by the fact that there was a Ῥύτιον in Crete (see 10. 4. 14).
[5] Καβείρου CD*hlnos*, Καβείρης B*k*.
[6] Κορυβαντεῖον, Meineke, for Κορυβάντιον.

21. Acusilaüs,[1] the Argive, calls Cadmilus the son of Cabeiro and Hephaestus, and Cadmilus the father of three Cabeiri, and these the fathers of the nymphs called Cabeirides. Pherecydes[2] says that nine Cyrbantes were sprung from Apollo and Rhetia, and that they took up their abode in Samothrace; and that three Cabeiri and three nymphs called Cabeirides were the children of Cabeiro, the daughter of Proteus, and Hephaestus, and that sacred rites were instituted in honour of each triad. Now it has so happened that the Cabeiri are most honoured in Imbros and Lemnos, but they are also honoured in separate cities of the Troad; their names, however, are kept secret. Herodotus[3] says that there were temples of the Cabeiri in Memphis, as also of Hephaestus, but that Cambyses destroyed them. The places where these deities were worshipped are uninhabited, both the Corybanteium in Hamaxitia in the territory now belonging to the Alexandreians near Sminthium,[4] and Corybissa in Scepsia in the neighbourhood of the river Eurëeis and of the village which bears the same name and also of the winter-torrent Aethalöeis. The Scepsian says that it is probable that the Curetes and the Corybantes were the same, being those who had been accepted as young men, or "youths," for the war-dance in connection with the holy rites of the Mother of the gods, and also as "corybantes" from the fact that they

[1] Acusilaüs (fl. fifth century B.C.) wrote works entitled *History* and *Genealogies*. Only fragments remain.

[2] Pherecydes (fl. in the fifth century B.C.) wrote a mythological and historical work in ten books. Only fragments remain.

[3] 3. 37.

[4] 13. 1. 48.

κορύβαντες δὲ ἀπὸ τοῦ κορύπτοντας βαίνειν ὀρχηστικῶς, οὓς καὶ βητάρμονας λέγει ὁ ποιητής·

δεῦτ' ἄγε Φαιήκων βητάρμονες, ὅσσοι ἄριστοι.

τῶν δὲ Κορυβάντων ὀρχηστικῶν καὶ ἐνθουσιαστικῶν ὄντων, καὶ τοὺς μανικῶς κινουμένους κορυβαντιᾶν φαμέν.

22. Δακτύλους δ' Ἰδαίους φασί τινες κεκλῆσθαι τοὺς πρώτους οἰκήτορας τῆς κατὰ τὴν Ἴδην ὑπωρείας· πόδας μὲν γὰρ λέγεσθαι τὰς ὑπωρείας, κορυφὰς δὲ τὰ ἄκρα τῶν ὀρῶν· αἱ οὖν κατὰ μέρος ἐσχατιαί (καὶ πᾶσαι τῆς μητρὸς τῶν θεῶν ἱεραί) περὶ τὴν Ἴδην δάκτυλοι ἐκαλοῦντο.[1] Σοφοκλῆς δὲ οἴεται πέντε τοὺς πρώτους ἄρσενας γενέσθαι, οἳ σίδηρόν τε ἐξεῦρον καὶ εἰργάσαντο πρῶτοι κ[αὶ] ἄλλα πολλὰ τῶν πρὸς τὸν βίον χρησίμων, πέν[τε] δὲ καὶ ἀδελφὰς τούτων, ἀπὸ δὲ τοῦ ἀριθμ[οῦ] Δακτύλους κληθῆναι. ἄλλοι δ' ἄλλως μυθεύο[υ]σιν, ἀπόροις ἄπορα συνάπτοντες, διαφόροις δὲ κα[ὶ] τοῖς ὀνόμασι καὶ τοῖς ἀριθμοῖς χρῶνται, ὧν Κέλμιν[2] ὀνομάζουσί τινα καὶ Δαμναμενέα[3] καὶ Ἡρακλέα καὶ Ἄκμονα· καὶ οἱ μὲν ἐπιχωρίους τῆς Ἴδης, οἱ δὲ ἐποίκους, πάντες δὲ σίδηρον εἰργάσθαι ὑπὸ τούτων ἐν Ἴδῃ πρῶτόν φασι, πάντες δὲ καὶ γόητας ὑπειλήφασι καὶ περὶ τὴν μητέρα τῶν θεῶν καὶ ἐν Φρυγίᾳ ᾠκηκότας περὶ τὴν Ἴδην, Φρυγίαν τὴν Τρωάδα καλοῦντες διὰ τὸ τοὺς

[1] Certain words must have been omitted from the text after Ἴδην. *x* adds δάκτυλοι, Jones also ἐκαλοῦντο. Others merely indicate a lacuna.

[2] Κέλμιν, Tzschucke, for Σαλαμῖνον; so the later editors

[3] Δαμναμενέα, Tzschucke, for Δαμνέα *x*, Δαμνανέα other MSS.

"walked with a butting of their heads" in a dancing way.[1] These are called by the poet "betarmones":[2] "Come now, all ye that are the best 'betarmones' of the Phaeacians."[3] And because the Corybantes are inclined to dancing and to religious frenzy, we say of those who are stirred with frenzy that they are "corybantising."

22. Some writers say that the name "Idaean Dactyli" was given to the first settlers of the lower slopes of Mt. Ida, for the lower slopes of mountains are called "feet," and the summits "heads"; accordingly, the several extremities of Ida (all of which are sacred to the Mother of the gods) were called Dactyli.[4] Sophocles[5] thinks that the first male Dactyli were five in number, who were the first to discover and to work iron, as well as many other things which are useful for the purposes of life, and that their sisters were five in number, and that they were called Dactyli from their number. But different writers tell the myth in different ways, joining difficulty to difficulty; and both the names and numbers they use are different; and they name one of them "Celmis" and others "Damnameneus" and "Heracles" and "Acmon." Some call them natives of Ida, others settlers; but all agree that iron was first worked by these on Ida; and all have assumed that they were wizards and attendants of the Mother of the gods, and that they lived in Phrygia about Ida; and they use the term Phrygia for the Troad

[1] *i.e.* "Cory-bant-es" is here derived from the two verbs "coryptein" ("butt with the head") and "bainein" ("walk" or "go").

[2] "Harmony-walkers."

[3] *Od.* 8. 250.

[4] "Dactyli" means either "fingers" or "toes."

[5] In his *Cophi Satyri*, now lost. *Frag.* 337 (Nauck).

Φρύγας ἐπικρατῆσαι πλησιοχώρους ὄντας, τῆς Τροίας ἐκπεπορθημένης. ὑπονοοῦσι δὲ τῶν Ἰδαίων Δακτύλων ἐκγόνους εἶναι τούς τε Κουρῆτας καὶ τοὺς Κορύβαντας· τοὺς γοῦν πρώτους γεννηθέντας ἐν Κρήνῃ ἑκατὸν ἄνδρας Ἰδαίους Δακτύλους κληθῆναι, τούτων δ' ἀπογόνους φασὶ Κουρῆτας ἐννέα γενέσθαι, τούτων δ' ἕκαστον δέκα παῖδας τεκνῶσαι τοὺς Ἰδαίους καλουμένους Δακτύλους.

C 474 23. Προήχθημεν δὲ διὰ πλειόνων εἰπεῖν περὶ τούτων, καίπερ ἥκιστα φιλομυθοῦντες, ὅτι τοῦ θεολογικοῦ γένους ἐφάπτεται τὰ πράγματα ταῦτα. πᾶς δὲ ὁ περὶ τῶν θεῶν λόγος ἀρχαίας ἐξετάζει δόξας καὶ μύθους, αἰνιττομένων[1] τῶν παλαιῶν ἃς εἶχον ἐννοίας φυσικὰς περὶ τῶν πραγμάτων καὶ προστιθέντων ἀεὶ τοῖς λόγοις τὸν μῦθον. ἅπαντα μὲν οὖν τὰ αἰνίγματα λύειν ἐπ' ἀκριβὲς οὐ ῥᾴδιον, τοῦ δὲ πλήθους τῶν μυθευομένων ἐκτεθέντος εἰς τὸ μέσον, τῶν μὲν ὁμολογούντων ἀλλήλοις, τῶν δ' ἐναντιουμένων, εὐπορώτερον ἄν τις δύναιτο[2] εἰκάζειν ἐξ αὐτῶν τἀληθές· οἷον τὰς ὀρειβασίας τῶν περὶ τὸ θεῖον σπουδαζόντων καὶ αὐτῶν τῶν θεῶν καὶ τοὺς ἐνθουσιασμοὺς εἰκότως μυθεύουσι κατὰ τὴν αὐτὴν αἰτίαν, καθ' ἣν καὶ οὐρανίους νομίζουσι τοὺς θεοὺς καὶ προνοητικοὺς τῶν τε ἄλλων καὶ τῶν προσημασιῶν· τῇ μὲν οὖν ὀρειβασίᾳ τὸ μεταλλευτικὸν καὶ τὸ θηρευτικὸν καὶ[3] ζητητικὸν τῶν πρὸς τὸν βίον χρησίμων ἐφάνη

[1] αἰνιττομένων, Xylander, for αἰνιττομένους; so the later editors.

[2] ἄν τις δύναιτο, Kramer, from conj. of Tyrwhitt, for ἀντιδοῦναι τό BCl*hil*, ἄν τι δοῦναι τό D, ἂν δοῦναί τι *no*, ἄν τις ἐξ αὐτῶν εἰκάσειε *x*, Tzschucke, Corais; so the later editors.

because, after Troy was sacked, the Phrygians, whose territory bordered on the Troad, got the mastery over it. And they suspect that both the Curetes and the Corybantes were offspring of the Idaean Dactyli; at any rate, the first hundred men born in Crete were called Idaean Dactyli, they say, and as offspring of these were born nine Curetes, and each of these begot ten children who were called Idaean Dactyli.

23. I have been led on to discuss these people rather at length, although I am not in the least fond of myths, because the facts in their case border on the province of theology. And theology as a whole must examine early opinions and myths, since the ancients expressed enigmatically the physical notions which they entertained concerning the facts and always added the mythical element to their accounts. Now it is not easy to solve with accuracy all the enigmas, but if the multitude of myths be set before us, some agreeing and others contradicting one another, one might be able more readily to conjecture out of them what the truth is. For instance, men probably speak in their myths about the "mountain-roaming" of religious zealots and of gods themselves, and about their "religious frenzies," for the same reason that they are prompted to believe that the gods dwell in the skies and show forethought, among their other interests, for prognostication by signs. Now seeking for metals, and hunting, and searching for the things that are useful for the purposes of life, are manifestly closely related to

---

[a] *καί*, Kramer inserts; so the later editors.

συγγενές, τῶν δ' ἐνθουσιασμῶν καὶ θρησκείας καὶ μαντικῆς τὸ ἀγυρτικὸν καὶ γοητεία ἐγγύς. τοιοῦτον δὲ καὶ τὸ φιλότεχνον μάλιστα τὸ περὶ τὰς Διονυσιακὰς τέχνας[1] καὶ τὰς Ὀρφικάς. ἀλλ' ἀπόχρη περὶ αὐτῶν.

IV

1. Ἐπεὶ δὲ πρῶτον περὶ τῶν τῆς Πελοποννήσου νήσων τῶν τε ἄλλων διῆλθον καὶ τῶν ἐν τῷ Κορινθιακῷ κόλπῳ καὶ τῶν πρὸ αὐτοῦ, περὶ τῆς Κρήτης ἐφεξῆς ῥητέον (καὶ γὰρ αὐτὴ[2] τῆς Πελοποννήσου ἐστί) καὶ εἴ τις περὶ τὴν Κρήτην. ἐν δὲ ταύταις αἵ τε Κυκλάδες εἰσὶ καὶ αἱ Σποράδες, αἱ μὲν ἄξιαι μνήμης, αἱ δ' ἀσημότεραι.

2. Νυνὶ δὲ περὶ τῆς Κρήτης πρῶτον λέγωμεν. Εὔδοξος μὲν οὖν ἐν τῷ Αἰγαίῳ φησὶν αὐτὴν ἱδρῦσθαι, δεῖ δὲ μὴ οὕτως, ἀλλὰ κεῖσθαι μὲν μεταξὺ τῆς Κυρηναίας καὶ τῆς Ἑλλάδος τῆς ἀπὸ Σουνίου μέχρι τῆς Λακωνικῆς, ἐπὶ μῆκος ταύταις ταῖς χώραις παράλληλον ἀπὸ τῆς ἑσπέρας ἐπὶ τὴν ἕω· κλύζεσθαι δὲ ἀπὸ μὲν τῶν ἄρκτων τῷ Αἰγαίῳ πελάγει καὶ τῷ Κρητικῷ, ἀπὸ δὲ τοῦ νότου τῷ Λιβυκῷ τῷ συνάπτοντι πρὸς τὸ Αἰγύπτιον πέλαγος. τῶν δὲ ἄκρων τὸ μὲν ἑσπέριόν ἐστι τὸ περὶ Φαλάσαρνα,[3] πλάτος ἔχον διακοσίων που σταδίων καὶ εἰς δύο ἀκρωτήρια μεριζόμενον (ὧν τὸ μὲν νότιον καλεῖται Κριοῦ μέτωπον, τὸ δ' ἀρκτικὸν Κίμαρος), τὸ δ' ἑῷον τὸ Σαμώνιόν ἐστιν, ὑπέρπιπτον τοῦ Σουνίου οὐ πολὺ πρὸς ἕω.

[1] For τέχνας, Jones conjectures τελετάς.

[2] αὐτή, Corais, and later editors (except Meineke αὕτη), for αὐτῆς. Corais inserts πρό after αὐτή.

[3] Φαλάσαρνα, Corais, for Φάλαρνα; so the later editors.

mountain-roaming, whereas juggling and magic are closely related to religious frenzies, worship, and divination. And such also is devotion to the arts, in particular to the Dionysiac and Orphic arts. But enough on this subject.

IV

1. SINCE I have already described the islands of the Peloponnesus in detail, not only the others, but also those in the Corinthian Gulf and those in front of it, I must next discuss Crete (for it, too, belongs to the Peloponnesus) and any islands that are in the neighbourhood of Crete. Among these are the Cyclades and the Sporades, some worthy of mention, others of less significance.

2. But at present let me first discuss Crete.[1] Now although Eudoxus says that it is situated in the Aegaean Sea, one should not so state, but rather that it lies between Cyrenaea and that part of Greece which extends from Sunium to Laconia, stretching lengthwise parallel with these countries from west to east, and that it is washed on the north by the Aegaean and the Cretan Seas, and on the south by the Libyan Sea, which borders on the Aegyptian. As for its two extremities, the western is in the neighbourhood of Phalasarna; it has a breadth of about two hundred stadia and is divided into two promontories (of these the southern is called Criumetopon,[2] the northern Cimarus), whereas the eastern is Samonium, which falls toward the east not much farther than Sunium.

[1] For map of Crete, see Insert in Map VIII at end of Vol. IV.
[2] "Ram's Forehead."

3. Μέγεθος δὲ Σωσικράτης μέν, ὃν φησιν ἀκριβοῦν Ἀπολλόδωρος τὰ περὶ τὴν νῆσον, ἀφορίζεται C 475 μήκει μὲν πλειόνων ἢ δισχιλίων σταδίων καὶ τριακοσίων, πλάτει δὲ ὑπὸ τὸ μέγεθος,[1] ὥσθ' ὁ κύκλος κατὰ τοῦτον γίνοιτ' ἂν πλέον ἢ πεντακισχίλιοι στάδιοι· Ἀρτεμίδωρος δὲ τετρακισχιλίους καὶ ἑκατόν φησιν. Ἱερώνυμος δέ, μῆκος δισχιλίων φήσας, τὸ δὲ πλάτος ἀνώμαλον, πλειόνων ἂν εἴη λέγων τὸν κύκλον, ἢ ὅσων Ἀρτεμίδωρος. κατὰ δὲ[2] τὸ τρίτον μέρος τοῦ μήκους. . . . .[3] τὸ δὲ ἔνθεν ἰσθμός ἐστιν ὡς ἑκατὸν σταδίων, ἔχων κατοικίαν πρὸς μὲν τῇ βορείῳ θαλάττῃ Ἀμφίμαλλαν,[4] πρὸς δὲ τῇ νοτίῳ Φοίνικα τὸν Λαμπέων·[5] πλατυτάτη δὲ κατὰ τὸ μέσον ἐστί. πάλιν δ' ἐντεῦθεν εἰς στενώτερον τοῦ προτέρου συμπίπτουσιν ἰσθμὸν αἱ ἠιόνες περὶ ἑξήκοντα σταδίων, τὸν[6] ἀπὸ Μινώας τῆς Λυκτίων εἰς Ἱεράπυτναν καὶ τὸ Λιβυκὸν πέλαγος· ἐν κόλπῳ δ' ἐστὶν ἡ πόλις. εἶτα πρόεισιν εἰς ὀξὺ ἀκρωτήριον τὸ Σαμώνιον ἐπὶ τὴν Αἴγυπτον νεῦον καὶ τὰς Ῥοδίων νήσους.

[1] ὑπὸ τὸ μέγεθος is corrupt. B has οὔπω τὸ μέγεθος; *kno* and *h* (between lines) and editors before Kramer read οὐ κατὰ τὸ μέγεθος. Groskurd conj. ὅσον διακοσίων (σ′ = 200); Kramer τετρακοσίων (υ′ = 400) or τριακοσίων (τ′ = 300), Meineke τετρακοσίων (υ′), Jones τετρακοσίων ὀγδοήκοντα (υ′ π′), omitting τὸ μέγεθος. υ′ π′ (480) is more in proportion to Strabo's number for the maximum length (2400).

[2] δέ, Corais, for τε; so the later editors.

[3] Something has fallen out after μήκους. Jones conj. διακοσίων (σ′ = 200). Others suggest a number of words, but these contain no number (see Müller, *Ind. Var. Lect.*, p. 1011).

[4] Ἀμφίμαλλαν, Casaubon, for Ἀμφιπαλίαν; so the later editors.

[5] Λαμπέων, Tzschucke, for Λαμπέω: so the later editors.

3. As for its size, Sosicrates, whose account of the island, according to Apollodorus, is exact, defines it as follows: In length, more than two thousand three hundred stadia, and in breadth, . . .,[1] so that its circuit, according to him, would amount to more than five thousand stadia; but Artemidorus says it is four thousand one hundred. Hieronymus[2] says that its length is two thousand stadia and its breadth irregular, and therefore might mean that the circuit is greater than Artemidorus says. For about a third of its length . . .;[3] and then comes an isthmus of about one hundred stadia, which, on the northern sea, has a settlement called Amphimalla, and, on the southern, Phoenix, belonging to the Lampians. The island is broadest near the middle. And from here the shores again converge to an isthmus narrower than the former, about sixty stadia in width, which extends from Minoa, city of the Lyctians, to Hierapytna and the Libyan Sea; the city is situated on the gulf. Then the island projects into a sharp promontory, Samonium, which slopes in the direction of Aegypt and the islands of the Rhodians.

[1] The text is corrupt (see critical note), and no known MS. contains a number for the breadth of the island. Moreover, the Greek words (either three or four) contained in the MSS. at this point are generally unintelligible. According to measurements on Kiepert's wall map, however, the maximum dimensions are 1400 × 310 stadia.

[2] On Hieronymus, see notes on 8. 6. 21 and 9. 5. 22.

[3] All MSS. omit something here (see critical note). Jones conjectures "(it is) about two hundred stadia" in breadth (the breadth of the western end as given in 10. 4. 2).

---

[6] τόν, Corais, for τῶν; so the later editors.

4. Ἔστι δ' ὀρεινὴ καὶ δασεῖα ἡ νῆσος, ἔχει δ' αὐλῶνας εὐκάρπους. τῶν δ' ὀρῶν τὰ μὲν πρὸς δύσιν καλεῖται Λευκά, οὐ λειπόμενα τοῦ Ταϋγέτου κατὰ τὸ ὕψος, ἐπὶ τὸ μῆκος δ' ἐκτεταμένα ὅσον τριακοσίων σταδίων, καὶ ποιοῦντα ῥάχιν, τελευτῶσάν πως ἐπὶ τὰ στενά. ἐν μέσῳ δ' ἐστὶ κατὰ τὸ εὐρυχωρότατον τῆς νήσου τὸ Ἰδαῖον ὄρος, ὑψηλότατον τῶν ἐκεῖ, περιφερὲς δ' ἐν κύκλῳ σταδίων ἑξακοσίων· περιοικεῖται δ' ὑπὸ τῶν ἀρίστων πόλεων. ἄλλα δ' ἐστὶ πάρισα τοῖς Λευκοῖς, τὰ μὲν ἐπὶ νότον, τὰ δ' ἐπὶ τὴν ἕω λήγοντα.

5. Ἔστι δ' ἀπὸ τῆς Κυρηναίας ἐπὶ τὸ Κριοῦ μέτωπον δυεῖν ἡμερῶν καὶ νυκτῶν πλοῦς, ἀπὸ δὲ Κιμάρου ἐπὶ Ταίναρόν[1] εἰσι στάδιοι ἑπτακόσιοι (μεταξὺ δὲ Κύθηρα), ἀπὸ δὲ τοῦ Σαμωνίου πρὸς Αἴγυπτον τεττάρων ἡμερῶν καὶ νυκτῶν πλοῦς, οἱ δὲ τριῶν φασί· σταδίων δ' εἶναι τοῦτόν τινες πεντακισχιλίων εἰρήκασιν, οἱ δὲ ἔτι ἐλαττόνων. Ἐρατοσθένης δ' ἀπὸ μὲν τῆς Κυρηναίας μέχρι Κριοῦ μετώπου δισχιλίους φησίν, ἔνθεν δ' εἰς Πελοπόννησον ἐλάττους. . . .[2]

6. Ἄλλη δ' ἄλλων γλῶσσα μεμιγμένη, φησὶν ὁ ποιητής,

> ἐν μὲν Ἀχαιοί,
>
> ἐν δ' Ἐτεόκρητες μεγαλήτορες, ἐν δὲ Κύδωνες,
> Δωριέες τε τριχάϊκες δῖοί τε Πελασγοί.

[1] ἐπὶ Ταίναρον, Meineke, from conj. of Kramer, inserts; others, ἐπὶ Μαλέα(ς).

[2] After ἐλάττους probably χιλίων (,α) has fallen out, as Groskurd suggests.

4. The island is mountainous and thickly wooded, but it has fruitful glens. Of the mountains, those towards the west are called Leuca;[1] they do not fall short of Taÿgetus in height, extend in length about three hundred stadia, and form a ridge which terminates approximately at the narrows. In the middle, in the most spacious part of the island, is Mount Ida, loftiest of the mountains of Crete and circular in shape, with a circuit of six hundred stadia; and around it are the best cities. There are other mountains in Crete that are about as high as the Leuca, some terminating towards the south and others towards the east.

5. The voyage from Cyrenaea to Criumetopon takes two days and nights, and the distance from Cimarus to Taenarum is seven hundred stadia,[2] Cythera lying between them; and the voyage from Samonium to Aegypt takes four days and nights, though some say three. Some state that this is a voyage of five thousand stadia, but others still less. Eratosthenes says that the distance from Cyrenaea to Criumetopon is two thousand, and from there to the Peloponnesus less. . .[3]

6. "But one tongue with others is mixed," the poet says; "there dwell Achaeans, there Eteo-Cretans[4] proud of heart, there Cydonians and Dorians, too, of waving plumes, and goodly Pelasgians."[5] Of these

[1] "White."

[2] A very close estimate (for the same estimate, see 8. 5. 1).

[3] Eratosthenes probably said "a *thousand* less," but no number is given in the MSS. (see critical note).

[4] "Cretans of the old stock."

[5] See 5. 2. 4, where the same passage (*Od.* 19. 175) is quoted.

τούτων φησὶ Στάφυλος[1] τὸ μὲν πρὸς ἕω Δωριεῖς κατέχειν, τὸ δὲ δυσμικὸν Κύδωνας, τὸ δὲ νότιον Ἐτεόκρητας, ὧν εἶναι πολίχνιον Πρᾶσον, ὅπου τὸ τοῦ Δικταίου Διὸς ἱερόν· τοὺς δ' ἄλλους, ἰσχύοντας πλέον, οἰκῆσαι τὰ πεδία. τοὺς μὲν οὖν Ἐτεόκρητας καὶ τοὺς Κύδωνας αὐτόχθονας ὑπάρξαι εἰκός, τοὺς δὲ λοιποὺς ἐπήλυδας, οὓς ἐκ Θετταλίας φησὶν ἐλθεῖν Ἄνδρων[2] τῆς Δωρίδος μὲν πρότερον, νῦν δὲ Ἑστιαιώτιδος λεγομένης· ἐξ ἧς ὡρμήθησαν, ὥς φησιν, οἱ περὶ τὸν Παρνασ-
C 476 σὸν οἰκήσαντες Δωριεῖς καὶ ἔκτισαν τήν τε Ἐρινεὸν καὶ Βοῖον καὶ Κυτίνιον, ἀφ' οὗ καὶ τριχάϊκες ὑπὸ τοῦ ποιητοῦ λέγονται. οὐ πάνυ δὲ τὸν τοῦ Ἄνδρωνος λόγον ἀποδέχονται, τὴν μὲν τετράπολιν Δωρίδα τρίπολιν ἀποφαίνοντος, τὴν δὲ μητρόπολιν τῶν Δωριέων ἄποικον Θετταλῶν· τριχάϊκας δὲ δέχονται ἤτοι ἀπὸ τῆς τριλοφίας ἢ ἀπὸ τοῦ τριχίνους[1] εἶναι τοὺς λόφους.[2]

7. Πόλεις δ' εἰσὶν ἐν τῇ Κρήτῃ πλείους μέν, μέγισται δὲ καὶ ἐπιφανέσταται τρεῖς, Κνωσσός, Γόρτυνα, Κυδωνία. διαφερόντως δὲ τὴν Κνωσσὸν

[1] τριχίνους, Xylander (from Eustath., note on *Od.* 19. 176) for τριχινίου; so the later editors.

[2] After λόφους CD*hi* have ευαμισολοφος (εὐαμίλλους added above in *h*), ἐφαμισολόφος B, εὐαμισολόφους *gl*, καὶ ἡμισολόφος *s*, ἐφαμίλλους *nok* and editors before Corais (who brackets it). Kramer and Meineke omit, following Eustathius (*l.c.*).

---

[1] Staphylus of Naucratis wrote historical works on Thessaly, Athens, Aeolia, and Arcadia, but only a few fragments are preserved. The translator does not know when he lived.

[2] Andron (fl. apparently in the fourth century B.C.) wrote a work entitled *Kinships*, of which only a few fragments

peoples, according to Staphylus,[1] the Dorians occupy the part towards the east, the Cydonians the western part, the Eteo-Cretans the southern; and to these last belongs the town Prasus, where is the temple of the Dictaean Zeus; whereas the other peoples, since they were more powerful, dwelt in the plains. Now it is reasonable to suppose that the Eteo-Cretans and the Cydonians were autochthonous, and that the others were foreigners, who, according to Andron,[2] came from Thessaly, from the country which in earlier times was called Doris, but is now called Hestiaeotis;[3] it was from this country that the Dorians who lived in the neighbourhood of Parnassus set out, as he says, and founded Erineüs, Boeüm, and Cytinium, and hence by Homer[4] are called "trichaïces."[5] However, writers do not accept the account of Andron at all, since he represents the Tetrapolis Doris as being a Tripolis,[6] and the metropolis of the Dorians as a mere colony of Thessalians; and they derive the meaning of "trichaïces" either from the "trilophia,"[7] or from the fact that the crests were "trichini."[8]

7. There are several cities in Crete, but the greatest and most famous are three: Cnossus, Gortyna and Cydonia. The praises of Cnossus are

remain. It treated the genealogical relationships between the Greek tribes and cities, and appears to have been an able work.

[3] See foot-note 2, p. 397, in Vol. IV. [4] *Odyssey*, 19. 177.

[5] Andron fancifully connects this adjective with "tricha" ("in three parts"), making it mean "three-fold" (so Liddell and Scott *q.v.*), but it is surely a compound of θρίξ and ἀΐσσω (cp. κορυθάϊξ), and means "hair-shaking," or, as translated in the above passage from Homer, "of waving plumes."

[6] *i.e.* as composed of three cities instead of four.

[7] "Triple-crest" (of a helmet). [8] "Made of hair."

καὶ Ὅμηρος ὑμνεῖ, μεγάλην καλῶν καὶ βασίλειον τοῦ Μίνω, καὶ οἱ ὕστερον. καὶ δὴ καὶ διετέλεσε μέχρι πολλοῦ φερομένη τὰ πρῶτα, εἶτα ἐταπεινώθη καὶ πολλὰ τῶν νομίμων[1] ἀφῃρέθη, μετέστη δὲ τὸ ἀξίωμα εἴς τε Γόρτυναν καὶ Λύκτον, ὕστερον δ' ἀνέλαβε πάλιν τὸ παλαιὸν σχῆμα τὸ τῆς μητροπόλεως. κεῖται δ' ἐν πεδίῳ κύκλον ἔχουσα ἡ Κνωσσὸς τὸν ἀρχαῖον τριάκοντα σταδίων μεταξὺ τῆς Λυκτίας καὶ τῆς Γορτυνίας, διέχουσα τῆς μὲν Γορτύνης[2] σταδίους διακοσίους, τῆς δὲ Λύττου,[3] ἣν ὁ ποιητὴς Λύκτον ὠνόμασεν, ἑκατὸν εἴκοσι· τῆς δὲ θαλάττης Κνωσσὸς μὲν τῆς βορείου πέντε καὶ εἴκοσι, Γόρτυνα δὲ τῆς Λιβυκῆς ἐνενήκοντα, Λύκτος δὲ καὶ αὐτὴ τῆς Λιβυκῆς ὀγδοήκοντα. ἔχει δ' ἐπίνειον τὸ Ἡράκλειον ἡ Κνωσσός.

8. Μίνω δέ φασιν ἐπινείῳ χρήσασθαι τῷ Ἀμνισῷ, ὅπου τὸ τῆς Εἰλειθυίας ἱερόν. ἐκαλεῖτο δ' ἡ Κνωσσὸς Καίρατος[4] πρότερον, ὁμώνυμος τῷ παραρρέοντι ποταμῷ. ἱστόρηται δ' ὁ Μίνως νομοθέτης γενέσθαι σπουδαῖος θαλαττοκρατῆσαί τε πρῶτος, τριχῇ δὲ διελὼν τὴν νῆσον ἐν ἑκάστῳ τῷ μέρει κτίσαι πόλιν, τὴν μὲν Κνωσσὸν ἐν τῷ. . . . . .[5] καταντικρὺ τῆς Πελοποννήσου· καὶ αὐτὴ δ' ἐστὶ προσβόρειος. ὡς δ' εἴρηκεν Ἔφορος,

[1] νόμων CDghlsx.

[2] διέχουσα τῆς μὲν Γορτύνης, Meineke inserts, from conj. of Tyrwhitt.

[3] Λύττου, Xylander, for Λύκτου; so Meineke.

[4] Καίρατος, Casaubon, for Κέρατος; so the later editors.

[5] After ἐν τῷ Müller-Dübner insert from Diod. Sic. (5. 78): πρὸς βορρᾶν καὶ τὴν Ἀσίαν νεύοντι μέρει τῆς νήσου, Φαιστὸν δ' ἐπὶ θαλάσσης ἐστραμμένην ἐπὶ μεσημβρίαν, Κυδωνίαν δ' ἐν τοῖς πρὸς ἑσπέραν κεκλιμένοις τόποις.

hymned above the rest both by Homer, who calls it "great" and "the kingdom of Minos,"[1] and by the later poets. Furthermore, it continued for a long time to win the first honours; then it was humbled and deprived of many of its prerogatives, and its superior rank passed over to Gortyna and Lyctus; but later it again recovered its olden dignity as the metropolis. Cnossus is situated in a plain, its original circuit being thirty stadia, between the Lyctian and Gortynian territories, being two hundred stadia distant from Gortyna, and a hundred and twenty from Lyttus, which the poet named Lyctus.[2] Cnossus is twenty-five stadia from the northern sea, Gortyna is ninety from the Libyan Sea, and Lyctus itself is eighty from the Libyan. And Cnossus has Heracleium as its seaport.

8. But Minos is said to have used as seaport Amnisus, where is the temple of Eileithuia.[3] In earlier times Cnossus was called Caeratus, bearing the same name as the river which flows past it. According to history, Minos was an excellent law-giver, and also the first to gain the mastery of the sea;[4] and he divided the island into three parts and founded a city in each part, Cnossus in the . . .[5] opposite the Peloponnesus. And it, too,[6] lies to the north. As Ephorus

[1] *Od.* 19. 178. [2] *Iliad* 2. 647 and 17. 611.

[3] The goddess of child-birth.

[4] So Diodorus Siculus (*l.c.*), but see Herodotus 3. 122.

[5] The thought, if not the actual Greek words, of the passage here omitted from the Greek MSS. can be supplied from Diodorus Siculus (5. 78), who, like Strabo, depends much upon Ephorus for historical material: "(Cnossus in the) part of the island which inclines towards Asia, Phaestus on the sea, turned towards the south, and Cydonia in the region which lies towards the west, opposite the Peloponnesus". [6] Cydonia, as well as Cnossus.

ζηλωτὴς ὁ Μίνως ἀρχαίου τινὸς Ῥαδαμάνθυος, δικαιοτάτου ἀνδρός, ὁμωνύμου τοῦ ἀδελφοῦ αὐτοῦ, ὃς πρῶτος τὴν νῆσον ἐξημερῶσαι δοκεῖ νομίμοις καὶ συνοικισμοῖς πόλεων καὶ πολιτείαις, σκηψάμενος παρὰ Διὸς φέρειν ἕκαστα τῶν τιθεμένων δογμάτων εἰς μέσον. τοῦτον δὴ μιμούμενος καὶ ὁ Μίνως δι' ἐννέα ἐτῶν, ὡς ἔοικεν, ἀναβαίνων ἐπὶ τὸ τοῦ Διὸς ἄντρον καὶ διατρίβων ἐνθάδε, ἀπῄει συντεταγμένα ἔχων παραγγέλματά τινα, ἃ ἔφασκεν εἶναι προστάγματα τοῦ Διός· ἀφ' ἧς αἰτίας καὶ τὸν ποιητὴν οὕτως εἰρηκέναι·

ἐνθάδε Μίνως
ἐννέωρος βασίλευε Διὸς μεγάλου ὀαριστής.

τοιαῦτα δ' εἰπόντος, οἱ ἀρχαῖοι περὶ αὐτοῦ πάλιν ἄλλους εἰρήκασι λόγους ὑπεναντίους τούτοις, ὡς C 477 τυραννικός τε γένοιτο καὶ βίαιος καὶ δασμολόγος, τραγῳδοῦντες τὰ περὶ τὸν Μινώταυρον καὶ τὸν Λαβύρινθον καὶ τὰ Θησεῖ συμβάντα καὶ Δαιδάλῳ.

9. Ταῦτα μὲν οὖν ὁποτέρως ἔχει, χαλεπὸν εἰπεῖν. ἔστι δὲ καὶ ἄλλος λόγος οὐχ ὁμολο-

---

[1] See 10. 4. 14.

[2] We should say "every *eight* years," or "every ninth year."

[3] Five different interpretations of this passage have been set forth, dependent on the meaning and syntax of ἐννέωρος: that Minos (1) reigned as king for nine years, (2) was nine years old when he became king, (3) for nine years held converse with Zeus, (4) every nine years held converse with Zeus, and (5) reigned as king when he had come to mature age. Frazer (*Pausanias* 3. 2. 4) adopts the first. Butcher and Lang, and A. T. Murray, adopt the second. Heracleides of Pontus (*On the Cretan Constitutions* 3) seems to have

states, Minos was an emulator of a certain Rhadamanthys of early times, a man most just and bearing the same name as Minos's brother, who is reputed to have been the first to civilise the island by establishing laws and by uniting cities under one city as metropolis[1] and by setting up constitutions, alleging that he brought from Zeus the several decrees which he promulgated. So, in imitation of Rhadamanthys, Minos would go up every nine years,[2] as it appears, to the cave of Zeus, tarry there, and come back with commandments drawn up in writing, which he alleged were ordinances of Zeus; and it was for this reason that the poet says, "there Minos reigned as king, who held converse with great Zeus every ninth year.'[3] Such is the statement of Ephorus; but again the early writers have given a different account of Minos, which is contrary to that of Ephorus, saying that he was tyrannical, harsh, and an exactor of tribute, representing in tragedy the story of the Minotaur and the Labyrinth, and the adventures of Theseus and Daedalus.

9. Now, as for these two accounts, it is hard to say which is true; and there is another subject

adopted the third, saying that Minos spent nine years formulating his laws. But Plato (*Minos* 319 C and *Laws* 624 D) says that Minos visited the cave of his father "every ninth year" (δι' ἐνάτου ἔτους); and Strabo (as 16. 2. 38 shows) expressly follows Plato. Hence the above rendering of the Homeric passage. Apart from the above interpretations, Eustathius (note on *Od.* 10. 19, on a different passage) suggests that ἐννέωρος might pertain to "nine seasons, that is, two years and one month" (the "one month," however, instead of "one season," seems incongruous). This suggests that the present passage might mean that Minos held converse with Zeus during a period of one season every other year.

γούμενος, τῶν μὲν ξένον τῆς νήσου τὸν Μίνω λεγόντων, τῶν δ' ἐπιχώριον. ὁ μέντοι ποιητὴς τῇ δευτέρᾳ δοκεῖ μᾶλλον συνηγορεῖν ἀποφάσει, ὅταν φῇ, ὅτι

πρῶτον Μίνωα τέκε Κρήτῃ ἐπίουρον.

ὑπὲρ δὲ τῆς Κρήτης ὁμολογεῖται, διότι κατὰ τοὺς παλαιοὺς χρόνους ἐτύγχανεν εὐνομουμένη καὶ ζηλωτὰς ἑαυτῆς τοὺς ἀρίστους τῶν Ἑλλήνων ἀπέφηνεν, ἐν δὲ τοῖς πρώτοις Λακεδαιμονίους, καθάπερ Πλάτων τε ἐν τοῖς Νόμοις δηλοῖ καὶ Ἔφορος ὃς[1] ἐν τῇ Εὐρώπῃ τὴν πολιτείαν[2] [3] ἀναγέγραφεν· ὕστερον δὲ πρὸς τὸ χεῖρον μετέβαλεν ἐπὶ πλεῖστον. μετὰ γὰρ τοὺς Τυρρηνούς,[4] οἳ μάλιστα ἐδῄωσαν τὴν καθ' ἡμᾶς θάλατταν, οὗτοί εἰσιν οἱ διαδεξάμενοι τὰ λῃστήρια· τούτους δ' ἐπόρθησαν ὕστερον οἱ Κίλικες· κατέλυσαν δὲ πάντας Ῥωμαῖοι, τήν τε Κρήτην ἐκπολεμήσαντες καὶ τὰ πειρατικὰ τῶν Κιλίκων φρούρια. νῦν δὲ Κνωσσὸς καὶ Ῥωμαίων ἀποικίαν ἔχει.

10. Περὶ μὲν οὖν Κνωσσοῦ ταῦτα, πόλεως οὐκ ἀλλοτρίας ἡμῖν, διὰ δὲ τἀνθρώπινα καὶ τὰς ἐν αὐτοῖς μεταβολὰς καὶ συντυχίας ἐκλελειμμένων τῶν συμβολαίων τῶν ὑπαρξάντων ἡμῖν πρὸς τὴν πόλιν. Δορύλαος γὰρ ἦν ἀνὴρ τακτικός, τῶν Μιθριδάτου τοῦ Εὐεργέτου φίλων· οὗτος διὰ τὴν ἐν τοῖς πολεμικοῖς ἐμπειρίαν ξενολογεῖν ἀποδειχθείς, πολὺς ἦν ἔν τε τῇ Ἑλλάδι καὶ τῇ Θρᾴκῃ, πολὺς δὲ καὶ τοῖς παρὰ τῆς Κρήτης ἰοῦσιν, οὔπω τὴν νῆσον ἐχόντων Ῥωμαίων, συχνοῦ δὲ ὄντος ἐν

[1] ὅς, Jones inserts, from conj. of C. Müller.
[2] τὴν πολιτείαν, Jones inserts, from conj. of C. Müller.

that is not agreed upon by all, some saying that Minos was a foreigner, but others that he was a native of the island. The poet, however, seems rather to advocate the second view when he says, "Zeus first begot Minos, guardian o'er Crete." In regard to Crete, writers agree that in ancient times it had good laws, and rendered the best of the Greeks its emulators, and in particular the Lacedaemonians, as is shown, for instance, by Plato in his *Laws*,[1] and also by Ephorus, who in his *Europe*[2] has described its constitution. But later it changed very much for the worse; for after the Tyrrhenians, who more than any other people ravaged Our Sea,[3] the Cretans succeeded to the business of piracy; their piracy was later destroyed by the Cilicians; but all piracy was broken up by the Romans, who reduced Crete by war and also the piratical strongholds of the Cilicians. And at the present time Cnossus has even a colony of Romans.

10. So much for Cnossus, a city to which I myself am not alien, although, on account of man's fortune and of the changes and issues therein, the bonds which at first connected me with the city have disappeared: Dorylaüs was a military expert and one of the friends of Mithridates Euergetes. He, because of his experience in military affairs, was appointed to enlist mercenaries, and often visited not only Greece and Thrace, but also the mercenaries of Crete, that is, before the Romans were

[1] 631 B, 693 E, 751 D ff., 950.
[2] The fourth book of his history was so entitled.
[3] The Mediterranean.

[3] Before ἀναγέγραφεν C. Müller would insert αὐτῶν.
[4] Τυρρηνούς, Tzschucke, for τυράννους; so the later editors.

αὐτῇ τοῦ μισθοφορικοῦ καὶ στρατιωτικοῦ πλήθους, ἐξ οὗ καὶ τὰ λῃστήρια πληροῦσθαι συνέβαινεν. ἐπιδημοῦντος δὲ τοῦ Δορυλάου, κατὰ τύχην ἐνέστη πόλεμος τοῖς Κνωσσίοις πρὸς τοὺς Γορτυνίους· αἱρεθεὶς δὲ στρατηγὸς καὶ κατορθώσας διὰ ταχέων ἤρατο τιμὰς τὰς μεγίστας, καὶ ἐπειδὴ μικρὸν ὕστερον ἐξ ἐπιβουλῆς δολοφονηθέντα ἔγνω τὸν Εὐεργέτην ὑπὸ τῶν φίλων ἐν Σινώπῃ, τὴν διαδοχὴν δὲ εἰς γυναῖκα καὶ παιδία ἥκουσαν, ἀπογνοὺς τῶν ἐκεῖ κατέμεινεν ἐν[1] τῇ Κνωσσῷ· τεκνοποιεῖται δ' ἐκ Μακέτιδος[2] γυναικός, Στερόπης τοὔνομα, δύο μὲν υἱεῖς, Λαγέταν καὶ Στρατάρχαν, ὧν τὸν Στρατάρχαν ἐσχατογήρων καὶ ἡμεῖς ἤδη εἴδομεν, θυγατέρα δὲ μίαν. δυεῖν δὲ ὄντων υἱῶν τοῦ Εὐεργέτου, διεδέξατο τὴν βασιλείαν Μιθριδάτης ὁ προσαγορευθεὶς Εὐπάτωρ, ἕνδεκα ἔτη γεγονώς· τούτῳ σύντροφος ὑπῆρξεν ὁ
C 478 τοῦ Φιλεταίρου Δορύλαος· ἦν δ' ὁ Φιλέταιρος ἀδελφὸς τοῦ τακτικοῦ Δορυλάου. ἀνδρωθεὶς δ' ὁ βασιλεὺς ἐπὶ τοσοῦτο ᾕρητο τῇ συντροφίᾳ τῇ πρὸς τὸν Δορύλαον, ὥστ' οὐκ ἐκεῖνον μόνον εἰς τιμὰς ἦγε τὰς μεγίστας, ἀλλὰ καὶ τῶν συγγενῶν ἐπεμελεῖτο καὶ τοὺς ἐν Κνωσσῷ μετεπέμπετο· ἦσαν δ' οἱ περὶ Λαγέταν, τοῦ μὲν πατρὸς ἤδη τετελευτηκότος, αὐτοὶ δ' ἠνδρωμένοι, καὶ ἧκον ἀφέντες τὰ ἐν Κνωσσῷ· τοῦ δὲ Λαγέτα θυγάτηρ ἦν ἡ μήτηρ τῆς ἐμῆς μητρός. εὐτυχοῦντος μὲν δὴ ἐκείνου, συνευτυχεῖν καὶ τούτοις συνέβαινε, καταλυθέντος δέ (ἐφωράθη γὰρ ἀφιστὰς τοῖς Ῥωμαίοις

[1] ἐν is omitted except in *Bkl.* [2] Μαμέτιδος *Bk.*

yet in possession of the island and while the number of mercenary soldiers in the island, from whom the piratical bands were also wont to be recruited, was large. Now when Dorylaüs was sojourning there war happened to break out between the Cnossians and the Gortynians, and he was appointed general, finished the war successfully, and speedily won the greatest honours. But when, a little later, he learned that Euergetes, as the result of a plot, had been treacherously slain in Sinopê by his closest associates, and heard that the succession had passed to his wife and young children, he despaired of the situation there and stayed on at Cnossus. There, by a Macetan woman, Steropê by name, he begot two sons, Lagetas and Stratarchas (the latter of whom I myself saw when he was an extremely old man), and also one daughter. Now Euergetes had two sons, one of whom, Mithridates, surnamed Eupâtor, succeeded to the rule when he was eleven years old. Dorylaüs, the son of Philetaerus, was his foster brother; and Philotaerus was a brother of Dorylaüs the military expert. And when the king Mithridates reached manhood, he was so infatuated with the companionship of his foster brother Dorylaüs that he not only conferred upon him the greatest honours, but also cared for his kinsmen and summoned those who lived at Cnossus. These were the household of Lagetas and his brother, their father having already died, and they themselves having reached manhood; and they quit Cnossus and went home. My mother's mother was the sister of Lagetas. Now when Lagetas prospered, these others shared in his prosperity, but when he was ruined (for he was caught in the act of trying to cause the kingdom to revolt

τὴν βασιλείαν, ἐφ᾿ ᾧ αὐτὸς εἰς τὴν ἀρχὴν καταστήσεται), συγκατελύθη καὶ τὰ τούτων καὶ ἐταπεινώθησαν· ὠλιγωρήθη δὲ καὶ τὰ πρὸς τοὺς Κνωσσίους συμβόλαια, καὶ αὐτοὺς μυρίας μεταβολὰς δεξαμένους. ἀλλὰ γὰρ ὁ μὲν περὶ τῆς Κνωσσοῦ λόγος τοιοῦτος.

11. Μετὰ δὲ ταύτην δευτερεῦσαι δοκεῖ κατὰ τὴν δύναμιν ἡ τῶν Γορτυνίων πόλις. συμπράττουσαί τε γὰρ ἀλλήλαις ἅπαντας ὑπηκόους εἶχον αὗται τοὺς ἄλλους, στασιάσασαί τε διέστησαν τὰ κατὰ τὴν νῆσον· προσθήκη δ᾿ ἦν ἡ Κυδωνία μεγίστη ὁποτέροις προσγένοιτο. κεῖται δ᾿ ἐν πεδίῳ καὶ ἡ τῶν Γορτυνίων πόλις, τὸ παλαιὸν μὲν ἴσως τετειχισμένη (καθάπερ καὶ Ὅμηρος εἴρηκε·

Γόρτυνά τε τειχήεσσαν)

ὕστερον δ᾿ ἀποβαλοῦσα τὸ τεῖχος ἐκ θεμελίων καὶ πάντα τὸν χρόνον μείνασα ἀτείχιστος· καὶ γὰρ ὁ Φιλοπάτωρ Πτολεμαῖος ἀρξάμενος τειχίζειν ὅσον ἐπὶ ὀγδοήκοντα[1] σταδίους παρῆλθε μόνον· ἀξιόλογον δ᾿ οὖν ἐξεπλήρου ποτὲ κύκλον ἡ οἴκησις, ὅσον πεντήκοντα σταδίων· διέχει δὲ τῆς Λιβυκῆς θαλάττης κατὰ[2] Λεβῆνα, τὸ ἐμπόριον αὐτῆς, ἐνενήκοντα· ἔχει δέ τι καὶ ἄλλο ἐπίνειον, τὸ Μάταλον,[3] διέχει δ᾿ αὐτῆς ἑκατὸν τριάκοντα. διαρρεῖ δ᾿ αὐτὴν ὅλην ὁ Ληθαῖος ποταμός.

12. Ἐκ δὲ Λεβῆνος ἦν Λευκοκόμας τε καὶ ὁ

[1] For ὀγδοήκοντα (MSS., Eustath. on *Iliad* 2. 645, Phrantzes *Chron.* 1. 34), Tzschucke and Corais, from conj. of Casaubon, read ὀκτώ, following *x*, which has in the margin ἢ ὄκτω.

[2] κατά, Casaubon, for καί; so the later editors.

to the Romans, on the understanding that he was to be established at the head of the government), their fortunes were also ruined at the same time, and they were reduced to humility; and the bonds which connected them with the Cnossians, who themselves had undergone countless changes, fell into neglect. But enough for my account of Cnossus.

11. After Cnossus, the city of the Gortynians seems to have ranked second in power; for when these two co-operated they held in subjection all the rest of the inhabitants, and when they had a quarrel there was dissension throughout the island. But Cydonia was the greatest addition to whichever side it attached itself. The city of the Gortynians also lies in a plain; and in ancient times, perhaps, it was walled, as Homer states, "and well-walled Gortyn,"[1] but later it lost its walls from their very foundations, and has remained unwalled ever since; for although Ptolemy Philopator began to build a wall, he proceeded with it only about eighty[2] stadia; at any rate, it is worth mentioning that the settlement once filled out a circuit of about fifty stadia. It is ninety stadia distant from the Libyan Sea at Leben, which is its trading-centre; it also has another seaport, Matalum, from which it is a hundred and thirty stadia distant. The Lethaeus River flows through the whole of its territory.

12. From Leben came Leucocomas and his lover

[1] *Iliad* 2. 646.
[2] "Eighty" seems to be an error for "eight."

---

[3] Μάταλον, Corais and later editors, from conj. of Villebrun, for Μέταλλον.

ἐραστὴς αὐτοῦ Εὐξύνθετος,[1] οὗς ἱστορεῖ Θεόφραστος ἐν τῷ Περὶ Ἔρωτος λόγῳ·[2] ἄθλων δ',[3] ὧν ὁ Λευκοκόμας τῷ Εὐξυνθέτῳ προσέταξεν, ἕνα φησὶν εἶναι τοῦτον, τὸν ἐν Πράσῳ[4] κύνα ἀναγαγεῖν αὐτῷ· ὅμοροι δ' εἰσὶν αὐτοῖς οἱ Πράσιοι, τῆς μὲν θαλάττης ἑβδομήκοντα,[5] Γόρτυνος δὲ διέχοντες ἑκατὸν καὶ ὀγδοήκοντα. εἴρηται δέ, ὅτι τῶν Ἐτεοκρήτων ὑπῆρχεν ἡ Πρᾶσος, καὶ διότι ἐνταῦθα τὸ τοῦ Δικταίου Διὸς ἱερόν· καὶ γὰρ ἡ Δίκτη πλησίον, οὐχ, ὡς Ἄρατος, ὄρεος σχεδὸν Ἰδαίοιο· καὶ γὰρ χιλίους ἡ Δίκτη τῆς Ἴδης ἀπέχει, πρὸς ἀνίσχοντα ἥλιον ἀπ' αὐτῆς κειμένη, τοῦ δὲ Σαμωνίου ἑκατόν. μεταξὺ δὲ τοῦ Σαμωνίου καὶ τῆς Χερρονήσου ἡ Πρᾶσος C 479 ἵδρυτο, ὑπὲρ τῆς θαλάττης ἑξήκοντα σταδίοις· κατέσκαψαν δ' Ἱεραπύτνιοι. οὐκ εὖ δὲ οὐδὲ τὸν Καλλίμαχον λέγειν φασίν, ὡς ἡ Βριτόμαρτις, φεύγουσα τὴν Μίνω βίαν, ἀπὸ τῆς Δίκτης ἅλοιτο εἰς ἁλιέων δίκτυα, καὶ διὰ τοῦτο αὐτὴ μὲν Δίκτυννα ὑπὸ τῶν Κυδωνιατῶν προσαγορευθείη, Δίκτη δὲ τὸ ὄρος· οὐδὲ γὰρ ὅλως ἐκ γειτόνων ἐστὶ τοῖς τόποις τούτοις ἡ Κυδωνία, πρὸς δὲ τοῖς ἑσπερίοις κεῖται τῆς νήσου πέρασι. τῆς μέντοι Κυδωνίας ὄρος ἐστὶ Τίτυρος, ἐν ᾧ ἱερόν ἐστιν, οὐ Δικταῖον, ἀλλὰ Δικτύνναιον.

13. Κυδωνία δ' ἐπὶ θαλάττῃ μὲν ἵδρυται, βλέπουσα πρὸς τὴν Λακωνικήν. διέχει δ' ἑκατέρας

[1] Εὐξύνθεος *k*, Εὐσύνθεος *i*, Εὐξύνθεος other MSS.; emended by all editors.
[2] *hi* add εἶναι before ἄθλων.
[3] δ', after ἄθλων, Jones inserts, from conj. of Kramer.
[4] Πράσκῳ *k*, Πραίσῳ Tzschucke and Corais.

Euxynthetus, the story of whom is told by Theophrastus in his treatise *On Love.* Of the tasks which Leucocomas assigned to Euxynthetus, one, he says, was this—to bring back his dog from Prasus. The country of the Prasians borders on that of the Lebenians, being seventy stadia distant from the sea and a hundred and eighty from Gortyn. As I have said,[1] Prasus belonged to the Eteo-Cretans; and the temple of the Dictaean Zeus was there; for Dictê is near it, not "close to the Idaean Mountain," as Aratus says,[2] for Dictê is a thousand stadia distant from Ida, being situated at that distance from it towards the rising sun, and a hundred from Samonium. Prasus was situated between Samonium and the Cherronesus, sixty stadia above the sea; it was rased to the ground by the Hierapytnians. And neither is Callimachus right, they say, when he says that Britomartis, in her flight from the violence of Minos, leaped from Dictê into fishermen's "nets,"[3] and that because of this she herself was called Dictynna by the Cydoniatae, and the mountain Dictê; for Cydonia is not in the neighbourhood of these places at all, but lies near the western limits of the island. However, there is a mountain called Tityrus in Cydonia, on which is a temple, not the "Dictaean" temple, but the "Dictynnaean."

13. Cydonia is situated on the sea, facing Laconia, and is equidistant, about eight hundred stadia, from

[1] 10. 4. 6. [2] *Phaenomena* 33. [3] "Dictya."

[5] On ἑβδομήκοντα (ο'), see Kramer (*ad loc.*) and C. Müller, *Ind. Var. Lect.* p. 101[illegible]. D*h* have ο', *h* has διακοσίους (σ'), added above, *i* has διακοσίους and the other MSS. δ.

τὸ ἴσον, τῆς τε Κνωσσοῦ καὶ τῆς Γόρτυνος,[1] οἷον ὀκτακοσίους σταδίους, Ἀπτέρας δὲ ὀγδοήκοντα, τῆς ταύτῃ δὲ θαλάττης τετταράκοντα. Ἀπτέρας δ' ἐπίνειόν ἐστι Κίσαμος· πρὸς ἑσπέραν δ' ὅμοροι τοῖς Κυδωνιάταις Πολυρρήνιοι, παρ' οἷς ἐστι τὸ τῆς Δικτύννης ἱερόν· ἀπέχουσι δὲ τῆς θαλάττης ὡς τριάκοντα σταδίους, Φαλασάρνης δὲ ἑξήκοντα. κωμηδὸν δ' ᾤκουν πρότερον· εἶτ' Ἀχαιοὶ καὶ Λάκωνες συνῴκησαν, τειχίσαντες ἐρυμνὸν χωρίον βλέπον πρὸς μεσημβρίαν.

14. Τῶν δ' ὑπὸ Μίνω συνῳκισμένων τριῶν τὴν λοιπὴν (Φαιστὸς δ' ἦν αὕτη)[2] κατέσκαψαν Γορτύνιοι, τῆς μὲν Γόρτυνος[3] διέχουσαν ἑξήκοντα, τῆς δὲ θαλάττης εἴκοσι, τοῦ δὲ Ματάλου[4] τοῦ ἐπινείου τετταράκοντα· τὴν δὲ χώραν ἔχουσιν οἱ κατασκάψαντες. Γορτυνίων δ' ἐστὶ καὶ τὸ Ῥύτιον σὺν τῇ Φαιστῷ·

Φαιστόν τε Ῥύτιόν τε.

ἐκ δὲ τῆς Φαιστοῦ τὸν τοὺς καθαρμοὺς ποιήσαντα διὰ τῶν ἐπῶν Ἐπιμενίδην φασὶν εἶναι. καὶ ὁ Λισσὴν[5] δὲ τῆς Φαιστίας. Λύκτου[6] δέ, ἧς

[1] Γορτύνης *ikx*, Corais.
[2] ἦν, before κατέσκαψαν, Xylander omits; so the later editors.
[3] Γορτύνης *ix*.
[4] Ματάλου B (by corr.) *o*, Μαρτάλου BCD*ghlxy*, Μετάλου *n*.
[5] ὁ Λισσήν (Stephanus ὁ Λισσής), Corais, for Ὀλύσσην; so Meineke.
[6] Λύκτου B*hiklno*, and D (corr. second hand); Λύτου B (first hand)*x*. Kramer and Meineke avoid the Homeric spelling, reading Λύττου.

---

[1] Strabo refers, respectively, to the distance by land to Aptera and by sea, but his estimates are erroneous (see Pauly-Wissowa *s.v.* "Aptera").

the two cities Cnossus and Gortyn, and is eighty stadia distant from Aptera, and forty from the sea in that region.[1] The seaport of Aptera is Cisamus. The territory of the Polyrrhenians borders on that of the Cydoniatae towards the west, and the temple of Dictynna is in their territory. They are about thirty stadia distant from the sea, and sixty from Phalasarna. They lived in villages in earlier times; and then Achaeans and Laconians made a common settlement, building a wall round a place that was naturally strong and faced towards the south.

14. Of the three cities that were united under one metropolis by Minos, the third, which was Phaestus, was rased to the ground by the Gortynians; it is sixty stadia distant from Gortyn, twenty from the sea, and forty from the seaport Matalum; and the country is held by those who rased it. Rhytium, also, together with Phaestus, belongs to the Gortynians: "and Phaestus and Rhytium."[2] Epimenides,[3] who performed the purifications by means of his verses, is said to have been from Phaestus. And Lissen also is in the Phaestian territory. Of Lyctus, which I have mentioned

[2] *Iliad* 2. 648.

[3] Epimenides was a wizard, an ancient "Rip Van Winkle," who, according to Suidas, slept for sixty of his one hundred and fifty years. According to Diogenes Laertius (1. 110), he went to Athens in "the forty-sixth Olympiad" (596–593 B.C.) "and purified the city, and put a stop to the plague" (see Plutarch's account of his visit in Solon's time, *Solon* 12). According to Plato (*Laws* 642 D) he went to Athens "ten years before the Persian War" (*i.e.* 500 B.C.), and uttered the prophecy that the Persians would not come for ten years, and would get the worst of it when they came. But see Pauly-Wissowa *s.v.* "Epimenides."

ἐμνήσθημεν καὶ πρότερον, ἐπίνειόν ἐστιν ἡ λεγομένη Χερρόνησος, ἐν ᾗ τὸ τῆς Βριτομάρτεως ἱερόν· αἱ δὲ συγκαταλεχθεῖσαι πόλεις οὐκέτ' εἰσί, Μίλητός τε καὶ Λύκαστος, τὴν δὲ χώραν, τὴν μὲν ἐνείμαντο Λύκτιοι,[1] τὴν δὲ Κνώσσιοι, κατασκάψαντες τὴν πόλιν.

15. Τοῦ δὲ ποιητοῦ τὸ μὲν ἑκατόμπολιν λέγοντος τὴν Κρήτην, τὸ δὲ ἐνενηκοντάπολιν, Ἔφορος μὲν ὕστερον ἐπικτισθῆναι τὰς δέκα φησὶ μετὰ τὰ Τρωικὰ ὑπὸ τῶν Ἀλθαιμένει τῷ Ἀργείῳ συνακολουθησάντων Δωριέων· τὸν μὲν οὖν Ὀδυσσέα λέγει ἐνενηκοντάπολιν ὀνομάσαι· οὗτος μὲν οὖν πιθανός ἐστιν ὁ λόγος· ἄλλοι δ' ὑπὸ τῶν Ἰδομενέως ἐχθρῶν κατασκαφῆναί φασι τὰς δέκα. ἀλλ' οὔτε κατὰ τὰ Τρωικά φησιν ὁ ποιητὴς ἑκατοντάπολιν ὑπάρξαι τὴν Κρήτην, ἀλλὰ μᾶλλον κατ' αὐτόν (ἐκ γὰρ τοῦ ἰδίου προσώπου λέγει· εἰ C 480 δ' ἐκ τῶν τότε ὄντων τινὸς ἦν ὁ λόγος, καθάπερ ἐν τῇ Ὀδυσσείᾳ, ἡνίκα ἐνενηκοντάπολιν φράζει, καλῶς εἶχεν ἂν[2] οὕτω δέχεσθαι), οὔτ' εἰ[3] συγχωρήσαιμεν τοῦτό γε, ὁ ἑξῆς λόγος σώζοιτ' ἄν. οὔτε γὰρ κατὰ τὴν στρατείαν οὔτε μετὰ τὴν ἐπάνοδον τὴν ἐκεῖθεν τοῦ Ἰδομενέως[4] εἰκός ἐστιν ὑπὸ τῶν ἐχθρῶν αὐτοῦ τὰς πόλεις ἠφανίσθαι ταύτας· ὁ γὰρ ποιητὴς φήσας,[5]

[1] Λύκτιοι *Dhikln*, and B (first hand); Λύτιοι *kx*; Kramer and Meineke Λύττιοι.

[2] ἄν is omitted by all MSS. except *x*.

[3] For οὔτ' εἰ BCD*his* have ὅτι, *x* ὅτι εἰ, Tzschucke and Corais, from conj. of Tyrwhitt, ἀλλ' οὐδ' εἰ.

[4] Tzchucke, Corais, Meineke, and others omit ὡς, after Ἰδομενέως.

[5] φήσας, Meineke, from conj. of Kramer, for φησι.

before,[1] the seaport is Chersonesus, as it is called, where is the temple of Britomartis. But the cities Miletus and Lycastus, which are catalogued along with Lyctus,[2] no longer exist; and as for their territory, the Lyctians took one portion of it and the Cnossians the other, after they had rased the city to the ground.

15. Since the poet speaks of Crete at one time as "possessing a hundred cities,"[3] and also at another as "possessing ninety cities,"[4] Ephorus says that the ten were founded later than the others, after the Trojan War, by the Dorians who accompanied Althaemenes the Argive; he adds that it was Odysseus, however, who called it "Crete of the ninety cities." Now this statement is plausible, but others say that the ten cities were rased to the ground by the enemies of Idomeneus.[5] However, in the first place, the poet does not say that Crete had one hundred cities at the time of the Trojan War, but rather in his own time (for he is speaking in his own person, although, if the statement was made by some person who was living at the time of the Trojan War, as is the case in the *Odyssey*, when Odysseus says "of the ninety cities," then it would be well to interpret it accordingly). In the second place, if we should concede this,[6] the next statement[7] could not be maintained; for it is not likely that these cities were wiped out by the enemies of Idomeneus either during the expedition or after his return from Troy; for when

[1] 10. 4. 7. [2] *Iliad* 2. 647. [3] *Iliad* 2. 649.
[4] *Od.* 19. 174. [5] The grandson of Minos.
[6] *i.e.* that Homer was speaking of his own time.
[7] *i.e.* that ten were rased by the enemies of Idomeneus.

πάντας δ' Ἰδομενεὺς Κρήτην εἰσήγαγ' ἑταίρους,
οἵ φύγον ἐκ πολέμου, πόντος δέ οἱ οὔτιν' ἀπηύρα·

καὶ[1] τούτου τοῦ πάθους ἐμέμνητ' ἄν·[2] οὐ γὰρ δήπου Ὀδυσσεὺς μὲν ἔγνω τὸν ἀφανισμὸν τῶν πόλεων ὁ μηδενὶ συμμίξας τῶν Ἑλλήνων μήτε κατὰ τὴν πλάνην μήθ' ὕστερον. ὁ δὲ καὶ συστρατεύσας τῷ Ἰδομενεῖ καὶ συνανασωθεὶς οὐκ ἔγνω τὰ συμβάντα οἴκοι αὐτῷ οὔτε[3] κατὰ τὴν στρατείαν οὔτε τὴν ἐπάνοδον τὴν ἐκεῖθεν· ἀλλὰ μὴν οὐδὲ μετὰ τὴν ἐπάνοδον· εἰ γὰρ μετὰ πάντων ἐσώθη τῶν ἑταίρων, ἰσχυρὸς ἐπανῆλθεν, ὥστ' οὐκ ἔμελλον ἰσχύσειν οἱ ἐχθροὶ τοσοῦτον, ὅσον δέκα ἀφαιρεῖσθαι πόλεις αὐτόν.[4] τῆς μὲν οὖν χώρας τῶν Κρητῶν τοιαύτη τις ἡ περιοδεία.

16. Τῆς δὲ πολιτείας, ἧς Ἔφορος ἀνέγραψε, τὰ κυριώτατα ἐπιδραμεῖν ἀποχρώντως ἂν ἔχοι. δοκεῖ δέ, φησίν, ὁ νομοθέτης μέγιστον ὑποθέσθαι ταῖς πόλεσιν ἀγαθὸν τὴν ἐλευθερίαν· μόνην γὰρ ταύτην ἴδια ποιεῖν τῶν κτησαμένων τὰ ἀγαθά, τὰ δ' ἐν δουλείᾳ τῶν ἀρχόντων, ἀλλ' οὐχὶ τῶν ἀρχομένων εἶναι· τοῖς δ' ἔχουσι ταύτην φυλακῆς δεῖν· τὴν μὲν οὖν ὁμόνοιαν διχοστασίας αἱρομένης[5] ἀπαντᾶν, ἣ γίνεται διὰ πλεονεξίαν καὶ τρυφήν· σωφρόνως γὰρ καὶ λιτῶς ζῶσιν ἅπασιν οὔτε φθόνον οὔθ' ὕβριν οὔτε μῖσος ἀπαντᾶν πρὸς τοὺς ὁμοίους·

[1] Before καὶ τούτου B(by corr.)*kno* and the earlier editors insert ὥστε.

[2] ἐμέμνητ' ἄν B*no*, ἐμέμνητο other MSS.

[3] οὔτε, after αὐτῷ, Corais inserts: so Müller-Dübner and others. Meineke ejects κατὰ . . . ἐκεῖθεν.

the poet said, "and all his companions Idomeneus brought to Crete, all who escaped from the war, and the sea robbed him of none,"[1] he would also have mentioned this disaster; for of course Odysseus could not have known of the obliteration of the cities, since he came in contact with no Greeks either during his wanderings or later. And he[2] who accompanied Idomeneus on the expedition to Troy and returned safely home at the same time could not have known what occurred in the homeland of Idomeneus either during the expedition or the return from Troy, nor yet even after the return; for if Idomeneus escaped with all his companions, he returned home strong, and therefore his enemies were not likely to be strong enough to take ten cities away from him. Such, then, is my description of the country of the Cretans.

16. As for their constitution, which is described by Ephorus, it might suffice to tell in a cursory way its most important provisions. The lawgiver, he says, seems to take it for granted that liberty is a state's greatest good, for this alone makes property belong specifically to those who have acquired it, whereas in a condition of slavery everything belongs to the rulers and not to the ruled; but those who have liberty must guard it; now harmony ensues when dissension, which is the result of greed and luxury, is removed; for when all citizens live a self-restrained and simple life there arises neither envy nor arrogance nor hatred towards those who are like them; and this is

[1] *Od.* 3. 191 (Nestor speaking). [2] Nestor.

[4] αὐτόν, Corais, for αὐτῶν; so the later editors.
[5] αἰρομένης C; αἰρουμένης other MSS.

διόπερ τοὺς μὲν παῖδας εἰς τὰς ὀνομαζομένας ἀγέλας κελεῦσαι φοιτᾶν, τοὺς δὲ τελείους ἐν τοῖς συσσιτίοις, ἃ καλοῦσιν ἀνδρεῖα, συσσιτεῖν[1] ὅπως τῶν ἴσων μετάσχοιεν τοῖς εὐπόροις οἱ πενέστεροι, δημοσίᾳ τρεφόμενοι· πρὸς δὲ τὸ μὴ δειλίαν ἀλλ' ἀνδρείαν κρατεῖν ἐκ παίδων ὅπλοις καὶ πόνοις συντρέφειν, ὥστε καταφρονεῖν καύματος καὶ ψύχους καὶ τραχείας ὁδοῦ καὶ ἀνάντους καὶ πληγῶν τῶν ἐν γυμνασίοις καὶ μάχαις ταῖς κατὰ σύνταγμα· ἀσκεῖν δὲ καὶ τοξικῇ καὶ ἐνοπλίῳ ὀρχήσει, ἣν καταδεῖξαι Κουρῆτας[2] πρῶτον, ὕστερον δὲ καὶ τὸν[3] συντάξαντα τὴν κληθεῖσαν ἀπ' αὐτοῦ πυρρίχην, ὥστε μηδὲ τὴν παιδιὰν ἄμοιρον εἶναι τῶν πρὸς πόλεμον χρησίμων· ὡς δ' αὔτως καὶ τοῖς ῥυθμοῖς Κρητικοῖς χρῆσθαι κατὰ τὰς ᾠδὰς συντονωτάτοις οὖσιν, οὓς Θάλητα C 481 ἀνευρεῖν, ᾧ καὶ τοὺς παιᾶνας καὶ τὰς ἄλλας τὰς ἐπιχωρίους ᾠδὰς ἀνατιθέασι καὶ πολλὰ τῶν νομίμων, καὶ ἐσθῆτι δὲ καὶ ὑποδέσει πολεμικῇ χρῆσθαι, καὶ τῶν δώρων τιμιώτατα αὐτοῖς εἶναι τὰ ὅπλα.

17. Λέγεσθαι δ' ὑπό τινων, ὡς Λακωνικὰ εἴη τὰ πολλὰ τῶν νομιζομένων Κρητικῶν, τὸ δ' ἀληθές, εὑρῆσθαι μὲν ὑπ' ἐκείνων, ἠκριβωκέναι δὲ τοὺς Σπαρτιάτας, τοὺς δὲ Κρῆτας ὀλιγωρῆσαι, κακωθεισῶν τῶν πόλεων, καὶ μάλιστα τῆς Κνωσσίων, τῶν πολεμικῶν· μεῖναι δέ τινα τῶν νομίμων παρὰ

[1] συσσιτεῖν, Meineke, for συσσίτια.
[2] Κουρῆτας, Groskurd, for Κουρῆτα, Kramer approving.
[3] τόν, before συντάξαντα, Corais inserts; so Jones independently.

why the lawgiver commanded the boys to attend the "Troops,"[1] as they are called, and the full-grown men to eat together at the public messes which they call the "Andreia," so that the poorer, being fed at public expense, might be on an equality with the well-to-do; and in order that courage, and not cowardice, might prevail, he commanded that from boyhood they should grow up accustomed to arms and toils, so as to scorn heat, cold, marches over rugged and steep roads, and blows received in gymnasiums or regular battles; and that they should practise, not only archery, but also the war-dance, which was invented and made known by the Curetes at first, and later, also, by the man[2] who arranged the dance that was named after him, I mean the Pyrrhic dance, so that not even their sports were without a share in activities that were useful for warfare; and likewise that they should use in their songs the Cretic rhythms, which were very high-pitched, and were invented by Thales, to whom they ascribe, not only their Paeans and other local songs, but also many of their institutions; and that they should use military dress and shoes; and that arms should be to them the most valuable of gifts.

17. It is said by some writers, Ephorus continues, that most of the Cretan institutions are Laconian, but the truth is that they were invented by the Cretans and only perfected by the Spartans; and the Cretans, when their cities, and particularly that of the Cnossians, were devastated, neglected military affairs; but some of the institutions continued in

[1] Literally, "Herds" (cf. the Boy Scout "Troops").
[2] Pyrrhicus (see 10. 3. 8).

Λυκτίοις καὶ Γορτυνίοις καὶ ἄλλοις τισὶ πολιχνίοις μᾶλλον, ἢ παρ' ἐκείνοις· καὶ δὴ καὶ τὰ Λυκτίων νόμιμα ποιεῖσθαι μαρτύρια τοὺς τὰ Λακωνικὰ πρεσβύτερα ἀποφαίνοντας· ἀποίκους γὰρ ὄντας φυλάττειν τὰ τῆς μητροπόλεως ἔθη, ἐπεὶ ἄλλως γε εὔηθες εἶναι τὸ τοὺς βέλτιον συνεστῶτας καὶ πολιτευομένους τῶν χειρόνων ζηλωτὰς ἀποφαίνειν· οὐκ εὖ δὲ ταῦτα λέγεσθαι· οὔτε γὰρ ἐκ τῶν νῦν καθεστηκότων τὰ παλαιὰ τεκμηριοῦσθαι δεῖν, εἰς τἀναντία ἑκατέρων μεταπεπτωκότων· καὶ γὰρ ναυκρατεῖν πρότερον τοὺς Κρῆτας, ὥστε καὶ παροιμιάζεσθαι πρὸς τοὺς προσποιουμένους μὴ εἰδέναι ἃ ἴσασιν· Ὁ Κρὴς ἀγνοεῖ τὴν θάλατταν, νῦν δ' ἀποβεβληκέναι τὸ ναυτικόν· οὔτε ὅτι ἄποικοί τινες τῶν πόλεων γεγόνασι τῶν ἐν Κρήτῃ Σπαρτιατῶν, ἐν τοῖς ἐκείνων νομίμοις ἐπηναγκάσθαι· πολλὰς γοῦν τῶν ἀποικίδων μὴ φυλάττειν τὰ πάτρια, πολλὰς δὲ καὶ τῶν μὴ ἀποικίδων ἐν Κρήτῃ τὰ αὐτὰ ἔχειν τοῖς ἀποίκοις ἔθη.

18. Τῶν τε Σπαρτιατῶν τὸν νομοθέτην Λυκοῦργον πέντε γενεαῖς νεώτερον Ἀλθαιμένους εἶναι τοῦ στείλαντος τὴν εἰς Κρήτην ἀποικίαν· τὸν μὲν γὰρ ἱστορεῖσθαι Κίσσου παῖδα τοῦ τὸ Ἄργος κτίσαντος περὶ τὸν αὐτὸν χρόνον ἡνίκα Προκλῆς τὴν Σπάρτην συνῴκιζε, Λυκοῦργον δ' ὁμολογεῖσθαι παρὰ πάντων ἕκτον ἀπὸ Προκλέους γεγονέναι· τὰ δὲ μιμήματα μὴ εἶναι πρότερα τῶν παραδειγ-

---

[1] This Althaemenes, therefore, is not to be confused with the Althaemenes who was the grandson of Minos.
[2] *i.e.* of Laconia (see 8. 5. 4).

use among the Lyctians, Gortynians, and certain other small cities to a greater extent than among the Cnossians; in fact, the institutions of the Lyctians are cited as evidence by those who represent the Laconian as older; for, they argue, being colonists, they preserve the customs of the mother-city, since even on general grounds it is absurd to represent those who are better organised and governed as emulators of their inferiors; but this is not correct, Ephorus says, for, in the first place, one should not draw evidence as to antiquity from the present state of things, for both peoples have undergone a complete reversal; for instance, the Cretans in earlier times were masters of the sea, and hence the proverb, "The Cretan does not know the sea," is applied to those who pretend not to know what they do know, although now the Cretans have lost their fleet; and, in the second place, it does not follow that, because some of the cities in Crete were Spartan colonies, they were under compulsion to keep to the Spartan institutions; at any rate, many colonial cities do not observe their ancestral customs, and many, also, of those in Crete that are not colonial have the same customs as the colonists.

18. Lycurgus the Spartan law-giver, Ephorus continues, was five generations later than the Althaemenes who conducted the colony to Crete;[1] for historians say that Althaemenes was son of the Cissus who founded Argos about the same time when Procles was establishing Sparta as metropolis;[2] and Lycurgus, as is agreed by all, was sixth in descent from Procles; and copies are not earlier than their models, nor more recent things earlier

μάτων μηδὲ τὰ νεώτερα τῶν πρεσβυτέρων· τήν τε ὄρχησιν τὴν παρὰ τοῖς Λακεδαιμονίοις ἐπιχωριάζουσαν καὶ τοὺς ῥυθμοὺς καὶ παιᾶνας τοὺς κατὰ νόμον ᾀδομένους καὶ ἄλλα πολλὰ τῶν νομίμων Κρητικὰ καλεῖσθαι παρ' αὐτοῖς, ὡς ἂν ἐκεῖθεν ὁρμώμενα· τῶν δ' ἀρχείων τὰ μὲν καὶ τὰς διοικήσεις ἔχειν τὰς αὐτὰς καὶ τὰς ἐπωνυμίας, ὥσπερ καὶ τὴν τῶν γερόντων ἀρχὴν καὶ τὴν τῶν ἱππέων
C 482 (πλὴν ὅτι τοὺς ἐν Κρήτῃ ἱππέας καὶ ἵππους κεκτῆσθαι συμβέβηκεν· ἐξ οὗ τεκμαίρονται πρεσβυτέραν εἶναι τῶν ἐν Κρήτῃ ἱππέων τὴν ἀρχήν· σώζειν γὰρ τὴν ἐτυμότητα τῆς προσηγορίας· τοὺς δὲ μὴ ἱπποτροφεῖν), τοὺς ἐφόρους δὲ τὰ αὐτὰ τοῖς ἐν Κρήτῃ κόσμοις διοικοῦντας ἑτέρως ὠνομάσθαι· τὰ δὲ συσσίτια ἀνδρεῖα παρὰ μὲν τοῖς Κρησὶν καὶ νῦν ἔτι καλεῖσθαι, παρὰ δὲ τοῖς Σπαρτιάταις μὴ διαμεῖναι καλούμενα ὁμοίως ὡς[1] πρότερον· παρ' Ἀλκμᾶνι γοῦν οὕτω κεῖσθαι·

φοίναις δὲ καὶ ἐν θιάσοισιν
ἀνδρείων[2] παρὰ δαιτυμόνεσσι πρέπει[3] παιᾶνα
κατάρχειν.

19. Λέγεσθαι δ' ὑπὸ τῶν Κρητῶν, ὡς καὶ παρ' αὐτοὺς ἀφίκοιτο Λυκοῦργος κατὰ τοιαύτην αἰτίαν· ἀδελφὸς ἦν πρεσβύτερος τοῦ Λυκούργου Πολυδέκτης· οὗτος τελευτῶν ἔγκυον κατέλιπε τὴν γυναῖκα· τέως μὲν οὖν ἐβασίλευεν ὁ Λυκοῦργος ἀντὶ τοῦ ἀδελφοῦ, γενομένου δὲ παιδός, ἐπετρό-

[1] ὡς only *no*; ὁμοίως ὡς B (by corr.), and so Tzschucke and Corais; ὁμοίως only, other MSS. (except *k*, which has neither word), and so Müller-Dübner and Meineke.
[2] ἀνδρίων BCD*hi*.
[3] πρέπει, Kramer, from conj. of Ursinus, for πρέπε.

than older things; not only the dancing which is customary among the Lacedaemonians, but also the rhythms and paeans that are sung according to law, and many other Spartan institutions, are called "Cretan" among the Lacedaemonians, as though they originated in Crete; and some of the public offices are not only administered in the same way as in Crete, but also have the same names, as, for instance, the office of the "Gerontes,"[1] and that of the "Hippeis"[2] (except that the "Hippeis" in Crete actually possessed horses, and from this fact it is inferred that the office of the "Hippeis" in Crete is older, for they preserve the true meaning of the appellation, whereas the Lacedaemonian "Hippeis" do not keep horses); but though the Ephors have the same functions as the Cretan Cosmi, they have been named differently; and the public messes are, even to-day, still called "Andreia" among the Cretans, but among the Spartans they ceased to be called by the same name as in earlier times;[3] at any rate, the following is found in Alcman: "In feasts and festive gatherings, amongst the guests who partake of the Andreia, 'tis meet to begin the paean."[4]

19. It is said by the Cretans, Ephorus continues, that Lycurgus came to them for the following reason: Polydectes was the elder brother of Lycurgus; when he died he left his wife pregnant; now for a time Lycurgus reigned in his brother's place, but when a child was born he became the child's

[1] "Old Men," *i.e.* "Senators."

[2] "Horsemen," *i.e.* "Knights."

[3] The later Spartan name was "Syssitia" or "Philitia" (sometimes "Phiditia").

[4] *Frag.* 22 (Bergk).

πευεν ἐκεῖνον, εἰς ὃν ἡ ἀρχὴ καθήκουσα ἐτύγχανε· λοιδορούμενος δή τις αὐτῷ σαφῶς εἶπεν εἰδέναι, διότι βασιλεύσοι· λαβὼν δ' ὑπόνοιαν ἐκεῖνος, ὡς ἐκ τοῦ λόγου τούτου διαβάλλοιτο ἐπιβουλὴ ἐξ αὐτοῦ τοῦ παιδός, δείσας, μὴ ἐκ τύχης ἀποθανόντος αἰτίαν αὐτὸς ἔχοι παρὰ τῶν ἐχθρῶν, ἀπῆρεν εἰς Κρήτην· ταύτην μὲν δὴ λέγεσθαι τῆς ἀποδημίας αἰτίαν, ἐλθόντα δὲ πλησιάσαι Θάλητι μελοποιῷ ἀνδρὶ καὶ νομοθετικῷ, ἱστορήσαντα δὲ παρ' αὐτοῦ τὸν τρόπον, ὃν Ῥαδάμανθύς τε πρότερον καὶ ὕστερον Μίνως, ὡς παρὰ τοῦ Διὸς τοὺς νόμους ἐκφέροι εἰς ἀνθρώπους, γενόμενον δὲ καὶ ἐν Αἰγύπτῳ καὶ καταμαθόντα καὶ τὰ ἐκεῖ νόμιμα, ἐντυχόντα δ', ὥς φασί τινες, καὶ Ὁμήρῳ διατρίβοντι ἐν Χίῳ, κατᾶραι πάλιν εἰς τὴν οἰκείαν, καταλαβεῖν δὲ τὸν τοῦ ἀδελφοῦ υἱόν, τὸν Πολυδέκτου Χαρίλαον, βασιλεύοντα· εἶθ' ὁρμῆσαι διαθεῖναι τοὺς νόμους, φοιτῶντα ὡς τὸν θεὸν τὸν ἐν Δελφοῖς, κἀκεῖθεν κομίζοντα τὰ προστάγματα, καθάπερ οἱ περὶ Μίνω ἐκ τοῦ ἄντρου τοῦ Διός, παραπλήσια ἐκείνοις τὰ πλείω.

20. Τῶν Κρητικῶν τὰ κυριώτατα τῶν καθ' ἕκαστα τοιαῦτα εἴρηκε. γαμεῖν μὲν ἅμα πάντες ἀναγκάζονται παρ' αὐτοῖς οἱ κατὰ τὸν αὐτὸν χρόνον ἐκ τῆς τῶν παίδων ἀγέλης ἐκκριθέντες, οὐκ εὐθὺς δ' ἄγονται παρ' ἑαυτοὺς τὰς γαμηθείσας παῖδας, ἀλλ' ἐπὰν ἤδη διοικεῖν ἱκαναὶ ὦσι τὰ περὶ τοὺς οἴκους· φερνὴ δ' ἐστίν, ἂν ἀδελφοὶ ὦσι, τὸ ἥμισυ τῆς τοῦ ἀδελφοῦ μερίδος· παῖδας δὲ

guardian, since the office of king descended to the child, but some man, railing at Lycurgus, said that he knew for sure that Lycurgus would be king; and Lycurgus, suspecting that in consequence of such talk he himself might be falsely accused of plotting against the child, and fearing that, if by any chance the child should die, he himself might be blamed for it by his enemies, sailed away to Crete; this, then, is said to be the cause of his sojourn in Crete; and when he arrived he associated with Thales, a melic poet and an expert in lawgiving; and after learning from him the manner in which both Rhadamanthys in earlier times and Minos in later times published their laws to men as from Zeus, and after sojourning in Egypt also and learning among other things their institutions, and, according to some writers, after meeting Homer, who was living in Chios, he sailed back to his homeland, and found his brother's son, Charilaüs the son of Polydectes, reigning as king; and then he set out to frame the laws, making visits to the god at Delphi, and bringing thence the god's decrees, just as Minos and his house had brought their ordinances from the cave of Zeus, most of his being similar to theirs.

20. The following are the most important provisions in the Cretan institutions as stated by Ephorus. In Crete all those who are selected out of the "Troop" of boys at the same time are forced to marry at the same time, although they do not take the girls whom they have married to their own homes immediately, but as soon as the girls are qualified to manage the affairs of the house. A girl's dower, if she has brothers, is half of the brother's portion. The children must learn, not only

C 483 γράμματά τε μανθάνειν καὶ τὰς ἐκ τῶν νόμων ᾠδὰς καί τινα εἴδη τῆς μουσικῆς· τοὺς μὲν οὖν ἔτι νεωτέρους εἰς τὰ συσσίτια ἄγουσι τὰ ἀνδρεῖα· χαμαὶ δὲ καθήμενοι διαιτῶνται μετ' ἀλλήλων ἐν φαύλοις τριβωνίοις καὶ χειμῶνος καὶ θέρους τὰ αὐτά, διακονοῦσί τε καὶ ἑαυτοῖς καὶ τοῖς ἀνδράσι· συμβάλλουσι δ'[1] εἰς μάχην καὶ οἱ ἐκ τοῦ αὐτοῦ συσσιτίου πρὸς ἀλλήλους, καὶ πρὸς ἕτερα συσσίτια· καθ' ἕκαστον δὲ ἀνδρεῖον ἐφέστηκε παιδονόμος· οἱ δὲ μείζους εἰς τὰς ἀγέλας ἄγονται· τὰς δ' ἀγέλας συνάγουσιν οἱ ἐπιφανέστατοι τῶν παίδων καὶ δυνατώτατοι, ἕκαστος ὅσους πλείστους οἷός τέ ἐστιν ἀθροίζων· ἑκάστης δὲ τῆς ἀγέλης ἄρχων ἐστὶν ὡς τὸ πολὺ ὁ πατὴρ τοῦ συναγαγόντος, κύριος ὢν ἐξάγειν ἐπὶ θήραν καὶ δρόμους, τὸν δ' ἀπειθοῦντα κολάζειν· τρέφονται δὲ δημοσίᾳ· τακταῖς δέ τισιν ἡμέραις ἀγέλη πρὸς ἀγέλην συμβάλλει μετὰ αὐλοῦ καὶ λύρας εἰς μάχην ἐν ῥυθμῷ, ὥσπερ καὶ ἐν τοῖς πολεμικοῖς εἰώθασιν, ἐκφέρουσι δὲ καὶ τὰς πληγάς, τὰς μὲν διὰ χειρός, τὰς δὲ καὶ δι' ὅπλων σιδηρῶν.

21. Ἴδιον δ' αὐτοῖς τὸ περὶ τοὺς ἔρωτας νόμιμον· οὐ γὰρ πειθοῖ κατεργάζονται τοὺς ἐρωμένους, ἀλλ' ἁρπαγῇ· προλέγει τοῖς φίλοις πρὸ τριῶν ἢ πλειόνων ἡμερῶν ὁ ἐραστής, ὅτι μέλλει[2] τὴν ἁρπαγὴν ποιεῖσθαι· τοῖς δ' ἀποκρύπτειν μὲν τὸν παῖδα ἢ μὴ ἐᾶν πορεύεσθαι τὴν τεταγμένην ὁδὸν τῶν αἰσχίστων ἐστίν, ὡς

[1] δ', Casaubon inserts; so the later editors.
[2] μέλλοι BC*lno*.

---

[1] Others translate ἐκφέρουσι in the sense of *delivering* blows.

their letters, but also the songs prescribed in the laws and certain forms of music. Now those who are still younger are taken to the public messes, the "Andreia"; and they sit together on the ground as they eat their food, clad in shabby garments, the same both winter and summer, and they also wait on the men as well as on themselves. And those who eat together at the same mess join battle both with one another and with those from different messes. A boy-director presides over each mess. But the older boys are taken to the "Troops"; and the most conspicuous and influential of the boys assemble the "Troops," each collecting as many boys as he possibly can; the leader of each "Troop" is generally the father of the assembler, and he has authority to lead them forth to hunt and to run races, and to punish anyone who is disobedient; and they are fed at public expense; and on certain appointed days "Troop" contends with "Troop," marching rhythmically into battle, to the tune of flute and lyre, as is their custom in actual war; and they actually bear marks of[1] the blows received, some inflicted by the hand, others by iron[2] weapons.

21. They have a peculiar custom in regard to love affairs,[3] for they win the objects of their love, not by persuasion, but by abduction; the lover tells the friends of the boy three or four days beforehand that he is going to make the abduction; but for the friends to conceal the boy, or not to let him go forth by the appointed road, is indeed a most disgraceful thing,

[2] Possibly an error for "wooden."

[3] The discussion of "love affairs" is strangely limited to pederasty.

ἐξομολογουμένοις,[1] ὅτι ἀνάξιος ὁ παῖς εἴη τοιούτου ἐραστοῦ τυγχάνειν. συνιόντες δ', ἂν μὲν τῶν ἴσων ἢ τῶν ὑπερεχόντων τις ᾖ τοῦ παιδὸς τιμῇ καὶ τοῖς ἄλλοις ὁ ἁρπάζων, ἐπιδιώκοντες ἀνθήψαντο μόνον μετρίως, τὸ νόμιμον ἐκπληροῦντες, τἆλλα δ' ἐπιτρέπουσιν ἄγειν χαίροντες· ἂν δ' ἀνάξιος, ἀφαιροῦνται· πέρας δὲ τῆς ἐπιδιώξεώς[2] ἐστιν, ἕως ἂν ἀχθῇ ὁ παῖς εἰς τὸ τοῦ ἁρπάσαντος ἀνδρεῖον. ἐράσμιον δὲ νομίζουσιν οὐ τὸν κάλλει διαφέροντα, ἀλλὰ τὸν ἀνδρείᾳ καὶ κοσμιότητι·[3] καὶ δωρησάμενος ἀπάγει τὸν παῖδα τῆς χώρας εἰς ὃν βούλεται τόπον· ἐπακολουθοῦσι δὲ τῇ ἁρπαγῇ οἱ παραγενόμενοι, ἑστιαθέντες δὲ καὶ συνθηρεύσαντες δίμηνον (οὐ γὰρ ἔξεστι πλείω χρόνον κατέχειν τὸν παῖδα) εἰς τὴν πόλιν καταβαίνουσιν. ἀφίεται δ' ὁ παῖς, δῶρα λαβὼν στολὴν πολεμικὴν καὶ βοῦν καὶ ποτήριον (ταῦτα μὲν τὰ κατὰ τὸν νόμον δῶρα)[4] καὶ ἄλλα πλείω καὶ πολυτελῆ, ὥστε συνερανίζειν τοὺς φίλους διὰ τὸ πλῆθος τῶν ἀναλωμάτων. τὸν μὲν οὖν βοῦν θύει τῷ Διὶ καὶ ἑστιᾷ τοὺς συγκαταβαίνοντας· εἶτ' ἀποφαίνεται περὶ τῆς πρὸς τὸν ἐραστὴν ὁμιλίας, εἴτ' ἀσμενίζων τετύχηκεν, εἴτε μή, τοῦ νόμου τοῦτ' ἐπιτρέψαντος, C 484 ἵν', εἴ τις αὐτῷ βία προσενήνεκται κατὰ τὴν ἁρπαγήν, ἐνταῦθα παρῇ τιμωρεῖν[5] ἑαυτῷ καὶ

[1] ἐξομολογουμένοις, the editors, for ἐξομολογουμένους.

[2] ἐπιδιώξεως *no*, ἐπιδείξεως other MSS.

[3] Before καὶ δωρησάμενος Meineke, following Groskurd's conj., indicates a lacuna, suspecting that something like ὁ δ' ἐραστὴς ἀσπασάμενος has fallen out of the MSS.

[4] After δῶρα Meineke indicates a lacuna.

a confession, as it were, that the boy is unworthy to obtain such a lover; and when they meet, if the abductor is the boy's equal or superior in rank or other respects, the friends pursue him and lay hold of him, though only in a very gentle way, thus satisfying the custom; and after that they cheerfully turn the boy over to him to lead away; if, however, the abductor is unworthy, they take the boy away from him. And the pursuit does not end until the boy is taken to the "Andreium" of his abductor. They regard as a worthy object of love, not the boy who is exceptionally handsome, but the boy who is exceptionally manly and decorous. After giving the boy presents, the abductor takes him away to any place in the country he wishes; and those who were present at the abduction follow after them, and after feasting and hunting with them for two months (for it is not permitted to detain the boy for a longer time), they return to the city. The boy is released after receiving as presents a military habit, an ox, and a drinking-cup (these are the gifts required by law), and other things so numerous and costly that the friends, on account of the number of the expenses, make contributions thereto. Now the boy sacrifices the ox to Zeus and feasts those who returned with him; and then he makes known the facts about his intimacy with his lover, whether, perchance, it has pleased him or not, the law allowing him this privilege in order that, if any force was applied to him at the time of the abduction, he might be able at this feast to avenge himself and be rid of the lover. It is disgraceful

---

[5] παρῇ τιμωρεῖν, Corais, for παρατιμωρεῖν; so the later editors.

ἀπαλλάττεσθαι. τοῖς δὲ καλοῖς τὴν ἰδέαν καὶ προγόνων ἐπιφανῶν ἐραστῶν μὴ τυχεῖν αἰσχρόν,[1] ὡς διὰ τὸν τρόπον τοῦτο παθοῦσιν. ἔχουσι δὲ τιμὰς οἱ παρασταθέντες (οὕτω γὰρ καλοῦσι τοὺς ἁρπαγέντας)· ἔν τε γὰρ τοῖς χοροῖς[2] καὶ τοῖς δρόμοις ἔχουσι τὰς ἐντιμοτάτας χώρας, τῇ τε στολῇ κοσμεῖσθαι διαφερόντως τῶν ἄλλων ἐφίεται τῇ δοθείσῃ παρὰ τῶν ἐραστῶν, καὶ οὐ τότε μόνον, ἀλλὰ καὶ τέλειοι γενόμενοι διάσημον ἐσθῆτα φέρουσιν, ἀφ᾽ ἧς γνωσθήσεται ἕκαστος κλεινὸς γενόμενος· τὸν μὲν γὰρ ἐρώμενον καλοῦσι κλεινόν, τὸν δ᾽ ἐραστὴν φιλήτορα. ταῦτα μὲν τὰ περὶ τοὺς ἔρωτας νόμιμα.

22. Ἄρχοντας δὲ δέκα αἱροῦνται· περὶ δὲ τῶν μεγίστων συμβούλοις χρῶνται τοῖς γέρουσι καλουμένοις· καθίστανται δ᾽ εἰς τοῦτο τὸ συνέδριον οἱ τῆς τῶν κόσμων ἀρχῆς ἠξιωμένοι καὶ τἆλλα δόκιμοι κρινόμενοι. ἀξίαν δ᾽ ἀναγραφῆς τὴν τῶν Κρητῶν πολιτείαν ὑπέλαβον διά τε τὴν ἰδιότητα καὶ διὰ[3] τὴν δόξαν· οὐ πολλὰ δὲ διαμένει τούτων τῶν νομίμων, ἀλλὰ τοῖς Ῥωμαίων διατάγμασι τὰ πλεῖστα διοικεῖται, καθάπερ καὶ ἐν ταῖς ἄλλαις ἐπαρχίαις συμβαίνει.

[1] αἰσχρόν, Casaubon inserts; so the later editors.
[2] χρόνοις BCD*hil*, θρόνοις *hnox* and by corr. in B.
[3] διά is omitted by D*hik*, and the later editors.

for those who are handsome in appearance or descendants of illustrious ancestors to fail to obtain lovers, the presumption being that their character is responsible for such a fate. But the parastathentes[1] (for thus they call those who have been abducted) receive honours; for in both the dances and the races they have the positions of highest honour, and are allowed to dress in better clothes than the rest, that is, in the habit given them by their lovers; and not then only, but even after they have grown to manhood, they wear a distinctive dress, which is intended to make known the fact that each wearer has become "kleinos,"[2] for they call the loved one "kleinos" and the lover "philetor."[3] So much for their customs in regard to love affairs.

22. The Cretans choose ten Archons. Concerning the matters of greatest importance they use as counsellors the "Gerontes," as they are called. Those who have been thought worthy to hold the office of the "Cosmi" and are otherwise adjudged men of approved worth are appointed members of this Council. I have assumed that the constitution of the Cretans is worthy of description both on account of its peculiar character and on account of its fame. Not many, however, of these institutions endure, but the administration of affairs is carried on mostly by means of the decrees of the Romans, as is also the case in the other provinces.

[1] The *literal* meaning of the word seems to be "those who were chosen as *stand-bys*" by lovers.

[2] Famous.

[3] *i.e.* "lover" or "sweetheart."

V

1. Περὶ δὲ τὴν Κρήτην εἰσὶ νῆσοι, Θήρα μέν, ἡ τῶν Κυρηναίων μητρόπολις, ἄποικος Λακεδαιμονίων, καὶ πλησίον ταύτης Ἀνάφη, ἐν ᾗ τὸ τοῦ Αἰγλήτου Ἀπόλλωνος ἱερόν. λέγει δὲ καὶ Καλλίμαχος τοτὲ μὲν οὕτως·

Αἰγλήτην Ἀνάφην τε, Λακωνίδι γείτονα Θήρᾳ·

τοτὲ δὲ τῆς Θήρας μνησθείς·

μήτηρ εὐίππου πατρίδος ἡμετέρης,

ἔστι δὲ μακρὰ ἡ Θήρα, διακοσίων οὖσα τὴν περίμετρον σταδίων, κειμένη δὲ κατὰ Δίαν νῆσον τὴν πρὸς Ἡρακλείῳ τῷ Κνωσσίῳ, διέχει δὲ τῆς Κρήτης εἰς ἑπτακοσίους· πλησίον δ' αὐτῆς ἥ τε Ἀνάφη καὶ Θηρασία. ταύτης δ' εἰς ἑκατὸν[1] ἀπέχει νησίδιον Ἴος, ἐν ᾧ κεκηδεῦσθαί τινές φασι τὸν ποιητὴν Ὅμηρον· ἀπὸ δὲ τῆς Ἴου πρὸς ἑσπέραν ἰόντι Σίκινος[2] καὶ Λάγουσα καὶ Φολέγανδρος, ἣν Ἄρατος σιδηρείην ὀνομάζει διὰ τὴν τραχύτητα· ἐγγὺς δὲ τούτων Κίμωλος, ὅθεν ἡ γῆ ἡ Κιμωλία· ἔνθεν ἡ Σίφνος ἐν ὄψει ἐστίν, ἐφ' ᾗ λέγουσι Σίφνιον ἀστράγαλον διὰ τὴν εὐτέλειαν. ἔτι δ' ἐγγυτέρω καὶ τῆς Κιμώλου καὶ τῆς Κρήτης ἡ Μῆλος, ἀξιολογωτέρα τούτων, διέχουσα τοῦ Ἑρμιονικοῦ ἀκρωτηρίου, τοῦ Σκυλλαίου, σταδίους ἑπτακοσίους· τοσούτους δὲ

[1] ταύτης δ' εἰς ἑκατόν, Tzschucke, from conj. of Casaubon, for τούτων δ' ἴσον ἑκάστῃ *Bkno*, ἑκαστόν *CDghilsxy*; so the later editors.

[2] Σίκινος, Tzschucke, for Σίκηνος; so the later editors.

V

1. The islands near Crete are Thera, the metropolis of the Cyrenaeans, a colony of the Lacedaemonians, and, near Thera, Anaphê, where is the temple of the Aegletan Apollo. Callimachus speaks in one place as follows, "Aegletan Anaphê, neighbour to Laconian Thera," [1] and in another, mentioning only Thera, "mother of my fatherland, famed for its horses." [2] Thera is a long island, being two hundred stadia in perimeter; it lies opposite Dia,[3] an island near the Cnossian Heracleium,[4] but it is seven hundred stadia distant from Crete. Near it are both Anaphê and Therasia. One hundred stadia distant from the latter is the little island Ios, where, according to some writers, the poet Homer was buried. From Ios towards the west one comes to Sicinos and Lagusa and Pholegandros, which last Aratus calls "Iron" Island, because of its ruggedness. Near these is Cimolos, whence comes the Cimolian earth.[5] From Cimolos Siphnos is visible, in reference to which island, because of its worthlessness, people say "Siphnian knuckle-bone." [6] And still nearer both to Cimolos and to Crete is Melos, which is more notable than these and is seven hundred stadia from the Hermionic promontory, the Scyllaeum, and almost the same distance

[1] *Frag.* 113 (Schneider).
[2] *Frag.* 112 (Schneider).
[3] *i.e.* almost due north of Dia.
[4] Heracleium was the seaport of Cnossus (10. 4. 7).
[5] A hydrous silicate of aluminium, now called "cimolite."
[6] *i.e.* the phrase is a proverb applied to worthless people or things.

σχεδόν τι καὶ τοῦ Δικτυνναίου. Ἀθηναῖοι δέ ποτε πέμψαντες στρατείαν, ἡβηδὸν κατέσφαξαν τοὺς πλείους. αὗται μὲν οὖν ἐν τῷ Κρητικῷ πελάγει, ἐν δὲ τῷ Αἰγαίῳ μᾶλλον αὐτή τε ἡ Δῆλος καὶ αἱ περὶ αὐτὴν Κυκλάδες καὶ αἱ ταύταις προσκείμεναι[1] Σποράδες, ὧν εἰσὶ καὶ αἱ λεχθεῖσαι περὶ τὴν Κρήτην.

C 485

2. Ἡ μὲν οὖν Δῆλος ἐν πεδίῳ κειμένην ἔχει τὴν πόλιν καὶ τὸ ἱερὸν τοῦ Ἀπόλλωνος καὶ τὸ Λητῷον, ὑπέρκειται δὲ τῆς πόλεως ὄρος ψιλὸν[2] ὁ Κύνθος καὶ τραχύ, ποταμὸς δὲ διαρρεῖ τὴν νῆσον Ἰνωπὸς οὐ μέγας· καὶ γὰρ ἡ νῆσος μικρά. τετίμηται δὲ ἐκ παλαιοῦ διὰ τοὺς θεοὺς ἀπὸ τῶν ἡρωικῶν χρόνων ἀρξαμένη· μυθεύεται γὰρ ἐνταῦθα ἡ Λητὼ τὰς ὠδῖνας ἀποθέσθαι τοῦ τε Ἀπόλλωνος καὶ τῆς Ἀρτέμιδος·

ἦν γὰρ τοπάροιθε[3] φορητά,

φησὶν ὁ Πίνδαρος,

κυμάτεσσι παντοδαπῶν[4] ἀνέμων
ῥιπαῖσιν· ἀλλ᾽ ἁ Κοιογενὴς[5] ὁπότ᾽ ὠδίνεσσι[6]
θύοισ᾽[7]
ἀγχιτόκοις ἐπέβα[8] νιν, δὴ τότε τέσσαρες ὀρθαί
πρέμνων[9] ἀπώρουσαν χθονίων,

[1] προκείμεναι *lno*. [2] ψιλόν CD, ὑψηλόν other MSS.

[3] τοπάροιθε, Casaubon and later editors, instead of πάροιθεν οὐ (all MSS.). Eustathius omits the οὐ (note on *Od.* 10. 3).

[4] Before ἀνέμων Tzschucke and later editors insert τ᾽.

[5] ἀλλ᾽ ἁ Κοιογενής, Kramer and Meineke, from conj. of Porson, for ἀλλὰ Καιογενης D, ἀλλὰ καὶ ὁ γένης C*s*, ἀλλ᾽ ἀκαιογένης B*k*, ἀλλὰ καινογενῆς *hi*, ἀλλὰ καὶ ὁ γένος *l*, ἀλλὰ Κοίου γένος Schneider, Hermann, Tzschucke, Corais.

from the Dictynnaeum. The Athenians once sent an expedition to Melos and slaughtered most of the inhabitants from youth upwards.[1] Now these islands are indeed in the Cretan Sea, but Delos itself and the Cyclades in its neighbourhood and the Sporades which lie close to these, to which belong the aforesaid islands in the neighbourhood of Crete, are rather in the Aegaean Sea.

2. Now the city which belongs to Delos, as also the temple of Apollo, and the Letöum,[2] are situated in a plain; and above the city lies Cynthus, a bare and rugged mountain; and a river named Inopus flows through the island—not a large river, for the island itself is small. From olden times, beginning with the times of the heroes, Delos has been revered because of its gods, for the myth is told that there Leto was delivered of her travail by the birth of Apollo and Artemis: "for aforetime," says Pindar,[3] "it[4] was tossed by the billows, by the blasts of all manner of winds,[5] but when the daughter of Coeüs[6] in the frenzied pangs of childbirth set foot upon it, then did four pillars, resting on adamant, rise perpendicular from the roots of the earth, and

[1] 416 B.C. (see Thucydides 5. 115–116).
[2] Temple of Leto. [3] *Frag.* 58 (Bergk). [4] Delos.
[5] There was a tradition that Delos was a floating isle until Leto set foot on it.
[6] Leto.

---

[6] *ὠδίνεσι* BD*hios*, *ὠξύναισι* *k*, *ὀδύ αισι* editors before before Kramer.
[7] *θύοισ'*, Bergk, for *θυοις* CD*hl*, *θείαις* B*knos* and editors before Kramer.
[8] *ἐπέβα νιν*, Wilamowitz, for *ἐπιβαίνειν*.
[9] *πρέμνων*, Hermann, for *πρύμνων* CD*hilos*, *πρεμνῶν* B*k*.

ἂν δ' ἐπικράνοις σχέθον πέτραν ἀδαμαντο-
πέδιλοι
κίονες· ἔνθα τεκοῖσ' εὐδαίμον' ἐπόψατο γένναν.

ἔνδοξον δ' ἐποίησαν αὐτὴν αἱ περιοικίδες νῆσοι, καλούμεναι Κυκλάδες, κατὰ τιμὴν πέμπουσαι δημοσίᾳ θεωρούς τε καὶ θυσίας καὶ χοροὺς παρθένων πανηγύρεις τε ἐν αὐτῇ συνάγουσαι μεγάλας.

3. Κατ' ἀρχὰς μὲν οὖν δώδεκα λέγονται· προσεγένοντο δὲ καὶ πλείους. Ἀρτεμίδωρος γοῦν[1] πεντεκαίδεκα[2] διαριθμεῖται περὶ τῆς Ἑλένης εἰπών, ὅτι ἀπὸ Θορίκου μέχρι Σουνίου παράκειται, μακρά, σταδίων ὅσον ἑξήκοντα τὸ μῆκος· ἀπὸ ταύτης γάρ, φησίν, αἱ καλούμεναι Κυκλάδες εἰσίν· ὀνομάζει δὲ Κέω, τὴν ἐγγυτάτω τῇ Ἑλένῃ, καὶ μετὰ ταύτην Κύθνον καὶ Σέριφον καὶ Μῆλον καὶ Σίφνον καὶ Κίμωλον καὶ Πρεπέσινθον καὶ Ὠλίαρον[3] καὶ πρὸς ταύταις Πάρον, Νάξον, Σῦρον, Μύκονον, Τῆνον, Ἄνδρον, Γύαρον. τὰς μὲν οὖν ἄλλας τῶν δώδεκα νομίζω, τὴν δὲ Πρεπέσινθον καὶ Ὠλίαρον[4] καὶ Γύαρον ἧττον· ὧν τῇ Γυάρῳ προσορμισθεὶς ἔγνων κώμιον ὑπὸ ἁλιέων συνοικούμενον· ἀπαίροντες δ' ἐδεξάμεθα πρεσβευτὴν ἐνθένδε ὡς Καίσαρα προκεχειρισμένον, τῶν ἁλιέων τινά (ἦν δ' ἐν Κορίνθῳ Καῖσαρ, βαδίζων ἐπὶ τὸν θρίαμβον τὸν Ἀκτιακόν)· συμπλέων δὴ ἔλεγε πρὸς τοὺς πυθομένους, ὅτι πρεσβεύοι περὶ κουφισμοῦ τοῦ φόρου· τελοῖεν C 486 γὰρ δραχμὰς ἑκατὸν πεντήκοντα, καὶ τὰς ἑκατὸν

[1] γοῦν, Meineke, for δ' οὖν.
[2] πεντεκαίδεκα (ιε'), Corais inserts; so Meineke.
[3] Ἀλίαρον Dhil. [4] Ἀλίαρον BCDhix.

on their capitals sustain the rock. And there she gave birth to, and beheld, her blessed offspring." The neighbouring islands, called the Cyclades, made it famous, since in its honour they would send at public expense sacred envoys, sacrifices, and choruses composed of virgins, and would celebrate great general festivals there.[1]

3. Now at first the Cyclades are said to have been only twelve in number, but later several others were added. At any rate, Artemidorus enumerates fifteen, after saying of Helena that it stretches parallel to the coast from Thoricus to Sunium and is a long island, about sixty stadia in length; for it is from Helena, he says, that the Cyclades, as they are called, begin; and he names Ceos, the island nearest to Helena, and, after this island, Cythnos and Seriphos and Melos and Siphnos and Cimolos and Prepesinthos and Oliaros, and, in addition to these, Paros, Naxos, Syros, Myconos, Tenos, Andros, and Gyaros. Now I consider all of these among the twelve except Prepesinthos, Oliaros, and Gyaros. When our ship anchored at one of these, Gyaros, I saw a small village that was settled by fishermen; and when we sailed away we took on board one of the fishermen, who had been chosen to go from there to Caesar as ambassador (Caesar was at Corinth, on his way[2] to celebrate the Triumph after the victory at Actium[3]). While on the voyage he told enquirers that he had been sent as ambassador to request a reduction in their tribute; for, he said, they were paying one hundred and fifty drachmas when they could only with difficulty pay

[1] *i.e.* in honour of Apollo and Leto (see Thucydides 3. 104).
[2] *i.e.* back to Rome. [3] 31 B.C.

χαλεπῶς ἂν τελοῦντες. δηλοῖ δὲ τὰς ἀπορίας αὐτῶν καὶ Ἄρατος ἐν τοῖς κατὰ λεπτόν·

ὦ Λητοῖ, σὺ μὲν ἤ με σιδηρείῃ Φολεγάνδρῳ,
δειλῇ[1] ἢ Γυάρῳ παρελεύσεαι αὐτίχ᾽ ὁμοίην.

4. Τὴν μὲν οὖν Δῆλον ἔνδοξον γενομένην οὕτως ἔτι μᾶλλον ηὔξησε κατασκαφεῖσα ὑπὸ Ῥωμαίων Κόρινθος. ἐκεῖσε γὰρ μετεχώρησαν οἱ ἔμποροι, καὶ τῆς ἀτελείας τοῦ ἱεροῦ προκαλουμένης αὐτοὺς καὶ τῆς εὐκαιρίας τοῦ λιμένος· ἐν καλῷ γὰρ κεῖται τοῖς ἐκ τῆς Ἰταλίας καὶ τῆς Ἑλλάδος εἰς τὴν Ἀσίαν πλέουσιν· ἥ τε πανήγυρις ἐμπορικόν τι πρᾶγμά ἐστι, καὶ συνήθεις ἦσαν αὐτῇ καὶ Ῥωμαῖοι τῶν ἄλλων μάλιστα, καὶ ὅτε συνειστήκει ἡ Κόρινθος· Ἀθηναῖοί τε λαβόντες τὴν νῆσον καὶ τῶν ἱερῶν ἅμα καὶ τῶν ἐμπόρων ἐπεμελοῦντο ἱκανῶς· ἐπελθόντες δ᾽ οἱ τοῦ Μιθριδάτου στρατηγοὶ καὶ ὁ ἀποστήσας τύραννος αὐτὴν διελυμήναντο πάντα, καὶ παρέλαβον ἐρήμην οἱ Ῥωμαῖοι πάλιν τὴν νῆσον, ἀναχωρήσαντος εἰς τὴν οἰκείαν τοῦ βασιλέως, καὶ διετέλεσε μέχρι νῦν ἐνδεῶς πράττουσα. ἔχουσι δ᾽ αὐτὴν Ἀθηναῖοι.

5. Ῥήνεια[2] δ᾽ ἔρημον νησίδιόν ἐστιν ἐν τέτρασι τῆς Δήλου σταδίοις, ὅπου τὰ μνήματα τοῖς Δηλίοις ἐστίν. οὐ γὰρ ἔξεστιν ἐν αὐτῇ τῇ Δήλῳ θάπτειν οὐδὲ καίειν νεκρόν, οὐκ ἔξεστι δὲ οὐδὲ κύνα ἐν Δήλῳ τρέφειν. ὠνομάζετο δὲ καὶ Ὀρτυγία πρότερον.

[1] δειλῇ, Müller-Dübner, for δειλήν *s* (and Meineke), δειλή other MSS.
[2] Ῥήνεια B*kno*, Ῥήναια other MSS.

---

[1] *i.e. Trifles.* [2] 146 B.C.

one hundred. Aratus also points out the poverty of the island in his *Catalepton*:[1] "O Leto, shortly thou wilt pass by me, who am like either iron Pholegandros or worthless Gyaros."

4. Now although Delos had become so famous, yet the rasing of Corinth to the ground by the Romans[2] increased its fame still more; for the importers changed their business to Delos because they were attracted both by the immunity which the temple enjoyed and by the convenient situation of the harbour; for it is happily situated for those who are sailing from Italy and Greece to Asia. The general festival is a kind of commercial affair, and it was frequented by Romans more than by any other people, even when Corinth was still in existence.[3] And when the Athenians took the island they at the same time took good care of the importers as well as of the religious rites. But when the generals of Mithridates, and the tyrant[4] who caused it to revolt, visited Delos, they completely ruined it, and when the Romans again got the island, after the king withdrew to his homeland, it was desolate; and it has remained in an impoverished condition until the present time. It is now held by the Athenians.

5. Rheneia is a desert isle within four stadia from Delos, and there the Delians bury their dead;[5] for it is unlawful to bury, or even burn, a corpse in Delos itself, and it is unlawful even to keep a dog there. In earlier times it was called Ortygia.

[3] As many as ten thousand slaves were sold there in one day (14. 5. 2).

[4] Aristion, through the aid of Mithridates, made himself tyrant of Athens in 88 B.C. (cf. 9. 1. 20).

[5] This began in 426 B.C., when "all the sepulchres of the dead in Delos were removed" to Rheneia (Thucydides 3. 104).

6. Κέως δὲ τετράπολις μὲν ὑπῆρξε, λείπονται δὲ δύο, ἥ τε Ἰουλὶς καὶ ἡ Καρθαία, εἰς ἃς συνεπολίσθησαν αἱ λοιπαί, ἡ μὲν Ποιήεσσα εἰς τὴν Καρθαίαν, ἡ δὲ Κορησσία εἰς τὴν Ἰουλίδα. ἐκ δὲ τῆς Ἰουλίδος ὅ τε Σιμωνίδης ἦν ὁ μελοποιὸς καὶ Βακχυλίδης, ἀδελφιδοῦς ἐκείνου, καὶ μετὰ ταῦτα Ἐρασίστρατος ὁ ἰατρὸς καὶ τῶν ἐκ τοῦ περιπάτου φιλοσόφων Ἀρίστων, ὁ τοῦ Βορυσθενίτου Βίωνος ζηλωτής. παρὰ τούτοις δὲ δοκεῖ τεθῆναί ποτε νόμος, οὗ μέμνηται καὶ Μένανδρος·

καλὸν τὸ Κείων νόμιμόν ἐστι, Φανία·
ὁ μὴ δυνάμενος ζῆν καλῶς οὐ ζῇ κακῶς.

προσέταττε γάρ, ὡς ἔοικεν, ὁ νόμος τοὺς ὑπὲρ ἑξήκοντα ἔτη γεγονότας κωνειάζεσθαι,[1] [2] τοῦ διαρκεῖν τοῖς ἄλλοις τὴν τροφήν· καὶ πολιορκουμένους δέ ποτε ὑπ' Ἀθηναίων ψηφίσασθαί φασι τοὺς πρεσβυτάτους ἐξ αὐτῶν ἀποθανεῖν, ὁρισθέντος πλήθους ἐτῶν, τοὺς δὲ παύσασθαι πολιορκοῦντας. κεῖται δ' ἐν ὄρει τῆς θαλάττης διέχουσα ἡ πόλις ὅσον πέντε καὶ εἴκοσι σταδίους, ἐπίνειον δ' ἐστὶν C 487 αὐτῆς τὸ χωρίον, ἐν ᾧ ἵδρυτο ἡ Κορησσία, κατοικίαν οὐδὲ κώμης ἔχουσα. ἔστι δὲ καὶ πρὸς τῇ Κορησσίᾳ Σμινθέου Ἀπόλλωνος ἱερὸν καὶ πρὸς Ποιηέσσῃ, μεταξὺ δὲ τοῦ ἱεροῦ καὶ τῶν τῆς Ποιηέσσης ἐρειπίων τὸ τῆς Νεδουσίας Ἀθηνᾶς ἱερόν, ἱδρυσαμένου Νέστορος κατὰ τὴν ἐκ Τροίας ἐπάνοδον. ἔστι δὲ καὶ Ἕλιξος ποταμὸς περὶ τὴν Κορησσίαν.

7. Μετὰ δὲ ταύτην Νάξος καὶ Ἄνδρος ἀξιόλογοι καὶ Πάρος· ἐντεῦθεν ἦν Ἀρχίλοχος ὁ ποιητής. ὑπὸ δὲ Παρίων ἐκτίσθη Θάσος καὶ Πάριον

6. Ceos was at first a Tetrapolis, but only two cities are left, Iulis and Carthaea, into which the remaining two were incorporated, Poeëessa into Carthaea and Coressia into Iulis. Both Simonides the melic poet and his nephew Bacchylides were natives of Iulis, and also after their time Erasistratus the physician, and Ariston the peripatetic philosopher and emulator of Bion the Borysthenite. It is reputed that there was once a law among these people (it is mentioned by Menander, "Phanias, the law of the Ceians is good, that he who is unable to live well should not live wretchedly"), which appears to have ordered those who were over sixty years of age to drink hemlock, in order that the food might be sufficient for the rest. And it is said that once, when they were being besieged by the Athenians, they voted, setting a definite age, that the oldest among them should be put to death, but the Athenians raised the siege. The city lies on a mountain, about twenty-five stadia distant from the sea; and its seaport is the place on which Coressia was situated, which has not as great a population as even a village. Near Coressia, and also near Poeëessa, is a temple of Sminthian Apollo; and between the temple and the ruins of Poeëessa is the temple of Nedusian Athena, founded by Nestor when he was on his return from Troy. There is also a River Elixus in the neighbourhood of Coressia.

7. After Ceos one comes to Naxos and Andros, notable islands, and to Paros. Archilochus the poet was a native of Paros. Thasos was founded by the Parians, as also Parium, a city on the Propontis.

---

[1] κωνεάζεσθαι CD*ghlxy*, κονεάζεσθαι B*k*.
[2] καί, before τοῦ, omitted by *nox*.

ἐν τῇ Προποντίδι πόλις. ἐν ταύτῃ μὲν οὖν ὁ βωμὸς λέγεται θέας ἄξιος, σταδιαίας ἔχων τὰς πλευράς· ἐν δὲ τῇ Πάρῳ ἡ Παρία λίθος λεγομένη, ἀρίστη πρὸς τὴν μαρμαρογλυφίαν.

8. Σῦρος δ' ἐστί (μηκύνουσι τὴν πρώτην συλλαβήν), ἐξ ἧς Φερεκύδης ὁ Βάβυος[1] ἦν· νεώτερος δ' ἐστὶν ὁ Ἀθηναῖος ἐκείνου. ταύτης δοκεῖ μνημονεύειν ὁ ποιητής, Συρίην καλῶν·

νῆσός τις Συρίη κικλήσκεται
Ὀρτυγίης καθύπερθε.

9. Μύκονος δ' ἐστίν, ὑφ' ᾗ μυθεύουσι κεῖσθαι τῶν γιγάντων τοὺς ὑστάτους[2] ὑφ' Ἡρακλέους καταλυθέντας, ἀφ' ὧν ἡ παροιμία Πάνθ' ὑπὸ μίαν Μύκονον ἐπὶ τῶν ὑπὸ μίαν ἐπιγραφὴν ἀγόντων καὶ[3] τὰ διηρτημένα τῇ φύσει. καὶ τοὺς φαλακροὺς δέ τινες Μυκονίους καλοῦσιν ἀπὸ τοῦ τὸ πάθος τοῦτο ἐπιχωριάζειν[4] τῇ νήσῳ.

10. Σέριφος δ' ἐστίν, ἐν ᾗ τὰ περὶ τὸν Δίκτυν μεμύθευται, τὸν ἀνελκύσαντα τὴν λάρνακα τοῖς δικτύοις τὴν περιέχουσαν τὸν Περσέα καὶ τὴν μητέρα Δανάην, καταπεποντωμένους ὑπ' Ἀκρισίου τοῦ πατρὸς τῆς Δανάης· τραφῆναί τε γὰρ ἐνταῦθα τὸν Περσέα φασί, καὶ κομίσαντα τὴν τῆς Γοργόνος[5] κεφαλήν, δείξαντα τοῖς Σεριφίοις ἀπολιθῶσαι πάντας· τοῦτο δὲ πρᾶξαι τιμωροῦντα τῇ μητρί, ὅτι αὐτὴν Πολυδέκτης ὁ βασιλεὺς ἄκουσαν ἄγεσθαι προείλετο πρὸς γάμον, συμπραττόντων

[1] Except D the MSS. have Βάβιος.
[2] ὑγιεινοτάτους Stephanus (*s.v.* Μύκονος) and Eustathius (note on *Dionysius* 525).
[3] καί omitted by B*knox*.
[4] Before τῇ BCD have ἐν.
[5] Γοργόνης BCD.

Now the altar in this city is said to be a spectacle worth seeing, its sides being a stadium in length; and so is the Parian stone, as it is called, in Paros, the best for sculpture in marble.

8. And there is Syros (the first syllable is pronounced long), where Pherecydes[1] the son of Babys was born. The Athenian Pherecydes is later than he.[2] The poet seems to mention this island, though he calls it Syria: "There is an island called Syria, above Ortygia."[3]

9. And there is Myconos, beneath which, according to the myth, lie the last of the giants that were destroyed by Heracles. Whence the proverb, "all beneath Myconos alone," applied to those who bring under one title even those things which are by nature separate. And further, some call bald men Myconians, from the fact that baldness is prevalent in the island.

10. And there is Seriphos, the scene of the mythical story of Dictys, who with his net drew to land the chest in which were enclosed Perseus and his mother Danaê, who had been sunk in the sea by Acrisius the father of Danaê; for Perseus was reared there, it is said, and when he brought the Gorgon's head there, he showed it to the Seriphians and turned them all into stone. This he did to avenge his mother, because Polydectes the king, with their co-operation, intended to marry his mother against

[1] Fl. about 560 B.C.

[2] Pherecydes of Leros (fl. in the first half of the fifth century B.C.), often called "the Athenian," wrote, among other things, a work in ten books on the mythology and antiquities of Attica.

[3] *Od.* 15. 403.

ἐκείνων. οὕτω δ' ἐστὶ πετρώδης ἡ νῆσος, ὥστε ὑπὸ τῆς Γοργόνος τοῦτο παθεῖν αὐτήν φασιν οἱ κωμῳδοῦντες.

11. Τῆνος δὲ πόλιν μὲν οὐ μεγάλην ἔχει, τὸ δ' ἱερὸν τοῦ Ποσειδῶνος μέγα ἐν ἄλσει τῆς πόλεως ἔξω, θέας ἄξιον· ἐν ᾧ καὶ ἑστιατόρια πεποίηται μεγάλα, σημεῖον τοῦ συνέρχεσθαι πλῆθος ἱκανὸν τῶν συνθυόντων αὐτοῖς ἀστυγειτόνων τὰ Ποσειδώνια.

12. Ἔστι δὲ καὶ Ἀμοργὸς τῶν Σποράδων, ὅθεν ἦν Σιμωνίδης ὁ τῶν ἰάμβων ποιητής, καὶ Λέβινθος καὶ Λέρος·[1]

καὶ τόδε Φωκυλίδου· Λέριοι κακοί, οὐχ ὁ μέν, ὃς δ' οὔ,
πάντες, πλὴν Προκλέους· καὶ Προκλέης Λέριος.

C 488 διεβέβλητο γὰρ ὡς κακοήθεις οἱ ἐνθένδε ἄνθρωποι.

13. Πλησίον δ' ἐστὶ καὶ ἡ Πάτμος καὶ Κορασσίαι, πρὸς δύσιν κείμεναι τῇ Ἰκαρίᾳ, αὕτη δὲ Σάμῳ. ἡ μὲν οὖν Ἰκαρία ἔρημός ἐστι, νομὰς δ' ἔχει, καὶ χρῶνται αὐταῖς Σάμιοι· τοιαύτη δ' οὖσα ἔνδοξος ὅμως ἐστί, καὶ ἀπ' αὐτῆς Ἰκάριον καλεῖται τὸ προκείμενον πέλαγος, ἐν ᾧ καὶ αὐτὴ καὶ Σάμος καὶ Κῶς ἐστί, καὶ αἱ ἄρτι λεχθεῖσαι Κορασσίαι καὶ Πάτμος καὶ Λέρος. ἔνδοξον δὲ καὶ τὸ ἐν αὐτῇ ὄρος ὁ Κερκετεύς, μᾶλλον τῆς Ἀμπέλου·[2] αὕτη δ' ὑπέρκειται τῆς Σαμίων πόλεως. συνάπτει δὲ τῷ Ἰκαρίῳ τὸ Καρπάθιον πέλαγος πρὸς νότον, τούτῳ δὲ τὸ Αἰγύπτιον, πρὸς δὲ δύσιν τό τε Κρητικὸν καὶ τὸ Λιβυκόν.

[1] Λέρος, Groskurd, for Λερία; so Meineke.
[2] Meineke ejects the words ἔνδοξον . . . Ἀμπέλου.

her will. The island is so rocky that the comedians say that it was made thus by the Gorgon.

11. Tenos has no large city, but it has the temple of Poseidon, a great temple in a sacred precinct outside the city, a spectacle worth seeing. In it have been built great banquet-halls—an indication of the multitude of neighbours who congregate there and take part with the inhabitants of Tenos in celebrating the Poseidonian festival.

12. And there is Amorgos, one of the Sporades, the home of Simonides the iambic poet; and also Lebinthos, and Leros: "And thus saith Phocylides, 'the Lerians are bad, not one, but every one, all except Procles; and Procles is a Lerian.'"[1] For the natives of the island were reproached with being unprincipled.

13. Near by are both Patmos and the Corassiae; these are situated to the west of Icaria, and Icaria to the west of Samos. Now Icaria is deserted, though it has pastures, which are used by the Samians. But although it is such an isle as it is, still it is famous, and after it is named the sea that lies in front of it, in which are itself and Samos and Cos and the islands just mentioned—the Corassiae and Patmos and Leros. Famous, also, is the mountain in it, Cerceteus, more famous than the Ampelus,[2] which is situated above the city of Samians.[3] The Icarian Sea connects with the Carpathian Sea on the south, and the Carpathian with the Aegyptian, and on the west with the Cretan and the Libyan.

[1] *Frag.* 1 (Bergk). [2] See 14. 1. 15.

[3] But *both* of these mountains are in Samos (Pliny, in 5. 37, spells the former "Cercetius"). Hence the sentence seems to be a gloss that has crept in from the margin of the text.

14. Καὶ ἐν τῷ Καρπαθίῳ δ' εἰσὶ πολλαὶ τῶν Σποράδων μεταξὺ τῆς Κῶ μάλιστα καὶ Ῥόδου καὶ Κρήτης· ὧν εἰσὶν Ἀστυπάλαιά τε καὶ Τῆλος καὶ Χαλκία, καὶ ἃς Ὅμηρος ὀνομάζει ἐν τῷ Καταλόγῳ·

> οἳ δ' ἄρα Νίσυρόν τ' εἶχον Κράπαθόν τε Κάσον τε,
> καὶ Κῶν, Εὐρυπύλοιο πόλιν, νήσους τε Καλύδνας.

ἔξω γὰρ τῆς Κῶ καὶ τῆς Ῥόδου, περὶ ὧν ἐροῦμεν ὕστερον, τάς τε ἄλλας ἐν ταῖς Σποράσι τίθεμεν, καὶ δὴ καὶ ἐνταῦθα μεμνήμεθα αὐτῶν, καίπερ τῆς Ἀσίας, οὐ τῆς Εὐρώπης, ἐγγὺς οὐσῶν, ἐπειδὴ τῇ Κρήτῃ καὶ ταῖς Κυκλάσι καὶ τὰς Σποράδας συμπεριλαβεῖν ἠπείγετό[1] πως ὁ λόγος· ἐν δὲ τῇ τῆς Ἀσίας περιοδείᾳ τὰς προσεχεῖς αὐτῇ τῶν ἀξιολόγων νήσων προσπεριοδεύσομεν, Κύπρον καὶ Ῥόδον καὶ Κῶν καὶ τὰς ἐν τῇ ἐφεξῆς παραλίᾳ κειμένας, Σάμον, Χίον, Λέσβον, Τένεδον· νῦν δὲ τὰς Σποράδας, ὧν ἄξιον μνησθῆναι λοιπόν, ἔπιμεν.

15. Ἡ μὲν οὖν Ἀστυπάλαια ἱκανῶς ἐστὶ πελαγία, πόλιν ἔχουσα. ἡ δὲ Τῆλος ἐκτέταται παρὰ τὴν Κνιδίαν, μακρά, ὑψηλή, στενή, τὴν περίμετρον ὅσον ἑκατὸν καὶ τετταράκοντα σταδίων, ἔχουσα ὕφορμον. ἡ δὲ Χαλκία[2] τῆς Τήλου διέχει σταδίους ὀγδοήκοντα, Καρπάθου δὲ τετρακοσίους, Ἀστυπαλαίας δὲ περὶ διπλασίους, ἔχει δὲ καὶ κατοικίαν ὁμώνυμον καὶ ἱερὸν Ἀπόλλωνος καὶ λιμένα.

[1] ἠπείγετο, Kramer, for ἐπείγετο BCD*hikl*, ἐπείγεται *nox*; so Müller-Dübner and Meineke.

14. In the Carpathian Sea, also, are many of the Sporades, and in particular between Cos and Rhodes and Crete. Among these are Astypalaea, Telos, Chalcia, and those which Homer names in the *Catalogue*: "And those who held the islands Nisyros and Crapathos and Casos and Cos, the city of Eurypylus, and the Calydnian Islands";[1] for, excepting Cos and Rhodes, which I shall discuss later,[2] I place them all among the Sporades, and in fact, even though they are near Asia and not Europe, I make mention of them here because my argument has somehow impelled me to include the Sporades with Crete and the Cyclades. But in my geographical description of Asia I shall add a description of such islands that lie close to it as are worthy of note, Cyprus, Rhodes, Cos, and those that lie on the seaboard next thereafter, Samos, Chios, Lesbos, and Tenedos. But now I shall traverse the remainder of the Sporades that are worth mentioning.

15. Now Astypalaea lies far out in the high sea, and has a city. Telos extends alongside Cnidia, is long, high, narrow, has a perimeter of about one hundred and forty stadia, and has an anchoring-place. Chalcia is eighty stadia distant from Telos, four hundred from Carpathos, about twice as far from Astypalaea, and has also a settlement of the same name and a temple of Apollo and a harbour.

[1] *Iliad* 2. 676. Cf. the interpretation of this passage in 10. 5. 19.
[2] 14. 2. 5–13, 19.

---

[2] Χαλκεία BC*ksx*.

16. Νίσυρος δὲ πρὸς ἄρκτον μέν ἐστι Τήλου, διέχουσα αὐτῆς ὅσον ἑξήκοντα σταδίους, ὅσους καὶ Κῶ διέχει, στρογγύλη δὲ καὶ ὑψηλὴ καὶ πετρώδης τοῦ μυλίου λίθου· τοῖς γοῦν ἀστυγείτοσιν ἐκεῖθέν ἐστιν ἡ τῶν μύλων εὐπορία. ἔχει δὲ καὶ πόλιν ὁμώνυμον καὶ λιμένα καὶ θερμὰ καὶ Ποσειδῶνος ἱερόν· περίμετρον δὲ αὐτῆς ὀγδοήκοντα

C 489 στάδιοι. ἔστι δὲ καὶ νησία πρὸς αὐτῇ Νισυρίων λεγόμενα. φασὶ δὲ τὴν Νίσυρον ἀπόθραυσμα εἶναι τῆς Κῶ, προσθέντες καὶ μῦθον, ὅτι Ποσειδῶν διώκων ἕνα τῶν Γιγάντων, Πολυβώτην, ἀποθραύσας τῇ τριαίνῃ τρύφος τῆς Κῶ ἐπ᾽ αὐτὸν βάλοι, καὶ γένοιτο νῆσος τὸ βληθὲν ἡ Νίσυρος, ὑποκείμενον ἔχουσα ἐν αὐτῇ τὸν Γίγαντα· τινὲς δὲ αὐτὸν ὑποκεῖσθαι τῇ Κῴ φασίν.

17. Ἡ δὲ Κάρπαθος, ἣν Κράπαθον εἶπεν ὁ ποιητής, ὑψηλή ἐστι, κύκλον ἔχουσα σταδίων διακοσίων. τετράπολις δ᾽ ὑπῆρξε καὶ ὄνομα εἶχεν ἀξιόλογον· ἀφ᾽ οὗ καὶ τῷ πελάγει τοὔνομα ἐγένετο. μία δὲ τῶν πόλεων ἐκαλεῖτο Νίσυρος, ὁμώνυμος τῇ τῶν Νισυρίων[1] νήσῳ. κεῖται δὲ τῆς Λιβύης κατὰ Λευκὴν ἀκτήν, ἣ τῆς μὲν Ἀλεξανδρείας περὶ χιλίους διέχει σταδίους, τῆς δὲ Καρπάθου περὶ τετρακισχιλίους.

18. Κάσος[2] δὲ ταύτης μὲν ἀπὸ ἑβδομήκοντά ἐστι σταδίων, τοῦ δὲ Σαμωνίου[3] τοῦ ἄκρου τῆς Κρήτης διακοσίων πεντήκοντα· κύκλον δὲ ἔχει σταδίων ὀγδοήκοντα. ἔστι δ᾽ ἐν αὐτῇ καὶ πόλις ὁμώνυμος, καὶ Κασίων νῆσοι καλούμεναι πλείους περὶ αὐτήν.

19. Νήσους δὲ Καλύδνας τὰς Σποράδας λέγειν φασὶ τὸν ποιητήν, ὧν μίαν εἶναι Κάλυμναν· εἰκὸς

16. Nisyros lies to the north of Telos, and is about sixty stadia distant both from it and from Cos. It is round and high and rocky, the rock being that of which millstones are made; at any rate, the neighbouring peoples are well supplied with millstones from there. It has also a city of the same name and a harbour and hot springs and a temple of Poseidon. Its perimeter is eighty stadia. Close to it are also isles called Isles of the Nisyrians. They say that Nisyros is a fragment of Cos, and they add the myth that Poseidon, when he was pursuing one of the giants, Polybotes, broke off a fragment of Cos with his trident and hurled it upon him, and the missile became an island, Nisyros, with the giant lying beneath it. But some say that he lies beneath Cos.

17. Carpathos, which the poet calls Crapathos, is high, and has a circuit of two hundred stadia. At first it was a Tetrapolis, and it had a renown which is worth noting; and it was from this fact that the sea got the name Carpathian. One of the cities was called Nisyros, the same name as that of the island of the Nisyrians. It lies opposite Leucê Actê in Libya, which is about one thousand stadia distant from Alexandreia and about four thousand from Carpathos.

18. Casos is seventy stadia from Carpathos, and two hundred and fifty from Cape Samonium in Crete. It has a circuit of eighty stadia. In it there is also a city of the same name, and round it are several islands called Islands of the Casians.

19. They say that the poet calls the Sporades "Calydnian Islands," one of which, they say, is Calymna. But it is reasonable to suppose that, as

[1] Νισυρίων, Corais, for Νισύρων; so the later editors.
[2] νῆσος BCDklsx.
[3] Σαλμωνίου BChkno.

δ', ὡς ἐκ τῶν Νισυρίων λέγονται καὶ Κασίων[1] αἱ ἐγγὺς καὶ ὑπήκοοι, οὕτως καὶ τὰς τῇ Καλύμνῃ περικειμένας, ἴσως τότε λεγομένῃ Καλύδνῃ· τινὲς δὲ δύο εἶναι Καλύδνας φασί, Λέρον καὶ Κάλυμναν, ὥσπερ καὶ λέγειν τὸν ποιητήν. ὁ δὲ Σκήψιος πληθυντικῶς ὠνομάσθαι τὴν νῆσον Καλύμνας φησίν, ὡς Ἀθήνας καὶ Θήβας, δεῖν δὲ ὑπερβατῶς δέξασθαι τὸ τοῦ ποιητοῦ· οὐ γὰρ νήσους Καλύδνας λέγειν, ἀλλ' οἳ[2] δ' ἄρα νήσους Νίσυρόν τ' εἶχον Κράπαθόν τε Κάσον τε καὶ Κῶν, Εὐρυπύλοιο πόλιν, Καλύδνας τε. ἅπαν μὲν οὖν τὸ νησιωτικὸν μέλι ὡς ἐπὶ τὸ πολὺ ἀστεῖόν ἐστι καὶ ἐνάμιλλον τῷ Ἀττικῷ, τὸ δ' ἐν ταῖσδε ταῖς νήσοις διαφερόντως, μάλιστα δὲ τὸ Καλύμνιον.

[1] Κασσίων BDhklno. [2] ἀλλ' οἵ, the editors, for ἄλλοι.

the islands which are near, and subject to, Nisyros and Casos are called "Islands of the Nisyrians" and "Islands of the Casians," so also those which lie round Calymna were called "Islands of the Calymnians"—Calymna at that time, perhaps, being called Calydna. But some say that there are only two Calydnian islands, Leros and Calymna, the two mentioned by the poet. The Scepsian[1] says that the name of the island was used in the plural, "Calymnae," like "Athenae" and "Thebae"; but, he adds, the words of the poet should be interpreted as a case of hyperbaton, for he does not say, "Calydnian Islands," but "those who held the islands Nisyros and Crapathos and Casos and Cos, the city of Eurypylus, and Calydnae." Now all the honey produced in the islands is, for the most part, good, and rivals that of Attica, but the honey produced in the islands in question is exceptionally good, and in particular the Calymnian.

[1] Demetrius of Scepsis.

BOOK XI

IA′

I

C 490 1. Τῇ δ' Εὐρώπῃ συνεχής ἐστιν ἡ Ἀσία, κατὰ τὸν Τάναϊν συνάπτουσα αὐτῇ· περὶ ταύτης οὖν ἐφεξῆς ῥητέον, διελόντας φυσικοῖς τισὶν ὅροις τοῦ σαφοῦς χάριν. ὅπερ οὖν Ἐρατοσθένης ἐφ' ὅλης τῆς οἰκουμένης ἐποίησε, τοῦθ' ἡμῖν ἐπὶ τῆς Ἀσίας ποιητέον.

2. Ὁ γὰρ Ταῦρος μέσην πως διέζωκε ταύτην τὴν ἤπειρον, ἀπὸ τῆς ἑσπέρας ἐπὶ τὴν ἕω τεταμένος,[1] τὸ μὲν αὐτῆς ἀπολείπων πρὸς βορρᾶν, τὸ δὲ μεσημβρινόν. καλοῦσι δὲ αὐτῶν οἱ Ἕλληνες τὸ μὲν ἐντὸς τοῦ Ταύρου, τὸ δὲ ἐκτός. εἴρηται δὲ ταῦθ' ἡμῖν καὶ πρότερον, ἀλλ' εἰρήσθω καὶ νῦν ὑπομνήσεως χάριν.

3. Πλάτος μὲν οὖν ἔχει τὸ ὄρος πολλαχοῦ καὶ τρισχιλίων σταδίων, μῆκος δ' ὅσον καὶ τὸ τῆς Ἀσίας, τεττάρων που μυριάδων καὶ πεντακισχιλίων, ἀπὸ τῆς Ῥοδίων περαίας ἐπὶ τὰ ἄκρα τῆς Ἰνδικῆς καὶ Σκυθίας πρὸς τὰς ἀνατολάς.

4. Διῄρηται δ' εἰς μέρη πολλὰ καὶ ὀνόματα περιγραφαῖς καὶ μείζοσι καὶ ἐλάττοσιν ἀφωρισμένα. ἐπεὶ δ' ἐν τῷ τοσούτῳ πλάτει τοῦ ὄρους

[1] τετμημένος *Cgloxwz*, τετραμένος Eustath. (note on *Dionys.* 647).

---

[1] The Don. [2] See 2. 1. 1.

BOOK XI

I

1. Asia is adjacent to Europe, bordering thereon along the Tanaïs[1] River. I must therefore describe this country next, first dividing it, for the sake of clearness, by means of certain natural boundaries. That is, I must do for Asia precisely what Eratosthenes did for the inhabited world as a whole.[2]

2. The Taurus forms a partition approximately through the middle of this continent, extending from the west towards the east, leaving one portion of it on the north and the other on the south. Of these portions, the Greeks call the one the "Cis-Tauran" Asia and the other "Trans-Tauran." I have said this before,[3] but let me repeat it by way of reminder.

3. Now the mountain has in many places as great a breadth as three thousand stadia, and a length as great as that of Asia itself, that is, about forty-five thousand stadia, reckoning from the coast opposite Rhodes to the eastern extremities of India and Scythia.

4. It has been divided into many parts with many names, determined by boundaries that circumscribe areas both large and small. But since certain tribes are comprised within the vast width of the mountain,

[3] *i.e.* "Asia this side Taurus and Asia outside Taurus." (Cp. 2. 5. 31.)

ἀπολαμβανεταί τινα ἔθνη, τὰ μὲν ἀσημότερα, τὰ δὲ καὶ παντελῶς γνώριμα (καθάπερ ἡ Παρθυαία καὶ Μηδία καὶ Ἀρμενία καὶ Καππαδοκῶν τινὲς καὶ Κίλικες καὶ Πισίδαι), τὰ μὲν πλεονάζοντα[1] ἐν[2] τοῖς προσβόροις μέρεσιν ἐνταῦθα τακτέον, τὰ δ' ἐν τοῖς νοτίοις εἰς τὰ νότια, καὶ τὰ ἐν μέσῳ δὲ τῶν ὀρῶν κείμενα διὰ τὰς τῶν ἀέρων ὁμοιότητας πρὸς βορρᾶν πως θετέον· ψυχροὶ γάρ εἰσιν, οἱ δὲ νότιοι θερμοί. καὶ τῶν ποταμῶν δὲ αἱ ῥύσεις ἐνθένδε οὖσαι πᾶσαι σχεδόν τι εἰς τἀναντία, αἱ μὲν εἰς τὰ βόρεια, αἱ δ' εἰς τὰ νότια μέρη (τά γε[3] πρῶτα, κἂν ὕστερόν τινες ἐπιστρέφωσι πρὸς ἀνατολὰς ἢ δύσεις), ἔχουσί τι εὐφυὲς πρὸς τὸ τοῖς ὄρεσιν ὁρίοις χρῆσθαι κατὰ τὴν εἰς δύο μέρη διαίρεσιν τῆς Ἀσίας· καθάπερ καὶ ἡ θάλαττα ἡ ἐντὸς Στηλῶν, ἐπ' εὐθείας πως οὖσα ἡ πλείστη τοῖς ὄρεσι τούτοις, ἐπιτηδεία γεγένηται πρὸς τὸ δύο ποιεῖν ἠπείρους, τήν τε Εὐρώπην καὶ τὴν Λιβύην, ὅριον ἀμφοῖν οὖσα ἀξιόλογον.

C 491

5. Τοῖς δὲ μεταβαίνουσιν ἀπὸ τῆς Εὐρώπης ἐπὶ τὴν Ἀσίαν ἐν τῇ γεωγραφίᾳ τὰ πρὸς βορρᾶν ἐστὶ πρῶτα τῆς εἰς δύο διαιρέσεως· ὥστε ἀπὸ τούτων ἀρκτέον. αὐτῶν δὲ τούτων πρῶτά ἐστι τὰ περὶ τὸν Τάναϊν, ὅνπερ τῆς Εὐρώπης καὶ τῆς Ἀσίας ὅριον ὑπεθέμεθα. ἔστι δὲ ταῦτα τρόπον τινὰ χερρονησίζοντα, περιέχεται γὰρ ἐκ μὲν τῆς ἑσπέρας τῷ ποταμῷ τῷ Τανάϊδι καὶ

[1] πλησιάζοντα *hi* and Xylander, instead of πλεονάζοντα.
[2] ἐν, before τοῖς, Groskurd inserts; so C. Müller.
[3] γε D, τε other MSS.

some rather insignificant, but others extremely well known (as, for instance, the Parthians, the Medes, the Armenians, a part of the Cappadocians, the Cilicians, and the Pisidians), those which lie for the most part in its northerly parts must be assigned there,[1] and those in its southern parts to the southern,[2] while those which are situated in the middle of the mountains should, because of the likeness of their climate, be assigned to the north, for the climate in the middle is cold, whereas that in the south is hot. Further, almost all the rivers that rise in the Taurus flow in contrary directions, that is, some into the northern region and others into the southern (they do so at first, at least, although later some of them bend towards the east or west), and they therefore are naturally helpful in our use of these mountains as boundaries in the two-fold division of Asia—just as the sea inside the Pillars,[3] which for the most part is approximately in a straight line with these mountains, has proved convenient in the forming of two continents, Europe and Libya, it being the noteworthy boundary between the two.

5. As we pass from Europe to Asia in our geography, the northern division is the first of the two divisions to which we come; and therefore we must begin with this. Of this division the first portion is that in the region of the Tanaïs River, which I have taken as the boundary between Europe and Asia. This portion forms, in a way, a peninsula, for it is surrounded on the west by the Tanaïs River

[1] *i.e.* to the Cis-Tauran Asia. [2] *i.e.* Trans-Tauran.
[3] *i.e.* the Mediterranean (see 2. 1. 1).

τῇ Μαιώτιδι μέχρι τοῦ Βοσπόρου καὶ τῆς τοῦ Εὐξείνου παραλίας τῆς τελευτώσης εἰς τὴν Κολχίδα· ἐκ δὲ τῶν ἄρκτων τῷ Ὠκεανῷ μέχρι τοῦ στόματος τῆς Κασπίας θαλάττης· ἕωθεν δὲ αὐτῇ ταύτῃ τῇ θαλάττῃ μέχρι τῶν μεθορίων τῆς τε Ἀλβανίας καὶ τῆς Ἀρμενίας, καθ᾽ ἃ ὁ Κῦρος καὶ ὁ Ἀράξης ἐκδιδοῦσι ποταμοί, ῥέοντες ὁ μὲν διὰ τῆς Ἀρμενίας, Κῦρος δὲ διὰ τῆς Ἰβηρίας καὶ τῆς Ἀλβανίας· ἐκ νότου δὲ τῇ[1] ἀπὸ τῆς ἐκβολῆς τοῦ Κύρου μέχρι τῆς Κολχίδος, ὅσον τρισχιλίων οὔσῃ[2] σταδίων ἀπὸ θαλάττης ἐπὶ θάλατταν, δι᾽ Ἀλβανῶν καὶ Ἰβήρων, ὥστε ἰσθμοῦ λόγον ἔχειν. οἱ δ᾽ ἐπὶ τοσοῦτον συναγαγόντες τὸν ἰσθμόν, ἐφ᾽ ὅσον Κλείταρχος, ἐπίκλυστον φήσας ἐξ ἑκατέρου τοῦ πελάγους, οὐδ᾽ ἂν λόγου ἀξιοῖντο. Ποσειδώνιος δὲ χιλίων καὶ πεντακοσίων εἴρηκε τὸν ἰσθμόν, ὅσον καὶ τὸν ἀπὸ Πηλουσίου ἰσθμὸν ἐς τὴν Ἐρυθράν· δοκῶ δέ, φησί, μὴ πολὺ διαφέρειν μηδὲ τὸν ἀπὸ τῆς Μαιώτιδος εἰς τὸν Ὠκεανόν.

6. Οὐκ οἶδα δέ, πῶς ἄν τις περὶ τῶν ἀδήλων αὐτῷ πιστεύσειε, μηδὲν εἰκὸς ἔχοντι εἰπεῖν περὶ αὐτῶν, ὅταν περὶ τῶν φανερῶν οὕτω παραλόγως λέγῃ, καὶ ταῦτα φίλος Πομπηίῳ γεγονὼς τῷ στρατεύσαντι ἐπὶ τοὺς Ἴβηρας καὶ τοὺς C 492 Ἀλβανοὺς μέχρι τῆς ἐφ᾽ ἑκάτερα θαλάττης, τῆς τε Κασπίας καὶ τῆς Κολχικῆς. φασὶ γοῦν

[1] τῇ, Corais, for ἡ; so the later editors.
[2] οὔσῃ, Corais, for οὖσα; so the later editors.

[1] The Cimmerian Bosporus.

and Lake Maeotis as far as the Bosporus[1] and that part of the coast of the Euxine Sea which terminates at Colchis; and then on the north by the Ocean as far as the mouth of the Caspian Sea;[2] and then on the east by this same sea as far as the boundary between Albania and Armenia, where empty the rivers Cyrus and Araxes, the Araxes flowing through Armenia and the Cyrus through Iberia and Albania; and lastly, on the south by the tract of country which extends from the outlet of the Cyrus River to Colchis, which is about three thousand stadia from sea to sea, across the territory of the Albanians and the Iberians, and therefore is described as an isthmus. But those writers who have reduced the width of the isthmus as much as Cleitarchus[3] has, who says that it is subject to inundation from either sea, should not be considered even worthy of mention. Poseidonius states that the isthmus is fifteen hundred stadia across, as wide as the isthmus from Pelusium to the Red Sea.[4] "And in my opinion," he says, "the isthmus from Lake Maeotis to the Ocean does not differ much therefrom."

6. But I do not know how anyone can trust him concerning things that are uncertain if he has nothing plausible to say about them, when he reasons so illogically about things that are obvious; and this too, although he was a friend of Pompey, who made an expedition against the Iberians and the Albanians, from sea to sea on either side, both the Caspian and the Colchian[5] Seas. At any rate, it is

[2] Strabo thought that the Caspian (Hyrcanian) Sea was an inlet of the Northern Sea (2. 5. 14).
[3] See *Dictionary* in Vol. II.
[4] Cf. 17. 1. 21.
[5] The Euxine.

ἐν Ῥόδῳ γενόμενον τὸν Πομπήιον, ἡνίκα ἐπὶ τὸν ληστρικὸν πόλεμον ἐξῆλθεν (εὐθὺς δ' ἔμελλε καὶ ἐπὶ Μιθριδάτην ὁρμήσειν καὶ τὰ μέχρι τῆς Κασπίας ἔθνη), παρατυχεῖν διαλεγομένῳ τῷ Ποσειδωνίῳ, ἀπιόντα δ' ἐρέσθαι, εἴ τι προστάττει, τὸν δ' εἰπεῖν·

αἰὲν ἀριστεύειν καὶ ὑπείροχον ἔμμεναι ἄλλων.

προστίθει[1] δὲ τούτοις, ὅτι καὶ τὴν ἱστορίαν συνέγραψε τὴν περὶ αὐτόν. διὰ δὴ ταῦτα ἐχρῆν φροντίσαι τἀληθοῦς πλέον τι.

7. Δεύτερον δ' ἂν εἴη μέρος τὸ ὑπὲρ τῆς Ὑρκανίας θαλάττης, ἣν Κασπίαν καλοῦμεν, μέχρι τῶν κατ' Ἰνδοὺς Σκυθῶν. τρίτον δὲ μέρος τὸ συνεχὲς τῷ λεχθέντι ἰσθμῷ καὶ τὰ ἑξῆς τούτῳ καὶ ταῖς Κασπίαις πύλαις, τῶν ἐντὸς τοῦ Ταύρου καὶ τῆς Εὐρώπης ἐγγυτάτω· ταῦτα δ' ἐστὶ Μηδία καὶ Ἀρμενία καὶ Καππαδοκία καὶ τὰ μεταξύ. τέταρτον δ' ἡ ἐντὸς Ἅλυος γῆ καὶ τὰ ἐν αὐτῷ τῷ Ταύρῳ καὶ ἐκτὸς ὅσα εἰς τὴν χερρόνησον ἐμπίπτει ἣν ποιεῖ ὁ διείργων ἰσθμὸς τήν τε Ποντικὴν καὶ τὴν Κιλικίαν θάλασσαν. τῶν δὲ ἄλλων, τῶν ἔξω τοῦ Ταύρου, τήν τε Ἰνδικὴν τίθεμεν καὶ τὴν Ἀριανὴν μέχρι τῶν ἐθνῶν τῶν καθηκόντων πρός τε τὴν κατὰ Πέρσας θάλατταν καὶ τὸν Ἀράβιον κόλπον καὶ τὸν Νεῖλον καὶ πρὸς τὸ Αἰγύπτιον πέλαγος καὶ τὸ Ἰσσικόν.

[1] προστίθει, Corais, for προσετίθει; so the later editors.

said that Pompey, upon arriving at Rhodes on his expedition against the pirates (immediately thereafter he was to set out against both Mithridates and the tribes which extended as far as the Caspian Sea), happened to attend one of the lectures of Poseidonius, and that when he went out he asked Poseidonius whether he had any orders to give, and that Poseidonius replied: "Ever bravest be, and preeminent o'er others." Add to this that among other works he wrote also the history of Pompey. So for this reason he should have been more regardful of the truth.

7. The second portion would be that beyond the Hyrcanian Sea, which we call the Caspian Sea, as far as the Scythians near India. The third portion would consist of the part which is adjacent to the isthmus above mentioned and of those parts of the region inside Taurus[1] and nearest Europe which come next after this isthmus and the Caspian Gates, I mean Media and Armenia and Cappadocia and the intervening regions. The fourth portion is the land inside[2] the Halys River, and all the region in the Taurus itself and outside thereof which falls within the limits of the peninsula which is formed by the isthmus that separates the Pontic and the Cilician Seas. As for the other countries, I mean the Trans-Tauran, I place among them not only India, but also Ariana as far as the tribes that extend to the Persian Sea and the Arabian Gulf and the Nile and the Egyptian and Issic Seas.

[1] Cis-Tauran. [2] *i.e.* "west of."

II

1. Οὕτω δὲ διακειμένων, τὸ πρῶτον μέρος οἰκοῦσιν ἐκ μὲν τῶν πρὸς ἄρκτον μερῶν καὶ τὸν Ὠκεανὸν Σκυθῶν τινὲς νομάδες καὶ ἁμάξοικοι,[1] ἐνδοτέρω δὲ τούτων Σαρμάται, καὶ οὗτοι Σκύθαι, Ἄορσοι καὶ Σιρακοί, μέχρι τῶν Καυκασίων ὀρῶν ἐπὶ μεσημβρίαν τείνοντες, οἱ μὲν νομάδες, οἱ δὲ καὶ σκηνῖται καὶ γεωργοί· περὶ δὲ τὴν λίμνην Μαιῶται· πρὸς δὲ τῇ θαλάττῃ τοῦ Βοσπόρου τὰ κατὰ τὴν Ἀσίαν ἐστὶ καὶ ἡ Σινδική· μετὰ δὲ ταύτην Ἀχαιοὶ καὶ Ζυγοὶ καὶ Ἡνίοχοι, Κερκέται τε καὶ Μακροπώγωνες. ὑπέρκεινται δὲ τούτων καὶ τὰ τῶν Φθειροφάγων στενά· μετὰ δὲ τοὺς Ἡνιόχους ἡ Κολχίς, ὑπὸ τοῖς Καυκασίοις ὄρεσι κειμένη καὶ τοῖς Μοσχικοῖς. ἐπεὶ δ' ὅριον ὑπόκειται τῆς Εὐρώπης καὶ τῆς Ἀσίας ὁ Τάναϊς ποταμός, ἐντεῦθεν ἀρξάμενοι τὰ καθ' ἕκαστα ὑπογράψομεν.

2. Φέρεται μὲν οὖν ἀπὸ τῶν ἀρκτικῶν μερῶν, οὐ μὴν ὡς ἂν κατὰ διάμετρον ἀντίρρους τῷ Νείλῳ, καθάπερ νομίζουσιν οἱ πολλοί, ἀλλὰ C 493 ἑωθινώτερος ἐκείνου, παραπλησίως ἐκείνῳ τὰς ἀρχὰς ἀδήλους ἔχων· ἀλλὰ τοῦ μὲν πολὺ τὸ φανερόν, χώραν διεξιόντος πᾶσαν εὐεπίμικτον καὶ μακροὺς ἀνάπλους ἔχοντος· τοῦ δὲ Ταναΐδος τὰς μὲν ἐκβολὰς ἴσμεν (δύο δ' εἰσὶν εἰς τὰ ἀρκτικώτατα μέρη τῆς Μαιώτιδος, ἑξήκοντα

[1] ἁμάξοικοι, Corais, for ἁμάξικοι; so the later editors.

---

[1] Also spelled "Siraces." See 11. 5. 8.

II

1. Of the portions thus divided, the first is inhabited, in the region toward the north and the ocean, by Scythian nomads and waggon-dwellers, and south of these, by Sarmatians, these too being Scythians, and by Aorsi and Siraci,[1] who extend towards the south as far as the Caucasian Mountains, some being nomads and others tent-dwellers and farmers. About Lake Maeotis live the Maeotae. And on the sea lies the Asiatic side of the Bosporus, or the Sindic territory. After this latter, one comes to the Achaei and the Zygi and the Heniochi, and also the Cercetae and the Macropogones.[2] And above these are situated the narrow passes of the Phtheirophagi;[3] and after the Heniochi the Colchian country, which lies at the foot of the Caucasian, or Moschian, Mountains. But since I have taken the Tanaïs River as the boundary between Europe and Asia, I shall begin my detailed description therewith.

2. Now the Tanaïs flows from the northerly region,—not, however, as most people think, in a course diametrically opposite to that of the Nile, but more to the east than the Nile—and like the Nile its sources are unknown. Yet a considerable part of the Nile is well known, since it traverses a country which is everywhere easily accessible and since it is navigable for a great distance inland. But as for the Tanaïs, although we know its outlets (they are two in number and are in the most northerly region of Lake Maeotis, being sixty stadia

[2] "Long-beards." [3] "Lice-eaters."

σταδίους ἀλλήλων διέχουσαι), τοῦ[1] δ' ὑπὲρ τῶν ἐκβολῶν ὀλίγον τὸ γνώριμόν ἐστι διὰ τὰ ψύχη καὶ τὰς ἀπορίας τῆς χώρας, ἃς οἱ μὲν αὐτόχθονες δύνανται φέρειν, σαρξὶ καὶ γάλακτι τρεφόμενοι νομαδικῶς, οἱ δ' ἀλλοεθνεῖς οὐχ ὑπομένουσιν. ἄλλως τε[2] οἱ νομάδες δυσεπίμικτοι τοῖς ἄλλοις ὄντες καὶ πλήθει καὶ βίᾳ διαφέροντες ἀποκεκλείκασιν, εἰ καί τι πορεύσιμον τῆς χώρας ἐστὶν ἢ εἴ τινας τετύχηκεν ἀνάπλους ἔχων ὁ ποταμός. ἀπὸ δὲ τῆς αἰτίας ταύτης οἱ μὲν ὑπέλαβον τὰς πηγὰς ἔχειν αὐτὸν ἐν τοῖς Καυκασίοις ὄρεσι, πολὺν δ' ἐνεχθέντα ἐπὶ τὰς ἄρκτους, εἶτ' ἀναστρέψαντα ἐκβάλλειν εἰς τὴν Μαιῶτιν· τούτοις δὲ ὁμοδοξεῖ καὶ Θεοφάνης ὁ Μιτυληναῖος· οἱ δ' ἀπὸ τῶν ἄνω μερῶν τοῦ Ἴστρου φέρεσθαι, σημεῖον δὲ φέρουσιν οὐδὲν τῆς πόρρωθεν οὕτω ῥύσεως καὶ ἀπ' ἄλλων κλιμάτων, ὥσπερ οὐ δυνατὸν ὂν καὶ ἐγγύθεν καὶ ἀπὸ τῶν ἄρκτων.

3. Ἐπὶ δὲ τῷ ποταμῷ καὶ τῇ λίμνῃ πόλις ὁμώνυμος οἰκεῖται Τάναϊς, κτίσμα τῶν τὸν Βόσπορον ἐχόντων Ἑλλήνων· νεωστὶ μὲν οὖν ἐξεπόρθησεν αὐτὴν Πολέμων ὁ βασιλεὺς ἀπειθοῦσαν. ἦν δ' ἐμπόριον κοινὸν τῶν τε Ἀσιανῶν καὶ τῶν Εὐρωπαίων νομάδων καὶ τῶν ἐκ τοῦ Βοσπόρου τὴν λίμνην πλεόντων, τῶν μὲν ἀνδράποδα ἀγόντων καὶ δέρματα καὶ εἴ τι ἄλλο τῶν νομαδικῶν, τῶν

[1] τοῦ, Corais, for τό; so the later editors.
[2] τε, Corais, for δέ; so the later editors.

---

[1] Intimate friend of Pompey; wrote a history of his campaigns.
[2] See Vol. I, p. 22, foot-note 2.

distant from one another), yet but little of the part that is beyond its outlets is known to us, because of the coldness and the poverty of the country. This poverty can indeed be endured by the indigenous peoples, who, in nomadic fashion, live on flesh and milk, but people from other tribes cannot stand it. And besides, the nomads, being disinclined to intercourse with any other people and being superior both in numbers and in might, have blocked off whatever parts of the country are passable, or whatever parts of the river happen to be navigable. This is what has caused some to assume that the Tanaïs has its sources in the Caucasian Mountains, flows in great volume towards the north, and then, making a bend, empties into Lake Maeotis (Theophanes of Mitylenê[1] has the same opinion as these), and others to assume that it flows from the upper region of the Ister, although they produce no evidence of its flowing from so great a distance or from other "climata,"[2] as though it were impossible for the river to flow both from a near-by source and from the north.

3. On the river and the lake is an inhabited city bearing the same name, Tanaïs; it was founded by the Greeks who held the Bosporus. Recently, however, it was sacked by King Polemon[3] because it would not obey him. It was a common emporium, partly of the Asiatic and the European nomads, and partly of those who navigated the lake from the Bosporus, the former bringing slaves, hides, and such other things as nomads possess, and the latter

[3] Polemon I. He became king of the Bosporus about 16 B.C. (Dio Cassius 54. 24).

δ' ἐσθῆτα καὶ οἶνον καὶ τἆλλα, ὅσα τῆς ἡμέρου διαίτης οἰκεῖα, ἀντιφορτιζομένων. πρόκειται δ' ἐν ἑκατὸν σταδίοις τοῦ ἐμπορίου νῆσος Ἀλωπεκία, κατοικία μιγάδων ἀνθρώπων· ἔστι δὲ καὶ ἄλλα νησίδια πλησίον ἐν τῇ λίμνῃ. διέχει δὲ τοῦ στόματος τῆς Μαιώτιδος εὐθυπλοοῦσι ἐπὶ τὰ βόρεια δισχιλίους καὶ διακοσίους σταδίους ὁ Τάναϊς, οὐ πολὺ δὲ πλείους εἰσὶ παραλεγομένῳ τὴν γῆν.

4. Ἐν δὲ τῷ παράπλῳ τῷ παρὰ γῆν πρῶτον μέν ἐστιν ἀπὸ τοῦ Ταναΐδος προϊοῦσιν ἐν ὀκτακοσίοις ὁ μέγας καλούμενος Ῥομβίτης, ἐν ᾧ τὰ πλεῖστα ἁλιεύματα τῶν εἰς ταριχείας ἰχθύων· ἔπειτα ἐν ἄλλοις ὀκτακοσίοις ὁ ἐλάσσων Ῥομβίτης καὶ[1] ἄκρα, ἔχουσα καὶ αὐτὴ ἁλιείας ἐλάττους· ἔχουσι δὲ οἱ μὲν περὶ τὸν[2] πρότερον νησία ὁρμητήρια, οἱ δ' ἐν τῷ μικρῷ Ῥομβίτῃ αὐτοί εἰσιν οἱ Μαιῶται ἐργαζόμενοι· οἰκοῦσι γὰρ ἐν τῷ παράπλῳ C 494 τούτῳ παντὶ οἱ Μαιῶται, γεωργοὶ μέν, οὐχ ἧττον δὲ τῶν νομάδων πολεμισταί. διῄρηνται δὲ εἰς ἔθνη πλείω, τὰ μὲν πλησίον τοῦ Ταναΐδος ἀγριώτερα, τὰ δὲ συνάπτοντα τῷ Βοσπόρῳ χειροήθη μᾶλλον. ἀπὸ δὲ τοῦ μικροῦ Ῥομβίτου στάδιοί εἰσιν ἑξακόσιοι ἐπὶ Τυράμβην καὶ τὸν Ἀντικείτην ποταμόν· εἶθ' ἑκατὸν καὶ εἴκοσιν ἐπὶ τὴν κώμην τὴν Κιμμερικήν,[3] ἥτις ἐστὶν ἀφετήριον τοῖς τὴν λίμνην πλέουσιν· ἐν δὲ τῷ παράπλῳ τούτῳ καὶ σκοπαί τινες λέγονται Κλαζομενίων.

[1] καί, before ἄκρα, Corais inserts ; so the later editors.

[2] περὶ τόν, before πρότερον, Groskurd inserts ; so Müller-Dübner, but Meineke merely indicates a lacuna.

[3] Κιμμερικήν, Xylander, for Κιμβρικήν ; so the later editors.

giving in exchange clothing, wine, and the other things that belong to civilised life. At a distance of one hundred stadia off the emporium lies an island called Alopecia, a settlement of promiscuous people. There are also other small islands near by in the lake. The Tanaïs[1] is two thousand two hundred stadia distant from the mouth of Lake Maeotis by a direct voyage towards the north; but it is not much farther by a voyage along the coast.

4. In the voyage along the coast, one comes first, at a distance of eight hundred stadia from the Tanaïs, to the Greater Rhombites River, as it is called, where are made the greatest catches of the fish that are suitable for salting. Then, at a distance of eight hundred more, to the Lesser Rhombites and a cape, which latter also has fisheries, although they are smaller. The people who live about the Greater Rhombites have small islands as bases for their fishing; but the people who carry on the business at the Lesser Rhombites are the Maeotae themselves, for the Maeotae live along the whole of this coast; and though farmers, they are no less warlike than the nomads. They are divided into several tribes, those who live near the Tanaïs being rather ferocious, but those whose territory borders on the Bosporus being more tractable. It is six hundred stadia from the Lesser Rhombites to Tyrambê and the Anticeites River; then a hundred and twenty to the Cimmerian village, which is a place of departure for those who navigate the lake; and on this coast are said to be some look-out places[2] belonging to the Clazomenians.

[1] *i.e.* the *mouth* of the Tanaïs.

[2] *i.e.* for the observation of fish.

5. Τὸ δὲ Κιμμερικὸν πόλις ἦν πρότερον ἐπὶ χερρονήσου ἱδρυμένη, τὸν ἰσθμὸν τάφρῳ καὶ χώματι κλείουσα· ἐκέκτηντο δ' οἱ Κιμμέριοι μεγάλην ποτὲ ἐν τῷ Βοσπόρῳ δύναμιν, διόπερ καὶ Κιμμερικὸς Βόσπορος ὠνομάσθη. οὗτοι δ' εἰσὶν οἱ τοὺς τὴν μεσόγαιαν οἰκοῦντας ἐν τοῖς δεξιοῖς μέρεσι τοῦ Πόντου μέχρι Ἰωνίας ἐπιδραμόντες. τούτους μὲν οὖν ἐξήλασαν ἐκ τῶν τόπων Σκύθαι, τοὺς δὲ Σκύθας Ἕλληνες οἱ Παντικάπαιον καὶ τὰς ἄλλας οἰκίσαντες πόλεις τὰς ἐν Βοσπόρῳ.

6. Εἶτ' ἐπὶ τὴν Ἀχίλλειον κώμην εἴκοσιν, ἐν ᾗ τὸ Ἀχιλλέως ἱερόν· ἐνταῦθα δ' ἐστὶν ὁ στενώτατος πορθμὸς τοῦ στόματος τῆς Μαιώτιδος, ὅσον εἴκοσι σταδίων ἢ πλειόνων, ἔχων ἐν τῇ περαίᾳ κώμην τὸ Μυρμήκιον· πλησίον δ' ἐστὶ τὸ Ἡρακλεῖον [1] καὶ τὸ Παρθένιον.

7. Εντεῦθεν δ' ἐπὶ τὸ Σατύρου μνῆμα ἐνενήκοντα στάδιοι· τοῦτο δ' ἐστὶν ἐπ' ἄκρας τινὸς χωστὸν ἀνδρὸς τῶν ἐπιφανῶς δυναστευσάντων τοῦ Βοσπόρου.

8. Πλησίον δὲ κώμη Πατραεύς, ἀφ' ἧς ἐπὶ κώμην Κοροκονδάμην ἑκατὸν τριάκοντα· αὕτη δ' ἐστὶ τοῦ Κιμμερικοῦ καλουμένου Βοσπόρου πέρας. καλεῖται δὲ οὕτως [2] ὁ στενωπὸς ἐπὶ [3] τοῦ στόματος τῆς Μαιώτιδος ἀπὸ τῶν κατὰ τὸ Ἀχίλλειον καὶ τὸ Μυρμήκιον στενῶν διατείνων μέχρι πρὸς τὴν Κοροκονδάμην καὶ τὸ ἀντικείμενον αὐτῇ κώμιον τῆς Παντικαπαίων γῆς, ὄνομα Ἄκραν,[4] ἑβδομή-

[1] τὸ Ἡρακλεῖον, Jones, following conj. of Kramer; so C. Müller.

[2] οὕτως, Xylander, for οὗτος; so the later editors.

[3] ἐπί, Xylander, for ἀπό: so the later editors.

5. Cimmericum was in earlier times a city situated on a peninsula, and it closed the isthmus by means of a trench and a mound. The Cimmerians once possessed great power in the Bosporus, and this is why it was named Cimmerian Bosporus. These are the people who overran the country of those who lived in the interior on the right side of the Pontus as far as Ionia. However, these were driven out of the region by the Scythians; and then the Scythians were driven out by the Greeks who founded Panticapaeum and the other cities on the Bosporus.

6. Then, twenty stadia distant, one comes to the village Achilleium, where is the temple of Achilles. Here is the narrowest passage across the mouth of Lake Maeotis, about twenty stadia or more; and on the opposite shore is a village, Myrmecium; and near by are Heracleium and Parthenium.[1]

7. Thence ninety stadia to the monument of Satyrus, which consists of a mound thrown up on a certain cape in memory of one of the illustrious potentates of the Bosporus.[2]

8. Near by is a village, Patraeus, from which the distance to a village Corocondamê is one hundred and thirty stadia; and this village constitutes the limit of the Cimmerian Bosporus, as it is called. The Narrows at the mouth of the Maeotis are so called from the narrow passage at Achilleium and Myrmecium; they extend as far as Corocondamê and the small village named Acra, which lies opposite to it in the land of the Panticapaeans, this village

[1] Cf. 7. 4. 5. [2] See 7. 4. 4.

---

[4] ῎Ακραν, Meineke, for ῎Ακρα; Corais and others insert ᾧ before ὄνομα.

κοντα σταδίων διειργόμενον πορθμῷ· μέχρι γὰρ δεῦρο καὶ ὁ κρύσταλλος διατείνει, πηττομένης τῆς Μαιώτιδος κατὰ τοὺς κρυμούς, ὥστε πεζεύεσθαι. ἅπας δ' ἐστὶν εὐλίμενος ὁ στενωπὸς οὗτος.

9. Ὑπέρκειται δὲ τῆς Κοροκονδάμης εὐμεγέθης λίμνη, ἣν καλοῦσιν ἀπ' αὐτῆς Κοροκονδαμῖτιν· ἐκδίδωσι δ' ἀπὸ δέκα σταδίων τῆς κώμης εἰς τὴν θάλατταν· ἐμβάλλει δὲ εἰς τὴν λίμνην ἀπορρώξ τις τοῦ Ἀντικείτου ποταμοῦ, καὶ ποιεῖ νῆσον περίκλυστόν τινα ταύτῃ τε τῇ λίμνῃ καὶ τῇ Μαιώτιδι καὶ τῷ ποταμῷ. τινὲς δὲ καὶ τοῦτον τὸν ποταμὸν Ὕπανιν προσαγορεύουσι, καθάπερ καὶ τὸν πρὸς τῷ Βορυσθένει.

10. Εἰσπλεύσαντι δ' εἰς τὴν Κοροκονδαμῖτιν ἥ C 495 τε Φαναγόρειά ἐστι, πόλις ἀξιόλογος, καὶ Κῆποι καὶ Ἑρμώνασσα καὶ τὸ Ἀπάτουρον, τὸ τῆς Ἀφροδίτης ἱερόν· ὧν ἡ Φαναγόρεια καὶ οἱ Κῆποι κατὰ τὴν λεχθεῖσαν νῆσον ἵδρυνται, εἰσπλέοντι ἐν ἀριστερᾷ, αἱ δὲ λοιπαὶ πόλεις ἐν δεξιᾷ πέραν Ὑπάνιος ἐν τῇ Σινδικῇ. ἔστι δὲ καὶ Γοργιπία[1] ἐν τῇ Σινδικῇ, τὸ βασίλειον τῶν Σινδῶν, πλησίον θαλάττης, καὶ Ἀβοράκη. τοῖς δὲ τοῦ Βοσπόρου δυνάσταις ὑπήκοοι ὄντες ἅπαντες Βοσπορανοὶ καλοῦνται· καὶ ἔστι τῶν μὲν Εὐρωπαίων Βοσπορανῶν μητρόπολις τὸ Παντικάπαιον, τῶν δ' Ἀσιανῶν τὸ Φαναγόρειον (καλεῖται γὰρ καὶ οὕτως ἡ πόλις), καὶ δοκεῖ τῶν μὲν ἐκ τῆς Μαιώτιδος καὶ τῆς ὑπερκειμένης βαρβάρου κατακομιζομένων ἐμπόριον εἶναι ἡ[2] Φαναγόρεια, τῶν δ' ἐκ τῆς θαλάττης

[1] Γοργίπια, Kramer, for Γοργίππια.
[2] ἡ, *xz* and Corais (ἡ Φαναγορία), instead of τά.

being separated from it by a strait seventy stadia wide; for the ice, also,[1] extends as far as this, the Maeotis being so frozen at the time of frosts that it can be crossed on foot. And these Narrows have good harbours everywhere.

9. Above Corocondamê lies a lake of considerable size, which derives its name, Corocondamitis, from that of the village. It empties into the sea at a distance of ten stadia from the village. A branch of the Anticeites empties into the lake and forms a kind of island which is surrounded by this lake and the Maeotis and the river. Some apply the name Hypanis to this river, just as they do to the river near the Borysthenes.

10. Sailing into Lake Corocondamitis one comes to Phanagoreia, a noteworthy city, and to Cepi, and to Hermonassa, and to Apaturum, the sanctuary of Aphroditê. Of these, Phanagoreia and Cepi are situated on the island above-mentioned, on the left as one sails in, but the other cities are on the right, across the Hypanis, in the Sindic territory. There is also a place called Gorgipia in the Sindic territory, the royal residence of the Sindi, near the sea; and also a place called Aboracê. All the people who are subject to the potentates of the Bosporus are called Bosporians; and Panticapaeum is the metropolis of the European Bosporians, while Phanagoreium (for the name of the city is also spelled thus) is the metropolis of the Asiatic Bosporians. Phanagoreia is reputed to be the emporium for the commodities that are brought down from the Maeotis and the barbarian country that lies above it, and Panti-

[1] *i.e.* as well as the Narrows.

ἀναφερομένων ἐκεῖσε τὸ Παντικάπαιον. ἔστι δὲ καὶ ἐν τῇ Φαναγορείᾳ τῆς Ἀφροδίτης ἱερὸν ἐπίσημον τῆς Ἀπατούρου· ἐτυμολογοῦσι δὲ τὸ ἐπίθετον τῆς θεοῦ μῦθόν τινα προστησάμενοι, ὡς, ἐπιθεμένων ἐνταῦθα τῇ θεῷ τῶν Γιγάντων, ἐπικαλεσαμένη τὸν Ἡρακλέα κρύψειεν[1] ἐν κευθμῶνί τινι, εἶτα τῶν Γιγάντων ἕκαστον δεχομένη καθ' ἕνα τῷ Ἡρακλεῖ παραδιδοίη δολοφονεῖν ἐξ ἀπάτης.

11. Τῶν Μαιωτῶν δ' εἰσὶν αὐτοί τε οἱ Σινδοὶ καὶ Δανδάριοι καὶ Τορεάται[2] καὶ Ἄγροι καὶ Ἀρρηχοί, ἔτι δὲ Τάρπητες, Ὀβιδιακηνοί, Σιττακηνοί, Δόσκοι, ἄλλοι πλείους· τούτων δ' εἰσὶ καὶ οἱ Ἀσπουργιανοί, μεταξὺ Φαναγορείας[3] οἰκοῦντες καὶ Γοργιπίας ἐν πεντακοσίοις σταδίοις, οἷς ἐπιθέμενος Πολέμων ὁ βασιλεὺς ἐπὶ προσποιήσει φιλίας, οὐ λαθὼν ἀντεστρατηγήθη καὶ ζωγρίᾳ ληφθεὶς ἀπέθανε. τῶν τε συμπάντων Μαιωτῶν τῶν Ἀσιανῶν οἱ μὲν ὑπήκουον τῶν τὸ ἐμπόριον ἐχόντων τὸ ἐν τῷ Ταναΐδι, οἱ δὲ τῶν Βοσπορανῶν· τότε δ' ἀφίσταντο ἄλλοτ' ἄλλοι. πολλάκις δ' οἱ τῶν Βοσπορανῶν ἡγεμόνες καὶ τὰ μέχρι τοῦ Ταναΐδος κατεῖχον, καὶ μάλιστα οἱ ὕστατοι, Φαρνάκης καὶ Ἄσανδρος καὶ Πολέμων. Φαρνάκης δέ ποτε καὶ τὸν Ὕπανιν τοῖς Δανδαρίοις ἐπαγαγεῖν λέγεται διά τινος παλαιᾶς διώρυγος, ἀνακαθάρας αὐτήν, καὶ[4] κατακλύσαι τὴν χώραν.

12. Μετὰ δὲ τὴν Σινδικὴν καὶ τὴν Γοργιπίαν

[1] κρύψειεν z, instead of κρύψει, κρύψοι, κρύψαι, κρύψι other MSS.

[2] Τορεάται is probably an error for Τορέται.

[3] Φαναγορείας, Meineke, for Φαναγορίας.

capaeum for those which are carried up thither from the sea. There is also in Phanagoreia a notable temple of Aphroditê Apaturus. Critics derive the etymology of the epithet of the goddess by adducing a certain myth, according to which the Giants attacked the goddess there; but she called upon Heracles for help and hid him in a cave, and then, admitting the Giants one by one, gave them over to Heracles to be murdered through "treachery."[1]

11. Among the Maeotae are the Sindi themselves, Dandarii, Toreatae, Agri, and Arrechi, and also the Tarpetes, Obidiaceni, Sittaceni, Dosci, and several others. Among these belong also the Aspurgiani, who live between Phanagoreia and Gorgipia, within a stretch of five hundred stadia; these were attacked by King Polemon under a pretence of friendship, but they discovered his pretence, outgeneralled him, and taking him alive killed him. As for the Asiatic Maeotae in general, some of them were subjects of those who possessed the emporium on the Tanaïs, and the others of the Bosporians; but in those days different peoples at different times were wont to revolt. And often the rulers of the Bosporians held possession of the region as far as the Tanaïs, and particularly the latest rulers, Pharnaces, Asander, and Polemon. Pharnaces is said at one time actually to have conducted the Hypanis River over the country of the Dandarii through an old canal which he cleared out, and to have inundated the country.

12. After the Sindic territory and Gorgipia, on

[1] In Greek, "apatê."

---

[4] καί, before κατακλύσαι, Casaubon inserts; so the later editors.

ἐπὶ τῇ θαλάττῃ ἡ[1] τῶν Ἀχαιῶν καὶ Ζυγῶν καὶ Ἡνιόχων παραλία, τὸ πλέον ἀλίμενος καὶ ὀρεινή, τοῦ Καυκάσου μέρος οὖσα. ζῶσι δὲ ἀπὸ τῶν κατὰ θάλατταν λῃστηρίων, ἀκάτια ἔχοντες λεπτά, στενὰ καὶ κοῦφα, ὅσον ἀνθρώπους πέντε καὶ εἴκοσι δεχόμενα, σπάνιον δὲ τριάκοντα δέξασθαι τοὺς πάντας δυνάμενα· καλοῦσι δ' αὐτὰ οἱ Ἕλληνες καμάρας. φασὶ δ' ἀπὸ τῆς Ἰάσονος στρατιᾶς τοὺς μὲν Φθιώτας Ἀχαιοὺς τὴν ἐνθάδε Ἀχαΐαν οἰκίσαι, Λάκωνας δὲ τὴν Ἡνιοχίαν, ὧν C 496 ἦρχον Ῥέκας[2] καὶ Ἀμφίστρατος, οἱ τῶν Διοσκούρων ἡνίοχοι, καὶ τοὺς Ἡνιόχους ἀπὸ τούτων εἰκὸς ὠνομάσθαι. τῶν δ' οὖν καμαρῶν στόλους κατασκευαζόμενοι καὶ ἐπιπλέοντες τοτὲ μὲν ταῖς ὁλκάσι, τοτὲ δὲ χώρᾳ τινὶ[3] ἢ καὶ πόλει θαλαττοκρατοῦσι. προσλαμβάνουσι δ' ἔσθ' ὅτε καὶ οἱ τὸν Βόσπορον ἔχοντες, ὑφόρμους χορηγοῦντες καὶ ἀγορὰν καὶ διάθεσιν τῶν ἁρπαζομένων· ἐπανιόντες δὲ εἰς τὰ οἰκεῖα χωρία, ναυλοχεῖν οὐκ ἔχοντες, ἀναθέμενοι τοῖς ὤμοις τὰς καμάρας ἀναφέρουσιν ἐπὶ τοὺς δρυμούς, ἐν οἷσπερ καὶ οἰκοῦσι, λυπρὰν ἀροῦντες γῆν· καταφέρουσι δὲ πάλιν, ὅταν ᾖ καιρὸς τοῦ πλεῖν. τὸ δ' αὐτὸ ποιοῦσι καὶ ἐν τῇ ἀλλοτρίᾳ, γνώριμα ἔχοντες ὑλώδη χωρία, ἐν οἷς ἀποκρύψαντες τὰς καμάρας αὐτοὶ πλανῶνται πεζῇ[4] νύκτωρ καὶ μεθ' ἡμέραν ἀνδραποδισμοῦ

[1] ἡ, after θαλάττῃ, Xylander, for τῇ; so the later editors.
[2] Meineke emends Ῥέκας to Κρέκας (see critical notes of Kramer and C. Müller).
[3] τινί is found only in *Clowz*.
[4] *lowz* have πεζοί instead of πεζῇ.

the sea, one comes to the coast of the Achaei and the Zygi and the Heniochi, which for the most part is harbourless and mountainous, being a part of the Caucasus. These peoples live by robberies at sea. Their boats are slender, narrow, and light, holding only about twenty-five people, though in rare cases they can hold thirty in all; the Greeks call them "camarae."[1] They say that the Phthiotic Achaei[2] in Jason's crew settled in this Achaea, but the Laconians in Heniochia, the leaders of the latter being Rhecas[3] and Amphistratus, the "heniochi"[4] of the Dioscuri,[5] and that in all probability the Heniochi were named after these. At any rate, by equipping fleets of "camarae" and sailing sometimes against merchant-vessels and sometimes against a country or even a city, they hold the mastery of the sea. And they are sometimes assisted even by those who hold the Bosporus, the latter supplying them with mooring-places, with market-place, and with means of disposing of their booty. And since, when they return to their own land, they have no anchorage, they put the "camarae" on their shoulders and carry them to the forests where they live and where they till a poor soil. And they bring the "camarae" down to the shore again when the time for navigation comes. And they do the same thing in the countries of others, for they are well acquainted with wooded places; and in these they first hide their "camarae" and then themselves wander on foot night and day for the sake of kidnapping

[1] *i.e.* "covered boats" (cf. Lat. and English "camera"). See the description of Tacitus (*Hist.* 3. 47).

[2] Cf. 9. 5. 10.

[3] Apparently an error for "Crecas."

[4] "charioteers."

[5] Castor and Pollux.

χάριν. ἃ δ' ἂν λάβωσιν ἐπίλυτρα ποιοῦσι ῥᾳδίως, μετὰ τοὺς ἀνάπλους μηνύοντες τοῖς ἀπολέσασιν. ἐν μὲν οὖν τοῖς δυναστευομένοις τόποις ἐστί τις βοήθεια ἐκ τῶν ἡγεμόνων τοῖς ἀδικουμένοις· ἀντεπιτίθενται γὰρ πολλάκις καὶ κατάγουσιν αὐτάνδρους τὰς καμάρας· ἡ δ' ὑπὸ Ῥωμαίοις ἀβοηθητοτέρα ἐστὶ διὰ τὴν ὀλιγωρίαν τῶν πεμπομένων.

13. Τοιοῦτος μὲν ὁ τούτων βίος· δυναστεύονται δὲ καὶ οὗτοι ὑπὸ τῶν καλουμένων σκηπτούχων· καὶ αὐτοὶ δὲ οὗτοι ὑπὸ τυράννοις ἢ βασιλεῦσίν εἰσιν. οἱ γοῦν Ἡνίοχοι τέτταρας εἶχον βασιλέας, ἡνίκα Μιθριδάτης ὁ Εὐπάτωρ, φεύγων ἐκ τῆς προγονικῆς εἰς Βόσπορον, διῄει τὴν χώραν αὐτῶν· καὶ αὕτη μὲν ἦν πορεύσιμος αὐτῷ, τῆς δὲ τῶν Ζυγῶν[1] ἀπογνοὺς διά τε δυσχερείας καὶ ἀγριότητας τῇ παραλίᾳ χαλεπῶς ᾔει, τὰ[2] πολλὰ ἐμβαίνων ἐπὶ τὴν θάλατταν, ἕως ἐπὶ τὴν τῶν Ἀχαιῶν ἧκε· καὶ προσλαβόντων τούτων ἐξετέλεσε τὴν ὁδὸν τὴν ἐκ Φάσιδος, οὐ πολὺ τῶν τετρακισχιλίων λείπουσαν σταδίων.

14. Εὐθὺς δ' οὖν ἀπὸ τῆς Κοροκονδάμης πρὸς ἕω μὲν ὁ πλοῦς ἐστίν. ἐν δὲ σταδίοις ἑκατὸν ὀγδοήκοντα ὁ Σινδικός ἐστι λιμὴν καὶ πόλις, εἶτα ἐν τετρακοσίοις τὰ καλούμενα Βατά, κώμη καὶ λιμήν, καθ' ὃ μάλιστα ἀντικεῖσθαι δοκεῖ πρὸς νότον ἡ Σινώπη ταύτῃ τῇ παραλίᾳ, καθάπερ ἡ Κάραμβις εἴρηται τοῦ Κριοῦ μετώπῳ· ἀπὸ δὲ

[1] Ζυγῶν (as spelled elsewhere by Strabo), Meineke, for Ζυγίων.
[2] τά should probably be ejected from the text.

people. But they readily offer to release their captives for ransom, informing their relatives after they have put out to sea. Now in those places which are ruled by local chieftains the rulers go to the aid of those who are wronged, often attacking and bringing back the "camarae," men and all. But the territory that is subject to the Romans affords but little aid, because of the negligence of the governors who are sent there.

13. Such is the life of these people. They are governed by chieftains called "sceptuchi,"[1] but the "sceptuchi" themselves are subject to tyrants or kings. For instance, the Heniochi had four kings at the time when Mithridates Eupator,[2] in flight from the country of his ancestors to the Bosporus, passed through their country; and while he found this country passable, yet he despaired of going through that of the Zygi, both because of the ruggedness of it and because of the ferocity of the inhabitants; and only with difficulty could he go along the coast, most of the way marching on the edge of the sea, until he arrived at the country of the Achaei; and, welcomed by these, he completed his journey from Phasis, a journey not far short of four thousand stadia.

14. Now the voyage from Corocondamê is straight towards the east; and at a distance of one hundred and eighty stadia is the Sindic harbour and city; and then, at a distance of four hundred stadia, one comes to Bata, as it is called, a village and harbour, at which place Sinopê on the south is thought to lie almost directly opposite this coast, just as Carambis has been referred to as opposite Criume-

[1] "Sceptre-bearers" (see note on "sceptuchies," § 18 below).

[2] See *Dictionary* in Vol. I.

τῶν Βατῶν ὁ μὲν Ἀρτεμίδωρος τὴν Κερκετῶν λέγει παραλίαν, ὑφόρμους ἔχουσαν καὶ κώμας, ὅσον ἐπὶ σταδίους ὀκτακοσίους καὶ πεντήκοντα· εἶτα τὴν τῶν Ἀχαιῶν σταδίων πεντακοσίων, εἶτα τὴν τῶν Ἡνιόχων χιλίων, εἶτα τὸν Πιτυοῦντα
C 497 τὸν μέγαν τριακοσίων ἑξήκοντα μέχρι Διοσκουριάδος. οἱ δὲ τὰ Μιθριδατικὰ συγγράψαντες, οἷς μᾶλλον προσεκτέον, Ἀχαιοὺς λέγουσι πρώτους, εἶτα Ζυγούς, εἶτα Ἡνιόχους, εἶτα Κερκέτας καὶ Μόσχους καὶ Κόλχους καὶ τοὺς ὑπὲρ τούτων Φθειροφάγους καὶ Σοάνας [1] καὶ ἄλλα μικρὰ ἔθνη τὰ περὶ τὸν Καύκασον. κατ᾽ ἀρχὰς μὲν οὖν ἡ παραλία, καθάπερ εἶπον, ἐπὶ τὴν ἕω τείνει καὶ βλέπει πρὸς νότον, ἀπὸ δὲ τῶν Βατῶν ἐπιστροφὴν λαμβάνει κατὰ μικρόν, εἶτ᾽ ἀντιπρόσωπος γίνεται τῇ δύσει καὶ τελευτᾷ πρὸς τὸν Πιτυοῦντα καὶ τὴν Διοσκουριάδα· ταῦτα γὰρ τὰ χωρία τῆς Κολχίδος συνάπτει τῇ λεχθείσῃ παραλίᾳ. μετὰ δὲ τὴν Διοσκουριάδα ἡ λοιπὴ τῆς Κολχίδος ἐστὶ παραλία καὶ ἡ συνεχὴς Τραπεζοῦς, καμπὴν ἀξιόλογον ποιήσασα· εἶτα εἰς εὐθεῖαν ταθεῖσά πως πλευρὰν τὴν τὰ δεξιὰ τοῦ Πόντου ποιοῦσαν, τὰ βλέποντα πρὸς ἄρκτον. ἅπασα δ᾽ ἡ τῶν Ἀχαιῶν καὶ τῶν ἄλλων παραλία μέχρι Διοσκουριάδος καὶ τῶν ἐπ᾽ εὐθείας πρὸς νότον ἐν τῇ μεσογαίᾳ τόπων ὑποπέπτωκε τῷ Καυκάσῳ.

15. Ἔστι δ᾽ ὄρος τοῦτο ὑπερκείμενον τοῦ πελάγους ἑκατέρου, τοῦ τε Ποντικοῦ καὶ τοῦ Κασπίου, διατείχιζον τὸν ἰσθμὸν τὸν διείργοντα αὐτά. ἀφορίζει δὲ πρὸς νότον μὲν τήν τε Ἀλβανίαν καὶ τὴν Ἰβηρίαν, πρὸς ἄρκτον δὲ τὰ τῶν Σαρματῶν πεδία· εὔδενδρον δ᾽ ἐστὶν ὕλῃ παντο-

topon.[1] After Bata Artemidorus[2] mentions the coast of the Cercetae, with its mooring-places and villages, extending thence about eight hundred and fifty stadia; and then the coast of the Achaei, five hundred stadia; and then that of the Heniochi, one thousand; and then Greater Pityus, extending three hundred and sixty stadia to Dioscurias. The more trustworthy historians of the Mithridatic wars name the Achaei first, then the Zygi, then the Heniochi, and then the Cercetae and Moschi and Colchi, and the Phtheirophagi who live above these three peoples and the Soanes, and other small tribes that live in the neighbourhood of the Caucasus. Now at first the coast, as I have said, stretches towards the east and faces the south, but from Bata it gradually takes a turn, and then faces the west and ends at Pityus and Dioscurias; for these places border on the above-mentioned coast of Colchis. After Dioscurias comes the remaining coast of Colchis and the adjacent coast of Trapezus, which makes a considerable bend, and then, extending approximately in a straight line, forms the right-hand side of the Pontus, which faces the north. The whole of the coast of the Achaei and of the other peoples as far as Dioscurias and of the places that lie in a straight line towards the south in the interior lie at the foot of the Caucasus.

15. This mountain lies above both seas, both the Pontic and the Caspian, and forms a wall across the isthmus that separates the two seas. It marks the boundary, on the south, of Albania and Iberia, and, on the north, of the plains of the Sarmatae. It is

[1] See 2. 5. 22 and 7. 4. 3. [2] See *Dictionary* in Vol. II.

[1] Σοάνας, Tzschucke from conj. of Casaubon, for Θοάνας; so the later editors.

δαπῇ, τῇ τε ἄλλῃ καὶ τῇ ναυπηγησίμῳ. φησὶ δ' Ἐρατοσθένης ὑπὸ τῶν ἐπιχωρίων καλεῖσθαι Κάσπιον τὸν Καύκασον, ἴσως ἀπὸ τῶν Κασπίων παρονομασθέντα. ἀγκῶνες δέ τινες αὐτοῦ προπίπτουσιν ἐπὶ τὴν μεσημβρίαν, οἳ τήν τε Ἰβηρίαν περιλαμβάνουσι μέσην καὶ τοῖς Ἀρμενίων ὄρεσι συνάπτουσι καὶ τοῖς Μοσχικοῖς καλουμένοις, ἔτι δὲ τῷ Σκυδίσῃ καὶ τῷ Παρυάδρῃ· ταῦτα δ' ἐστὶ μέρη τοῦ Ταύρου πάντα, τοῦ ποιοῦντος τὸ νότιον τῆς Ἀρμενίας πλευρόν, ἀπερρωγότα πως ἐκεῖθεν πρὸς ἄρκτον καὶ προπίπτοντα[1] μέχρι τοῦ Καυκάσου καὶ τῆς τοῦ Εὐξείνου παραλίας, τῆς ἐπὶ Θεμίσκυραν διατεινούσης ἀπὸ τῆς Κολχίδος.

16. Ἡ δ' οὖν Διοσκουριὰς ἐν κόλπῳ τοιούτῳ κειμένη καὶ τὸ ἑωθινώτατον σημεῖον ἐπέχουσα τοῦ σύμπαντος πελάγους, μυχός τε τοῦ Εὐξείνου λέγεται καὶ ἔσχατος πλοῦς· τό τε παροιμιακῶς λεχθὲν

εἰς Φᾶσιν, ἔνθα ναυσὶν ἔσχατος δρόμος,

οὕτω δεῖ δέξασθαι, οὐχ ὡς τὸν ποταμὸν λέγοντος τοῦ ποιήσαντος τὸ ἰάμβειον, οὐδὲ δὴ ὡς τὴν ὁμώνυμον αὐτῷ πόλιν κειμένην ἐπὶ τῷ ποταμῷ, ἀλλ' ὡς τὴν Κολχίδα ἀπὸ μέρους, ἐπεὶ ἀπό γε τοῦ ποταμοῦ καὶ τῆς πόλεως οὐκ ἐλάττων ἑξακοσίων σταδίων λείπεται πλοῦς ἐπ' εὐθείας εἰς C 498 τὸν μυχόν. ἡ δ' αὐτὴ Διοσκουριάς ἐστι καὶ ἀρχὴ τοῦ ἰσθμοῦ τοῦ μεταξὺ τῆς Κασπίας καὶ τοῦ Πόντου καὶ ἐμπόριον τῶν ὑπερκειμένων καὶ σύνεγγυς ἐθνῶν κοινόν· συνέρχεσθαι γοῦν εἰς αὐτὴν ἑβδομήκοντα, οἱ δὲ καὶ τριακόσια ἔθνη

[1] προπίπτοντα, Niese, for προσπίπτοντα; so Meineke.

well wooded with all kinds of timber, and especially the kind suitable for ship-building. According to Eratosthenes, the Caucasus is called "Caspius" by the natives the name being derived perhaps from the "Caspi." Branches of it project towards the south; and these not only comprise the middle of Albania but also join the mountains of Armenia and the Moschian Mountains, as they are called, and also the Scydises and the Paryadres Mountains. All these are parts of the Taurus, which forms the southern side of Armenia,—parts broken off, as it were, from that mountain on the north and projecting as far as the Caucasus and that part of the coast of the Euxine which stretches from Colchis to Themiscyra.

16. Be this as it may, since Dioscurias is situated in such a gulf and occupies the most easterly point of the whole sea, it is called not only the recess of the Euxine but also the "farthermost" voyage. And the proverbial verse, "To Phasis, where for ships is the farthermost run," must be interpreted thus, not as though the author[1] of the iambic verse meant the river, much less the city of the same name situated on the river, but as meaning by a part of Colchis the whole of it since from the river and the city of that name there is left a straight voyage into the recess of not less than six hundred stadia. The same Dioscurias is the beginning of the isthmus between the Caspian Sea and the Euxine, and also the common emporium of the tribes who are situated above it and in its vicinity; at any rate, seventy tribes come together in it, though others, who care nothing for the facts, actually say three hundred.

[1] An unknown tragic poet (*Adesp.* 559, Nauck).

φασίν, οἷς οὐδὲν τῶν ὄντων μέλει. πάντα δὲ ἑτερόγλωττα διὰ τὸ σποράδην καὶ ἀμίκτως οἰκεῖν ὑπὸ αὐθαδείας καὶ ἀγριότητος· Σαρμάται δ' εἰσὶν οἱ πλείους, πάντες δὲ Καυκάσιοι. ταῦτα μὲν δὴ τὰ περὶ τὴν Διοσκουριάδα.

17. Καὶ ἡ λοιπὴ δὲ Κολχὶς ἐπὶ τῇ θαλάττῃ ἡ πλείων ἐστί· διαρρεῖ δ' αὐτὴν ὁ Φᾶσις, μέγας ποταμὸς ἐξ Ἀρμενίας τὰς ἀρχὰς ἔχων, δεχόμενος τόν τε Γλαῦκον καὶ τὸν Ἵππον, ἐκ τῶν πλησίον ὀρῶν ἐκπίπτοντας· ἀναπλεῖται δὲ μέχρι Σαραπανῶν, ἐρύματος δυναμένου δέξασθαι καὶ πόλεως συνοικισμόν, ὅθεν πεζεύουσιν ἐπὶ τὸν Κῦρον ἡμέραις τέτταρσι δι' ἁμαξιτοῦ. ἐπίκειται δὲ τῷ Φάσιδι ὁμώνυμος πόλις, ἐμπόριον τῶν Κόλχων, τῇ μὲν προβεβλημένη τὸν ποταμόν, τῇ δὲ λίμνην, τῇ δὲ τὴν θάλατταν. ἐντεῦθεν δὲ πλοῦς ἐπ' Ἀμισοῦ καὶ Σινώπης τριῶν ἡμερῶν ἢ δύο[1] διὰ τὸ τοὺς αἰγιαλοὺς μαλακοὺς εἶναι καὶ τὰς τῶν ποταμῶν ἐκβολάς. ἀγαθὴ δ' ἐστὶν ἡ χώρα καὶ καρποῖς πλὴν τοῦ μέλιτος (πικρίζει γὰρ τὸ πλέον) καὶ τοῖς πρὸς ναυπηγίαν πᾶσι· πολλήν τε γὰρ ὕλην[2] φύει καὶ ποταμοῖς κατακομίζει, λινόν τε ποιεῖ πολὺ καὶ κάνναβιν καὶ κηρὸν καὶ πίτταν. ἡ δὲ λινουργία καὶ τεθρύληται· καὶ γὰρ εἰς τοὺς ἔξω τόπους ἐπεκόμιζον, καί τινες βουλόμενοι συγγένειάν τινα τοῖς Κόλχοις πρὸς τοὺς Αἰγυπτίους

[1] τριῶν ἡμερῶν ἢ δύο ("three or two days") cannot be right, since, according to Strabo (12. 3 17) the distance from Phasis to Amisus is 3600 stadia. Gosselin, Groskurd, and Kramer think that the copyists confused γ' (3) and β' (2) with η' (8) and θ (9). C. Müller thinks that the β' has been confused with δ' (4), and would emend ἡμερῶν to νυχθημερῶν.

All speak different languages because of the fact that, by reason of their obstinacy and ferocity, they live in scattered groups and without intercourse with one another. The greater part of them are Sarmatae, but they are all Caucasii. So much, then, for the region of Dioscurias.

17. Further, the greater part of the remainder of Colchis is on the sea. Through it flows the Phasis, a large river having its sources in Armenia and receiving the waters of the Glaucus and the Hippus, which issue from the neighbouring mountains. It is navigated as far as Sarapana, a fortress capable of admitting the population even of a city. From here people go by land to the Cyrus in four days by a wagon-road. On the Phasis is situated a city bearing the same name, an emporium of the Colchi, which is protected on one side by the river, on another by a lake, and on another by the sea. Thence people go to Amisus and Sinopê by sea (a voyage of two or three days), because the shores are soft and because of the outlets of the rivers. The country is excellent both in respect to its produce—except its honey, which is generally bitter—and in respect to everything that pertains to ship-building; for it not only produces quantities of timber but also brings it down on rivers. And the people make linen in quantities, and hemp, wax, and pitch. Their linen industry has been famed far and wide; for they used to export linen to outside places; and some writers, wishing to show forth a kinship between the Colchians and the

---

[2] ὕλην, Jones inserts, following conj. of Kramer, and also, following *x*, omits καὶ before φύει.

ἐμφανίζειν ἀπὸ τούτων πιστοῦνται. ὑπέρκειται δὲ τῶν λεχθέντων ποταμῶν ἐν τῇ Μοσχικῇ τὸ τῆς Λευκοθέας ἱερόν, Φρίξου ἵδρυμα, καὶ μαντεῖον ἐκείνου, ὅπου κριὸς οὐ θύεται, πλούσιόν ποτε ὑπάρξαν, συληθὲν δὲ ὑπὸ Φαρνάκου καθ' ἡμᾶς, καὶ μικρὸν ὕστερον ὑπὸ Μιθριδάτου τοῦ Περγαμηνοῦ· κακωθείσης γὰρ χώρας,

νοσεῖ τὰ τῶν θεῶν, οὐδὲ τιμᾶσθαι θέλει,

φησὶν Εὐριπίδης.[1]

18. Τὸ μὲν γὰρ παλαιὸν ὅσην ἐπιφάνειαν ἔσχεν ἡ χώρα αὕτη, δηλοῦσιν οἱ μῦθοι, τὴν Ἰάσονος στρατείαν αἰνιττόμενοι προελθόντος μέχρι καὶ Μηδίας, ἔτι δὲ πρότερον τὴν Φρίξου. μετὰ δὲ ταῦτα διαδεξάμενοι βασιλεῖς εἰς σκηπτουχίας διῃρημένην ἔχοντες τὴν χώραν μέσως ἔπραττον· αὐξηθέντος δὲ ἐπὶ πολὺ Μιθριδάτου τοῦ Εὐπάτορος, εἰς ἐκεῖνον ἡ χώρα περιέστη· ἐπέμπετο C 499 δ' ἀεί τις τῶν φίλων ὕπαρχος καὶ διοικητὴς τῆς χώρας. τούτων δὲ ἦν καὶ Μοαφέρνης, ὁ τῆς μητρὸς ἡμῶν θεῖος πρὸς πατρός· ἦν δ' ἔνθεν ἡ πλείστη τῷ βασιλεῖ πρὸς τὰς ναυτικὰς δυνάμεις ὑπουργία. καταλυθέντος δὲ Μιθριδάτου, συγκατελύθη καὶ ἡ ὑπ' αὐτῷ πᾶσα καὶ διενεμήθη πολλοῖς· ὕστατα δὲ Πολέμων ἔσχε τὴν Κολχίδα, κἀκείνου τελευτήσαντος ἡ γυνὴ Πυθοδωρὶς κρατεῖ, βασιλεύουσα καὶ Κόλχων καὶ Τραπεζοῦντος καὶ Φαρνακίας καὶ τῶν ὑπερκειμένων βαρβάρων, περὶ ὧν ἐροῦμεν ἐν τοῖς ὕστερον. ἡ

---

[1] *Troades* 26.

Egyptians, confirm their belief by this. Above the aforesaid rivers in the Moschian country lies the temple of Leucothea, founded by Phrixus, and the oracle of Phrixus, where a ram is never sacrificed; it was once rich, but it was robbed in our time by Pharnaces, and a little later by Mithridates of Pergamum. For when a country is devastated, "things divine are in sickly plight and wont not even to be respected," says Euripides.[1]

18. The great fame this country had in early times is disclosed by the myths, which refer in an obscure way to the expedition of Jason as having proceeded as far even as Media, and also, before that time, to that of Phrixus. After this, when kings succeeded to power, the country being divided into "sceptuchies,"[2] they were only moderately prosperous; but when Mithridates Eupator[3] grew powerful, the country fell into his hands; and he would always send one of his friends as sub-governor or administrator of the country. Among these was Moaphernes, my mother's uncle on her father's side. And it was from this country that the king received most aid in the equipment of his naval forces. But when the power of Mithridates had been broken up, all the territory subject to him was also broken up and distributed among many persons. At last Polemon got Colchis; and since his death his wife Pythodoris has been in power, being queen, not only of the Colchians, but also of Trapezus and Pharnacia and of the barbarians who live above these places, concerning whom I shall speak later on.[4] Now the Moschian country, in

[2] *i.e.* divisions corresponding to the rank of Persian "sceptuchi" ("sceptre-bearers").

[3] See *Dictionary* in Vol. I.

[4] 12. 3. 28 ff.

δ' οὖν Μοσχική, ἐν ᾗ τὸ ἱερόν, τριμερής ἐστι· τὸ μὲν γὰρ ἔχουσιν αὐτῆς Κόλχοι, τὸ δὲ Ἴβηρες, τὸ δὲ Ἀρμένιοι. ἔστι δὲ καὶ πολίχνιον ἐν τῇ Ἰβηρίᾳ, Φρίξου πόλις, ἡ νῦν Ἰδήεσσα, εὐερκὲς χωρίον, ἐν μεθορίοις τῆς Κολχίδος. περὶ δὲ[1] τὴν Διοσκουριάδα ῥεῖ ὁ Χάρης[2] ποταμός.

19. Τῶν δὲ συνερχομένων ἐθνῶν εἰς τὴν Διοσκουριάδα καὶ οἱ Φθειροφάγοι εἰσίν, ἀπὸ τοῦ αὐχμοῦ καὶ τοῦ πίνου λαβόντες τοὔνομα. πλησίον δὲ καὶ οἱ Σοάνες, οὐδὲν βελτίους τούτων τῷ πίνῳ, δυνάμει δὲ βελτίους, σχεδὸν δέ τι καὶ κράτιστοι κατὰ ἀλκὴν καὶ δύναμιν· δυναστεύουσι γοῦν τῶν[3] κύκλῳ, τὰ ἄκρα τοῦ Καυκάσου κατέχοντες τὰ ὑπὲρ τῆς Διοσκουριάδος. βασιλέα δ' ἔχουσι καὶ συνέδριον ἀνδρῶν τριακοσίων, συνάγουσι δ', ὥς φασι, στρατιὰν[4] καὶ εἴκοσι μυριάδων· ἅπαν γάρ ἐστι τὸ πλῆθος μάχιμον, οὐ συντεταγμένον· παρὰ τούτοις δὲ λέγεται καὶ χρυσὸν καταφέρειν τοὺς χειμάρρους, ὑποδέχεσθαι δ' αὐτὸν τοὺς βαρβάρους φάτναις κατατετρημέναις καὶ μαλλωταῖς δοραῖς· ἀφ' οὗ δὴ μεμυθεῦσθαι καὶ τὸ χρυσόμαλλον δέρος· εἰ μὴ[5] καὶ Ἴβηρας ὁμωνύμως τοῖς ἑσπερίοις καλοῦσιν ἀπὸ τῶν ἑκατέρωθι χρυσείων. χρῶνται δ' οἱ Σοάνες φαρμάκοις πρὸς τὰς ἀκίδας θαυμαστοῖς,[6]

[1] δέ, after περί, Casaubon adds from *rw*; so the later editors in general.

[2] CD*hi* have ῥιοχάρης instead of ῥεῖ ὁ Χάρης; but Meineke ejects the whole sentence.

[3] τῶν, Casaubon, for τῷ MSS., except C, which has τά; so the later editors.

[4] στρατιάν, Corais, for στρατείαν; so the later editors.

[5] εἰ μή seems to be corrupt. Kramer proposes ἔνιοι.

which is situated the temple,[1] is divided into three parts: one part is held by the Colchians, another by the Iberians, and another by the Armenians. There is also a small city in Iberia, the city of Phrixus,[2] the present Ideëssa, well fortified, on the confines of Colchis. And near Dioscurias flows the Chares River.

19. Among the tribes which come together at Dioscurias are the Phtheirophagi,[3] who have received their name from their squalor and their filthiness. Near them are the Soanes, who are no less filthy, but superior to them in power,—indeed, one might almost say that they are foremost in courage and power. At any rate, they are masters of the peoples around them, and hold possession of the heights of the Caucasus above Dioscurias. They have a king and a council of three hundred men; and they assemble, according to report, an army of two hundred thousand; for the whole of the people are a fighting force, though unorganised. It is said that in their country gold is carried down by the mountain-torrents, and that the barbarians obtain it by means of perforated troughs and fleecy skins, and that this is the origin of the myth of the golden fleece—unless they call them Iberians, by the same name as the western Iberians, from the gold mines in both countries. The Soanes use remarkable poisons for the points of their missiles; and even people who

[1] Of Leucothea § 17 above).
[2] Phrixopolis.
[3] "Lice-eaters."

---

[6] θαυμαστοῖς, Casaubon, for θαυμαστῶς; so Kramer and Müller-Dübner.

ἃ[1] καὶ τοὺς μὴ[2] φαρμακτοῖς[3] τετρωμένους βέλεσι λυπεῖ κατὰ τὴν ὀσμήν. τὰ μὲν οὖν ἄλλα ἔθνη τὰ πλησίον τὰ περὶ τὸν Καύκασον λυπρὰ καὶ μικρόχωρα, τὸ δὲ τῶν Ἀλβανῶν ἔθνος καὶ τὸ τῶν Ἰβήρων, ἃ δὴ πληροῖ μάλιστα τὸν λεχθέντα ἰσθμόν, Καυκάσια καὶ αὐτὰ λέγοιτ' ἄν, εὐδαίμονα δὲ χώραν ἔχει καὶ σφόδρα καλῶς οἰκεῖσθαι δυναμένην.

III

1. Καὶ δὴ καὶ ἥ γε Ἰβηρία κατοικεῖται[4] καλῶς τὸ πλέον πόλεσί τε καὶ ἐποικίοις, ὥστε καὶ κεραμωτὰς εἶναι στέγας καὶ ἀρχιτεκτονικὴν τὴν τῶν οἰκήσεων κατασκευὴν καὶ ἀγορὰς καὶ τἆλλα κοινά.

2. Τῆς δὲ χώρας τὰ μὲν κύκλῳ τοῖς Καυκασίοις C 500 ὄρεσι περιέχεται. προπεπτώκασι γάρ, ὡς εἶπον, ἀγκῶνες ἐπὶ τὴν μεσημβρίαν εὔκαρποι, περιλαμβάνοντες τὴν σύμπασαν Ἰβηρίαν καὶ συνάπτοντες πρός τε τὴν Ἀρμενίαν καὶ τὴν Κολχίδα· ἐν μέσῳ δ' ἐστὶ πεδίον ποταμοῖς διάρρυτον, μεγίστῳ δὲ τῷ Κύρῳ· ὃς τὴν ἀρχὴν ἔχων ἀπὸ τῆς Ἀρμενίας, εἰσβαλὼν εὐθὺς εἰς τὸ πεδίον τὸ λεχθέν, παραλαβὼν καὶ τὸν Ἄραγον, ἐκ[5] τοῦ Καυκάσου ῥέοντα, καὶ ἄλλα ὕδατα, διὰ στενῆς ποταμίας εἰς τὴν Ἀλβανίαν ἐκπίπτει· μεταξὺ δὲ ταύτης τε καὶ τῆς Ἀρμενίας ἐνεχθεὶς πολὺς

[1] ἅ, Casaubon inserts; so Kramer and Müller-Dübner.
[2] μή, Jones inserts, on suggestion of Professor Capps.
[3] φαρμακτοῖς, Corais, for ἀφαρμακτοῖς; so Kramer and Müller-Dübner.

are not wounded by the poisoned missiles suffer from their odour. Now in general the tribes in the neighbourhood of the Caucasus occupy barren and cramped territories, but the tribes of the Albanians and the Iberians, which occupy nearly all the isthmus above-mentioned, might also be called Caucasian tribes; and they possess territory that is fertile and capable of affording an exceedingly good livelihood.

III

1. FURTHERMORE, the greater part of Iberia is so well built up in respect to cities and farmsteads that their roofs are tiled, and their houses as well as their market-places and other public buildings are constructed with architectural skill.

2. Parts of the country are surrounded by the Caucasian Mountains; for branches of these mountains, as I said before,[1] project towards the south; they are fruitful, comprise the whole of Iberia, and border on both Armenia and Colchis. In the middle is a plain intersected by rivers, the largest being the Cyrus. This river has its beginning in Armenia, flows immediately into the plain above-mentioned, receives both the Aragus, which flows from the Caucasus, and other streams, and empties through a narrow valley into Albania; and between the valley and Armenia it flows in great volume

[1] 11. 2. 15.

[4] κατοικεῖται, Meineke, for καὶ οἰκεῖται; earlier editors merely omit the καί.

[5] Ἄραγον (see § 5 following) ἐκ, Corais, for Ἀραγῶνα κάτω; so Meineke.

διὰ πεδίων εὐβοτουμένων σφόδρα, δεξάμενος καὶ πλείους ποταμούς, ὧν ἐστὶν ὅ τε Ἀλαζόνιος καὶ ὁ Σανδοβάνης καὶ ὁ Ῥοιτάκης καὶ Χάνης, πλωτοὶ πάντες, εἰς τὴν Κασπίαν ἐμβάλλει[1] θάλατταν. ἐκαλεῖτο δὲ πρότερον Κόρος.

3. Τὸ μὲν οὖν πεδίον τῶν Ἰβήρων οἱ γεωργικώτεροι καὶ πρὸς εἰρήνην νενευκότες οἰκοῦσιν, Ἀρμενιστί τε καὶ Μηδιστὶ ἐσκευασμένοι, τὴν δ' ὀρεινὴν οἱ πλείους καὶ μάχιμοι κατέχουσι, Σκυθῶν δίκην ζῶντες καὶ Σαρματῶν, ὧνπερ καὶ ὅμοροι καὶ συγγενεῖς εἰσίν· ἅπτονται δ' ὅμως καὶ γεωργίας, πολλάς τε μυριάδας συνάγουσιν καὶ ἐξ ἑαυτῶν καὶ ἐξ ἐκείνων, ἐπειδάν τι συμπέσῃ θορυβῶδες.

4. Τέτταρες δ' εἰσὶν εἰς τὴν χώραν εἰσβολαί· μία μὲν διὰ Σαραπανῶν, φρουρίου Κολχικοῦ, καὶ τῶν κατ' αὐτὸ στενῶν, δι' ὧν ὁ Φᾶσις γεφύραις ἑκατὸν καὶ εἴκοσι περατὸς γενόμενος διὰ τὴν σκολιότητα καταρρεῖ τραχὺς καὶ βίαιος εἰς τὴν Κολχίδα, πολλοῖς χειμάρροις κατὰ τὰς ἐπομβρίας ἐκχαραδρουμένων τῶν τόπων. γεννᾶται δ' ἐκ τῶν ὑπερκειμένων ὀρῶν πολλαῖς συμπληρούμενος πηγαῖς, ἐν δὲ τοῖς πεδίοις καὶ ἄλλους προσλαμβάνει ποταμούς, ὧν ἐστὶν ὅ τε Γλαῦκος καὶ ὁ Ἵππος· πληρωθεὶς δὲ καὶ γενόμενος πλωτὸς ἐξίησιν εἰς τὸν Πόντον καὶ ἔχει πόλιν ὁμώνυμον ἐπ' αὐτῷ καὶ λίμνην πλησίον. ἡ μὲν οὖν ἐκ τῆς Κολχίδος εἰς τὴν Ἰβηρίαν ἐμβολὴ τοιαύτη, πέτραις καὶ ἐρύμασι καὶ ποταμοῖς χαραδρώδεσι διακεκλεισμένη.

[1] ἐμβάλλει *oz Epit.*; ἐμβάλλουσι other MSS.

through plains that have exceedingly good pasture, receives still more rivers, among which are the Alazonius, Sandobanes, Rhoetaces, and Chanes, all navigable, and empties into the Caspian Sea. It was formerly called Corus.

3. Now the plain of the Iberians is inhabited by people who are rather inclined to farming and to peace, and they dress after both the Armenian and the Median fashion; but the major, or warlike, portion occupy the mountainous territory, living like the Scythians and the Sarmatians, of whom they are both neighbours and kinsmen; however, they engage also in farming. And they assemble many tens of thousands, both from their own people and from the Scythians and Sarmatians, whenever anything alarming occurs.

4. There are four passes leading into their country; one through Sarapana, a Colchian stronghold, and through the narrow defiles there. Through these defiles the Phasis, which has been made passable by one hundred and twenty bridges because of the windings of its course, flows down into Colchis with rough and violent stream, the region being cut into ravines by many torrents at the time of the heavy rains. The Phasis rises in the mountains that lie above it, where it is supplied by many springs; and in the plains it receives still other rivers, among which are the Glaucus and the Hippus. Thus filled and having by now become navigable, it issues forth into the Pontus; and it has on its banks a city bearing the same name; and near it is a lake. Such, then, is the pass that leads from Colchis into Iberia, being shut in by rocks, by strongholds, and by rivers that run through ravines.

5. Ἐκ δὲ τῶν πρὸς ἄρκτον νομάδων ἐπὶ τρεῖς ἡμέρας ἀνάβασις χαλεπή, καὶ μετὰ ταύτην ποταμία στενὴ ἐπὶ τοῦ Ἀράγου ποταμοῦ τεττάρων ἡμερῶν ὁδὸν ἔχουσα ἐφ' ἕνα, φρουρεῖ δὲ τὸ πέρας τῆς ὁδοῦ τεῖχος δύσμαχον· ἀπὸ δὲ τῆς Ἀλβανίας διὰ πέτρας πρῶτον λατομητὴ εἴσοδος, εἶτα διὰ τέλματος, ὃ ποιεῖ ὁ ποταμὸς Ἀλαζόνιος[1] ἐκ τοῦ Καυκάσου καταπίπτων· ἀπὸ δὲ τῆς Ἀρμενίας τὰ ἐπὶ τῷ Κύρῳ στενὰ καὶ τὰ ἐπὶ τῷ Ἀράγῳ. πρὶν γὰρ εἰς ἀλλήλους συμπεσεῖν, ἔχουσιν ἐπικειμένας πόλεις ἐρυμνὰς ἐπὶ πέτραις, C 501 διεχούσαις ἀλλήλων ὅσον ἑκκαίδεκα σταδίους, ἐπὶ μὲν τῷ Κύρῳ τὴν Ἁρμοζικήν, ἐπὶ δὲ θατέρῳ Σευσάμορα. ταύταις δὲ ἐχρήσατο ταῖς εἰσβολαῖς πρότερον Πομπήιος ἐκ τῶν Ἀρμενίων ὁρμηθείς, καὶ μετὰ ταῦτα Κανίδιος.

6. Τέτταρα δὲ καὶ γένη τῶν ἀνθρώπων οἰκεῖ τὴν χώραν· ἓν μὲν καὶ πρῶτον, ἐξ οὗ τοὺς βασιλέας καθιστᾶσι, κατ' ἀγχιστείαν τε καὶ ἡλικίαν τὸν πρεσβύτατον, ὁ δὲ δεύτερος δικαιοδοτεῖ καὶ στρατηλατεῖ· δεύτερον δὲ τὸ τῶν ἱερέων,[2] οἳ ἐπιμελοῦνται καὶ τῶν πρὸς τοὺς ὁμόρους δικαίων· τρίτον δὲ τὸ τῶν στρατευομένων καὶ γεωργούντων· τέταρτον δὲ τὸ τῶν λαῶν, οἳ βασιλικοὶ δοῦλοί εἰσι καὶ πάντα διακονοῦνται τὰ πρὸς τὸν βίον. κοιναὶ δ' εἰσὶν αὐτοῖς αἱ κτήσεις κατὰ συγγένειαν, ἄρχει δὲ καὶ ταμιεύει ἑκάστην ὁ πρεσβύτατος. τοιοῦτοι μὲν οἱ Ἴβηρες καὶ ἡ χώρα αὐτῶν.

[1] Ἀλαζόνιος, Groskurd inserts; so the later editors.
[2] ἱερέων, Xylander, for ἱερῶν; so the later editors.

---

[1] Crassus the Triumvir.
[2] *i.e.* as well as four passes leading into the country (see § 4, beginning).

5. From the country of the nomads on the north there is a difficult ascent into Iberia requiring three days' travel; and after this ascent comes a narrow valley on the Aragus River, with a single-file road requiring a four days' journey. The end of the road is guarded by a fortress which is hard to capture. The pass leading from Albania into Iberia is at first hewn through rock, and then leads through a marsh formed by the River Alazonius, which falls from the Caucasus. The passes from Armenia into Iberia are the defiles on the Cyrus and those on the Aragus. For, before the two rivers meet, they have on their banks fortified cities that are situated upon rocks, these being about sixteen stadia distant from each other—I mean Harmozicê on the Cyrus and Seusamora on the other river. These passes were used first by Pompey when he set out from the country of the Armenians, and afterwards by Canidius.[1]

6. There are also[2] four castes among the inhabitants of Iberia. One, and the first of all, is that from which they appoint their kings, the appointee being both the nearest of kin to his predecessor and the eldest, whereas the second in line administers justice and commands the army. The second caste is that of the priests, who among other things attend to all matters of controversy with the neighbouring peoples. The third is that of the soldiers and the farmers. And the fourth is that of the common people, who are slaves of the king and perform all the services that pertain to human livelihood. Their possessions are held in common by them according to families, although the eldest is ruler and steward of each estate. Such are the Iberians and their country.

IV

1. Ἀλβανοὶ δὲ ποιμενικώτεροι καὶ τοῦ νομαδικοῦ γένους ἐγγυτέρω, πλὴν ἀλλ' οὐκ ἄγριοι·[1] ταύτῃ δὲ καὶ πολεμικοὶ μετρίως. οἰκοῦσι δὲ μεταξὺ τῶν Ἰβήρων καὶ τῆς Κασπίας θαλάττης, πρὸς ἕω μὲν ἁπτόμενοι τῆς θαλάττης, πρὸς δύσιν δὲ ὁμοροῦντες τοῖς Ἴβηρσι· τῶν δὲ λοιπῶν πλευρῶν τὸ μὲν βόρειον φρουρεῖται τοῖς Καυκασίοις ὄρεσι (ταῦτα γὰρ ὑπέρκειται τῶν πεδίων, καλεῖται δὲ τὰ πρὸς τῇ θαλάττῃ μάλιστα Κεραύνια), τὸ δὲ νότιον ποιεῖ ἡ Ἀρμενία παρήκουσα, πολλὴ μὲν πεδιάς, πολλὴ δὲ καὶ ὀρεινή, καθάπερ ἡ Καμβυσηνή, καθ' ἣν ἅμα καὶ τοῖς Ἴβηρσι καὶ τοῖς Ἀλβανοῖς οἱ Ἀρμένιοι συνάπτουσιν.

2. Ὁ δὲ Κῦρος ὁ διαρρέων τὴν Ἀλβανίαν καὶ οἱ ἄλλοι ποταμοὶ οἱ πληροῦντες ἐκεῖνον ταῖς μὲν τῆς γῆς ἀρεταῖς προσλαμβάνουσι, τὴν δὲ θάλατταν ἀλλοτριοῦσιν, ἡ γὰρ χοῦς προσπίπτουσα πολλὴ πληροῖ τὸν πόρον, ὥστε καὶ τὰς ἐπικειμένας νησῖδας ἐξηπειροῦσθαι καὶ τενάγη ποιεῖν ἀνώμαλα καὶ δυσφύλακτα, τὴν δ' ἀνωμαλίαν ἐπιτείνουσιν αἱ ἐκ τῶν πλημμυρίδων ἀνακοπαί. καὶ δὴ καὶ εἰς στόματα δώδεκά φασι μεμερίσθαι τὰς ἐκβολάς, τὰ μὲν τυφλά, τὰ δὲ παντελῶς ἐπίπεδα ὄντα[2] καὶ μηδὲ[3] ὕφορμον ἀπολείποντα· ἐπὶ πλείους γοῦν ἢ ἑξήκοντα σταδίους ἀμφι-

[1] ἀλλ' οὐκ ἄγριοι, Meineke from conj. of Kramer, for ἀλλότριοι.

[2] For ἐπιγελῶντα Meineke and C. Müller conj. ἐπίπεδα ὄντα. ἐπίγεια ὄντα conj. Tyrwhitt, ἐπιπόλαια ὄντα Corais, ἐπίπλεα ὄντα Kramer.

IV

1. The Albanians are more inclined to the shepherd's life than the Iberians and closer akin to the nomadic people, except that they are not ferocious; and for this reason they are only moderately warlike. They live between the Iberians and the Caspian Sea, their country bordering on the sea towards the east and on the country of the Iberians towards the west. Of the remaining sides the northern is protected by the Caucasian Mountains (for these mountains lie above the plains, though their parts next to the sea are generally called Ceraunian), whereas the southern side is formed by Armenia, which stretches alongside it; and much of Armenia consists of plains, though much of it is mountainous, like Cambysenê, where the Armenians border on both the Iberians and the Albanians.

2. The Cyrus, which flows through Albania, and the other rivers by which it is supplied, contribute to the excellent qualities of the land; and yet they thrust back the sea, for the silt, being carried forward in great quantities, fills the channel, and consequently even the adjacent isles are joined to the mainland and form shoals that are uneven and difficult to avoid; and their unevenness is made worse by the back-wash of the flood-tides. Moreover, they say that the outlet of the river is divided into twelve mouths, of which some are choked with silt, while the others are altogether shallow and leave not even a mooring-place. At any rate, they add, although the shore is washed on all sides by the sea

---

[3] μηδέ, Kramer, for μηδέν; so the later editors.

κλύστου τῆς ᾐόνος οὔσης τῇ θαλάττῃ καὶ τοῖς ποταμοῖς, ἅπαν εἶναι μέρος αὐτῆς ἀπροσπέλαστον, τὴν δὲ χοῦν καὶ μέχρι πεντακοσίων παρήκειν σταδίων, θινώδη ποιοῦσαν τὸν αἰγιαλόν. πλησίον δὲ καὶ ὁ Ἀράξης ἐμβάλλει, τραχὺς ἐκ τῆς Ἀρμενίας ἐκπίπτων· ἣν δὲ ἐκεῖνος προωθεῖ χοῦν, πορευτὸν ποιῶν τὸ ῥεῖθρον, ταύτην ὁ Κῦρος ἀναπληροῖ.

3. Τάχα μὲν οὖν τῷ τοιούτῳ γένει τῶν ἀνθρώ-
C 502 πων οὐδὲν δεῖ θαλάττης· οὐδὲ γὰρ τῇ γῇ χρῶνται κατ' ἀξίαν, πάντα μὲν[1] ἐκφερούσῃ καρπόν, καὶ τὸν ἡμερώτατον, πᾶν δὲ φυτόν· καὶ γὰρ τὰ ἀειθαλῆ φέρει· τυγχάνει δ' ἐπιμελείας οὐδὲ μικρᾶς, ἀλλὰ τἀγαθὰ ἄσπαρτα καὶ ἀνήροτα ἅπαντα φύονται, καθάπερ οἱ στρατεύσαντές φασι, Κυκλώπειόν τινα διηγούμενοι βίον· πολλαχοῦ γοῦν σπαρεῖσαν ἅπαξ δὶς ἐκφέρειν καρπὸν ἢ καὶ τρίς, τὸν δὲ πρῶτον καὶ πεντηκοντάχουν, ἀνέαστον καὶ ταῦτα, οὐδὲ σιδήρῳ τμηθεῖσαν, ἀλλ' αὐτοξύλῳ ἀρότρῳ. ποτίζεται δὲ πᾶν τὸ πεδίον τοῦ Βαβυλωνίου καὶ τοῦ Αἰγυπτίου μᾶλλον τοῖς ποταμοῖς καὶ τοῖς ἄλλοις ὕδασιν, ὥστ' ἀεὶ ποώδη φυλάττειν τὴν ὄψιν· διὰ δὲ τοῦτο καὶ εὔβοτόν ἐστι· πρόσεστι δὲ καὶ τὸ εὐάερον ἐκείνῳ μᾶλλον. ἄσκαφοι δὲ ἄμπελοι μένουσαι διὰ τέλους, τεμνόμεναι δὲ[2] διὰ πενταετηρίδος, νέαι μὲν διετεῖς

[1] γάρ, after μέν, is omitted by *oxz* and the later editors.
[2] δέ, D *man. pr.* inserts after τεμνόμεναι; so Meineke.

---

[1] *i.e.* the excessive amount of silt deposited by the Cyrus compensates for the failure of the Araxes in this respect. On these rivers see Tozer, *Selections*, pp. 262–263.

and the rivers for a distance of more than sixty stadia, every part of it is inaccessible; and the silt extends even as far as five hundred stadia, making the shore sandy. Near by is also the mouth of the Araxes, a turbulent stream that flows down from Armenia. But the silt which this river pushes before it, thus making the channel passable for its stream, is compensated for by the Cyrus.[1]

3. Now perhaps a people of this kind have no need of a sea; indeed, they do not make appropriate use of their land either, which produces, not only every kind of fruit, even the most highly cultivated kind, but also every plant, for it bears even the evergreens. It receives not even slight attention, yet the good things all "spring up for them without sowing and ploughing,"[2] according to those who have made expeditions there,[3] who describe the mode of life there as "Cyclopeian." In many places, at any rate, they say, the land when sown only once produces two crops or even three, the first a crop of even fifty-fold, and that too without being ploughed between crops; and even when it is ploughed, it is not ploughed with an iron share, but with a wooden plough shaped by nature. The plain as a whole is better watered by its rivers and other waters than the Babylonian and the Egyptian plains; consequently it always keeps a grassy appearance, and therefore is also good for pasturage. In addition to this, the climate here is better than there. And the people never dig about the vines, although they prune them every fifth year;[4] the new vines begin

[2] *Odyssey* 9. 109.
[3] In particular Theophanes of Mitylenê (already mentioned in 11. 2. 2). [4] *i.e.* every *four* years.

ἐκφέρουσιν ἤδη καρπόν, τέλειαι δ' ἀποδιδόασι τοσοῦτον, ὥστ' ἀφιᾶσιν ἐν τοῖς κλήμασι πολὺ μέρος. εὐερνῆ δ' ἐστὶ καὶ τὰ βοσκήματα παρ' αὐτοῖς τά τε ἥμερα καὶ τὰ ἄγρια.

4. Καὶ οἱ ἄνθρωποι κάλλει καὶ μεγέθει διαφέροντες, ἁπλοῖ δὲ καὶ οὐ καπηλικοί· οὐδὲ γὰρ νομίσματι τὰ πολλὰ χρῶνται, οὐδὲ ἀριθμὸν ἴσασι μείζω[1] τῶν ἑκατόν, ἀλλὰ φορτίοις τὰς ἀμοιβὰς ποιοῦνται, καὶ πρὸς τἆλλα δὲ τὰ τοῦ βίου ῥᾳθύμως ἔχουσιν. ἄπειροι δ' εἰσὶ καὶ μέτρων τῶν ἐπ' ἀκριβὲς καὶ σταθμῶν, καὶ πολέμου δὲ καὶ πολιτείας καὶ γεωργίας ἀπρονοήτως ἔχουσιν· ὅμως δὲ καὶ πεζοὶ καὶ ἀφ' ἵππων ἀγωνίζονται, ψιλοί τε καὶ κατάφρακτοι, καθάπερ Ἀρμένιοι.

5. Στέλλουσι δὲ μείζω τῆς Ἰβήρων στρατιάν·[2] ὁπλίζουσι γὰρ ἓξ μυριάδας πεζῶν,[3] ἱππέας δὲ δισμυρίους[4] καὶ δισχιλίους, ὅσοις πρὸς Πομπήιον διεκινδύνευσαν. καὶ τούτοις δὲ συμπολεμοῦσιν οἱ νομάδες πρὸς τοὺς ἔξωθεν, ὥσπερ τοῖς Ἴβηρσι κατὰ τὰς αὐτὰς αἰτίας· ἄλλως δ' ἐπιχειροῦσι τοῖς ἀνθρώποις πολλάκις, ὥστε καὶ γεωργεῖν κωλύουσιν. ἀκοντισταὶ δέ εἰσι καὶ τοξόται, θώρακας ἔχοντες καὶ θυρεούς, περίκρανα δὲ θήρεια παραπλησίως τοῖς Ἴβηρσιν. ἔστι δὲ τῆς Ἀλβανῶν χώρας καὶ ἡ Κασπιανή, τοῦ Κασπίου

[1] E, and Eustath. (*ad Dion.* 730), have πλείω instead of μείζω.
[2] στρατιάν, Meineke, foll. conj. of Villebrun, for στρατιᾶς.
[3] πεζῶν E*g*, ἀνδρῶν other MSS.
[4] Plutarch has μυρίους (*Pomp.* 35).

---

[1] See § 8 following.

to produce fruit the second year, and when mature they yield so much that the people leave a large part of the fruit on the branches. Also the cattle in their country thrive, both the tame and the wild.

4. The inhabitants of this country are unusually handsome and large. And they are frank in their dealings, and not mercenary;[1] for they do not in general use coined money, nor do they know any number greater than one hundred, but carry on business by means of barter, and otherwise live an easy-going life. They are also unacquainted with accurate measures and weights, and they take no forethought for war or government or farming. But still they fight both on foot and on horseback, both in light armour and in full armour,[2] like the Armenians.[3]

5. They send forth a greater army than that of the Iberians; for they equip sixty thousand infantry and twenty-two thousand[4] horsemen, the number with which they risked their all against Pompey. Against outsiders the nomads join with the Albanians in war, just as they do with the Iberians, and for the same reasons; and besides, they often attack the people, and consequently prevent them from farming. The Albanians use javelins and bows; and they wear breastplates and large oblong shields, and helmets made of the skins of wild animals, similar to those worn by the Iberians. To the country of the Albanians belongs also the territory called Caspianê, which was named after the Caspian

[2] For a description of this heavy armour, see Tacitus, *Hist.* 1. 79.

[3] Cf. 11. 14. 9.

[4] Plutarch, *Pompey* 35, says twelve thousand.

ἔθνους ἐπώνυμος, οὗπερ καὶ ἡ θάλαττα, ἀφανοῦς ὄντος νυνί. ἡ δ' ἐκ τῆς Ἰβηρίας εἰς τὴν Ἀλβανίαν εἰσβολὴ διὰ τῆς Καμβυσηνῆς ἀνύδρου τε καὶ τραχείας ἐπὶ τὸν Ἀλαζόνιον ποταμόν. θηρευτικοὶ δὲ καὶ αὐτοὶ καὶ οἱ κύνες αὐτῶν εἰς ὑπερβολήν, οὐ τέχνῃ μᾶλλον ἢ σπουδῇ τῇ περὶ τοῦτο.

C 503 6. Διαφέρουσι δὲ καὶ οἱ βασιλεῖς· νυνὶ μὲν οὖν εἷς ἁπάντων ἄρχει, πρότερον δὲ καὶ καθ' ἑκάστην γλῶτταν ἰδίᾳ ἐβασιλεύοντο ἕκαστοι. γλῶτται δ' εἰσὶν ἓξ καὶ εἴκοσι αὐτοῖς διὰ τὸ μὴ εὐεπίμικτον πρὸς ἀλλήλους. φέρει δ' ἡ γῆ καὶ τῶν ἑρπετῶν ἔνια τῶν θανασίμων καὶ σκορπίους καὶ φαλάγγια· τῶν δὲ φαλαγγίων τὰ μὲν ποιεῖ γελῶντας ἀποθνήσκειν, τὰ δὲ κλαίοντας πόθῳ τῶν οἰκείων.

7. Θεοὺς δὲ τιμῶσιν Ἥλιον καὶ Δία καὶ Σελήνην, διαφερόντως δὲ τὴν Σελήνην. ἔστι δ' αὐτῆς τὸ ἱερὸν τῆς Ἰβηρίας πλησίον· ἱερᾶται δ' ἀνὴρ ἐντιμότατος μετά γε τὸν βασιλέα, προεστὼς τῆς ἱερᾶς χώρας, πολλῆς καὶ εὐάνδρου, καὶ αὐτῆς καὶ τῶν ἱεροδούλων, ὧν ἐνθουσιῶσι πολλοὶ καὶ προφητεύουσιν· ὃς δ' ἂν αὐτῶν ἐπὶ πλέον κατάσχετος γενόμενος πλανᾶται κατὰ τὰς ὕλας μόνος, τοῦτον συλλαβὼν ὁ ἱερεὺς ἁλύσει δήσας ἱερᾷ τρέφει πολυτελῶς τὸν ἐνιαυτὸν ἐκεῖνον, ἔπειτα προαχθεὶς εἰς τὴν θυσίαν τῆς θεοῦ, σὺν ἄλλοις ἱερείοις θύεται μυρισθείς. τῆς δὲ θυσίας ὁ τρόπος οὗτος· ἔχων τις ἱερὰν λόγχην, ᾗπερ

---

[1] Members of the spider family; but here, apparently, tarantulas (see Tozer, *op. cit.*, p. 265).

[2] The Sun. [3] The Moon.

[4] Cf. 12. 3. 31.

tribe, as was also the sea; but the tribe has now disappeared. The pass from Iberia into Albania leads through Cambysenê, a waterless and rugged country, to the Alazonius River. Both the people and their dogs are surpassingly fond of hunting, engaging in it not so much because of their skill in it as because of their love for it.

6. Their kings, also, are excellent. At the present time, indeed, one king rules all the tribes, but formerly the several tribes were ruled separately by kings of their own according to their several languages. They have twenty-six languages, because of the fact that they have no easy means of intercourse with one another. The country produces also certain of the deadly reptiles, and scorpions and phalangia.[1] Some of the phalangia cause people to die laughing, while others cause people to die weeping over the loss of their deceased kindred.

7. As for gods, they honour Helius,[2] Zeus, and Selenê,[3] but especially Selenê;[4] her temple is near Iberia. The office of priest is held by the man who, after the king, is held in highest honour; he has charge of the sacred land, which is extensive and well-populated, and also of the temple slaves, many of whom are subject to religious frenzy and utter prophecies. And any one of those who, becoming violently possessed, wanders alone in the forests, is by the priest arrested, bound with sacred fetters, and sumptuously maintained during that year, and then led forth to the sacrifice that is performed in honour of the goddess, and, being anointed, is sacrificed along with other victims. The sacrifice is performed as follows: Some person holding a sacred lance, with which it is the custom to sacrifice human

ἐστι νομος ἀνθρωποθυτεῖν, παρελθὼν ἐκ τοῦ πλήθους, παίει διὰ τῆς πλευρᾶς εἰς τὴν καρδίαν, οὐκ ἄπειρος τοιούτου· πεσόντος δὲ σημειοῦνται μαντεῖά τινα ἐκ τοῦ πτώματος καὶ εἰς τὸ κοινὸν ἀποφαίνουσι· κομισθέντος δὲ τοῦ σώματος εἴς τι χωρίον, ἐπιβαίνουσιν ἅπαντες καθαρσίῳ χρώμενοι.

8. Ὑπερβαλλόντως δὲ καὶ[1] τὸ γῆρας τιμῶσιν Ἀλβανοί, καὶ τὸ τῶν ἄλλων, οὐ τῶν γονέων μόνον· τεθνηκότων δὲ οὐχ ὅσιον φροντίζειν οὐδὲ μεμνῆσθαι. συγκατορύττουσι μέντοι τὰ χρήματα αὐτοῖς, καὶ διὰ τοῦτο πένητες ζῶσιν, οὐδὲν πατρῷον ἔχοντες. ταῦτα μὲν περὶ Ἀλβανῶν. λέγεται δ' Ἰάσονα μετὰ Ἀρμένου[2] τοῦ Θετταλοῦ κατὰ τὸν πλοῦν τὸν ἐπὶ τοὺς Κόλχους ὁρμῆσαι μέχρι τῆς Κασπίας θαλάττης, καὶ τήν τε Ἰβηρίαν καὶ τὴν Ἀλβανίαν ἐπελθεῖν καὶ πολλὰ τῆς Ἀρμενίας καὶ τῆς Μηδίας, ὡς μαρτυρεῖ τά τε Ἰασόνια καὶ ἄλλα ὑπομνήματα πλείω. τὸν δὲ Ἄρμενον[3] εἶναι ἐξ Ἀρμενίου πόλεως, τῶν περὶ τὴν Βοιβηίδα λίμνην μεταξὺ Φερῶν καὶ Λαρίσης· τοὺς σὺν αὐτῷ τε οἰκίσαι τήν τε Ἀκιλισηνὴν καὶ τὴν Συσπιρῖτιν ἕως Καλαχανῆς καὶ Ἀδιαβηνῆς, καὶ δὴ καὶ τὴν Ἀρμενίαν ἐπώνυμον καταλιπεῖν.

[1] Corais and Meineke eject the καί before τὸ γῆρας.
[2] Ἀρμένου, the editors, for Ἀρμενίου (cp. 11. 14. 12), and so five lines below.
[3] Ἀρμένου, Tzschucke and later editors (Eustath. on *Iliad* 2. 734 reads Ὀρμένου), for Ἀρμενίου.

victims, comes forward out of the crowd and strikes the victim through the side into the heart, he being not without experience in such a task; and when the victim falls, they draw auguries from his fall[1] and declare them before the public; and when the body is carried to a certain place, they all trample upon it, thus using it as a means of purification.

8. The Albanians are surpassingly respectful to old age, not merely to their parents, but to all other old people. And when people die it is impious to be concerned about them or even to mention them. Indeed, they bury their money with them, and therefore live in poverty, having no patrimony. So much for the Albanians. It is said that Jason, together with Armenus the Thessalian, on his voyage to the country of the Colchians, pressed on from there as far as the Caspian Sea, and visited, not only Iberia and Albania, but also many parts of Armenia and Media, as both the Jasonia[2] and several other memorials testify. And it is said that Armenus was a native of Armenium, one of the cities on Lake Boebeïs between Pherae and Larisa, and that he and his followers took up their abode in Acilisenê and Syspiritis, occupying the country as far as Calachanê and Adiabenê; and indeed that he left Armenia named after himself.

[1] As among the Lusitanians (3. 3. 6) and the Gauls (4. 4. 5).

[2] *i.e.* temples dedicated to Jason (see 11. 14. 12).

V

1. Ἐν δὲ τοῖς ὑπὲρ τῆς Ἀλβανίας ὄρεσι καὶ τὰς Ἀμαζόνας οἰκεῖν φασί. Θεοφάνης μὲν οὖν ὁ συστρατεύσας τῷ Πομπηίῳ καὶ γενόμενος ἐν τοῖς Ἀλβανοῖς, μεταξὺ τῶν Ἀμαζόνων καὶ τῶν Ἀλβανῶν φησὶ Γήλας οἰκεῖν καὶ Λήγας Σκύθας, καὶ ῥεῖν ἐνταῦθα τὸν Μερμάδαλιν ποταμὸν τούτων τε καὶ τῶν Ἀμαζόνων ἀνὰ μέσον. ἄλλοι δέ, C 504 ὧν καὶ ὁ Σκήψιος Μητρόδωρος καὶ Ὑψικράτης, οὐδὲ αὐτοὶ ἄπειροι τῶν τόπων γεγονότες, Γαργαρεῦσιν ὁμόρους αὐτὰς οἰκεῖν φασὶν ἐν ταῖς ὑπωρείαις ταῖς πρὸς ἄρκτον τῶν Καυκασίων ὀρῶν ἃ καλεῖται Κεραύνια· τὸν μὲν ἄλλον χρόνον καθ' αὐτὰς αὐτουργούσας ἕκαστα, τά τε πρὸς ἄροτον καὶ φυτουργίαν καὶ τὰ πρὸς τὰς νομάς, καὶ μάλιστα τῶν ἵππων, τὰς δ' ἀλκιμωτάτας ἐφ'[1] ἵππων κυνηγεσίαις πλεονάζειν καὶ τὰ πολέμια ἀσκεῖν· ἁπάσας δ' ἐπικεκαῦσθαι τὸν δεξιὸν μαστὸν ἐκ νηπίων, ὥστε εὐπετῶς χρῆσθαι τῷ βραχίονι πρὸς ἑκάστην χρείαν, ἐν δὲ τοῖς πρώτοις πρὸς ἀκοντισμόν· χρῆσθαι δὲ καὶ τόξῳ καὶ σαγάρι καὶ πέλτῃ, δορὰς δὲ θηρίων ποιεῖσθαι περίκρανά τε καὶ σκεπάσματα καὶ διαζώματα· δύο δὲ μῆνας ἐξαιρέτους ἔχειν τοῦ ἔαρος, καθ' οὓς ἀναβαίνουσιν εἰς τὸ πλησίον ὄρος τὸ διόριζον αὐτάς τε καὶ τοὺς Γαργαρέας. ἀναβαίνουσι δὲ κἀκεῖνοι κατὰ ἔθος τι παλαιόν, συνθύσοντές τε

[1] ἐφ' *l*(?)*oz* and the earlier editors for τῶν; Meineke ejects τῶν ἵππων.

---

[1] Cnaeus Pompeius Theophanes of Mytilenê.
[2] See 13. 1. 55. [3] See 11. 4. 1.

V

1. The Amazons, also, are said to live in the mountains above Albania. Now Theophanes,[1] who made the expedition with Pompey and was in the country of the Albanians, says that the Gelae and the Legae, Scythian people, live between the Amazons and the Albanians, and that the Mermadalis River flows there, midway between these people and the Amazons. But others, among whom are Metrodorus of Scepsis[2] and Hypsicrates, who themselves, likewise, were not unacquainted with the region in question, say that the Amazons live on the borders of the Gargarians, in the northerly foothills of those parts of the Caucasian Mountains which are called Ceraunian;[3] that the Amazons spend the rest of their time[4] off to themselves, performing their several individual tasks, such as ploughing, planting, pasturing cattle, and particularly in training horses, though the bravest engage mostly in hunting on horseback and practise warlike exercises; that the right breasts of all are seared when they are infants, so that they can easily use their right arm for every needed purpose, and especially that of throwing the javelin; that they also use bow and sagaris[5] and light shield, and make the skins of wild animals serve as helmets, clothing, and girdles; but that they have two special months in the spring in which they go up into the neighbouring mountain which separates them and the Gargarians. The Gargarians also, in accordance with an ancient custom, go up

[4] *i.e.* ten months of the year.
[5] Apparently some sort of single-edged weapon (see Hesychius *s.v.*).

καὶ συνεσόμενοι ταῖς γυναιξὶ τεκνοποιίας χάριν, ἀφανῶς τε καὶ ἐν σκότει, ὁ τυχὼν τῇ τυχούσῃ, ἐγκύμονας δὲ ποιήσαντες ἀποπέμπουσιν· αἱ δ' ὅ τι μὲν ἂν θῆλυ τέκωσι κατέχουσιν αὐταί, τὰ δ' ἄρρενα κομίζουσιν ἐκείνοις ἐκτρέφειν· ᾠκείωται δ' ἕκαστος πρὸς ἕκαστον, νομίζων υἱὸν διὰ τὴν ἄγνοιαν.

2. Ὁ δὲ Μερμόδας, καταράττων ἀπὸ τῶν ὀρῶν διὰ τῆς τῶν Ἀμαζόνων καὶ τῆς Σιρακηνῆς καὶ ὅση μεταξὺ ἔρημος, εἰς τὴν Μαιῶτιν ἐκδίδωσι. τοὺς δὲ Γαργαρέας συναναβῆναι μὲν ἐκ Θεμισκύρας φασὶ ταῖς Ἀμαζόσιν εἰς τούσδε τοὺς τόπους, εἶτ' ἀποστάντας αὐτῶν πολεμεῖν μετὰ Θρᾳκῶν καὶ Εὐβοέων τινῶν πλανηθέντων μέχρι δεῦρο πρὸς αὐτάς, ὕστερον δὲ καταλυσαμένους τὸν πρὸς αὐτὰς πόλεμον ἐπὶ τοῖς λεχθεῖσι ποιήσασθαι συμβάσεις, ὥστε τέκνων συγκοινωνεῖν μόνον, ζῆν δὲ καθ' αὑτοὺς ἑκατέρους.

3. Ἴδιον δέ τι συμβέβηκε τῷ λόγῳ περὶ τῶν Ἀμαζόνων· οἱ μὲν γὰρ ἄλλοι τὸ μυθῶδες καὶ τὸ ἱστορικὸν διωρισμένον ἔχουσι· τὰ γὰρ παλαιὰ καὶ ψευδῆ καὶ τερατώδη μῦθοι καλοῦνται, ἡ δ' ἱστορία βούλεται τἀληθές, ἄν τε παλαιὸν ἄν τε νέον, καὶ τὸ τερατῶδες ἢ οὐκ ἔχει ἢ σπάνιον· περὶ δὲ τῶν Ἀμαζόνων τὰ αὐτὰ λέγεται καὶ νῦν καὶ πάλαι,

---

[1] Apparently the same river as that called Mermadalis in the preceding paragraph.

thither to offer sacrifice with the Amazons and also to have intercourse with them for the sake of begetting children, doing this in secrecy and darkness, any Gargarian at random with any Amazon; and after making them pregnant they send them away; and the females that are born are retained by the Amazons themselves, but the males are taken to the Gargarians to be brought up; and each Gargarian to whom a child is brought adopts the child as his own, regarding the child as his son because of his uncertainty.

2. The Mermodas[1] dashes down from the mountains through the country of the Amazons and through Siracenê and the intervening desert and then empties into Lake Maeotis. It is said that the Gargarians went up from Themiscyra into this region with the Amazons, then revolted from them and in company with some Thracians and Euboeans who had wandered thus far carried on war against them, and that they later ended the war against them and made a compact on the conditions abovementioned, that is, that they should have dealings with one another only in the matter of children, and that each people should live independent of the other.

3. A peculiar thing has happened in the case of the account we have of the Amazons; for our accounts of other peoples keep a distinction between the mythical and the historical elements; for the things that are ancient and false and monstrous are called myths, but history wishes for the truth, whether ancient or recent, and contains no monstrous element, or else only rarely. But as regards the Amazons, the same stories are told now as in early

τερατώδη τε ὄντα καὶ πίστεως πόρρω. τίς γὰρ ἂν πιστεύσειεν ὡς γυναικῶν στρατὸς ἢ πόλις ἢ ἔθνος συσταίη ἄν ποτε χωρὶς ἀνδρῶν; καὶ οὐ μόνον γε συσταίη, ἀλλὰ καὶ ἐφόδους ποιήσαιτο ἐπὶ τὴν ἀλλοτρίαν καὶ κρατήσειεν οὐ τῶν ἐγγὺς μόνον, C 505 ὥστε καὶ μέχρι τῆς νῦν Ἰωνίας προελθεῖν, ἀλλὰ καὶ διαπόντιον στείλαιτο στρατείαν μέχρι τῆς Ἀττικῆς; τοῦτο γὰρ ὅμοιον, ὡς ἂν εἴ τις λέγοι, τοὺς μὲν ἄνδρας γυναῖκας γεγονέναι τοὺς τότε, τὰς δὲ γυναῖκας ἄνδρας. ἀλλὰ μὴν ταῦτά γε αὐτὰ καὶ νῦν λέγεται περὶ αὐτῶν, ἐπιτείνει δὲ τὴν ἰδιότητα καὶ τὸ πιστεύεσθαι τὰ παλαιὰ μᾶλλον ἢ τὰ νῦν.

4. Κτίσεις γοῦν πόλεων καὶ ἐπωνυμίαι λέγονται, καθάπερ Ἐφέσου καὶ Σμύρνης καὶ Κύμης καὶ Μυρίνης, καὶ τάφοι[1] καὶ ἄλλα ὑπομνήματα· τὴν δὲ Θεμίσκυραν καὶ τὰ περὶ τὸν Θερμώδοντα πεδία καὶ τὰ ὑπερκείμενα ὄρη ἅπαντες Ἀμαζόνων καλοῦσι, καί φασιν ἐξελαθῆναι αὐτὰς ἐνθένδε. ὅπου δὲ νῦν εἰσίν, ὀλίγοι τε καὶ ἀναποδείκτως καὶ ἀπίστως ἀποφαίνονται· καθάπερ καὶ περὶ Θαληστρίας, ἣν Ἀλεξάνδρῳ συμμῖξαί φασιν ἐν τῇ Ὑρκανίᾳ καὶ συγγενέσθαι τεκνοποιίας χάριν, δυναστεύουσαν[2] τῶν Ἀμαζόνων· οὐ γὰρ ὁμολογεῖται τοῦτο· ἀλλὰ τῶν συγγραφέων τοσούτων ὄντων, οἱ μάλιστα τῆς ἀληθείας φροντίσαντες οὐκ εἰρήκασιν, οὐδ' οἱ πιστευόμενοι μάλιστα οὐδενὸς μέμνηνται τοιούτου, οὐδ' οἱ εἰπόντες τὰ

[1] Instead of τάφοι, *Dhilrwx* have πάφου, *oz* πάφος, C πάφαι.

[2] δυναστεύουσαν, Casaubon, for δυναστεῦσαι *oxyz*, δυναστευσάντων other MSS.

times, though they are marvellous and beyond belief. For instance, who could believe that an army of women, or a city, or a tribe, could ever be organised without men, and not only be organised, but even make inroads upon the territory of other people, and not only overpower the peoples near them to the extent of advancing as far as what is now Ionia, but even send an expedition across the sea as far as Attica? For this is the same as saying that the men of those times were women and that the women were men. Nevertheless, even at the present time these very stories are told about the Amazons, and they intensify the peculiarity above-mentioned and our belief in the ancient accounts rather than those of the present time.

4. At any rate, the founding of cities and the giving of names to them are ascribed to the Amazons, as, for instance, Ephesus and Smyrna and Cymê and Myrinê; and so are tombs and other monuments; and Themiscyra and the plains about Thermodon and the mountains that lie above them are by all writers mentioned as having belonged to the Amazons; but they say that the Amazons were driven out of these places. Only a few writers make assertions as to where they are at the present time, but their assertions are without proof and beyond belief, as in the case of Thalestria, queen of the Amazons, with whom, they say, Alexander associated in Hyrcania and had intercourse for the sake of offspring; for this assertion is not generally accepted. Indeed, of the numerous historians, those who care most for the truth do not make the assertion, nor do those who are most trustworthy mention any such thing, nor do those

αὐτὰ εἰρήκασι· Κλείταρχος δέ[1] φησι τὴν Θαληστρίαν ἀπὸ Κασπίων πυλῶν καὶ Θερμώδοντος ὁρμηθεῖσαν ἐλθεῖν πρὸς Ἀλέξανδρον, εἰσὶ δ' ἀπὸ Κασπίας εἰς Θερμώδοντα στάδιοι πλείους ἑξακισχιλίων.

5. Καὶ τὰ πρὸς τὸ ἔνδοξον θρυληθέντα οὐκ ἀνωμολόγηται[2] παρὰ πάντων, οἱ δὲ πλάσαντες ἦσαν οἱ κολακείας μᾶλλον ἢ ἀληθείας φροντίζοντες· οἷον τὸ τὸν Καύκασον μετενεγκεῖν εἰς τὰ Ἰνδικὰ ὄρη καὶ τὴν πλησιάζουσαν ἐκείνοις ἑῴαν θάλατταν ἀπὸ τῶν ὑπερκειμένων τῆς Κολχίδος καὶ τοῦ Εὐξείνου ὀρῶν· ταῦτα γὰρ οἱ Ἕλληνες καὶ Καύκασον ὠνόμαζον, διέχοντα τῆς Ἰνδικῆς πλείους ἢ τρισμυρίους σταδίους, καὶ ἐνταῦθα ἐμύθευσαν τὰ περὶ Προμηθέα καὶ τὸν δεσμὸν αὐτοῦ· ταῦτα γὰρ τὰ ὕστατα πρὸς ἕω ἐγνώριζον οἱ τότε. ἡ δὲ ἐπὶ Ἰνδοὺς στρατεία Διονύσου καὶ Ἡρακλέους ὑστερογενῆ τὴν μυθοποιίαν ἐμφαίνει, ἅτε τοῦ Ἡρακλέους καὶ τὸν Προμηθέα λῦσαι λεγομένου χιλιάσιν ἐτῶν ὕστερον. καὶ ἦν μὲν ἐνδοξότερον τὸ τὸν Ἀλέξανδρον μέχρι τῶν Ἰνδικῶν ὀρῶν καταστρέψασθαι τὴν Ἀσίαν ἢ μέχρι τοῦ μυχοῦ τοῦ Εὐξείνου καὶ τοῦ Καυκάσου, ἀλλ' ἡ δόξα τοῦ ὄρους καὶ τοὔνομα καὶ τὸ τοὺς περὶ Ἰάσονα δοκεῖν μακροτάτην στρατείαν τελέσαι τὴν μέχρι τῶν πλησίον Καυκάσου καὶ τὸ τὸν Προμηθέα παραδεδόσθαι δεδεμένον ἐπὶ τοῖς ἐσχάτοις τῆς γῆς ἐν τῷ Καυκάσῳ,[3]

C 506

[1] δέ before φησί is found only in E.

[2] ἀνωμολόγηται E, instead of κἂν ὡμολόγηται; so Meineke, and Müller-Dübner.

[3] Meineke indicates a lacuna after Καυκάσῳ; but it is probably merely a case of anacolouthon.

who tell the story agree in their statements. Cleitarchus[1] says that Thalestria set out from the Caspian Gates and Thermodon and visited Alexander; but the distance from the Caspian country to Thermodon is more than six thousand stadia.

5. The stories that have been spread far and wide with a view to glorifying Alexander are not accepted by all; and their fabricators were men who cared for flattery rather than truth. For instance: they transferred the Caucasus into the region of the Indian mountains and of the eastern sea which lies near those mountains from the mountains which lie above Colchis and the Euxine; for these are the mountains which the Greeks named Caucasus, which is more than thirty thousand stadia distant from India; and here it was that they laid the scene of the story of Prometheus and of his being put in bonds; for these were the farthermost mountains towards the east that were known to writers of that time. And the expedition of Dionysus and Heracles to the country of the Indians looks like a mythical story of later date, because Heracles is said to have released Prometheus one thousand years later. And although it was a more glorious thing for Alexander to subdue Asia as far as the Indian mountains than merely to the recess of the Euxine and to the Caucasus, yet the glory of the mountain, and its name, and the belief that Jason and his followers had accomplished the longest of all expeditions, reaching as far as the neighbourhood of the Caucasus, and the tradition that Prometheus was bound at the ends of the earth on the Caucasus, led writers to suppose that they

[1] See *Dictionary* in Vol. II.

χαριεῖσθαί τι τῷ βασιλεῖ ὑπέλαβον, τοὔνομα τοῦ ὄρους μετενέγκαντες εἰς τὴν Ἰνδικήν.

6. Τὰ μὲν οὖν ὑψηλότατα τοῦ ὄντως Καυκάσου τὰ νοτιώτατά ἐστι, τὰ πρὸς Ἀλβανίᾳ καὶ Ἰβηρίᾳ καὶ Κόλχοις καὶ Ἡνιόχοις· οἰκοῦσι δὲ οὓς εἶπον τοὺς συνερχομένους εἰς τὴν Διοσκουριάδα· συνέρχονται δὲ τὸ πλεῖστον ἁλῶν χάριν. τούτων δ' οἱ μὲν τὰς ἀκρωρείας κατέχουσιν, οἱ δὲ ἐν νάπαις αὐλίζονται καὶ ζῶσιν ἀπὸ θηρείων σαρκῶν τὸ πλέον καὶ καρπῶν ἀγρίων καὶ γάλακτος. αἱ δὲ κορυφαὶ χειμῶνος μὲν ἄβατοι, θέρους δὲ προσβαίνουσιν ὑποδούμενοι κεντρωτὰ ὠμοβόϊνα δίκην τυμπάνων πλατεῖα διὰ τὰς χιόνας καὶ τοὺς κρυστάλλους. καταβαίνουσι δ' ἐπὶ δορᾶς κείμενοι σὺν τοῖς φορτίοις καὶ κατολισθαίνοντες, ὅπερ καὶ κατὰ τὴν Ἀτροπατίαν Μηδίαν καὶ κατὰ τὸ Μάσιον ὄρος τὸ ἐν Ἀρμενίᾳ συμβαίνει· ἐνταῦθα δὲ καὶ τροχίσκοι ξύλινοι κεντρωτοὶ τοῖς πέλμασιν ὑποτίθενται. τοῦ γοῦν Καυκάσου τὰ μὲν ἄκρα τοιαῦτα.

7. Καταβαίνοντι δ' εἰς τὰς ὑπωρείας ἀρκτικώτερα μέν ἐστι τὰ κλίματα, ἡμερώτερα δέ· ἤδη γὰρ συνάπτει τοῖς πεδίοις τῶν Σιράκων. εἰσὶ δὲ καὶ Τρωγλοδύται τινὲς ἐν φωλεοῖς οἰκοῦντες διὰ τὰ ψύχη, παρ' οἷς ἤδη καὶ ἀλφίτων ἐστὶν εὐπορία· μετὰ δὲ τοὺς Τρωγλοδύτας καὶ Χαμαικοῖται[1] καὶ Πολυφάγοι τινὲς καλούμενοι καὶ αἱ τῶν Εἰσαδίκων[2] κῶμαι, δυναμένων γεωργεῖν διὰ τὸ μὴ παντελῶς ὑποπεπτωκέναι ταῖς ἄρκτοις.

[1] Χαμαικοῖται, Du Theil, for χαινανοῖται (for other variants see C. Müller); so Meineke.

[2] Εἰσαδίκων is doubtful (see C. Müller).

would be doing the king a favour if they transferred the name Caucasus to India.

6. Now the highest parts of the real Caucasus are the most southerly—those next to Albania, Iberia, and the Colchians, and the Heniochians. They are inhabited by the peoples who, as I have said,[1] assemble at Dioscurias; and they assemble there mostly in order to get salt. Of these tribes, some occupy the ridges of the mountains, while the others have their abodes in glens and live mostly on the flesh of wild animals, and on wild fruits and milk. The summits of the mountains are impassable in winter, but the people ascend them in summer by fastening to their feet broad shoes made of raw ox-hide, like drums, and furnished with spikes, on account of the snow and the ice. They descend with their loads by sliding down seated upon skins, as is the custom in Atropatian Media and on Mount Masius in Armenia; there, however, the people also fasten wooden discs furnished with spikes to the soles of their shoes. Such, then, are the heights of the Caucasus.

7. As one descends into the foothills, the country inclines more towards the north, but its climate is milder, for there it borders on the plains of the Siraces. And here are also some Troglodytae, who, on account of the cold, live in caves; but even in their country there is plenty of barley. After the Troglodytae one comes to certain Chamaecoetae[2] and Polyphagi,[3] as they are called, and to the villages of the Eisadici, who are able to farm because they are not altogether exposed to the north.

[1] 11. 2. 16. [2] *i.e.* "People who sleep on the ground."
[3] *i.e.* "Heavy-eaters."

8. Οἱ δ' ἐφεξῆς ἤδη νομάδες οἱ μεταξὺ τῆς Μαιώτιδος καὶ τῆς Κασπίας Ναβιανοὶ καὶ Πανξανοὶ[1] καὶ ἤδη τὰ τῶν Σιράκων καὶ Ἀόρσων φῦλα. δοκοῦσι δ' οἱ Ἄορσοι καὶ οἱ Σίρακες φυγάδες εἶναι τῶν ἀνωτέρω καὶ προσάρκτιοι μᾶλλον Ἄορσοι.[2] Ἀβέακος μὲν οὖν, ὁ τῶν Σιράκων βασιλεύς, ἡνίκα Φαρνάκης τὸν Βόσπορον εἶχε, δύο μυριάδας ἱππέων ἔστειλε, Σπαδίνης δ', ὁ τῶν Ἀόρσων, καὶ εἴκοσιν, οἱ δὲ ἄνω Ἄορσοι καὶ πλείονας· καὶ γὰρ ἐπεκράτουν πλείονος γῆς, καὶ σχεδόν τι τῆς Κασπίων παραλίας τῆς πλείστης ἦρχον, ὥστε καὶ ἐνεπορεύοντο καμήλοις τὸν Ἰνδικὸν φόρτον καὶ τὸν Βαβυλώνιον, παρά τε Ἀρμενίων καὶ Μήδων διαδεχόμενοι· ἐχρυσοφόρουν δὲ διὰ τὴν εὐπορίαν. οἱ μὲν οὖν Ἄορσοι τὸν Τάναϊν παροικοῦσιν, οἱ Σίρακες δὲ τὸν Ἀχαρδέον, ὃς ἐκ τοῦ Καυκάσου ῥέων ἐκδίδωσιν εἰς τὴν Μαιῶτιν.

VI

1. Η δὲ δευτέρα μερὶς ἄρχεται μὲν ἀπὸ τῆς C 507 Κασπίας θαλάττης, εἰς ἣν κατέπαυεν ἡ προτέρα· καλεῖται δ' ἡ αὐτὴ θάλαττα καὶ Ὑρκανία. δεῖ δὲ περὶ τῆς θαλάττης εἰπεῖν πρότερον ταύτης καὶ τῶν προσοίκων ἐθνῶν.

Ἔστι δ' ὁ κόλπος ἀνέχων ἐκ τοῦ ὠκεανοῦ πρὸς

[1] The spelling of this name varies (see C. Müller).

[2] Ἄορσοι, Groskurd, for Ἀόρσων; so Müller-Dübner's Latin trans.

8. The next peoples to which one comes between Lake Maeotis and the Caspian Sea are nomads, the Nabiani and the Panxani, and then next the tribes of the Siraces and the Aorsi. The Aorsi and the Siraces are thought to be fugitives from the upper tribes of those names[1] and the Aorsi are more to the north than the Siraces. Now Abeacus, king of the Siraces, sent forth twenty thousand horsemen at the time when Pharnaces held the Bosporus; and Spadines, king of the Aorsi, two hundred thousand; but the upper Aorsi sent a still larger number, for they held dominion over more land, and, one may almost say, ruled over most of the Caspian coast; and consequently they could import on camels the Indian and Babylonian merchandise, receiving it in their turn from the Armenians and the Medes, and also, owing to their wealth, could wear golden ornaments. Now the Aorsi live along the Tanaïs, but the Siraces live along the Achardeüs, which flows from the Caucasus and empties into Lake Maeotis.

VI

1. The second[2] portion begins at the Caspian Sea, at which the first portion ends. The same sea is also called Hyrcanian. But I must first describe this sea and the tribes which live about it.

This sea is the gulf which extends from the

[1] *i.e.* the southern tribes. The tribes of the Aorsi and Siraces (also spelt Syraci, 11. 2. 1) extended towards the south as far as the Caucasian Mountains (11. 2. 1).
[2] *i.e.* of the First Division (see 11. 1. 5).

μεσημβρίαν κατ᾽ ἀρχὰς μὲν ἱκανῶς στενός, ἐνδοτέρω δὲ πλατύνεται προϊών, καὶ μάλιστα κατὰ τὸν μυχὸν ἐπὶ σταδίους που καὶ πεντακισχιλίους· ὁ δ᾽ εἴσπλους μέχρι τοῦ μυχοῦ μικρῷ πλειόνων[1] ἂν εἴη, συνάπτων πως ἤδη τῇ ἀοικήτῳ. φησὶ δ᾽ Ἐρατοσθένης τὸν ὑπὸ τῶν Ἑλλήνων γνώριμον περίπλουν τῆς θαλάττης ταύτης, τὸν μὲν παρὰ τοὺς Ἀλβανοὺς καὶ τοὺς Καδουσίους[2] εἶναι πεντακισχιλίων καὶ τετρακοσίων, τὸν δὲ παρὰ τὴν Ἀναριακῶν[3] καὶ Μάρδων καὶ Ὑρκανῶν μέχρι τοῦ στόματος τοῦ Ὤξου ποταμοῦ τετρακισχιλίων καὶ ὀκτακοσίων· ἔνθεν δ᾽ ἐπὶ τοῦ Ἰαξάρτου δισχιλίων τετρακοσίων. δεῖ δὲ περὶ τῶν ἐν τῇ μερίδι ταύτῃ καὶ τοῖς ἐπὶ τοσοῦτον ἐκτετοπισμένοις ἁπλούστερον ἀκούειν, καὶ μάλιστα περὶ τῶν διαστημάτων.

2. Εἰσπλέοντι δ᾽ ἐν δεξιᾷ μὲν τοῖς Εὐρωπαίοις οἱ συνεχεῖς Σκύθαι νέμονται καὶ Σαρμάται οἱ μεταξὺ τοῦ Ταναΐδος καὶ τῆς θαλάττης ταύτης, νομάδες οἱ πλείους, περὶ ὧν εἰρήκαμεν· ἐν ἀριστερᾷ δ᾽ οἱ πρὸς ἕω Σκύθαι, νομάδες καὶ οὗτοι, μέχρι τῆς ἑῴας θαλάττης καὶ τῆς Ἰνδικῆς παρατείνοντες. ἅπαντας μὲν δὴ τοὺς προσβόρους κοινῶς οἱ παλαιοὶ τῶν Ἑλλήνων συγγραφεῖς Σκύθας καὶ Κελτοσκύθας ἐκάλουν· οἱ δ᾽ ἔτι πρότερον διελόντες τοὺς μὲν ὑπὲρ τοῦ Εὐξείνου καὶ Ἴστρου καὶ τοῦ Ἀδρίου κατοικοῦντας Ὑπερβορέους ἔλεγον καὶ Σαυρομάτας καὶ Ἀριμασπούς,

[1] πλειόνων, Kramer, for πλεῖον C, πλείων other MSS.; so the later editors.

[2] Καδουσίους *Epit.*, for Κλουσίους MSS.

[3] Ἀναριακῶν, Tzschucke, for Ἀριάκων CD, Ἀναρίσκων *oz.*

ocean[1] towards the south; it is rather narrow at its entrance, but it widens out as it advances inland, and especially in the region of its recess, where its width is approximately five thousand stadia. The length of the voyage from its entrance to its recess might be slightly more than that, since its entrance is approximately on the borders of the uninhabited world. Eratosthenes says that the circuit of this sea was known to the Greeks; that the part along the coast of the Albanians and the Cadusians is five thousand four hundred stadia; and that the part along the coast of the Anariaci and Mardi and Hyrcani to the mouth of the Oxus River is four thousand eight hundred, and thence to the Iaxartes, two thousand four hundred. But we must understand in a more general sense the accounts of this portion and the regions that lie so far removed, particularly in the matter of distances.

2. On the right, as one sails into the Caspian Sea, are those Scythians, or Sarmatians,[2] who live in the country contiguous to Europe between the Tanaïs River and this sea; the greater part of them are nomads, of whom I have already spoken.[3] On the left are the eastern Scythians, also nomads, who extend as far as the Eastern Sea and India. Now all the peoples towards the north were by the ancient Greek historians given the general name "Scythians" or "Celtoscythians"; but the writers of still earlier times, making distinctions between them, called those who lived above the Euxine and the Ister and the Adriatic "Hyperboreans," "Sauromatians," and "Arimaspians," and they called those

[1] See note on "Caspian Sea" (11. 1. 5).
[2] See 11. 2. 1.
[3] 11. 2. 1.

τοὺς δὲ πέραν τῆς Κασπίας θαλάττης τοὺς μὲν Σάκας, τοὺς δὲ Μασσαγέτας ἐκάλουν, οὐκ ἔχοντες ἀκριβῶς[1] λέγειν περὶ αὐτῶν οὐδέν, καίπερ πρὸς Μασσαγέτας τοῦ Κύρου πόλεμον ἱστοροῦντες. ἀλλ' οὔτε περὶ τούτων οὐδὲν ἠκρίβωτο πρὸς ἀλήθειαν, οὔτε τὰ παλαιὰ τῶν Περσικῶν οὔτε τῶν Μηδικῶν ἢ Συριακῶν ἐς πίστιν ἀφικνεῖτο μεγάλην διὰ τὴν τῶν συγγραφέων ἁπλότητα καὶ τὴν φιλομυθίαν.

3. Ὁρῶντες γὰρ τοὺς φανερῶς μυθογράφους εὐδοκιμοῦντας ᾠήθησαν καὶ αὐτοὶ παρέξεσθαι τὴν γραφὴν ἡδεῖαν, ἐὰν ἐν ἱστορίας σχήματι λέγωσιν, ἃ μηδέποτε εἶδον μηδὲ[2] ἤκουσαν, ἢ οὐ παρά γε εἰδότων,[3] σκοποῦντες[4] αὐτὸ[5] μόνον τοῦτο, ὅ τι ἀκρόασιν ἡδεῖαν ἔχει καὶ θαυμαστήν. ῥᾷον δ' C 508 ἄν τις Ἡσιόδῳ καὶ Ὁμήρῳ πιστεύσειεν ἡρωολογοῦσι καὶ τοῖς τραγικοῖς ποιηταῖς ἢ Κτησίᾳ τε καὶ Ἡροδότῳ καὶ Ἑλλανίκῳ καὶ ἄλλοις τοιούτοις.

4. Οὐδὲ τοῖς περὶ Ἀλεξάνδρου δὲ συγγράψασιν οὐ[6] ῥᾴδιον πιστεύειν τοῖς πολλοῖς· καὶ γὰρ οὗτοι ῥᾳδιουργοῦσι διά τε τὴν δόξαν τὴν Ἀλεξάνδρου καὶ διὰ τὸ τὴν στρατείαν πρὸς τὰς ἐσχατιὰς γεγονέναι τῆς Ἀσίας πόρρω ἀφ' ἡμῶν· τὸ δὲ πόρρω δυσέλεγκτον. ἡ δὲ τῶν Ῥωμαίων ἐπικράτεια καὶ ἡ τῶν Παρθυαίων πλεῖόν τι προσεκκαλύπτει τῶν παραδεδομένων πρότερον· οἱ γὰρ

[1] ἀκριβές E, Meineke.
[2] μηδέ, Jones, for μήτε, from conj. of C. Müller.
[3] εἰδότων, Meineke emends to ἰδόντων.
[4] δι', before αὐτό, Corais omits.
[5] δέ, after αὐτό, Corais omits.
[6] οὐ is omitted by *oz* and some of the editors.

who lived across the Caspian Sea in part "Sacians" and in part "Massagetans," but they were unable to give any accurate account of them, although they reported a war between Cyrus[1] and the Massagetans. However, neither have the historians given an accurate and truthful account of these peoples, nor has much credit been given to the ancient history of the Persians or Medes or Syrians, on account of the credulity of the historians and their fondness for myths.

3. For, seeing that those who were professedly writers of myths enjoyed repute, they thought that they too would make their writings pleasing if they told in the guise of history what they had never seen, nor even heard—or at least not from persons who knew the facts—with this object alone in view, to tell what afforded their hearers pleasure and amazement. One could more easily believe Hesiod and Homer in their stories of the heroes, or the tragic poets, than Ctesias, Herodotus, Hellanicus,[2] and other writers of this kind.

4. Neither is it easy to believe most of those who have written the history of Alexander; for these toy with facts, both because of the glory of Alexander and because his expedition reached the ends of Asia, far away from us; and statements about things that are far away are hard to refute. But the supremacy of the Romans and that of the Parthians has disclosed considerably more knowledge than that which had previously come down to us by tradition;

[1] Cyrus the Elder. For an account of this war, see Herodotus 1. 201 ff.

[2] On their writings, see *Dictionary* in Vol. I.

περὶ ἐκείνων συγγράφοντες καὶ τὰ χωρία καὶ τὰ ἔθνη, ἐν οἷς αἱ πράξεις, πιστότερον λέγουσιν ἢ οἱ πρὸ αὐτῶν· μᾶλλον γὰρ κατωπτεύκασι.

VII

1. Τοὺς δ' οὖν ἐν ἀριστερᾷ εἰσπλέοντι τὸ Κάσπιον πέλαγος παροικοῦντας νομάδας Δάας οἱ νῦν προσαγορεύουσι τοὺς ἐπονομαζομένους Ἀπάρνους·[1] εἶτ' ἔρημος πρόκειται μεταξύ, καὶ ἐφεξῆς ἡ Ὑρκανία, καθ' ἣν ἤδη πελαγίζει μέχρι τοῦ συνάψαι τοῖς Μηδικοῖς ὄρεσι καὶ τοῖς Ἀρμενίων. τούτων δ' ἐστὶ μηνοειδὲς τὸ σχῆμα κατὰ τὰς ὑπωρείας, αἳ τελευτῶσαι πρὸς θάλατταν ποιοῦσι τὸν μυχὸν τοῦ κόλπου. οἰκεῖ δὲ τὴν παρώρειαν ταύτην μέχρι τῶν ἄκρων ἀπὸ θαλάττης ἀρξαμένοις ἐπὶ μικρὸν μὲν τῶν Ἀλβανῶν τι μέρος καὶ τῶν Ἀρμενίων, τὸ δὲ πλέον Γῆλαι καὶ Καδούσιοι καὶ Ἄμαρδοι καὶ Οὐίτιοι[2] καὶ Ἀναριάκαι. φασὶ δὲ Παρρασίων τινὰς συνοικῆσαι τοῖς Ἀναριάκαις, οὓς καλεῖσθαι νῦν Παρσίους·[3] Αἰνιᾶνας δ' ἐν τῇ Οὐιτίᾳ τειχίσαι πόλιν, ἣν Αἰνιάνα καλεῖσθαι, καὶ δείκνυσθαι[4] ὅπλα τε Ἑλληνικὰ ἐνταῦθα καὶ σκεύη χαλκᾶ καὶ ταφάς· ἐνταῦθα δὲ καὶ πόλιν Ἀναριάκην,[5] ἐν ᾗ,[6] φασί,

[1] Ἀπάρνους (so spelled in 11. 8. 2 (twice)), Jones, for Σπάρνους; others Πάρνους (as in MSS. 11. 9. 2, 3 *q.v.*).

[2] Οὐίτιοι E, Κουίτιοι other MSS. C. Müller conj. Κύρτιοι (see *Ind. Var. Lect.*, p. 1014).

[3] Παρσίους, Corais, for Παρρασίους; so the later editors.

for those who write about those distant regions tell a more trustworthy story than their predecessors, both of the places and of the tribes among which the activities took place, for they have looked into the matter more closely.

VII

1. Those nomads, however, who live along the coast on the left as one sails into the Caspian Sea are by the writers of to-day called Däae, I mean, those who are surnamed Aparni; then, in front of them, intervenes a desert country; and next comes Hyrcania, where the Caspian resembles an open sea to the point where it borders on the Median and Armenian mountains. The shape of these mountains is crescent-like along the foot-hills, which end at the sea and form the recess of the gulf. This side of the mountains, beginning at the sea, is inhabited as far as their heights for a short stretch by a part of the Albanians and the Armenians, but for the most part by Gelae, Cadusii, Amardi, Vitii, and Anariacae. They say that some of the Parrhasii took up their abode with the Anariacae, who, they say, are now called Parsii; and that the Aenianes built a walled city in the Vitian territory, which, they say, is called Aeniana; and that Greek armour, brazen vessels, and burial-places are to be seen there; and that there is also a city Anariacê there, in which, they

[4] δείκνυσθαι, Corais, for δείκνυται; so the later editors.
[5] Ἀναριάκην, Tzschucke, for Ἀβάρκην *Dh*, Ναβάρκην other MSS.; so the later editors.
[6] ᾗ, Tzschucke, for ᾧ; so the later editors.

δείκνυται μαντεῖον ἐγκοιμωμένων,[1] καὶ ἄλλα τινὰ ἔθνη λῃστρικὰ καὶ μάχιμα μᾶλλον ἢ γεωργικά·[2] ποιεῖ δὲ τοῦτο ἡ τραχύτης τῶν τόπων. τὸ μέντοι πλέον τῆς περὶ τὴν ὀρεινὴν παραλίας Καδούσιοι νέμονται, σχεδόν δέ τι ἐπὶ πεντακισχιλίους σταδίους, ὥς φησι Πατροκλῆς, ὃς καὶ πάρισον ἡγεῖται τὸ πέλαγος τοῦτο τῷ Ποντικῷ. ταῦτα μὲν οὖν τὰ χωρία λυπρά.

2. Ἡ δ' Ὑρκανία σφόδρα εὐδαίμων καὶ πολλὴ καὶ τὸ πλέον πεδιὰς πόλεσί τε ἀξιολόγοις διειλημμένη, ὧν ἐστὶ Ταλαβρόκη καὶ Σαμαριανὴ καὶ Κάρτα καὶ τὸ βασίλειον Τάπη· ὅ φασι μικρὸν ὑπὲρ τῆς θαλάττης ἱδρυμένον διέχειν τῶν Κασπίων πυλῶν σταδίους χιλίους τετρακοσίους, καὶ διὰ τὸ μὲν εἶδος[3] τῆς εὐδαιμονίας σημεῖα διηγοῦνται·[4] ἡ μὲν γὰρ ἄμπελος μετρητὴν οἴνου φέρει, ἡ δὲ συκῆ μεδίμνους ἑξήκοντα, ὁ δὲ σῖτος C 509 ἐκ τοῦ ἐκπεσόντος καρποῦ τῆς καλάμης φύεται, ἐν δὲ τοῖς δένδρεσι σμηνουργεῖται καὶ τῶν φύλλων ἀπορρεῖ μέλι· τοῦτο δὲ γίνεται καὶ τῆς Μηδίας ἐν τῇ Ματιανῇ καὶ τῆς Ἀρμενίας ἐν τῇ Σακασηνῇ καὶ τῇ Ἀραξηνῇ. τῆς μέντοι προσηκούσης ἐπιμελείας οὐκ ἔτυχεν οὔτε αὐτὴ οὔτε ἡ ἐπώνυμος αὐτῇ θάλαττα, ἄπλους τε οὖσα

[1] ἐγκοιμωμένων, Tzschucke, for ἐν κοιμωμένων; so the later editors.

[2] There appears to be an omission here. Groskurd suggests that Strabo wrote "and some other traces of Greek colonisation, and all these tribes are more inclined to brigandage and war."

[3] καὶ τοῦ μὲν εἴδους *owz*, καὶ ταῦτα μὲν τοῦ εἴδους *xy*. E omits the words, inserting δέ after σημεῖα. T. G. Tucker (*Classical Quarterly* 3. 101) proposes καὶ νὴ Δία τοῦ μεγέθους . . . διηγοῦνται.

say, is to be seen an oracle for sleepers,[1] [2] and some other tribes that are more inclined to brigandage and war than to farming; but this is due to the ruggedness of the region. However, the greater part of the seaboard round the mountainous country is occupied by Cadusii, for a stretch of almost five thousand stadia, according to Patrocles,[3] who considers this sea almost equal to the Pontic Sea. Now these regions have poor soil.

2. But Hyrcania is exceedingly fertile, extensive, and in general level; it is distinguished by notable cities, among which are Talabrocê, Samarianê, Carta, and the royal residence Tapê, which, they say, is situated slightly above the sea and at a distance of one thousand four hundred stadia from the Caspian Gates. And because of its particular kind of prosperity writers go on to relate evidences thereof: the vine produces one metretes[4] of wine, and the fig-tree sixty medimni;[5] the grain grows up from the seed that falls from the stalk; bees have their hives in the trees, and honey drips from the leaves; and this is also the case in Matianê in Media, and in Sacasenê and Araxenê in Armenia.[6] However, neither the country itself nor the sea that is named after it has received proper attention, the sea being both without vessels and unused. There

[1] *i.e.* people received oracles in their dreams while sleeping in the temple (cf. 16. 2. 35).
[2] See critical note
[3] See *Dictionary* in Vol. I.
[4] A little less than nine gallons.
[5] The medimnus was about a bushel and a half.
[6] Cf. 2. 1. 14.

---

[2] διηγοῦνται, Groskurd, for ἡγοῦνται, which E and Meineke omit.

καὶ ἀργός· νῆσοί τέ εἰσιν οἰκεῖσθαι δυνάμεναι, ὡς δ' εἰρήκασί τινες, καὶ χρυσῖτιν ἔχουσαι γῆν. αἴτιον δ', ὅτι καὶ οἱ ἡγεμόνες οἵ τ' ἐξαρχῆς ἐτύγχανον βάρβαροι ὄντες οἱ τῶν Ὑρκανῶν, Μῆδοί τε καὶ Πέρσαι, καὶ οἱ ὕστατοι Παρθυαῖοι, χείρους ἐκείνων ὄντες, καὶ ἡ γείτων ἅπασα χώρα λῃστῶν καὶ νομάδων μεστὴ καὶ ἐρημίας. Μακεδόνες δ' ὀλίγον μὲν χρόνον ἐπῆρξαν, καὶ ἐν πολέμοις ὄντες καὶ τὰ πόρρω σκοπεῖν οὐ δυνάμενοι. φησὶ δ' Ἀριστόβουλος ὑλώδη οὖσαν τὴν Ὑρκανίαν δρῦν ἔχειν, πεύκην δὲ καὶ ἐλάτην καὶ πίτυν μὴ φύειν, τὴν δ' Ἰνδικὴν πληθύειν τούτοις. τῆς δὲ Ὑρκανίας ἐστὶ καὶ ἡ Νησαία· τινὲς δὲ καὶ καθ' αὑτὴν τιθέασι τὴν Νησαίαν.

3. Διαρρεῖται δὲ καὶ ποταμοῖς ἡ Ὑρκανία τῷ τε Ὤχῳ καὶ τῷ Ὤξῳ μέχρι τῆς εἰς θάλατταν ἐκβολῆς, ὧν ὁ Ὦχος καὶ διὰ τῆς Νησαίας ῥεῖ· ἔνιοι δὲ τὸν Ὦχον εἰς τὸν Ὦξον ἐμβάλλειν φασίν. Ἀριστόβουλος δὲ καὶ μέγιστον ἀποφαίνει τὸν Ὦξον τῶν ἑωραμένων ὑφ' ἑαυτοῦ κατὰ τὴν Ἀσίαν, πλὴν τῶν Ἰνδικῶν· φησὶ δὲ καὶ εὔπλουν εἶναι (καὶ οὗτος καὶ Ἐρατοσθένης παρὰ Πατροκλέους λαβών) καὶ πολλὰ τῶν Ἰνδικῶν φορτίων κατάγειν εἰς τὴν Ὑρκανίαν θάλατταν, ἐντεῦθεν δ' εἰς τὴν Ἀλβανίαν περαιοῦσθαι, καὶ διὰ τοῦ Κύρου καὶ τῶν ἑξῆς τόπων εἰς τὸν Εὔξεινον καταφέρεσθαι. οὐ πάνυ δὲ ὑπὸ τῶν παλαιῶν ὁ Ὦχος ὀνομάζεται. Ἀπολ-

---

[1] *Pinus maritima.*
[2] *Pinus picea.*
[3] *Pinus pinea.*
[4] Cf. 11. 13. 7.
[5] This Aristobulus accompanied Alexander on his expedition and wrote a work of unknown title.

are islands in this sea which could afford a livelihood, and, according to some writers, contain gold ore. The cause of this lack of attention was the fact that the first governors of the Hyrcanians, I mean the Medes and Persians, as also the last, I mean the Parthians, who were inferior to the former, were barbarians, and also the fact that the whole of the neighbouring country was full of brigands and nomads and deserted regions. The Macedonians did indeed rule over the country for a short time, but they were so occupied with wars that they could not attend to their remote possessions. According to Aristobulus, Hyrcania, which is a wooded country, has the oak, but does not produce the torch-pine[1] or fir[2] or stone-pine,[3] though India abounds in these trees. Nesaea, also, belongs to Hyrcania, though some writers set it down as an independent district.[4]

3. Hyrcania is traversed by the rivers Ochus and Oxus to their outlets into the sea; and of these, the Ochus flows also through Nesaea, but some say that the Ochus empties into the Oxus. Aristobulus[5] declares that the Oxus is the largest of the rivers he has seen in Asia, except those in India. And he further says that it is navigable (both he and Eratosthenes taking this statement from Patrocles)[6] and that large quantities of Indian wares are brought down on it to the Hyrcanian Sea, and thence on that sea are transported to Albania and brought down on the Cyrus River and through the region that comes next after it to the Euxine. The Ochus is not mentioned at all by the ancient writers. Apollodorus,[7] however,

[6] See *Dictionary* in Vol. I. [7] Of Artemita.

λόδωρος μέντοι ὁ τὰ Παρθικὰ γράψας συνεχῶς αὐτὸν ὀνομάζει, ὡς ἐγγυτάτω τοῖς Παρθυαίοις ῥέοντα.

4. Προσεδοξάσθη δὲ καὶ περὶ τῆς θαλάττης ταύτης πολλὰ ψευδῆ διὰ τὴν Ἀλεξάνδρου φιλοτιμίαν· ἐπειδὴ γὰρ ὡμολόγητο ἐκ πάντων, ὅτι διείργει τὴν Ἀσίαν ἀπὸ τῆς Εὐρώπης ὁ Τάναϊς ποταμός, τὸ δὲ μεταξὺ τῆς θαλάττης καὶ τοῦ Ταναΐδος, πολὺ μέρος τῆς Ἀσίας ὄν, οὐχ ὑπέπιπτε τοῖς Μακεδόσι, στρατηγεῖν δ' ἔγνωστο, ὥστε τῇ φήμῃ γε κἀκείνων δόξαι τῶν μερῶν κρατεῖν τὸν Ἀλέξανδρον· εἰς ἓν οὖν συνῆγον τήν τε Μαιῶτιν λίμνην τὴν δεχομένην τὸν Τάναϊν καὶ τὴν Κασπίαν θάλατταν, λίμνην καὶ ταύτην καλοῦντες καὶ συντετρῆσθαι φάσκοντες πρὸς ἀλλήλας ἀμφοτέρας, ἑκατέραν δὲ εἶναι μέρος τῆς ἑτέρας. Πολύκλειτος δὲ καὶ πίστεις προσφέρεται περὶ τοῦ λίμνην εἶναι C 510 τὴν θάλατταν ταύτην (ὄφεις τε γὰρ ἐκτρέφειν καὶ ὑπόγλυκυ εἶναι τὸ ὕδωρ), ὅτι δὲ καὶ οὐχ ἑτέρα τῆς Μαιώτιδός ἐστι, τεκμαιρόμενος ἐκ τοῦ τὸν Τάναϊν εἰς αὐτὴν ἐμβάλλειν· ἐκ γὰρ τῶν αὐτῶν ὀρῶν τῶν Ἰνδικῶν, ἐξ ὧν ὅ τε Ὦχος καὶ ὁ Ὦξος καὶ ἄλλοι πλείους, φέρεται καὶ ὁ Ἰαξάρτης ἐκδίδωσί τε ὁμοίως ἐκείνοις εἰς τὸ Κάσπιον πέλαγος, πάντων ἀρκτικώτατος. τοῦτον οὖν ὠνόμασαν Τάναϊν, καὶ προσέθεσαν καὶ τούτῳ πίστιν, ὡς[1] εἴη Τάναϊς, ὃν εἴρηκεν ὁ Πολύκλειτος· τὴν γὰρ περαίαν τοῦ ποταμοῦ τούτου φέρειν ἐλάτην καὶ ὀϊστοῖς ἐλατίνοις χρῆσθαι τοὺς ταύτῃ Σκύθας, τοῦτο δὲ καὶ τεκμήριον τοῦ τὴν χώραν τὴν πέραν

[1] ὡς, Corais, for ὥστ'; so the later editors.

who wrote the *Parthica,* names it continually, implying that it flows very close to the country of the Parthians.

4. Many false notions were also added to the account of this sea because[1] of Alexander's love of glory; for, since it was agreed by all that the Tanaïs separated Asia from Europe, and that the region between the sea and the Tanaïs, being a considerable part of Asia, had not fallen under the power of the Macedonians, it was resolved to manipulate the account of Alexander's expedition so that in fame at least he might be credited with having conquered those parts of Asia too. They therefore united Lake Maeotis, which receives the Tanaïs, with the Caspian Sea, calling this too a lake and asserting that both were connected with one another by an underground passage and that each was a part of the other. Polycleitus goes on to adduce proofs in connection with his belief that the sea is a lake (for instance, he says that it produces serpents, and that its water is sweetish); and that it is no other than Maeotis he judges from the fact that the Tanaïs empties into it. From the same Indian mountains, where the Ochus and the Oxus and several other rivers rise, flows also the Iaxartes, which, like those rivers, empties into the Caspian Sea and is the most northerly of them all. This river, accordingly, they named Tanaïs; and in addition to so naming it they gave as proof that it was the Tanaïs mentioned by Polycleitus that the country on the far side of this river produces the fir-tree and that the Scythians in that region use arrows made of fir-wood; and they say that this is also evidence that the country on the

[1] See 11. 5. 5.

τῆς Εὐρώπης εἶναι, μὴ τῆς Ἀσίας· τὴν γὰρ Ἀσίαν τὴν ἄνω καὶ τὴν πρὸς ἕω μὴ φύειν ἐλάτην. Ἐρατοσθένης δέ φησι καὶ ἐν τῇ Ἰνδικῇ φύεσθαι ἐλάτην καὶ ἐντεῦθεν ναυπηγήσασθαι τὸν στόλον Ἀλέξανδρον· πολλὰ δὲ καὶ ἄλλα τοιαῦτα συγκρούειν Ἐρατοσθένης πειρᾶται, ἡμῖν δ' ἀποχρώντως εἰρήσθω περὶ αὐτῶν.

5. Καὶ τοῦτο δ' ἐκ τῶν κατὰ τὴν Ὑρκανίαν ἱστορουμένων παραδόξων ἐστὶν ὑπὸ Εὐδόξου καὶ ἄλλων, ὅτι πρόκεινταί τινες ἀκταὶ τῆς θαλάττης ὕπαντροι, τούτων δὲ μεταξὺ καὶ τῆς θαλάττης ὑπόκειται ταπεινὸς αἰγιαλός, ἐκ δὲ τῶν ὕπερθεν κρημνῶν ποταμοὶ ῥέοντες τοσαύτῃ προφέρονται βίᾳ, ὥστε ταῖς ἀκταῖς συνάψαντες ἐξακοντίζουσι τὸ ὕδωρ εἰς τὴν θάλατταν, ἄρραντον φυλάττοντες τὸν αἰγιαλόν, ὥστε καὶ στρατοπέδοις ὁδεύσιμον εἶναι, σκεπαζομένοις[1] τῷ ῥεύματι· οἱ δ' ἐπιχώριοι κατάγονται πολλάκις εὐωχίας καὶ θυσίας χάριν εἰς τὸν τόπον καὶ ποτὲ μὲν ὑπὸ τοῖς ἄντροις κατακλίνονται, ποτὲ δ' ὑπ' αὐτῷ τῷ ῥεύματι ἡλιαζόμενοι, ἄλλως[2] ἄλλοι τέρπονται, παραφαινομένης ἅμα καὶ τῆς θαλάττης ἑκατέρωθεν καὶ τῆς ἠιόνος, ποώδους καὶ ἀνθηρᾶς οὔσης διὰ τὴν ἰκμάδα.

VIII

1. Ἀπὸ δὲ τῆς Ὑρκανίας θαλάττης προϊόντι ἐπὶ τὴν ἕω δεξιὰ μέν ἐστι τὰ ὄρη μέχρι τῆς Ἰνδικῆς θαλάττης παρατείνοντα, ἅπερ οἱ Ἕλληνες

[1] σκεπαζομένοις *Epit.* for σκεπαζόμενον.
[2] δ', after ἄλλως, Meineke omits.

far side belongs to Europe and not to Asia, for, they add, Upper and Eastern Asia does not produce the fir-tree. But Eratosthenes says that the fir-tree grows also in India and that Alexander built his fleet out of fir-wood from there. Eratosthenes tries to reconcile many other differences of this kind, but as for me, let what I have said about them suffice.

5. This too, among the marvellous things recorded of Hyrcania, is related by Eudoxus[1] and others: that there are some cliffs facing the sea with caverns underneath, and between these and the sea, below the cliffs, is a low-lying shore; and that rivers flowing from the precipices above rush forward with so great force that when they reach the cliffs they hurl their waters out into the sea without wetting the shore, so that even armies can pass underneath sheltered by the stream above; and the natives often come down to the place for the sake of feasting and sacrifice, and sometimes they recline in the caverns down below and sometimes they enjoy themselves basking in the sunlight beneath the stream itself, different people enjoying themselves in different ways, having in sight at the same time on either side both the sea and the shore, which latter, because of the moisture, is grassy and abloom with flowers.

VIII

1. As one proceeds from the Hyrcanian Sea towards the east, one sees on the right the mountains that extend as far as the Indian Sea, which by

[1] Eudoxus of Cnidus (see *Dictionary* in Vol. I).

ὀνομάζουσι Ταῦρον, ἀρξάμενα[1] ἀπὸ τῆς Παμφυλίας καὶ τῆς Κιλικίας καὶ μέχρι δεῦρο προϊόντα ἀπὸ τῆς ἑσπέρας συνεχῆ καὶ τυγχάνοντα[2] ἄλλων καὶ ἄλλων ὀνομάτων. προσοικοῦσι δ' αὐτοῦ τὰ προσάρκτια μέρη πρῶτοι μὲν οἱ Γῆλαι καὶ Καδούσιοι καὶ Ἄμαρδοι, καθάπερ εἴρηται, καὶ τῶν Ὑρκανίων τινές, ἔπειτα τὸ τῶν Παρθυαίων ἔθνος καὶ τὸ τῶν Μαργιανῶν καὶ τῶν Ἀρίων καὶ
C 511 ἡ ἔρημος, ἣν ἀπὸ τῆς Ὑρκανίας ὁρίζει ὁ Σάρνιος ποταμὸς πρὸς ἕω βαδίζουσι καὶ ἐπὶ τὸν Ὦχον. καλεῖται δὲ τὸ μέχρι δεῦρο ἀπὸ τῆς Ἀρμενίας διατεῖνον, ἢ μικρὸν ἀπολεῖπον, Παραχοάθρας.[3] ἔστι δὲ ἀπὸ τῆς Ὑρκανίας θαλάττης εἰς τοὺς Ἀρίους περὶ ἑξακισχιλίους σταδίους, εἶθ' ἡ Βακτριανή ἐστι καὶ ἡ Σογδιανή, τελευταῖοι δὲ Σκύθαι νομάδες. τὰ δ' ὄρη Μακεδόνες μὲν ἅπαντα τὰ ἐφεξῆς ἀπὸ Ἀρίων Καύκασον ἐκάλεσαν, παρὰ δὲ τοῖς βαρβάροις τά τε ἄκρα κατὰ μέρος ὠνομάζετο ὁ Παροπάμισος τὰ προσβόρεια[4] καὶ τὰ Ἠμωδὰ καὶ τὸ Ἴμαον καὶ ἄλλα τοιαῦτα ὀνόματα ἑκάστοις μέρεσιν ἐπέκειτο.

2. Ἐν ἀριστερᾷ δὲ τούτοις ἀντιπαράκειται τὰ[5] Σκυθικὰ ἔθνη καὶ τὰ νομαδικά, ἅπασαν ἐκπληροῦντα τὴν βόρειον πλευράν. οἱ μὲν δὴ πλείους τῶν Σκυθῶν ἀπὸ τῆς Κασπίας θαλάττης ἀρξάμενοι Δάαι προσαγορεύονται, τοὺς δὲ προσεῴους τούτων

[1] ἀρξάμενα *Egxyz* (ἀρξάμενον other MSS.); so Tzschucke, Corais, Meineke.

[2] τυγχάνοντα E, τυγχανόντων other MSS.

[3] Παραχοάθρας, Tzschucke, for Παρωχοάρας; so the later editors.

[4] The reading of the MSS., τά τε ἄκρα καὶ τοῦ Παραπαμίσου τὰ προσβόρεια κτλ., is corrupt. Jones corrects the passage by

the Greeks are named the Taurus. Beginning at Pamphylia and Cilicia they extend thus far in a continuous line from the west and bear various different names. In the northerly parts of the range dwell first the Gelae and Cadusii and Amardi, as I have said,[1] and certain of the Hyrcanians, and after them the tribe of the Parthians and that of the Margianians and the Arians; and then comes the desert which is separated from Hyrcania by the Sarnius River as one goes eastwards and towards the Ochus River. The mountain which extends from Armenia to this point, or a little short of it, is called Parachoathras. The distance from the Hyrcanian Sea to the country of the Arians is about six thousand stadia. Then comes Bactriana, and Sogdiana, and finally the Scythian nomads. Now the Macedonians gave the name Caucasus to all the mountains which follow in order after the country of the Arians; but among the barbarians[2] the extremities[3] on the north were given the separate names "Paropamisus" and "Emoda" and "Imaus"; and other such names were applied to separate parts.

2. On the left and opposite these peoples are situated the Scythian or nomadic tribes, which cover the whole of the northern side. Now the greater part of the Scythians, beginning at the Caspian Sea, are called Däae, but those who are situated more to

[1] 11. 7. 1.
[2] *i.e.* the "natives," as referred to in 15. 1. 11.
[3] *i.e.* the "farthermost (or outermost) parts of the Taurus,' as mentioned in 15. 1. 11 (*q. v.*).

---

following the similar statement in 15. 1. 11 (but cp. Groskurd and C. Müller).
[5] τά, before Σκυθικά, Corais inserts; so the later editors.

μᾶλλον Μασσαγέτας καὶ Σάκας ὀνομάζουσι, τοὺς δ' ἄλλους κοινῶς μὲν Σκύθας ὀνομάζουσιν, ἰδίᾳ δ' ὡς ἑκάστους· ἅπαντες δ' ὡς ἐπὶ τὸ πολὺ νομάδες. μάλιστα δὲ γνώριμοι γεγόνασι τῶν νομάδων οἱ τοὺς Ἕλληνας ἀφελόμενοι τὴν Βακτριανήν, Ἄσιοι καὶ Πασιανοὶ καὶ Τόχαροι[1] καὶ Σακάραυλοι,[2] ὁρμηθέντες ἀπὸ τῆς περαίας τοῦ Ἰαξάρτου τῆς κατὰ Σάκας καὶ Σογδιανούς, ἣν κατεῖχον Σάκαι. καὶ τῶν Δαῶν οἱ μὲν προσαγορεύονται Ἄπαρνοι, οἱ δὲ Ξάνθιοι, οἱ δὲ Πίσσουροι· οἱ μὲν οὖν Ἄπαρνοι πλησιαίτατα τῇ Ὑρκανίᾳ παράκεινται καὶ τῇ κατ' αὐτὴν θαλάττῃ, οἱ δὲ λοιποὶ διατείνουσι[3] καὶ μέχρι τῆς ἀντιπαρηκούσης τῇ Ἀρίᾳ.

3. Μεταξὺ δ' αὐτῶν καὶ τῆς Ὑρκανίας καὶ τῆς Παρθυαίας μέχρι Ἀρίων ἔρημος πρόκειται πολλὴ καὶ ἄνυδρος, ἣν διεξιόντες μακραῖς ὁδοῖς κατέτρεχον τήν τε Ὑρκανίαν καὶ τὴν Νησαίαν[4] καὶ τὰ τῶν Παρθυαίων πεδία· οἱ δὲ συνέθεντο φόρους· φόρος δ' ἦν τὸ ἐπιτρέπειν τακτοῖς τισὶ χρόνοις τὴν χώραν κατατρέχειν καὶ φέρεσθαι λείαν. ἐπιπολαζόντων δ' αὐτῶν παρὰ τὰ συγκείμενα, ἐπολεμεῖτο, καὶ πάλιν διαλύσεις καὶ ἀναπολεμήσεις ὑπῆρχον. τοιοῦτος δὲ καὶ ὁ τῶν ἄλλων νομάδων βίος, ἀεὶ τοῖς πλησίον ἐπιτιθεμένων, τοτὲ δ' αὖ διαλλαττομένων.

4. Σάκαι μέντοι παραπλησίας ἐφόδους ἐποιή-

[1] Τόχαροι, the editors, for Τάχαροι.

[2] καί, before ὁρμηθέντες, Kramer omits; so the later editors.

[3] διατείνουσι, Corais, for διαμένουσι (but E omits the word); so the later editors.

[4] Νησαίαν, Xylander, for Ἰσαίαν; so the later editors.

the east than these are named Massagetae and Sacae, whereas all the rest are given the general name of Scythians, though each people is given a separate name of its own. They are all for the most part nomads. But the best known of the nomads are those who took away Bactriana from the Greeks, I mean the Asii, Pasiani, Tochari,[1] and Sacarauli, who originally came from the country on the other side of the Iaxartes River that adjoins that of the Sacae and the Sogdiani and was occupied by the Sacae. And as for the Däae, some of them are called Aparni, some Xanthii, and some Pissuri. Now of these the Aparni are situated closest to Hyrcania and the part of the sea that borders on it, but the remainder extend even as far as the country that stretches parallel to Aria.

3. Between them[2] and Hyrcania and Parthia and extending as far as the Arians is a great waterless desert, which they traversed by long marches and then overran Hyrcania, Nesaea, and the plains of the Parthians. And these people agreed to pay tribute, and the tribute was to allow the invaders at certain appointed times to overrun the country and carry off booty. But when the invaders overran their country more than the agreement allowed, war ensued, and in turn their quarrels were composed and new wars were begun. Such is the life of the other nomads also, who are always attacking their neighbours and then in turn settling their differences.

4. The Sacae, however, made raids like those of

[1] On the Tochari and their language, see the article by T. A. Sinclair in the *Classical Review*, xxxvii, Nov., Dec., 1923, p. 159.

[2] The Aparnian Däae (see 11. 9. 2).

σαντο τοῖς Κιμμερίοις καὶ Τρήρεσι,[1] τὰς μὲν μακροτέρας, τὰς δὲ καὶ ἐγγύθεν· καὶ γὰρ τὴν Βακτριανὴν κατέσχον καὶ τῆς Ἀρμενίας κατεκτήσαντο τὴν ἀρίστην γῆν, ἣν καὶ ἐπώνυμον ἑαυτῶν κατέλιπον τὴν Σακασηνήν, καὶ μέχρι Καππα-
C 512 δόκων, καὶ μάλιστα τῶν πρὸς Εὐξείνῳ, οὓς Ποντικοὺς νῦν καλοῦσι, προῆλθον. ἐπιθέμενοι δ' αὐτοῖς πανηγυρίζουσιν ἀπὸ τῶν λαφύρων οἱ ταύτῃ τότε τῶν Περσῶν στρατηγοί, νύκτωρ ἄρδην αὐτοὺς ἠφάνισαν. ἐν δὲ τῷ πεδίῳ πέτραν τινὰ προσχώματι συμπληρώσαντες εἰς βουνοειδὲς σχῆμα ἐπέθηκαν τεῖχος καὶ τὸ τῆς Ἀναΐτιδος καὶ τῶν συμβώμων θεῶν ἱερὸν ἱδρύσαντο, Ὠμανοῦ καὶ Ἀναδάτου, Περσικῶν δαιμόνων, ἀπέδειξάν τε πανήγυριν κατ' ἔτος ἱεράν, τὰ Σάκαια, ἣν μέχρι νῦν ἐπιτελοῦσιν οἱ τὰ Ζῆλα[2] ἔχοντες· οὕτω γὰρ καλοῦσι τὸν τόπον· ἔστι δὲ ἱεροδούλων πόλισμα τὸ πλέον· Πομπήιος δὲ προσθεὶς χώραν ἀξιόλογον καὶ τοὺς ἐν αὐτῇ συνοικίσας εἰς τὸ τεῖχος μίαν τῶν πόλεων ἀπέφηνεν, ὧν διέταξε μετὰ τὴν Μιθριδάτου κατάλυσιν.

5. Οἱ μὲν[3] οὕτω λέγουσι περὶ τῶν Σακῶν, οἱ δ', ὅτι Κῦρος ἐπιστρατεύσας τοῖς Σάκαις, ἡττηθεὶς τῇ μάχῃ φεύγει, στρατοπεδευσάμενος δ' ἐν ᾧ χωρίῳ τὰς παρασκευὰς ἀπελελοίπει[4] πλήρεις ἀφθονίας ἁπάσης, καὶ μάλιστα οἴνου, διαναπαύσας μικρὰ τὴν στρατιάν, ἤλαυνεν ἀφ' ἑσπέρας, ὡς φεύγων, πλήρεις ἀφεὶς τὰς σκηνάς· προελθὼν δ',

[1] Τρήρεσι, Xylander, for τριήρεσι; so the later editors.
[2] Ζῆλα, Tzschucke, for Σάκα; so the later editors.
[3] Corais, Meineke and others insert οὖν after μέν.
[4] ἀπελελοίπει, Jones, for ἀπολελοίπει.

Cimmerians and Treres,[1] some into regions close to their own country, others into regions farther away. For instance, they occupied Bactriana, and acquired possession of the best land in Armenia, which they left named after themselves, Sacasenê; and they advanced as far as the country of the Cappadocians, particularly those situated close to the Euxine, who are now called the Pontici. But when they were holding a general festival and enjoying their booty, they were attacked by night by the Persian generals who were then in that region and utterly wiped out. And these generals, heaping up a mound of earth over a certain rock in the plain, completed it in the form of a hill, and erected on it a wall, and established the temple of Anaïtis and the gods who share her altar—Omanus and Anadatus, Persian deities; and they instituted an annual sacred festival, the Sacaea, which the inhabitants of Zela (for thus the place is called) continue to celebrate to the present day. It is a small city belonging for the most part to the temple-slaves. But Pompey added considerable territory to it, settled the inhabitants thereof within the walls, and made it one of the cities which he organised after his overthrow of Mithridates.

5. Now this is the account which some writers give of the Sacae. Others say that Cyrus made an expedition against the Sacae, was defeated in the battle, and fled; but that he encamped in the place where he had left behind his supplies, which consisted of an abundance of everything and especially of wine, rested his army a short time, and set out at nightfall, as though he were in flight, leaving the tents full of supplies; and that he proceeded as far

[1] Cf. 1. 3. 21, 12. 3. 24, 12. 8. 7, 13. 1. 8, 13. 4. 8, 14. 1. 40.

ὅσον ἐδόκει συμφέρειν, ἱδρύθη· ἐπιόντες δ' ἐκεῖνοι καὶ καταλαβόντες ἔρημον ἀνδρῶν τὸ στρατόπεδον, τῶν δὲ πρὸς ἀπόλαυσιν μεστόν, ἀνέδην ἐνεπίμπλαντο· ὁ δ' ὑποστρέψας ἐξοίνους κατέλαβε καὶ παραπλῆγας, ὥσθ' οἱ μὲν ἐν κάρῳ κείμενοι καὶ ὕπνῳ κατεκόπτοντο, οἱ δ' ὀρχούμενοι καὶ βακχεύοντες γυμνοὶ περιέπιπτον τοῖς τῶν πολεμίων ὅπλοις, ὀλίγου δ' ἀπώλοντο ἅπαντες. ὁ δὲ θεῖον νομίσας τὸ εὐτύχημα, τὴν ἡμέραν ἐκείνην ἀνιερώσας τῇ πατρίῳ θεῷ προσηγόρευσε[1] Σάκαια· ὅπου δ' ἂν ᾖ τῆς θεοῦ ταύτης ἱερόν, ἐνταῦθα νομίζεται καὶ ἡ τῶν Σακαίων ἑορτὴ βακχεία τις[2] μεθ' ἡμέραν καὶ νύκτωρ, διεσκευασμένων Σκυθιστί, πινόντων ἅμα καὶ πληκτιζομένων πρὸς ἀλλήλους ἅμα τε καὶ τὰς συμπινούσας γυναῖκας.

6. Μασσαγέται δ' ἐδήλωσαν τὴν σφετέραν ἀρετὴν ἐν τῷ πρὸς Κῦρον πολέμῳ, περὶ ὧν[3] θρυλοῦσι πολλοί, καὶ δεῖ πυνθάνεσθαι παρ' ἐκείνων. λέγεται δὲ καὶ τοιαῦτα περὶ τῶν Μασσαγετῶν, ὅτι κατοικοῦσιν οἱ μὲν ὄρη, τινὲς δ' αὐτῶν πεδία, οἱ δὲ ἕλη, ἃ ποιοῦσιν οἱ ποταμοί, οἱ δὲ τὰς ἐν τοῖς ἕλεσι νήσους. μάλιστα δέ φασι τὸν Ἀράξην[4] ποταμὸν κατακλύζειν τὴν χώραν πολλαχῇ σχιζόμενον, ἐκπίπτοντα δὲ τοῖς μὲν C 513 ἄλλοις στόμασιν εἰς τὴν ἄλλην τὴν πρὸς ἄρκτοις θάλατταν, ἑνὶ δὲ μόνῳ πρὸς τὸν κόλπον τὸν Ὑρκάνιον. θεὸν δὲ ἥλιον μόνον ἡγοῦνται, τούτῳ δὲ ἱπποθυτοῦσι· γαμεῖ δ' ἕκαστος μίαν, χρῶνται

[1] προσηγόρευσε *oxz*, προσηγορεύσας other MSS.
[2] τις, Tzschucke, for τοῖς D, τῆς *Chilrwg*, τῶν *gxy*.
[3] For ὧν, Meineke, following conj. of Corais, reads οὗ.
[4] Ἀράξην *i*, Ἄραξον other MSS.

as he thought best and halted; and that the Sacae pursued, found the camp empty of men but full of things conducive to enjoyment, and filled themselves to the full; and that Cyrus turned back, and found them drunk and crazed, so that some were slain while lying stupefied and asleep, whereas others fell victims to the arms of the enemy while dancing and revelling naked, and almost all perished; and Cyrus, regarding the happy issue as of divine origin, consecrated that day to the goddess of his fathers and called it Sacaea; and that wherever there is a temple of this goddess, there the festival of the Sacaea, a kind of Bacchic festival, is the custom, at which men, dressed in the Scythian garb, pass day and night drinking and playing wantonly with one another, and also with the women who drink with them.

6. The Massagetae disclosed their valour in their war with Cyrus, to which many writers refer again and again; and it is from these that we must get our information. Statements to the following effect are made concerning the Massagetae: that some of them inhabit mountains, some plains, others marshes which are formed by the rivers, and others the islands in the marshes. But the country is inundated most of all, they say, by the Araxes River, which splits into numerous branches and empties by its other mouths into the other sea[1] on the north, though by one single mouth it reaches the Hyrcanian Gulf. They regard Helius[2] alone as god, and to him they sacrifice horses. Each man marries only one wife, but they use also the wives of

[1] The Northern Ocean. [2] The Sun.

δὲ καὶ ταῖς ἀλλήλων οὐκ ἀφανῶς, ὁ δὲ μιγνύμενος τῇ ἀλλοτρίᾳ, τὴν φαρέτραν ἐξαρτήσας ἐκ τῆς ἁμάξης, φανερῶς μίγνυται· θάνατος δὲ νομίζεται παρ' αὐτοῖς ἄριστος, ὅταν γηράσαντες κατακοπῶσι μετὰ προβατείων κρεῶν καὶ ἀναμὶξ βρωθῶσι· τοὺς δὲ νόσῳ θανόντας ῥίπτουσιν, ὡς ἀσεβεῖς καὶ ἀξίους ὑπὸ θηρίων βεβρῶσθαι. ἀγαθοὶ δὲ ἱππόται καὶ πεζοί, τόξοις δὲ χρῶνται καὶ μαχαίραις καὶ θώραξι καὶ σαγάρεσι χαλκαῖς, ζῶναι δὲ αὐτοῖς εἰσὶ χρυσαῖ καὶ διαδήματα ἐν ταῖς μάχαις· οἵ τε ἵπποι χρυσοχάλινοι, καὶ μασχαλιστῆρες δὲ χρυσοῖ· ἄργυρος δ' οὐ γίνεται παρ' αὐτοῖς, σίδηρος δ' ὀλίγος, χαλκὸς δὲ καὶ χρυσὸς ἄφθονος.

7. Οἱ μὲν οὖν ἐν ταῖς νήσοις, οὐκ ἔχοντες σπόριμα, ῥιζοφαγοῦσι καὶ ἀγρίοις χρῶνται καρποῖς, ἀμπέχονται δὲ τοὺς τῶν δένδρων φλοιούς (οὐδὲ γὰρ βοσκήματα ἔχουσι), πίνουσι δὲ τὸν ἐκ τῶν δένδρων καρπὸν ἐκθλίβοντες· οἱ δ' ἐν τοῖς ἕλεσιν ἰχθυοφαγοῦσιν, ἀμπέχονται δὲ τὰ τῶν φωκῶν δέρματα τῶν ἐκ θαλάττης ἀνατρεχουσῶν· οἱ δ' ὄρειοι τοῖς ἀγρίοις τρέφονται καὶ αὐτοὶ καρποῖς· ἔχουσι δὲ καὶ πρόβατα ὀλίγα, ὥστ' οὐδὲ κατακόπτουσι, φειδόμενοι τῶν ἐρίων χάριν καὶ τοῦ γάλακτος· τὴν δ' ἐσθῆτα ποικίλλουσιν ἐπιχρίστοις φαρμάκοις δυσεξίτηλον ἔχουσι τὸ ἄνθος. οἱ δὲ πεδινοί, καίπερ ἔχοντες χώραν, οὐ γεωργοῦσιν, ἀλλὰ ἀπὸ προβάτων καὶ ἰχθύων ζῶσι νομαδικῶς καὶ Σκυθικῶς, ἔτι γάρ τις καὶ κοινὴ ἡ δίαιτα πάντων τῶν τοιούτων, ἣν πολλάκις λέγω, καὶ ταφαὶ δ' εἰσὶ παραπλήσιαι καὶ ἤθη καὶ

one another; not in secret, however, for the man who is to have intercourse with the wife of another hangs up his quiver on the wagon and has intercourse with her openly. And they consider it the best kind of death when they are old to be chopped up with the flesh of cattle and eaten mixed up with that flesh. But those who die of disease are cast out as impious and worthy only to be eaten by wild beasts. They are good horsemen and foot-soldiers; they use bows, short swords, breastplates, and sagares[1] made of brass; and in their battles they wear head-bands and belts made of gold. And their horses have bits and girths made of gold. Silver is not found in their country, and only a little iron, but brass and gold in abundance.

7. Now those who live in the islands, since they have no grain to sow, use roots and wild fruits as food, and they clothe themselves with the bark of trees (for they have no cattle either), and they drink the juice squeezed out of the fruit of the trees. Those who live in the marshes eat fish, and clothe themselves in the skins of the seals that run up thither from the sea. The mountaineers themselves also live on wild fruits; but they have sheep also, though only a few, and therefore they do not butcher them, sparing them for their wool and milk; and they variegate the colour of their clothing by staining it with dyes whose colours do not easily fade. The inhabitants of the plains, although they possess land, do not till it, but in the nomadic or Scythian fashion live on sheep and fish. Indeed, there not only is a certain mode of life common to all such peoples, of which I often speak,[2] but their burials, customs, and their way of living as a whole,

[1] See note on "sagaris," 11. 5. 1. [2] *e.g.* 7. 3. 7—8.

ὁ σύμπας βίος, αὐθέκαστος μέν, σκαιὸς δὲ καὶ ἄγριος καὶ πολεμικός, πρὸς δὲ τὰ συμβόλαια ἁπλοῦς καὶ ἀκάπηλος.

8. Τοῦ δὲ τῶν Μασσαγετῶν καὶ τῶν Σακῶν ἔθνους καὶ οἱ Ἀττάσιοι[1] καὶ οἱ Χωράσμιοι, εἰς οὓς ἀπὸ τῶν Βακτριανῶν καὶ τῶν Σογδιανῶν ἔφυγε Σπιταμένης, εἷς ἐκ τῶν ἀποδράντων Περσῶν τὸν Ἀλέξανδρον, καθάπερ καὶ Βῆσσος· καὶ ὕστερον δὲ Ἀρσάκης τὸν Καλλίνικον φεύγων Σέλευκον εἰς τοὺς Ἀπασιάκας ἐχώρησε. φησὶ δ' Ἐρατοσθένης τοὺς Ἀραχωτοὺς καὶ Μασσαγέτας τοῖς Βακτρίοις παρακεῖσθαι πρὸς δύσιν παρὰ τὸν Ὦξον, καὶ Σάκας μὲν καὶ Σογδιανοὺς τοῖς ὅλοις ἐδάφεσιν ἀντικεῖσθαι τῇ Ἰνδικῇ, Βακτρίους δ' C 514 ἐπ' ὀλίγον· τὸ γὰρ πλέον τῷ Παροπαμισῷ παρακεῖσθαι· διείργειν δὲ Σάκας μὲν καὶ Σογδιανοὺς τὸν Ἰαξάρτην, καὶ Σογδιανοὺς δὲ καὶ Βακτριανοὺς τὸν Ὦξον, μεταξὺ δὲ Ὑρκανῶν καὶ Ἀρίων Ταπύρους οἰκεῖν· κύκλῳ δὲ περὶ τὴν θάλατταν μετὰ τοὺς Ὑρκανοὺς Ἀμάρδους[2] τε καὶ Ἀναριάκας[3] καὶ Καδουσίους καὶ Ἀλβανοὺς καὶ Κασπίους καὶ Οὐιτίους, τάχα δὲ καὶ ἑτέρους μέχρι Σκυθῶν, ἐπὶ θάτερα δὲ μέρη τῶν Ὑρκανῶν Δέρβικας, τοὺς δὲ Καδουσίους συμψαύειν Μήδων καὶ Ματιανῶν[4] ὑπὸ τὸν Παραχοάθραν.

9. Τὰ δὲ διαστήματα οὕτω λέγει· ἀπὸ μὲν τοῦ Κασπίου ἐπὶ τὸν Κῦρον ὡς χιλίους ὀκτακοσίους

[1] On Ἀττάσιοι, believed to be corrupt, see C. Müller, *Ind. Var. Lect.*, p. 1015.

[2] Ἀμάρδους, Xylander, for Ἀρμανούς E, Ἀμάρνους other MSS.; so the later editors.

[3] Ἀναριάκας, Xylander, for Ἀδριάκας E, Ἀνδριάκας other MSS.; so the later editors.

are alike, that is, they are self-assertive, uncouth, wild, and warlike, but, in their business dealings, straightforward and not given to deceit.

8. Belonging to the tribe of the Massagetae and the Sacae are also the Attasii and the Chorasmii, to whom Spitamenes[1] fled from the country of the Bactriani and the Sogdiani. He was one of the Persians who escaped from Alexander, as did also Bessus; and later Arsaces,[2] when he fled from Seleucus Callinicus,[3] withdrew into the country of the Apasiacae. Eratosthenes says that the Arachoti and Massagetae are situated alongside the Bactrians towards the west along the Oxus River, and that the Sacae and the Sogdiani, with the whole of their lands, are situated opposite India, but the Bactriani only for a slight distance; for, he says, they are situated for the most part alongside the Paropamisus, and the Sacae and the Sogdiani are separated from one another by the Iaxartes River, and the Sogdiani and the Bactriani by the Oxus River; and the Tapyri live between the Hyrcanians and the Arians; and in a circuit round the sea after the Hyrcanians one comes to the Amardi, Anariacae, Cadusii, Albani, Caspii, Vitii, and perhaps also other peoples, until one reaches the Scythians; and on the other side of the Hyrcanians are Derbices; and the Cadusii border on the Medi and Matiani below the Parachoathras.

9. Eratosthenes gives the distances as follows: From Mt. Caspius to the Cyrus River, about one

[1] See Arrian's *Expedition of Alexander*, 3. 28. 16, 29. 12, 30. 1.
[2] King of Parthia.
[3] King of Syria 246—226 B.C.

[4] E reads Μαντιανῶν (cp. Μαντιανή and note in 11. 14. 8).

σταδίους, ἔνθεν δ' ἐπὶ Κασπίας πύλας πεντακισχιλίους ἑξακοσίους, εἶτ' εἰς Ἀλεξάνδρειαν τὴν ἐν Ἀρίοις ἑξακισχιλίους τετρακοσίους, εἶτ' εἰς Βάκτραν τὴν πόλιν, ἣ καὶ Ζαριάσπα καλεῖται, τρισχιλίους ὀκτακοσίους ἑβδομήκοντα, εἶτ' ἐπὶ τὸν Ἰαξάρτην ποταμόν, ἐφ' ὃν Ἀλέξανδρος ἧκεν, ὡς πεντακισχιλίους· ὁμοῦ δισμύριοι δισχίλιοι ἑξακόσιοι ἑβδομήκοντα. λέγει δὲ καὶ οὕτω τὰ διαστήματα ἀπὸ Κασπίων πυλῶν εἰς Ἰνδούς, εἰς μὲν Ἑκατόμπυλον χιλίους ἐννακοσίους ἑξήκοντά φασιν, εἰς δ' Ἀλεξάνδρειαν τὴν ἐν Ἀρίοις τετρακισχιλίους πεντακοσίους τριάκοντα, εἶτ' εἰς Προφθασίαν τὴν ἐν Δραγγῇ[1] χιλίους ἑξακοσίους, οἱ δὲ πεντακοσίους, εἶτ' εἰς Ἀραχωτοὺς τὴν πόλιν τετρακισχιλίους ἑκατὸν εἴκοσιν, εἶτ' εἰς Ὀρτόσπανα, ἐπὶ τὴν ἐκ Βάκτρων τρίοδον, δισχιλίους, εἶτ' εἰς τὰ ὅρια τῆς Ἰνδικῆς χιλίους· ὁμοῦ μύριοι πεντακισχίλιοι τριακόσιοι.[2] ἐπ' εὐθείας δὲ τῷ διαστήματι τούτῳ[3] συνεχὲς δεῖ νοεῖν, τὸ ἀπὸ τοῦ Ἰνδοῦ μέχρι τῆς ἑῴας θαλάττης μῆκος τῆς Ἰνδικῆς. ταῦτα μὲν τὰ περὶ τοὺς Σάκας.

IX

1. Ἡ δὲ Παρθυαία πολλὴ μὲν οὐκ ἔστι· συνετέλει γοῦν μετὰ τῶν Ὑρκανῶν κατὰ[4] τὰ Περσικά, καὶ μετὰ ταῦτα, τῶν Μακεδόνων κρατούντων ἐπὶ

[1] Δραγγῇ, the editors, for Δράπῃ.
[2] τριακόσιοι, Kramer, for πεντακόσιοι; so the later editors.
[3] τό, before συνεχές, Jones deletes.
[4] κατά, before τά, Casaubon inserts; so the later editors.

thousand eight hundred stadia; thence to the Caspian Gates, five thousand six hundred; then to Alexandreia in the country of the Arians, six thousand four hundred; then to the city Bactra, also called Zariaspa, three thousand eight hundred and seventy; then to the Iaxartes River, to which Alexander came, about five thousand; a distance all told of twenty-two thousand six hundred and seventy stadia. He gives also the distance from the Caspian Gates to India as follows: To Hecatompylus, they say one thousand nine hundred and sixty stadia; to Alexandreia in the country of the Arians, four thousand five hundred and thirty; then to Prophthasia in Drangê, one thousand six hundred (others say one thousand five hundred); then to the city Arachoti four thousand one hundred and twenty; then to Ortospana, to the junction of the three roads leading from Bactra, two thousand; then to the borders of India, one thousand; a distance all told of fifteen thousand three hundred stadia.[1] We must conceive of the length of India, reckoned from the Indus River to the eastern sea, as continuous with this distance in a straight line. So much for the Sacae.

IX

1. As for the Parthian country, it is not large; at any rate, it paid its tribute along with the Hyrcanians in the Persian times, and also after this, when for a long time the Macedonians held the

[1] The sum total of the distances here given is 15,210 stadia, not 15,300 (15,500 MSS.). The total of 15,300 is again found in 15. 2. 8.

χρόνον πολύν. πρὸς δὲ τῇ σμικρότητι δασεῖα καὶ ὀρεινή ἐστι καὶ ἄπορος, ὥστε[1] διὰ τοῦτο δρόμῳ διεξιᾶσι τὸν ἑαυτῶν οἱ βασιλεῖς ὄχλον, οὐ δυναμένης τρέφειν τῆς χώρας οὐδ' ἐπὶ μικρόν· ἀλλὰ νῦν ηὔξηται. μέρη δ' ἐστὶ τῆς Παρθυηνῆς ἥ τε Κωμισηνὴ[2] καὶ ἡ Χωρηνή, σχεδὸν δέ τι καὶ τὰ μέχρι πυλῶν Κασπίων καὶ Ῥαγῶν καὶ Ταπύρων, ὄντα τῆς Μηδίας πρότερον. ἔστι δ' Ἀπάμεια καὶ Ἡράκλεια πόλεις περὶ τὰς Ῥάγας. εἰσὶ δ' ἀπὸ Κασπίων πυλῶν εἰς μὲν Ῥάγας στάδιοι πεντακόσιοι, ὥς φησιν Ἀπολλόδωρος, εἰς δ' Ἑκατόμπυλον, τὸ τῶν Παρθυαίων βασίλειον, χίλιοι διακόσιοι ἑξήκοντα· τοὔνομα δὲ ταῖς Ῥάγαις ἀπὸ τῶν γενομένων σεισμῶν γενέσθαι φασίν, ὑφ' ὧν πόλεις τε συχναὶ καὶ κῶμαι δισχίλιαι, ὡς Ποσειδώνιός φησι, ἀνετράπησαν. τοὺς δὲ Ταπύρους οἰκεῖν φασὶ μεταξὺ Δερβίκων τε καὶ
C 515 Ὑρκανῶν. ἱστοροῦσι δὲ περὶ τῶν Ταπύρων, ὅτι αὐτοῖς εἴη νόμιμον τὰς γυναῖκας ἐκδιδόναι τὰς γαμετὰς ἑτέροις ἀνδράσιν, ἐπειδὰν ἐξ αὐτῶν ἀνέλωνται δύο ἢ τρία τέκνα, καθάπερ καὶ Κάτων Ὁρτησίῳ δεηθέντι ἐξέδωκε τὴν Μαρκίαν ἐφ' ἡμῶν κατὰ παλαιὸν Ῥωμαίων ἔθος.

2. Νεωτερισθέντων δὲ τῶν ἔξω τοῦ Ταύρου διὰ τὸ πρὸς ἄλλοις[3] εἶναι τοὺς τῆς Συρίας καὶ τῆς Μηδίας βασιλέας τοὺς ἔχοντας καὶ ταῦτα, πρῶτον

[1] ὥστε *gixy*, ὡς other MSS. except E, which omits the word.
[2] Κωμισηνή, Tzschucke, for Κωμεισηνή CD*h*, Καμβυσηνή *y*, Καμεισηνή other MSS. ; so the later editors.
[3] ἄλλοις, Corais, from conj. of Tyrwhitt, for ἀλλήλους *loz*, ἀλλήλοις other MSS. (but see Kramer's note).

mastery. And, in addition to its smallness, it is thickly wooded and mountainous, and also poverty-stricken, so that on this account the kings send their own throngs through it in great haste, since the country is unable to support them even for a short time. At present, however, it has increased in extent. Parts of the Parthian country are Comisenê and Chorenê, and, one may almost say, the whole region that extends as far as the Caspian Gates and Rhagae and the Tapyri, which formerly belonged to Media. And in the neighbourhood of Rhagae are the cities Apameia and Heracleia. The distance from the Caspian Gates to Rhagae is five hundred stadia, as Apollodorus says, and to Hecatompylus, the royal seat of the Parthians, one thousand two hundred and sixty. Rhagae is said to have got its name from the earthquakes that took place in that country, by which numerous cities and two thousand villages, as Poseidonius says, were destroyed. The Tapyri are said to live between the Derbices and the Hyrcanians. It is reported of the Tapyri that it was a custom of theirs to give their wives in marriage to other husbands as soon as they had had two or three children by them; just as in our times, in accordance with an ancient custom of the Romans, Cato gave Marcia in marriage to Hortensius at the request of the latter.

2. But when revolutions were attempted by the countries outside the Taurus, because of the fact that the kings of Syria and Media, who were in possession also of these countries, were busily engaged with others, those who had been entrusted with their government first caused the revolt of

μὲν τὴν Βακτριανὴν ἀπέστησαν οἱ πεπιστευμένοι καὶ τὴν ἐγγὺς αὐτῆς πᾶσαν, οἱ περὶ Εὐθύδημον. ἔπειτ' Ἀρσάκης, ἀνὴρ Σκύθης, τῶν Δαῶν[1] τινὰς ἔχων, τοὺς Ἀπάρνους[2] καλουμένους νομάδας, παροικοῦντας τὸν Ὦχον, ἐπῆλθεν ἐπὶ τὴν Παρθυαίαν καὶ ἐκράτησεν αὐτῆς. κατ' ἀρχὰς μὲν οὖν ἀσθενὴς ἦν διαπολεμῶν πρὸς τοὺς ἀφαιρεθέντας τὴν χώραν καὶ αὐτὸς καὶ οἱ διαδεξάμενοι ἐκεῖνον, ἔπειθ' οὕτως ἴσχυσαν ἀφαιρούμενοι τὴν πλησίον ἀεὶ διὰ τὰς ἐν τοῖς πολέμοις κατορθώσεις, ὥστε τελευτῶντες ἁπάσης τῆς ἐντὸς Εὐφράτου κύριοι κατέστησαν. ἀφείλοντο δὲ καὶ τῆς Βακτριανῆς μέρος βιασάμενοι τοὺς Σκύθας καὶ ἔτι πρότερον τοὺς περὶ Εὐκρατίδαν, καὶ νῦν ἐπάρχουσι τοσαύτης γῆς καὶ τοσούτων ἐθνῶν, ὥστε ἀντίπαλοι τοῖς Ῥωμαίοις τρόπον τινὰ γεγόνασι κατὰ μέγεθος τῆς ἀρχῆς. αἴτιος δ' ὁ βίος αὐτῶν καὶ τὰ ἔθη τὰ ἔχοντα πολὺ μὲν τὸ βάρβαρον καὶ τὸ Σκυθικόν, πλέον μέντοι τὸ χρήσιμον πρὸς ἡγεμονίαν καὶ τὴν ἐν τοῖς πολέμοις κατόρθωσιν.

3. Φασὶ δὲ τοὺς Ἀπάρνους[3] Δάας μετανάστας εἶναι ἐκ τῶν ὑπὲρ τῆς Μαιώτιδος Δαῶν, οὓς Ξανδίους ἢ Παρίους καλοῦσιν· οὐ πάνυ δ' ὡμολόγηται Δάας εἶναί τινας τῶν ὑπὲρ τῆς Μαιώτιδος Σκυθῶν· ἀπὸ τούτων δ' οὖν ἕλκειν φασὶ τὸ γένος τὸν Ἀρσάκην, οἱ δὲ Βακτριανὸν λέγουσιν αὐτόν, φεύγοντα δὲ τὴν αὔξησιν τῶν περὶ Διόδοτον ἀποστῆσαι τὴν Παρθυαίαν. εἰρηκότες

[1] Δαῶν, Xylander, for Δατίων; so the later editors.
[2] Ἀπάρνους, Jones, for Πάρνους (see note on Ἀπάρνους, 11. 7. 1).

Bactriana and of all the country near it, I mean Euthydemus and his followers; and then Arsaces, a Scythian, with some of the Däae (I mean the Aparnians, as they were called, nomads who lived along the Ochus), invaded Parthia and conquered it. Now at the outset Arsaces was weak, being continually at war with those who had been deprived by him of their territory, both he himself and his successors, but later they grew so strong, always taking the neighbouring territory, through successes in warfare, that finally they established themselves as lords of the whole of the country inside the Euphrates. And they also took a part of Bactriana, having forced the Scythians, and still earlier Eucratides and his followers, to yield to them; and at the present time they rule over so much land and so many tribes that in the size of their empire they have become, in a way, rivals of the Romans. The cause of this is their mode of life, and also their customs, which contain much that is barbarian and Scythian in character, though more that is conducive to hegemony and success in war.

3. They say that the Aparnian Däae were emigrants from the Däae above Lake Maeotis, who are called Xandii or Parii. But the view is not altogether accepted that the Däae are a part of the Scythians who live about Maeotis. At any rate, some say that Arsaces derives his origin from the Scythians, whereas others say that he was a Bactrian, and that when in flight from the enlarged power of Diodotus and his followers he caused Parthia to revolt. But since I have said much

---

[3] Ἀπάρνους, Jones, for Πάρνους (see note on Ἀπάρνους, 11. 7. 1).

δὲ πολλὰ περὶ τῶν Παρθικῶν νομίμων ἐν τῇ ἕκτῃ τῶν ἱστορικῶν ὑπομνημάτων βίβλῳ, δευτέρᾳ δὲ τῶν μετὰ Πολύβιον, παραλείψομεν ἐνταῦθα, μὴ ταυτολογεῖν δόξωμεν, τοσοῦτον εἰπόντες μόνον, ὅτι τῶν Παρθυαίων συνέδριόν φησιν εἶναι Ποσειδώνιος διττόν, τὸ μὲν συγγενῶν, τὸ δὲ σοφῶν καὶ μάγων, ἐξ ὧν ἀμφοῖν τοὺς βασιλεῖς καθίστασθαι.

X

1. Ἡ δ' Ἀρία καὶ ἡ Μαργιανὴ[1] κράτιστα[2] χωρία ἐστὶ ταύτῃ, τῇ μὲν ὑπὸ τῶν ὀρῶν ἐγκλειόμενα, τῇ δ' ἐν πεδίοις τὰς οἰκήσεις ἔχοντα. τὰ μὲν οὖν ὄρη νέμονται σκηνῖταί τινες, τὰ δὲ πεδία ποταμοῖς διαρρεῖται ποτίζουσιν αὐτά, τὰ μὲν τῷ Ἀρίῳ, τὰ δὲ Μάργῳ. ὁμορεῖ δὲ ἡ Ἀρία τῇ C 516 Βακτριανῇ καὶ τὴν ὑποστᾶσαν ὄρει τῷ ἔχοντι τὴν Βακτριανήν·[3] διέχει δὲ τῆς Ὑρκανίας περὶ ἑξακισχιλίους σταδίους. συντελὴς δ' ἦν αὐτῇ καὶ ἡ Δραγγιανὴ μέχρι Καρμανίας, τὸ μὲν πλέον τοῖς νοτίοις μέρεσι τῶν ὀρῶν ὑποπεπτωκυῖα, ἔχουσα μέντοι τινὰ τῶν μερῶν[4] καὶ τοῖς ἀρκτικοῖς πλησιάζοντα τοῖς κατὰ τὴν Ἀρίαν· καὶ ἡ Ἀραχωσία δὲ οὐ πολὺ ἄπωθέν ἐστι, καὶ αὕτη

[1] Μαργιανή, Casaubon, for Ματιανή E, Μαντιανή *l*, Μαρτιανή other MSS.

[2] κράτιστα E, ἃ κράτιστα other MSS.

[3] The words καὶ τὴν ὑποστᾶσαν ὄρει τῷ ἔχοντι τὴν Βακτριανήν are unintelligible. For purely conjectural emendations see C. Müller, *Ind. Var. Lect.* p. 1016.

about the Parthian usages in the sixth book of my *Historical Sketches* and in the second book of my *History* of events after Polybius,[1] I shall omit discussion of that subject here, lest I may seem to be repeating what I have already said, though I shall mention this alone, that the Council of the Parthians, according to Poseidonius, consists of two groups, one that of kinsmen,[2] and the other that of wise men and Magi, from both of which groups the kings were appointed.[3]

X

1. Aria and Margiana are the most powerful districts in this part of Asia, these districts in part being enclosed by the mountains and in part having their habitations in the plains. Now the mountains are occupied by Tent-dwellers, and the plains are intersected by rivers that irrigate them, partly by the Arius and partly by the Margus. Aria borders on Margiana and . . . Bactriana;[4] it is about six thousand stadia distant from Hyrcania. And Drangiana, as far as Carmania, was joined with Aria in the payment of tribute—Drangiana, for the most part, lying below the southern parts of the mountains, though some parts of it approach the northern region opposite Aria. But Arachosia, also, is not far away, this country too lying below the

[1] See Vol. I, p. 47, note 1. [2] *i.e.* of the king.

[3] It appears that the kings were chosen from the first group by the members of the second (see Forbiger, Vol. III, p. 39, note 7).

[4] The text is corrupt (see critical note).

---

[4] Instead of μερῶν E reads ὀρῶν.

τοῖς νοτίοις μέρεσι τῶν ὀρῶν ὑποπεπτωκυῖα καὶ μέχρι τοῦ Ἰνδοῦ ποταμοῦ τεταμένη, μέρος οὖσα τῆς Ἀριανῆς. μῆκος δὲ τῆς Ἀρίας ὅσον δισχίλιοι στάδιοι, πλάτος δὲ τριακόσιοι τοῦ πεδίου· πόλεις δὲ Ἀρτακάηνα[1] καὶ Ἀλεξάνδρεια καὶ Ἀχαΐα, ἐπώνυμοι τῶν κτισάντων. εὐοινεῖ δὲ σφόδρα ἡ γῆ· καὶ γὰρ εἰς τριγονίαν παραμένει ἐν ἀπιτώττοις ἄγγεσι.

2. Παραπλησία δ' ἐστὶ καὶ ἡ Μαργιανή, ἐρημίαις δὲ περιέχεται τὸ πεδίον. θαυμάσας δὲ τὴν εὐφυΐαν ὁ Σωτὴρ Ἀντίοχος τείχει περιέβαλε κύκλον ἔχοντι χιλίων καὶ πεντακοσίων σταδίων, πόλιν δὲ ἔκτισεν Ἀντιόχειαν. εὐάμπελος δὲ καὶ αὕτη ἡ γῆ· φασὶ γοῦν τὸν πυθμένα εὑρίσκεσθαι πολλάκις δυσὶν ἀνδράσι περιληπτόν, τὸν δὲ βότρυν δίπηχυν.

XI

1. Τῆς δὲ Βακτρίας μέρη μέν τινα τῇ Ἀρίᾳ παραβέβληται πρὸς ἄρκτον, τὰ πολλὰ δ' ὑπέρκειται πρὸς ἕω· πολλὴ δ' ἐστὶ καὶ πάμφορος πλὴν ἐλαίου. τοσοῦτον δὲ ἴσχυσαν οἱ ἀποστήσαντες Ἕλληνες αὐτὴν διὰ τὴν ἀρετὴν τῆς χώρας, ὥστε τῆς τε Ἀριανῆς ἐπεκράτουν καὶ τῶν Ἰνδῶν, ὥς φησιν Ἀπολλόδωρος ὁ Ἀρτεμιτηνός,[2] καὶ πλείω ἔθνη κατεστρέψαντο ἢ Ἀλέξανδρος, καὶ μάλιστα Μένανδρος (εἴ γε καὶ τὸν Ὕπανιν διέβη

[1] For variant spellings see C. Müller, *Ind. Var. Lect.* p. 1016.

[2] Ἀρτεμιτηνός, Corais, for Ἀρταμιτηνός (cp. 2. 5. 12, 11. 11. 7, and 11. 13. 6).

southern parts of the mountains and extending as far as the Indus River, being a part of Ariana. The length of Aria is about two thousand stadia, and the breadth of the plain about three hundred. Its cities are Artacaëna and Alexandreia and Achaïa, all named after their founders. The land is exceedingly productive of wine, which keeps good for three generations in vessels not smeared with pitch.

2. Margiana is similar to this country, although its plain is surrounded by deserts. Admiring its fertility, Antiochus Soter[1] enclosed a circuit of fifteen hundred stadia with a wall and founded a city Antiocheia. The soil of the country is well suited to the vine; at any rate, they say that a stock of the vine is often found which would require two men to girth it,[2] and that the bunches of grapes are two cubits.[3]

XI

1. As for Bactria, a part of it lies alongside Aria towards the north, though most of it lies above Aria and to the east of it. And much of it produces everything except oil. The Greeks who caused Bactria to revolt grew so powerful on account of the fertility of the country that they became masters, not only of Ariana, but also of India, as Apollodorus of Artemita says: and more tribes were subdued by them than by Alexander—by Menander in particular (at least if he actually crossed the Hypanis towards

[1] King of Syria 280–261 B.C.
[2] *i.e.* about ten to eleven feet in circumference.
[3] *i.e.* about three feet; apparently in *length*, not in *circumference.*

πρὸς ἕω, καὶ μέχρι τοῦ Ἰμάου[1] προῆλθε), τὰ μὲν γὰρ αὐτός, τὰ δὲ Δημήτριος ὁ Εὐθυδήμου υἱός, τοῦ Βακτρίων βασιλέως· οὐ μόνον δὲ τὴν Παταληνὴν κατέσχον, ἀλλὰ καὶ τῆς ἄλλης παραλίας τήν τε Σαραόστου καλουμένην καὶ τὴν Σιγέρδιδος βασιλείαν. καθ' ὅλου δέ φησιν ἐκεῖνος τῆς συμπάσης Ἀριανῆς πρόσχημα εἶναι τὴν Βακτριανήν· καὶ δὴ καὶ μέχρι Σηρῶν καὶ Φρυνῶν[2] ἐξέτεινον τὴν ἀρχήν.

2. Πόλεις δ' εἶχον τά τε Βάκτρα, ἥνπερ καὶ Ζαριάσπαν καλοῦσιν, ἣν διαρρεῖ ὁμώνυμος ποταμὸς ἐκβάλλων εἰς τὸν Ὦξον, καὶ Δάραψα[3] καὶ ἄλλας πλείους· τούτων δ' ἦν καὶ ἡ Εὐκρατιδία, τοῦ ἄρξαντος ἐπώνυμος. οἱ δὲ κατασχόντες αὐτὴν Ἕλληνες καὶ εἰς σατραπείας διῃρήκασιν, C 517 ὧν τήν τε Ἀσπιώνου καὶ τὴν Τουριούαν[4] ἀφῄρηντο Εὐκρατίδην οἱ Παρθυαῖοι. ἔσχον δὲ καὶ τὴν Σογδιανὴν ὑπερκειμένην πρὸς ἕω τῆς Βακτριανῆς μεταξὺ τοῦ τε Ὤξου ποταμοῦ, ὃς ὁρίζει τήν τε τῶν Βακτρίων καὶ τὴν τῶν Σογδίων, καὶ τοῦ Ἰαξάρτου· οὗτος δὲ καὶ τοὺς Σογδίους ὁρίζει καὶ τοὺς νομάδας.

3. Τὸ μὲν οὖν παλαιὸν οὐ πολὺ διέφερον τοῖς βίοις καὶ τοῖς ἤθεσι[5] τῶν νομάδων οἵ τε Σογδιανοὶ καὶ οἱ Βακτριανοί, μικρὸν δ' ὅμως ἡμερώτερα ἦν τὰ τῶν Βακτριανῶν, ἀλλὰ καὶ περὶ τούτων οὐ τὰ βέλτιστα λέγουσιν οἱ περὶ Ὀνησίκριτον· τοὺς γὰρ ἀπειρηκότας διὰ γῆρας ἢ νόσον ζῶντας παρα-

[1] Ἰμάου, Meineke, from conj. of Casaubon, for Ἰσάμου.
[2] Φρυνῶν, Tzschucke, for Φαυνῶν.
[3] Δάραψα, Meineke emends to Ἄδραψα (cp. Ἀδράψα in 15. 2. 10), but the spelling is doubtful.

the east and advanced as far as the Imaüs), for some were subdued by him personally and others by Demetrius, the son of Euthydemus the king of the Bactrians; and they took possession, not only of Patalena, but also, on the rest of the coast, of what is called the kingdom of Saraostus and Sigerdis. In short, Apollodorus says that Bactriana is the ornament of Ariana as a whole; and, more than that, they extended their empire even as far as the Seres and the Phryni.

2. Their cities were Bactra (also called Zariaspa, through which flows a river bearing the same name and emptying into the Oxus), and Darapsa, and several others. Among these was Eucratidia, which was named after its ruler. The Greeks took possession of it and divided it into satrapies, of which the satrapy Turiva and that of Aspionus were taken away from Eucratides by the Parthians. And they also held Sogdiana, situated above Bactriana towards the east between the Oxus River, which forms the boundary between the Bactrians and the Sogdians, and the Iaxartes River. And the Iaxartes forms also the boundary between the Sogdians and the nomads.

3. Now in early times the Sogdians and Bactrians did not differ much from the nomads in their modes of life and customs, although the Bactrians were a little more civilised; however, of these, as of the others, Onesicritus[1] does not report their best traits, saying, for instance, that those who have become helpless because of old age or sickness are thrown out

[1] See *Dictionary* in Vol. I.

[4] Τουριοὺαν, Meineke emends to Ταπυρίαν, perhaps rightly.
[5] For ἤθεσι Meineke reads ἔθεσι.

βάλλεσθαι τρεφομένοις κυσὶν ἐπίτηδες πρὸς τοῦτο, οὓς ἐνταφιαστὰς καλεῖσθαι τῇ πατρῴᾳ γλώττῃ, καὶ ὁρᾶσθαι τὰ μὲν ἔξω τείχους τῆς μητροπόλεως τῶν Βάκτρων καθαρά, τῶν δ' ἐντὸς τὸ πλέον ὀστέων πλῆρες ἀνθρωπίνων· καταλῦσαι δὲ τὸν νόμον Ἀλέξανδρον. τοιαῦτα δέ πως καὶ τὰ περὶ τοὺς Κασπίους ἱστοροῦσι· τοὺς γὰρ γονέας, ἐπειδὰν ὑπὲρ ἑβδομήκοντα ἔτη γεγονότες τυγχάνωσιν, ἐγκλεισθέντας λιμοκτονεῖσθαι. τοῦτο μὲν οὖν ἀνεκτότερον καὶ τῷ Κείων[1] νόμῳ παραπλήσιον, καίπερ ὂν Σκυθικόν, πολὺ μέντοι Σκυθικώτερον τὸ τῶν Βακτριανῶν. καὶ δὴ εἰ[2] διαπορεῖν ἄξιον ἦν, ἡνίκα Ἀλέξανδρος τοιαῦτα κατελάμβανε τἀνταῦθα, τί χρὴ εἰπεῖν[3] τὰ ἐπὶ τῶν πρώτων Περσῶν καὶ τῶν ἔτι πρότερον ἡγεμόνων, ὁποῖα εἰκὸς ἦν παρ' αὐτοῖς νεμομίσθαι;

4. Φασὶ δ' οὖν ὀκτὼ πόλεις τὸν Ἀλέξανδρον ἔν τε τῇ Βακτριανῇ καὶ τῇ Σογδιανῇ κτίσαι, τινὰς δὲ κατασκάψαι, ὧν Καριάτας μὲν τῆς Βακτριανῆς, ἐν ᾗ Καλλισθένης συνελήφθη καὶ παρεδόθη φυλακῇ, Μαράκανδα δὲ τῆς Σογδιανῆς καὶ τὰ Κῦρα, ἔσχατον ὂν Κύρου κτίσμα, ἐπὶ τῷ Ἰαξάρτῃ ποταμῷ κείμενον, ὅπερ ἦν ὅριον τῆς Περσῶν ἀρχῆς· κατασκάψαι δὲ τὸ κτίσμα τοῦτο, καίπερ ὄντα φιλόκυρον, διὰ τὰς πυκνὰς ἀποστάσεις· ἑλεῖν δὲ καὶ πέτρας ἐρυμνὰς σφόδρα ἐκ προδοσίας, τήν τε ἐν τῇ Βακτριανῇ, τὴν Σισιμίθρου, ἐν ᾗ εἶχεν Ὀξυάρτης τὴν θυγατέρα Ῥωξάνην, καὶ τὴν ἐν τῇ

[1] Κείων, Kramer, for οἰκείῳ; so the later editors.
[2] εἰ, after δή, Jones inserts.
[3] εἰπεῖν, *o* and Corais, for ποιεῖν.

alive as prey to dogs kept expressly for this purpose, which in their native tongue are called "undertakers," and that while the land outside the walls of the metropolis of the Bactrians looks clean, yet most of the land inside the walls is full of human bones; but that Alexander broke up the custom. And the reports about the Caspians are similar, for instance, that when parents live beyond seventy years they are shut in and starved to death. Now this latter custom is more tolerable; and it is similar to that of the Ceians,[1] although it is of Scythian origin; that of the Bactrians, however, is much more like that of the Scythians. And so, if it was proper to be in doubt as to the facts at the time when Alexander was finding such customs there, what should one say as to what sort of customs were probably in vogue among them in the time of the earliest Persian rulers and the still earlier rulers?

4. Be this as it may, they say that Alexander founded eight cities in Bactriana and Sogdiana, and that he rased certain cities to the ground, among which was Cariatae in Bactriana, in which Callisthenes was seized and imprisoned, and Maracanda and Cyra in Sogdiana, Cyra being the last city founded by Cyrus[2] and being situated on the Iaxartes River, which was the boundary of the Persian empire; and that although this settlement was fond of Cyrus, he rased it to the ground because of its frequent revolts; and that through a betrayal he took also two strongly fortified rocks, one in Bactriana, that of Sisimithres, where Oxyartes kept his daughter

[1] Cf. 10. 5. 6.

[2] Cyrus the Elder.

Σογδιανῇ τὴν τοῦ Ὥξου, οἱ δ' Ἀριαμάζου φασί. τὴν μὲν οὖν Σισιμίθρου πεντεκαίδεκα σταδίων ἱστοροῦσι τὸ ὕψος, ὀγδοήκοντα δὲ τὸν κύκλον· ἄνω δ' ἐπίπεδον καὶ εὔγεων, ὅσον πεντακοσίους ἄνδρας τρέφειν δυναμένην, ἐν ᾗ καὶ ξενίας τυχεῖν πολυτελοῦς, καὶ γάμους ἀγαγεῖν Ῥωξάνης τῆς Ὀξυάρτου θυγατρὸς τὸν Ἀλέξανδρον· τὴν δὲ τῆς Σογδιανῆς διπλασίαν τὸ ὕψος φασί. περὶ τούτους δὲ τοὺς τόπους καὶ τὸ τῶν Βραγχιδῶν ἄστυ ἀνε-
C 518 λεῖν, οὓς Ξέρξην μὲν ἱδρῦσαι αὐτόθι, συναπάραντας αὐτῷ ἑκόντας ἐκ τῆς οἰκείας, διὰ τὸ παραδοῦναι τὰ χρήματα τοῦ θεοῦ τὰ ἐν Διδύμοις καὶ τοὺς θησαυρούς· ἐκεῖνον δ' ἀνελεῖν μυσαττόμενον τὴν ἱεροσυλίαν καὶ τὴν προδοσίαν.

5. Τὸν δὲ διὰ τῆς Σογδιανῆς ῥέοντα ποταμὸν καλεῖ[1] Πολυτίμητον Ἀριστόβουλος, τῶν Μακεδόνων ὄνομα[2] θεμένων (καθάπερ καὶ ἄλλα πολλὰ τὰ μὲν καινὰ ἔθεσαν, τὰ δὲ παρωνόμασαν), ἄρδοντα δὲ τὴν χώραν ἐκπίπτειν εἰς ἔρημον καὶ ἀμμώδη γῆν, καταπίνεσθαί τε εἰς τὴν ἄμμον, ὡς καὶ τὸν Ἄριον τὸν δι' Ἀρίων ῥέοντα. τοῦ δὲ Ὤχου ποταμοῦ πλησίον ὀρύττοντας εὑρεῖν ἐλαίου πηγὴν λέγουσιν· εἰκὸς δέ, ὥσπερ νιτρώδη τινὰ καὶ στύφοντα ὑγρὰ καὶ ἀσφαλτώδη καὶ θειώδη διαρρεῖ τὴν γῆν, οὕτω καὶ λιπαρὰ εὑρίσκεσθαι, τὸ δὲ σπάνιον ποιεῖ τὴν παραδοξίαν. ῥεῖν δὲ τὸν Ὦχον οἱ μὲν διὰ τῆς Βακτριανῆς φασίν, οἱ δὲ

[1] καλεῖ, Forbiger, from conj. of Casaubon, for καί. *ixy* insert λέγει after Ἀριστόβουλος. *xy* omit the καί, and so Tzschucke and Corais.

[2] ὄνομα, Jones inserts, from conj. of Kramer; others, τοὔνομα.

Rhoxana, and the other in Sogdiana, that of Oxus, though some call it the rock of Ariamazes. Now writers report that that of Sisimithres is fifteen stadia in height and eighty in circuit, and that on top it is level and has a fertile soil which can support five hundred men, and that here Alexander met with sumptuous hospitality and married Rhoxana, the daughter of Oxyartes; but the rock in Sogdiana, they say, is twice as high as that in Bactriana. And near these places, they say, Alexander destroyed also the city of the Branchidae, whom Xerxes had settled there—people who voluntarily accompanied him from their home-land—because of the fact that they had betrayed to him the riches and treasures of the god at Didyma. Alexander destroyed the city, they add, because he abominated the sacrilege and the betrayal.

5. Aristobulus[1] calls the river which flows through Sogdiana Polytimetus, a name imposed by the Macedonians (just as they imposed names on many other places, giving new names to some and slightly altering the spelling of the names of others); and watering the country it empties into a desert and sandy land, and is absorbed in the sand, like the Arius which flows through the country of the Arians. It is said that people digging near the Ochus River found a spring of oil. It is reasonable to suppose that, just as nitrous[2] and astringent and bituminous and sulphurous liquids flow through the earth, so also oily liquids are found; but the rarity causes surprise.[3] According to some, the Ochus flows through Bactriana; according to

[1] See 11. 7. 3 and foot-note.

[2] *i.e.* containing soda (see 11. 14. 8 and foot-note).

[3] *i.e.*, apparently, when one does happen to find them.

παρ' αὐτήν, καὶ οἱ μὲν ἕτερον τοῦ Ὤξου μέχρι τῶν ἐκβολῶν, νοτιώτερον ἐκείνου, ἀμφοτέρων δ' ἐν τῇ Ὑρκανίᾳ τὰς εἰς τὴν θάλατταν ὑπάρχειν ἐκρύσεις, οἱ δὲ κατ' ἀρχὰς μὲν ἕτερον, συμβάλλειν δ' εἰς ἓν τὸ τοῦ Ὤξου ῥεῖθρον, πολλαχοῦ καὶ ἓξ καὶ ἑπτὰ σταδίων ἔχοντα τὸ πλάτος. ὁ μέντοι Ἰαξάρτης ἀπ' ἀρχῆς μέχρι τέλους ἕτερός ἐστι τοῦ Ὤξου, καὶ εἰς μὲν τὴν αὐτὴν τελευτῶν θάλατταν, αἱ δ' ἐμβολαὶ διέχουσιν ἀλλήλων, ὥς φησι Πατροκλῆς, παρασάγγας ὡς ὀγδοήκοντα· τὸν δὲ παρασάγγην τὸν Περσικὸν οἱ μὲν ἑξήκοντα σταδίων φασίν, οἱ δὲ τριάκοντα ἢ[1] τετταράκοντα. ἀναπλεόντων δ' ἡμῶν τὸν Νεῖλον ἄλλοτ' ἄλλοις μέτροις χρώμενοι τὰς σχοίνους ὠνόμαζον ἀπὸ πόλεως ἐπὶ πόλιν, ὥστε τὸν αὐτὸν τῶν σχοίνων ἀριθμὸν ἀλλαχοῦ μὲν μείζω παρέχειν πλοῦν, ἀλλαχοῦ δὲ βραχύτερον· οὕτως ἐξ ἀρχῆς παραδεδομένον καὶ φυλαττόμενον μέχρι νῦν.

6. Μέχρι μὲν δὴ τῆς Σογδιανῆς πρὸς ἀνίσχοντα ἥλιον ἰόντι ἀπὸ τῆς Ὑρκανίας γνώριμα ὑπῆρξε τὰ ἔθνη καὶ τοῖς Πέρσαις πρότερον τὰ εἴσω[2] τοῦ Ταύρου καὶ τοῖς Μακεδόσι μετὰ ταῦτα καὶ τοῖς Παρθυαίοις. τὰ δ' ἐπέκεινα ἐπ' εὐθείας ὅτι μὲν Σκυθικά ἐστιν, ἐκ τῆς ὁμοειδείας εἰκάζεται, στρατεῖαι δ' οὐ γεγόνασιν ἐπ' αὐτοὺς ἡμῖν γνώριμοι, καθάπερ οὐδὲ ἐπὶ τοὺς βορειοτάτους τῶν νομάδων· ἐφ' οὓς ἐπεχείρησε μὲν ὁ Ἀλέξανδρος ἄγειν στρα-

[1] τριάκοντα ἤ, Xylander, for τριακοσίων; so the later editors.
[2] εἴσω, Du Theil, for ἔξω; so Meineke and others.

others, alongside it. And according to some, it is a different river from the Oxus as far as its mouths, being more to the south than the Oxus, although they both have their outlets into the Caspian Sea in Hyrcania, whereas others say that it is different at first, but unites with the Oxus, being in many places as much as six or seven stadia wide. The Iaxartes, however, from beginning to end, is a different river from the Oxus, and although it ends in the same sea, the mouths of the two, according to Patrocles, are about eighty parasangs distant from one another. The Persian parasang, according to some, is sixty stadia, but according to others thirty or forty. When I was sailing up the Nile, they used different measures when they named the distance in "schoeni" from city to city, so that in some places the same number of "schoeni" meant a longer voyage and in others a shorter;[1] and thus the variations have been preserved to this day as handed down from the beginning.

6. Now the tribes one encounters in going from Hyrcania towards the rising sun as far as Sogdiana became known at first to the Persians—I mean the tribes inside[2] Taurus—and afterwards to the Macedonians and to the Parthians; and the tribes situated on the far side of those tribes and in a straight line with them are supposed, from their identity in kind, to be Scythian, although no expeditions have been made against them that I know of, any more than against the most northerly of the nomads. Now Alexander did attempt to lead an expedition

[1] On the variations in the length of the "schoenus," see 17. 1. 24.

[2] *i.e.* "north of" Taurus (see 11. 1. 2).

τείαν ὅτε τὸν Βῆσσον μετῄει καὶ τὸν Σπιταμένην, ζωγρίᾳ δ' ἀναχθέντος τοῦ Βήσσου, τοῦ δὲ Σπιταμένους ὑπὸ τῶν βαρβάρων διαφθαρέντος, ἐπαύσατο τῆς ἐπιχειρήσεως. οὐχ ὁμολογοῦσι δ', ὅτι περιέπλευσάν τινες ἀπὸ τῆς Ἰνδικῆς ἐπὶ τὴν Ὑρκανίαν, ὅτι δὲ δυνατὸν Πατροκλῆς εἴρηκε.

C 519 7. Λέγεται δέ, διότι τοῦ Ταύρου τὸ τελευταῖον, ὃ καλοῦσιν Ἰμάϊον,[1] τῇ Ἰνδικῇ θαλάττῃ ξυνάπτον, οὐδὲν οὔτε προὔχει πρὸς ἕω τῆς Ἰνδικῆς μᾶλλον οὔτ' εἰσέχει· παριόντι δ' εἰς τὸ βόρειον πλευρόν, ἀεί τι τοῦ μήκους ὑφαιρεῖ καὶ τοῦ πλάτους ἡ θάλαττα, ὥστ' ἀποφαίνειν μείουρον[2] πρὸς ἕω τὴν νῦν ὑπογραφομένην μερίδα τῆς Ἀσίας, ἣν ὁ Ταῦρος ἀπολαμβάνει πρὸς τὸν ὠκεανὸν τὸν πληροῦντα τὸ Κάσπιον πέλαγος. μῆκος δ' ἐστὶ ταύτης τῆς μερίδος τὸ μέγιστον ἀπὸ τῆς Ὑρκανίας θαλάττης ἐπὶ τὸν ὠκεανὸν τὸν κατὰ τὸ Ἰμάϊον τρισμυρίων που σταδίων, παρὰ τὴν ὀρεινὴν τοῦ Ταύρου τῆς πορείας οὔσης, πλάτος δ' ἔλαττον τῶν μυρίων.[3] εἴρηται γάρ, ὅτι περὶ τετρακισμυρίους σταδίους ἐστὶ τὸ ἀπὸ τοῦ Ἰσσικοῦ κόλπου μέχρι τῆς ἑῴας θαλάττης τῆς κατὰ Ἰνδούς, ἐπὶ δ' Ἰσσὸν ἀπὸ τῶν ἑσπερίων ἄκρων τῶν κατὰ Στήλας ἄλλοι τρισμύριοι· ἔστι δὲ ὁ μυχὸς τοῦ Ἰσσικοῦ κόλπου μικρὸν ἢ οὐδὲν Ἀμισοῦ ἑωθινώτερος, τὸ δὲ ἀπὸ Ἀμισοῦ ἐπὶ τὴν Ὑρκανίαν γῆν περὶ μυρίους ἐστὶ σταδίους, παράλληλον ὂν τῷ ἀπὸ τοῦ Ἰσσοῦ λεχθέντι ἐπὶ τοὺς Ἰνδούς. λείπεται δὴ τὸ λεχθὲν μῆκος ἐπὶ τὴν

[1] Ἰμάϊον, Meineke, for Ἴμαιον E, Ἴμεον other MSS.
[2] E has μύουρον above μείουρον; Meineke so reads.
[3] See note of Groskurd, who would emend μυρίων to ἑξακισχιλίων; also Kramer's comment.

against these when he was in pursuit of Bessus[1] and Spitamenes, but when Bessus was captured alive and brought back, and Spitamenes was slain by the barbarians, he desisted from his undertaking. It is not generally agreed that persons have sailed around from India to Hyrcania, but Patrocles states that it is possible.

7. It is said that the last part of the Taurus, which is called Imaïus and borders on the Indian Sea, neither extends eastwards farther than India nor into it;[2] but that, as one passes to the northern side, the sea gradually reduces the length and breadth of the country, and therefore causes to taper towards the east the portion of Asia now being sketched, which is comprehended between the Taurus and the ocean that fills the Caspian Sea. The maximum length of this portion from the Hyrcanian Sea to the ocean that is opposite the Imaïus is about thirty thousand stadia, the route being along the mountainous tract of the Taurus, and the breadth less than ten thousand; for, as has been said,[3] the distance from the Gulf of Issus to the eastern sea at India is about forty thousand stadia, and to Issus from the western extremity at the Pillars of Heracles thirty thousand more.[4] The recess of the Gulf of Issus is only slightly, if at all, farther east than Amisus, and the distance from Amisus to the Hyrcanian land is about ten thousand stadia, being parallel to that of the above-mentioned distance from Issus to India. Accordingly, there remain thirty thousand stadia as the above-mentioned length

[1] Satrap of Bactria under Darius III.
[2] To understand this discussion, see Map in Vol. I.
[3] See 2. 1. 3 ff.
[4] See, and compare, 1. 4. 5, 2. 1. 35, 2. 4. 3, and 11. 1. 3.

ἕω τῆς περιωδευμένης νυνὶ μερίδος οἱ τρισμύριοι στάδιοι. πάλιν δὲ τοῦ πλάτους τοῦ μεγίστου[1] τῆς οἰκουμένης ὄντος περὶ τρισμυρίους σταδίους, χλαμυδειδοῦς οὔσης, τὸ διάστημα τοῦτο ἐγγὺς ἂν εἴη τοῦ μεσημβρινοῦ τοῦ διὰ τῆς Ὑρκανίας θαλάττης γραφομένου καὶ τῆς Περσικῆς, εἴπερ ἐστὶ τὸ μῆκος τῆς οἰκουμένης ἑπτὰ μυριάδες· εἰ οὖν ἀπὸ τῆς Ὑρκανίας ἐπὶ Ἀρτεμίταν τὴν ἐν τῇ Βαβυλωνίᾳ στάδιοί εἰσιν ὀκτακισχίλιοι, καθάπερ εἴρηκεν Ἀπολλόδωρος ἐκ τῆς Ἀρτεμίτας,[2] ἐκεῖθεν δ' ἐπὶ τὸ στόμα τῆς κατὰ Πέρσας θαλάττης ἄλλο τοσοῦτόν ἐστι, καὶ πάλιν τοσοῦτον ἢ μικρὸν ἀπολεῖπον εἰς τὰ ἀνταίροντα τοῖς ἄκροις τῆς Αἰθιοπίας, λοιπὸν ἂν εἴη τοῦ πλάτους τῆς οἰκουμένης τοῦ λεχθέντος ἀπὸ τοῦ μυχοῦ τῆς Ὑρκανίας θαλάττης ἐπὶ τοῦ στόματος αὐτῆς ὅσον εἰρήκαμεν. μειούρου δ' ὄντος τοῦ τμήματος τούτου τῆς γῆς ἐπὶ τὰ πρὸς ἕω μέρη, γίνοιτ' ἂν τὸ σχῆμα προσόμοιον μαγειρικῇ κοπίδι, τοῦ μὲν ὄρους ἐπ' εὐθείας ὄντος, καὶ νοουμένου κατὰ τὴν ἀκμὴν τῆς κοπίδος, τῆς δ' ἀπὸ τοῦ στόματος τοῦ Ὑρκανίου παραλίας ἐπὶ Τάμαρον κατὰ θάτερον πλευρὸν εἰς περιφερῆ καὶ μείουρον γραμμὴν ἀπολῆγον.

8. Ἐπιμνηστέον δὲ καὶ τῶν παραδόξων ἐνίων, ἃ θρυλοῦσι περὶ τῶν τελέως βαρβάρων, οἷον τῶν περὶ τὸν Καύκασον καὶ τὴν ἄλλην ὀρεινήν. τοῖς C 520 μὲν γὰρ νόμιμον εἶναί φασι τὸ τοῦ Εὐριπίδου,

> τὸν φύντα θρηνεῖν, εἰς ὅσ' ἔρχεται κακά,
> τὸν δ' αὖ θανόντα καὶ πόνων πεπαυμένον
> χαίροντας εὐφημοῦντας ἐκπέμπειν δόμων·

[1] τοῦ μεγίστου, Corais, for τῆς μεγίστης; so the later editors.
[2] Ἀρτεμίτας, Xylander, for Ἀρτεμησίας *Cx*, Ἀρτεμισίας other MSS.

towards the east of the portion now described. Again, since the maximum breadth of the inhabited world, which is chlamys-shaped,[1] is about thirty thousand stadia, this distance would be measured near the meridian line drawn through the Hyrcanian and Persian Seas, if it be true that the length of the inhabited world is seventy thousand stadia. Accordingly, if the distance from Hyrcania to Artemita in Babylonia is eight thousand stadia, as is stated by Apollodorus of Artemita, and the distance from there to the mouth of the Persian Sea another eight thousand, and again eight thousand, or a little less, to the places that lie on the same parallel as the extremities of Ethiopia, there would remain of the above-mentioned breadth of the inhabited world the distance which I have already given,[2] from the recess of the Hyrcanian Sea to the mouth of that sea. Since this segment of the earth tapers towards the eastern parts, its shape would be like a cook's knife, the mountain being in a straight line and conceived of as corresponding to the edge of the knife, and the coast from the mouth of the Hyrcanian Sea to Tamarum as corresponding to the other side of the knife, which ends in a line that curves sharply to the point.

8. I must also mention some strange customs, everywhere talked about, of the utterly barbarous tribes; for instance, the tribes round the Caucasus and the mountainous country in general. What Euripides refers to is said to be a custom among some of them, "to lament the new-born babe, in view of all the sorrows it will meet in life, but on the other hand to carry forth from their homes with joy and benedictions those who are dead and at rest from their

[1] See Vol. I, p. 435, note 3. [2] Six thousand (2. 1. 17).

ἑτέροις δὲ μηδένα ἀποκτείνειν τῶν ἐξαμαρτόντων τὰ μέγιστα, ἀλλ' ἐξορίζειν μόνον μετὰ τῶν τέκνων, ὑπεναντίως τοῖς Δέρβιξι· καὶ γὰρ ἐπὶ μικροῖς οὗτοι σφάττουσι. σέβονται δὲ Γῆν οἱ Δέρβικες· θύουσι δ' οὐδὲν θῆλυ οὐδὲ ἐσθίουσι· τοὺς δὲ ὑπὲρ ἑβδομήκοντα ἔτη γεγονότας σφάττουσι, ἀναλίσκουσι δὲ τὰς σάρκας οἱ ἄγχιστα γένους· τὰς δὲ γραίας ἀπάγχουσιν, εἶτα θάπτουσι· τοὺς δὲ ἐντὸς ἑβδομήκοντα ἐτῶν ἀποθανόντας οὐκ ἐσθίουσιν, ἀλλὰ θάπτουσι. Σίγιννοι δὲ τἆλλα μὲν περσίζουσιν, ἱππαρίοις δὲ χρῶνται μικροῖς, δασέσιν, ἅπερ ἱππότην ὀχεῖν μὲν οὐ δύνανται, τέθριππα δὲ ζευγνύουσιν· ἡνιοχοῦσι δὲ γυναῖκες, ἐκ παίδων ἠσκημέναι, ἡ δ' ἄριστα ἡνιοχοῦσα συνοικεῖ ᾧ βούλεται. τινὰς δ' ἐπιτηδεύειν φασίν, ὅπως ὡς μακροκεφαλώτατοι φανοῦνται, καὶ προπεπτωκότες τοῖς μετώποις, ὥσθ' ὑπερκύπτειν τῶν γενείων. Ταπύρων[1] δ' ἐστὶ καὶ τὸ τοὺς μὲν ἄνδρας μελανειμονεῖν καὶ μακροκομεῖν, τὰς δὲ γυναῖκας λευχειμονεῖν καὶ βραχυκομεῖν· οἰκοῦσι δὲ μεταξὺ Δερβίκων καὶ Ὑρκανῶν·[2] καὶ ὁ ἀνδρειότατος κριθεὶς γαμεῖ ἣν βούλεται. Κάσπιοι δὲ τοὺς ὑπὲρ ἑβδομήκοντα ἔτη λιμοκτονήσαντες εἰς τὴν ἐρημίαν ἐκτιθέασιν, ἄπωθεν δὲ σκοπεύοντες ἐὰν μὲν ὑπ' ὀρνίθων κατασπωμένους ἀπὸ τῆς κλίνης ἴδωσιν, εὐδαιμονίζουσι, ἐὰν δὲ ὑπὸ θηρίων ἢ κυνῶν, ἧττον, ἐὰν δ' ὑπὸ μηδενός, κακοδαιμονίζουσι.

[1] Ταπύρων, Corais, for Ταπυρίων; so Meineke.

[2] οἰκοῦσι δέ . . . Ὑρκανῶν appears to be a gloss from 11. 9. 1.

troubles";[1] and it is said to be a custom among others to put to death none of the greatest criminals, but only to cast them and their children out of their borders—a custom contrary to that of the Derbices, for these slaughter people even for slight offences. The Derbices worship Mother Earth; and they do not sacrifice, or eat, anything that is female; and when men become over seventy years of age they are slaughtered, and their flesh is consumed by their nearest of kin; but their old women are strangled and then buried. However, the men who die under seventy years of age are not eaten, but only buried. The Siginni imitate the Persians in all their customs, except that they use ponies that are small and shaggy, which, though unable to carry a horseman, are yoked together in a four-horse team and are driven by women trained thereto from childhood; and the woman who drives best cohabits with whomever she wishes. Others are said to practise making their heads appear as long as possible and making their foreheads project beyond their chins. It is a custom of the Tapyri for the men to dress in black and wear their hair long, and for the women to dress in white and wear their hair short. They live between the Derbices and the Hyrcanians. And he who is adjudged the bravest marries whomever he wishes. The Caspians starve to death those who are over seventy years of age and place their bodies out in the desert; and then they keep watch from a distance, and if they see them dragged from their biers by birds, they consider them fortunate, and if by wild beasts or dogs, less so, but if by nothing, they consider them cursed by fortune.

[1] *Frag. Cresphontes* 449 (Nauck).

XII

1. Ἐπεὶ δὲ τὰ βόρεια μέρη τῆς Ἀσίας ποιεῖ ὁ Ταῦρος, ἃ δὴ καὶ ἐντὸς τοῦ Ταύρου καλοῦσιν, εἰπεῖν προειλόμεθα πρῶτον περὶ τούτων·[1] τούτων δ᾽ ἐστὶ καὶ τὰ ἐν τοῖς ὄρεσιν αὐτοῖς ἢ ὅλα ἢ τὰ πλεῖστα. ὅσα μὲν τῶν Κασπίων πυλῶν ἑωθινώτερά ἐστιν, ἁπλουστέραν ἔχει τὴν περιήγησιν διὰ τὴν ἀγριότητα, οὐ πολύ τε ἂν διαφέροι τοῦδε ἢ τοῦδε τοῦ κλίματος συγκαταλεχθέντα· τὰ δ᾽ ἑσπέρια πάντα δίδωσιν εὐπορίαν τοῦ λέγειν περὶ αὐτῶν, ὥστε δεῖ προάγειν ἐπὶ τὰ παρακείμενα ταῖς Κασπίαις πύλαις. παράκειται δὲ ἡ Μηδία πρὸς δύσιν, χώρα καὶ πολλὴ καὶ δυναστεύσασά ποτε καὶ ἐν μέσῳ τῷ Ταύρῳ κειμένη, πολυσχιδεῖ κατὰ ταῦτα ὑπάρχοντι τὰ μέρη καὶ αὐλῶνας ἐμπεριλαμβάνοντι μεγάλους, καθάπερ καὶ τῇ Ἀρμενίᾳ τοῦτο συμβέβηκε.

2. Τὸ γὰρ ὄρος τοῦτο ἄρχεται μὲν ἀπὸ τῆς Καρίας καὶ Λυκίας, ἀλλ᾽ ἐνταῦθα μὲν οὔτε πλάτος οὔτε ὕψος ἀξιόλογον δείκνυσιν, ἐξαίρεται δὲ πολὺ πρῶτον κατὰ τὰς Χελιδονίας· αὗται δ᾽ εἰσὶ νῆσοι κατὰ τὴν ἀρχὴν τῆς Παμφύλων C 521 παραλίας· ἐπὶ δὲ τὰς ἀνατολὰς ἐκτεινόμενον[2] αὐλῶνας μακροὺς[3] ἀπολαμβάνει τοὺς τῶν Κιλίκων· εἶτα τῇ μὲν τὸ Ἀμανὸν ἀπ᾽ αὐτοῦ σχίζεται, τῇ δὲ ὁ Ἀντίταυρος, ἐν ᾧ τὰ Κόμανα ἵδρυται τὰ ἐν τοῖς ἄνω λεγομένοις Καππάδοξιν. οὗτος μὲν

[1] περὶ τούτων, Tzschucke, for περὶ τούτου *oz*; other MSS. omit the words.

[2] ἐκτεινόμενον, Meineke, for ἐκτεινόμενος, from correction in D.

XII

1. SINCE the northern parts of Asia are formed by the Taurus,—I mean the parts which are also called "Cis-Tauran" Asia,[1] I have chosen to describe these first. These include all or most of the regions in the mountains themselves. All that lie farther east than the Caspian Gates admit of a simpler description because of the wildness of their inhabitants; and it would not make much difference whether they were named as belonging to this "clima"[2] or that, whereas all that lie to the west afford abundant matter for description, and therefore I must proceed to the parts which are adjacent to the Caspian Gates. Adjacent to the Caspian Gates on the west is Media, a country at one time both extensive and powerful, and situated in the midst of the Taurus, which is split into many parts in the region of Media and contains large valleys, as is also the case in Armenia.

2. For this mountain has its beginning in Caria and Lycia; there, indeed, it has neither any considerable breadth nor height, but it first rises to a considerable height opposite the Chelidoniae, which are islands at the beginning of the coast of Pamphylia, and then stretching towards the east encloses long valleys, those in Cilicia, and then on one side the Amanus Mountain splits off it and on the other the Antitaurus Mountain, in which latter is situated Comana, in Upper Cappadocia, as it is called. Now

[1] See 11. 1. 1–5. [2] See Vol. I, p. 22, foot-note 2.

[3] μακρούς E, μικράς oz, μικρούς other MSS.

οὖν ἐν τῇ Καταονίᾳ τελευτᾷ, τὸ δὲ Ἀμανὸν ὄρος μέχρι τοῦ Εὐφράτου καὶ τῆς Μελιτηνῆς πρόεισι, καθ' ἣν ἡ Κομμαγηνὴ τῇ Καππαδοκίᾳ παράκειται· ἐκδέχεται δὲ τὰ πέραν τοῦ Εὐφράτου ὄρη, συνεχῆ μὲν τοῖς προειρημένοις, πλὴν ὅσον διακόπτει ῥέων διὰ μέσων ὁ ποταμός· πολλὴν δ' ἐπίδοσιν λαμβάνει εἰς τὸ ὕψος καὶ τὸ πλάτος καὶ τὸ πολυσχιδές. τὸ δ' οὖν νοτιώτατον μάλιστά ἐστιν ὁ Ταῦρος, ὁρίζων τὴν Ἀρμενίαν ἀπὸ τῆς Μεσοποταμίας.

3. Ἐντεῦθεν δὲ ἀμφότεροι ῥέουσιν οἱ τὴν Μεσοποταμίαν ἐγκυκλούμενοι ποταμοὶ καὶ συνάπτοντες ἀλλήλοις ἐγγὺς κατὰ τὴν Βαβυλωνίαν, εἶτα ἐκδιδόντες εἰς τὴν κατὰ Πέρσας θάλατταν, ὅ τε Εὐφράτης καὶ Τίγρις. ἔστι δὲ καὶ μείζων ὁ Εὐφράτης καὶ πλείω διέξεισι χώραν σκολιῷ τῷ ῥείθρῳ, τὰς πηγὰς ἔχων ἐν τῷ προσβόρῳ μέρει τοῦ Ταύρου, ῥέων δ' ἐπὶ δύσιν διὰ τῆς Ἀρμενίας τῆς μεγάλης καλουμένης μέχρι τῆς μικρᾶς, ἐν δεξιᾷ ἔχων ταύτην, ἐν ἀριστερᾷ δὲ τὴν Ἀκιλισηνήν·[1] εἶτ' ἐπιστρέφει πρὸς νότον, συνάπτει δὲ κατὰ τὴν ἐπιστροφὴν τοῖς Καππαδόκων ὁρίοις· δεξιᾷ δὲ ταῦτα ἀφεὶς καὶ τὰ τῶν Κομμαγηνῶν, ἀριστερᾷ δὲ τὴν Ἀκιλισηνὴν καὶ Σωφηνὴν τῆς μεγάλης Ἀρμενίας πρόεισιν ἐπὶ τὴν Συρίαν καὶ λαμβάνει πάλιν ἄλλην ἐπιστροφὴν εἰς τὴν Βαβυλωνίαν καὶ τὸν Περσικὸν κόλπον. ὁ δὲ Τίγρις ἐκ τοῦ νοτίου μέρους τοῦ αὐτοῦ ὄρους ἐνεχθεὶς ἐπὶ τὴν Σελεύκειαν συνάπτει τῷ Εὐφράτῃ πλησίον καὶ ποιεῖ τὴν Μεσοποταμίαν πρὸς αὐτόν, εἶτ' ἐκδίδωσι καὶ αὐτὸς εἰς τὸν αὐτὸν κόλπον. διέχουσι δὲ ἀλλήλων αἱ πηγαὶ τοῦ

the Antitaurus ends in Cataonia, whereas the mountain Amanus extends to the Euphrates River and Melitinê, where Commagenê lies adjacent to Cappadocia. And it is succeeded in turn by the mountains on the far side of the Euphrates, which are continuous with those aforementioned, except that they are cleft by the river that flows through the midst of them. Here its height and breadth greatly increase and its branches are more numerous. At all events, the most southerly part is the Taurus proper, which separates Armenia from Mesopotamia.

3. Thence flow both rivers, I mean the Euphrates and the Tigris, which encircle Mesopotamia and closely approach each other in Babylonia and then empty into the Persian Sea. The Euphrates is not only the larger of the two rivers, but also, with its winding stream, traverses more country, having its sources in the northerly region of the Taurus, and flowing towards the west through Greater Armenia, as it is called, to Lesser Armenia, having the latter on its right and Acilisenê on the left. It then bends towards the south, and at its bend joins the boundaries of Cappadocia; and leaving these and the region of Commagenê on the right, and Acilisenê and Sophenê in Greater Armenia on the left, it runs on to Syria and again makes another bend into Babylonia and the Persian Gulf. The Tigris, running from the southerly part of the same mountain to Seleuceia, approaches close to the Euphrates and with it forms Mesopotamia, and then flows into the same gulf as the Euphrates. The sources of the

[1] Ἀκιλισηνήν in margin of E, Λισηνήν MSS., Βασιλισηνήν *Epit.*, Casaubon and Corais.

τε Εὐφράτου καὶ τοῦ Τίγριος περὶ δισχιλίους καὶ πεντακοσίους σταδίους.

4. Ἀπὸ δ' οὖν τοῦ Ταύρου πρὸς ἄρκτον ἀποσχίδες πολλαὶ γεγόνασι, μία μὲν ἡ τοῦ καλουμένου Ἀντιταύρου· καὶ γὰρ ἐνταῦθα οὕτως ὠνομάζετο ὁ τὴν Σωφηνὴν ἀπολαμβάνων ἐν αὐλῶνι μεταξὺ κειμένῳ αὐτοῦ τε καὶ τοῦ Ταύρου. πέραν δὲ τοῦ Εὐφράτου κατὰ τὴν μικρὰν Ἀρμενίαν ἐφεξῆς τῷ Ἀντιταύρῳ πρὸς ἄρκτον ἐπεκτείνεται μέγα ὄρος καὶ πολυσχιδές· καλοῦσι δὲ τὸ μὲν αὐτοῦ Παρυάδρην,[1] τὸ δὲ Μοσχικὰ ὄρη, τὸ δ' ἄλλοις ὀνόμασι· ταῦτα δ' ἀπολαμβάνει τὴν Ἀρμενίαν ὅλην μέχρι Ἰβήρων καὶ Ἀλβανῶν. εἶτ' ἄλλ' ἐπανίσταται πρὸς ἕω, τὰ ὑπερκείμενα C 522 τῆς Κασπίας θαλάττης μέχρι Μηδίας, τῆς τε Ἀτροπατίου καὶ τῆς μεγάλης· καλοῦσι δὲ καὶ ταῦτα τὰ μέρη πάντα τῶν ὀρῶν Παραχοάθραν καὶ τὰ μέχρι τῶν Κασπίων πυλῶν καὶ ἐπέκεινα ἔτι πρὸς ταῖς ἀνατολαῖς τὰ συνάπτοντα τῇ Ἀρίᾳ. τὰ μὲν δὴ πρόσβορα ὄρη οὕτω καλοῦσι, τὰ δὲ νότια τὰ πέραν τοῦ Εὐφράτου, ἀπὸ[2] τῆς Καππαδοκίας καὶ τῆς Κομμαγηνῆς πρὸς ἕω τείνοντα, κατ' ἀρχὰς μὲν αὐτὸ τοῦτο καλεῖται Ταῦρος, διορίζων τὴν Σωφηνὴν καὶ τὴν ἄλλην Ἀρμενίαν ἀπὸ τῆς Μεσοποταμίας· τινὲς δὲ Γορδυαῖα ὄρη καλοῦσιν. ἐν δὲ τούτοις ἐστὶ καὶ τὸ Μάσιον, τὸ ὑπερκείμενον τῆς Νισίβιος ὄρος καὶ τῶν Τιγρανοκέρτων. ἔπειτα ἐξαίρεται πλέον καὶ καλεῖται Νιφάτης· ἐνταῦθα δέ που καὶ αἱ[3] τοῦ Τίγριος

[1] Παρυάδρην, Tzschucke, for Πολυάρρην; so the later editors.
[2] ἀπό, Groskurd inserts; *ουz* καί.
[3] αἱ, after καί, the editors insert.

Euphrates and the Tigris are about two thousand five hundred stadia distant from each other.

4. Now the Taurus has numerous branches towards the north, one of which is that of the Antitaurus, as it is called, for there too the mountain which encloses Sophenê in a valley situated between itself and the Taurus was so named. On the far side of the Euphrates, near Lesser Armenia and next to the Antitaurus towards the north, there stretches a large mountain with many branches, one of which is called Paryadres, another the Moschian Mountains, and another which is called by various names; and these comprehend the whole of Armenia as far as Iberia and Albania. Then other mountains rise towards the east, I mean those which lie above the Caspian Sea, extending as far as Media, not only the Atropatian Media but also the Greater Media. Not only all these parts of the mountains are called Parachoathras, but also those which extend to the Caspian Gates and those which extend still farther towards the east, I mean those which border on Aria. The mountains on the north, then, bear these names, whereas those on the south, on the far side of the Euphrates, in their extent towards the east from Cappadocia and Commagenê, are, at their beginning, called Taurus proper,[1] which separates Sophenê and the rest of Armenia from Mesopotamia; by some, however, these are called the Gordyaean Mountains, and among these belongs also Masius, the mountain which is situated above Nisibis and Tigranocerta. Then the Taurus rises higher and bears the name Niphates; and somewhere here are the sources of the Tigris, on

[1] Cf. 11. 12. 3.

πηγαὶ κατὰ τὸ νότιον τῆς ὀρεινῆς πλευρόν· εἶτ' ἀπὸ τοῦ Νιφάτου μᾶλλον ἔτι καὶ μᾶλλον ἡ ῥάχις ἐκτεινομένη τὸ Ζάγριον ὄρος ποιεῖ, τὸ διόριζον τὴν Μηδίαν καὶ τὴν Βαβυλωνίαν· μετὰ δὲ τὸ Ζάγριον ἐκδέχεται ὑπὲρ μὲν τῆς Βαβυλωνίας ἥ τε τῶν Ἐλυμαίων ὀρεινὴ καὶ ἡ τῶν Παραιτακηνῶν, ὑπὲρ δὲ τῆς Μηδίας ἡ τῶν Κοσσαίων· ἐν μέσῳ δ' ἐστὶν ἡ Μηδία καὶ ἡ Ἀρμενία, πολλὰ μὲν ὄρη περιλαμβάνουσα, πολλὰ δὲ ὀροπέδια, ὡσαύτως δὲ πεδία καὶ αὐλῶνας μεγάλους, συχνὰ δὲ καὶ ἔθνη τὰ περιοικοῦντα, μικρά, ὀρεινὰ καὶ ληστρικὰ τὰ πλείω. οὕτω μὲν τοίνυν τίθεμεν ἐντὸς τοῦ Ταύρου τήν τε Μηδίαν, ἧς εἰσὶ καὶ αἱ Κάσπιοι πύλαι, καὶ τὴν Ἀρμενίαν.

5. Καθ' ἡμᾶς μὲν τοίνυν προσάρκτια ἂν εἴη τὰ ἔθνη ταῦτα, ἐπειδὴ καὶ ἐντὸς τοῦ Ταύρου, Ἐρατοσθένης δέ, πεποιημένος τὴν διαίρεσιν εἰς τὰ νότια μέρη καὶ τὰ προσάρκτια καὶ τὰς ὑπ' αὐτοῦ λεγομένας σφραγῖδας, τὰς μὲν βορείους καλῶν, τὰς δὲ νοτίους, ὅρια ἀποφαίνει τῶν κλιμάτων ἀμφοῖν τὰς Κασπίους πύλας· εἰκότως οὖν τὰ νοτιώτερα, πρὸς ἕω τείνοντα,[1] τῶν Κασπίων πυλῶν νότια ἂν ἀποφαίνοι, ὧν ἐστὶ καὶ ἡ Μηδία καὶ ἡ Ἀρμενία, τὰ δὲ βορειότερα πρόσβορα, κατ' ἄλλην καὶ ἄλλην διάταξιν τούτου συμβαίνοντος. τάχα δὲ οὐκ ἐπέβαλε τούτῳ, διότι ἔξω τοῦ Ταύρου πρὸς νότον οὐδέν ἐστιν οὔτε τῆς Ἀρμενίας μέρος οὔτε τῆς Μηδίας.

[1] πρὸς ἕω τείνοντα, Kramer suspects, Meineke ejects.

---

[1] See 2. 1. 35 and note on "Sphragides."
[2] See Vol. I., p. 22, foot-note 2.

the southern side of the mountainous country. Then from the Niphates the mountain-chain extends still farther and farther and forms the mountain Zagrus which separates Media and Babylonia. After the Zagrus there follows, above Babylonia, the mountainous country of the Elymaei and that of the Paraetaceni, and also, above Media, that of the Cossaei. In the middle are Media and Armenia, which comprise many mountains, many plateaus, and likewise many low plains and large valleys, and also numerous tribes that live round among the mountains and are small in numbers and range the mountains and for the most part are given to brigandage. Thus, then, I am placing inside the Taurus both Media, to which the Caspian Gates belong, and Armenia.

5. According to the way in which I place them, then, these tribes would be towards the north, since they are inside the Taurus, but Eratosthenes, who is the author of the division of Asia into "Southern Asia" and "Northern Asia" and into "Sphragides,"[1] as he calls them, calling some of the "sphragides" "northern" and others "southern," represents the Caspian Gates as a boundary between the two "climata";[2] reasonably, therefore, he might represent as "southern" the parts that are more southerly, stretching towards the east,[3] than the Caspian Gates, among which are Media and Armenia, and the more northerly as "northern," since this is the case no matter what distribution into parts is otherwise made of the country. But perhaps it did not strike Eratosthenes that no part either of Armenia or of Media lay outside the Taurus.

[3] "Stretching towards the east" seems to be an interpolation (see critical note).

XIII

1. Ἡ δὲ Μηδία δίχα διῄρηται· καλοῦσι δὲ τὴν μὲν μεγάλην, ἧς μητρόπολις τὰ Ἐκβάτανα, μεγάλη πόλις καὶ τὸ βασίλειον ἔχουσα τῆς Μήδων ἀρχῆς (διατελοῦσι δὲ καὶ νῦν οἱ Παρθυαῖοι τούτῳ χρώμενοι βασιλείῳ, καὶ θερίζουσί γε ἐνταῦθα οἱ βασιλεῖς, ψυχρὰ γὰρ ἡ Μηδία· τὸ δὲ χειμάδιόν ἐστιν αὐτοῖς ἐν Σελευκείᾳ τῇ ἐπὶ τῷ Τίγριδι πλησίον Βαβυλῶνος), ἡ δ' ἑτέρα μερίς ἐστιν ἡ C 523 Ἀτροπάτιος Μηδία, τοὔνομα δ' ἔσχεν ἀπὸ τοῦ ἡγεμόνος Ἀτροπάτου, ὃς ἐκώλυσεν ὑπὸ τοῖς Μακεδόσι γίνεσθαι καὶ ταύτην, μέρος οὖσαν μεγάλης Μηδίας· καὶ δὴ καὶ βασιλεὺς ἀναγορευθεὶς ἰδίᾳ συνέταξε καθ' αὑτὴν τὴν χώραν ταύτην, καὶ ἡ διαδοχὴ σώζεται μέχρι νῦν ἐξ ἐκείνου, πρός τε τοὺς Ἀρμενίων βασιλέας ποιησαμένων ἐπιγαμίας τῶν ὕστερον καὶ Σύρων καὶ μετὰ ταῦτα Παρθυαίων.

2. Κεῖται δὲ ἡ χώρα τῇ μὲν Ἀρμενίᾳ καὶ τῇ Ματιανῇ πρὸς ἕω, τῇ δὲ μεγάλῃ Μηδίᾳ πρὸς δύσιν, πρὸς ἄρκτον δ' ἀμφοτέραις· τοῖς δὲ περὶ τὸν μυχὸν τῆς Ὑρκανίας θαλάττης καὶ τῇ Ματιανῇ[1] ἀπὸ νότου παράκειται. ἔστι δ' οὐ μικρὰ κατὰ τὴν δύναμιν, ὥς φησιν Ἀπολλωνίδης, ἥ γε καὶ[2] μυρίους ἱππέας δύναται παρέχεσθαι, πεζῶν δὲ τέτταρας μυριάδας. λίμνην δ' ἔχει τὴν Καπαῦτα,[3] ἐν ᾗ ἅλες ἐπανθοῦντες πήττονται· εἰσὶ

[1] τῇ Ματιανῇ, Kramer, for τῆς Ματιάνης; so Meineke.

[2] κατά before μυρίους, *z* and Corais omit.

[3] Καπαῦτα, conj. of C. Müller (Καπαῦταν, Kramer and others), for Σπαῦτα; so Tozer (see his note).

XIII

1. Media is divided into two parts. One part of it is called Greater Media, of which the metropolis is Ecbatana, a large city containing the royal residence of the Median empire (the Parthians continue to use this as a royal residence even now, and their kings spend at least their summers there, for Media is a cold country; but their winter residence is at Seleuceia, on the Tigris near Babylon). The other part is Atropatian Media, which got its name from the commander[1] Atropates, who prevented also this country, which was a part of Greater Media, from becoming subject to the Macedonians. Furthermore, after he was proclaimed king, he organised this country into a separate state by itself, and his succession of descendants is preserved to this day, and his successors have contracted marriages with the kings of the Armenians and Syrians and, in later times, with the kings of the Parthians.

2. This country lies east of Armenia and Matianê, west of Greater Media, and north of both; and it lies adjacent to the region round the recess of the Hyrcanian Sea and to Matianê on the south. It is no small country, considering its power, as Apollonides[2] says, since it can furnish as many as ten thousand horsemen and forty thousand foot-soldiers. It has a harbour, Capauta,[3] in which salts effloresce and solidify. These salts cause itching and are

[1] In the battle of Arbela, 331 B.C.
[2] Vol III., p. 234, foot-note 2.
[3] Now Lake Urmi (see 11. 14. 8 and note on "Blue").

δὲ κνησμώδεις καὶ ἐπαλγεῖς, ἔλαιον δὲ τοῦ πάθους ἄκος, ὕδωρ δὲ γλυκὺ τοῖς καπυρωθεῖσιν[1] ἱματίοις, εἴ τις κατ' ἄγνοιαν βάψειεν εἰς αὐτὴν πλύσεως χάριν. ἔχουσι δ' ἰσχυροὺς γείτονας τοὺς Ἀρμενίους καὶ τοὺς Παρθυαίους, ὑφ' ὧν περικόπτονται πολλάκις. ἀντέχουσι δ' ὅμως καὶ ἀπολαμβάνουσι τὰ ἀφαιρεθέντα, καθάπερ τὴν Συμβάκην ἀπέλαβον παρὰ τῶν Ἀρμενίων, ὑπὸ Ῥωμαίοις γεγονότων, καὶ αὐτοὶ προσεληλύθασι τῇ φιλίᾳ τῇ πρὸς Καίσαρα· θεραπεύουσι δ' ἅμα καὶ τοὺς Παρθυαίους.

3. Βασίλειον δ' αὐτῶν θερινὸν μὲν ἐν πεδίῳ ἱδρυμένον Γάζακα[2] χειμερινὸν δὲ[3] ἐν φρουρίῳ ἐρυμνῷ Οὔερα, ὅπερ Ἀντώνιος ἐπολιόρκησε κατὰ τὴν ἐπὶ Παρθυαίους στρατείαν. διέχει δὲ τοῦτο τοῦ Ἀράξου ποταμοῦ τοῦ ὁρίζοντος τήν τε Ἀρμενίαν καὶ τὴν Ἀτροπατηνὴν σταδίους δισχιλίους καὶ τετρακοσίους, ὥς φησιν ὁ Δέλλιος,[4] ὁ τοῦ Ἀντωνίου φίλος, συγγράψας τὴν ἐπὶ Παρθυαίους αὐτοῦ στρατείαν, ἐν ᾗ παρῆν καὶ αὐτὸς ἡγεμονίαν ἔχων. ἔστι δὲ τῆς χώρας ταύτης τὰ μὲν ἄλλα εὐδαίμονα χωρία, ἡ δὲ προσάρκτιος ὀρεινὴ καὶ τραχεῖα καὶ ψυχρά, Καδουσίων κατοικία τῶν ὀρεινῶν καὶ Ἀμάρδων καὶ Ταπύρων καὶ Κυρτίων καὶ ἄλλων τοιούτων, οἳ μετανάσται εἰσὶ καὶ ληστρικοί. καὶ γὰρ ὁ Ζάγρος καὶ ὁ Νιφάτης κατεσπαρμένα ἔχουσι τὰ ἔθνη ταῦτα, καὶ οἱ ἐν τῇ Περσίδι Κύρτιοι καὶ Μάρδοι (καὶ γὰρ οὕτω λέγονται οἱ Ἄμαρδοι) καὶ οἱ ἐν τῇ Ἀρμενίᾳ μέχρι νῦν ὁμωνύμως προσαγορευόμενοι τῆς αὐτῆς εἰσὶν ἰδέας.

[1] For καπυρωθεῖσιν, C. Müller conj. καταρρυπωθεῖσιν ("soiled").

painful, but this effect is relieved by olive-oil; and the water restores weathered garments, if perchance through ignorance one should dip them in it to wash them. They have powerful neighbours in the Armenians and the Parthians, by whom they are often plundered. But still they hold out against them and get back what has been taken away from them, as, for example, they got back Symbacê from the Armenians when the latter became subject to the Romans; and they themselves have attained to friendship with Caesar. But they are also paying court to the Parthians at the same time.

3. Their royal summer palace is situated in a plain at Gazaca, and their winter palace in a fortress called Vera, which was besieged by Antony on his expedition against the Parthians. This fortress is distant from the Araxes, which forms the boundary between Armenia and Atropatenê, two thousand four hundred stadia, according to Dellius, the friend of Antony, who wrote an account of Antony's expedition against the Parthians, on which he accompanied Antony and was himself a commander. All regions of this country are fertile except the part towards the north, which is mountainous and rugged and cold, the abode of the mountaineers called Cadusii, Amardi, Tapyri, Cyrtii and other such peoples, who are migrants and predatory; for the Zagrus and Niphates mountains keep these tribes scattered; and the Cyrtii in Persis, and the Mardi (for the Amardi are also thus called), and those in Armenia who to this day are called by the same name, are of the same character.

---

[2] Γάζακα, Groskurd, for Γάζα καί; so the later editors.
[3] χειμερινὸν δέ, Groskurd inserts; so Meineke.
[4] Δέλλιος, Casaubon, for Ἀδέλφιος; so the later editors.

4. Οἱ δ' οὖν Καδούσιοι πλήθει τῷ πεζῷ μικρὸν ἀπολείπονται τῶν Ἀριανῶν, ἀκοντισταὶ δ' εἰσὶν ἄριστοι, ἐν δὲ τοῖς τραχέσιν ἀνθ' ἱππέων πεζοὶ C 524 διαμάχονται. Ἀντωνίῳ δὲ χαλεπὴν τὴν στρατείαν ἐποίησεν οὐχ ἡ τῆς χώρας φύσις, ἀλλ' ὁ τῶν ὁδῶν ἡγεμών, ὁ τῶν Ἀρμενίων βασιλεὺς Ἀρταουάσδης, ὃν εἰκῇ[1] ἐκεῖνος, ἐπιβουλεύοντα αὐτῷ, σύμβουλον ἐποιεῖτο καὶ κύριον τῆς περὶ τοῦ πολέμου γνώμης· ἐτιμωρήσατο μὲν οὖν αὐτόν, ἀλλ' ὀψέ, ἡνίκα πολλῶν αἴτιος κατέστη κακῶν Ῥωμαίοις καὶ αὐτὸς καὶ ἐκεῖνος, ὅστις τὴν ἀπὸ τοῦ Ζεύγματος ὁδὸν τοῦ κατὰ τὸν Εὐφράτην μέχρι τοῦ ἅψασθαι τῆς Ἀτροπατηνῆς ὀκτακισχιλίων σταδίων ἐποίησε, πλέον ἢ διπλασίαν τῆς εὐθείας, διὰ ὀρῶν καὶ ἀνοδιῶν καὶ κυκλοπορίας.

5. Ἡ δὲ μεγάλη Μηδία τὸ μὲν παλαιὸν τῆς Ἀσίας ἡγήσατο πάσης, καταλύσασα τὴν τῶν Σύρων ἀρχήν· ὕστερον δ' ὑπὸ Κύρου καὶ Περσῶν ἀφαιρεθεῖσα τὴν τοσαύτην ἐξουσίαν ἐπὶ Ἀστυάγου, διεφύλαττεν ὅμως πολὺ τοῦ πατρίου ἀξιώματος, καὶ ἦν τὰ Ἐκβάτανα χειμάδιον[2] τοῖς Πέρσαις, ὁμοίως δὲ καὶ τοῖς ἐκείνους καταλύσασι Μακεδόσι τοῖς τὴν Συρίαν ἔχουσι καὶ νῦν ἔτι τοῖς Παρθυαίων βασιλεῦσι τὴν αὐτὴν παρέχεται χρείαν τε καὶ ἀσφάλειαν.

6. Ὁρίζεται δ' ἀπὸ μὲν τῆς ἕω τῇ τε Παρθυαίᾳ καὶ τοῖς Κοσσαίων ὄρεσι, λῃστρικῶν ἀνθρώπων, οἵ τοξότας μυρίους καὶ τρισχιλίους παρέσχοντό

[1] εἰκῇ, Meineke, for εἰκός, which *oz* omit.

[2] χειμάδιον must be an error for θερινὸν βασίλειον, or simply βασίλειον, unless certain words (see Corais) have fallen out of the text which make χειμάδιον apply to Seleuceia (see 11. 13. 1).

4. The Cadusii, however, are but little short of the Ariani in the number of their foot-soldiers; and their javelin-throwers are excellent; and in rugged places foot-soldiers instead of horsemen do the fighting. It was not the nature of the country that made the expedition difficult for Antony, but his guide Artavasdes, the king of the Armenians, whom, though plotting against him, Antony rashly made his counsellor and master of decisions respecting the war. Antony indeed punished him, but too late, when the latter had been proved guilty of numerous wrongs against the Romans, not only he himself, but also that other guide, who made the journey from the Zeugma on the Euphrates to the borders of Atropatenê eight thousand stadia long, more than twice the direct journey, guiding the army over mountains and roadless regions and circuitous routes.

5. In ancient times Greater Armenia ruled the whole of Asia, after it broke up the empire of the Syrians, but later, in the time of Astyages, it was deprived of that great authority by Cyrus and the Persians, although it continued to preserve much of its ancient dignity; and Ecbatana was winter residence[1] for the Persian kings, and likewise for the Macedonians who, after overthrowing the Persians, occupied Syria; and still to-day it affords the kings of the Parthians the same advantages and security.

6. Greater Media is bounded on the east by Parthia and the mountains of the Cossaei, a predatory people, who once supplied the Elymaei, with

[1] Apparently an error of the copyist for "summer residence" or "royal residence" (cf. § 1 above and § 6 below).

ποτε Ἐλυμαίοις, συμμαχοῦντες ἐπὶ Σουσίους καὶ Βαβυλωνίους. Νέαρχος δέ φησι, τεττάρων ὄντων λῃστρικῶν ἐθνῶν, ὧν Μάρδοι μὲν Πέρσαις προσεχεῖς ἦσαν, Οὔξιοι δὲ καὶ Ἐλυμαῖοι τούτοις τε καὶ Σουσίοις, Κοσσαῖοι δὲ Μήδοις, πάντας μὲν φόρους πράττεσθαι τοὺς βασιλέας, Κοσσαίους δὲ καὶ δῶρα λαμβάνειν, ἡνίκα ὁ βασιλεὺς θερίσας ἐν Ἐκβατάνοις εἰς τὴν Βαβυλωνίαν καταβαίνοι· καταλῦσαι δ' αὐτῶν τὴν πολλὴν τόλμαν Ἀλέξανδρον, ἐπιθέμενον χειμῶνος. τούτοις τε δὴ ἀφορίζεται πρὸς ἕω καὶ ἔτι τοῖς Παραιτακηνοῖς, οἳ συνάπτουσι Πέρσαις, ὀρεινοὶ καὶ αὐτοὶ καὶ λῃστρικοί· ἀπὸ δὲ τῶν ἄρκτων τοῖς ὑπεροικοῦσι τῆς Ὑρκανίας θαλάττης Καδουσίοις καὶ τοῖς ἄλλοις, οὓς ἄρτι διήλθομεν· πρὸς νότον[1] δὲ τῇ Ἀπολλωνιάτιδι, ἣν Σιτακηνὴν ἐκάλουν οἱ παλαιοί, καὶ τῷ Ζάγρῳ, καθ' ὃ ἡ Μασσαβατικὴ κεῖται, τῆς Μηδίας οὖσα, οἱ δὲ τῆς Ἐλυμαίας φασί· πρὸς δύσιν δὲ τοῖς Ἀτροπατίοις[2] καὶ τῶν Ἀρμενίων τισίν. εἰσὶ δὲ καὶ Ἑλληνίδες πόλεις, κτίσματα τῶν Μακεδόνων ἐν τῇ Μηδίᾳ, ὧν Λαοδίκειά τε καὶ Ἀπάμεια καὶ ἡ πρὸς Ῥάγαις[3] καὶ αὐτὴ Ῥάγα, τὸ τοῦ Νικάτορος κτίσμα· ὃ ἐκεῖνος μὲν Εὔρωπον ὠνόμασε, Πάρθοι δὲ Ἀρσακίαν, νοτιωτέραν οὖσαν τῶν Κασπίων πυλῶν πεντακοσίοις που σταδίοις, C 525 ὥς φησιν Ἀπολλόδωρος Ἀρτεμιτηνός.

[1] E has ἕω instead of νότον.

[2] Ἀτροπατίοις E, Ἀτραπίοις other MSS.

[3] Ἡρακλεία (the name of the city to which Strabo refers, see 11. 9. 1) is inserted after Ῥάγαις by Meineke, who follows conj. of Groskurd and Kramer.

whom they were allies in the war against the Susians and Babylonians, with thirteen thousand bowmen. Nearchus[1] says that there were four predatory tribes and that of these the Mardi were situated next to the Persians; the Uxii and Elymaei next to the Mardi and the Susians; and the Cossaei next to the Medians; and that whereas all four exacted tribute from the kings, the Cossaei also received gifts at the times when the king, after spending the summer in Ecbatana, went down into Babylonia; but that Alexander put an end to their great audacity when he attacked them in the winter time. So then, Greater Media is bounded on the east by these tribes, and also by the Paraetaceni, who border on the Persians and are themselves likewise mountaineers and predatory; on the north by the Cadusii who live above the Hyrcanian Sea, and by the other tribes which I have just described; on the south by Apollioniatis, which the ancients called Sitacenê, and by the mountain Zagrus, at the place where Massabaticê is situated, which belongs to Media, though some say that it belongs to Elymaea; and on the west by the Atropatii and certain of the Armenians. There are also some Greek cities in Media, founded by the Macedonians, among which are Laodiceia, Apameia and the city[2] near Rhagae, and Rhaga[3] itself, which was founded by Nicator.[4] By him it was named Europus, but by the Parthians Arsacia; it lies about five hundred stadia to the south of the Caspian Gates, according to Apollodorus of Artemita.

[1] See *Dictionary* in Vol. I. [2] Heracleia (see 11. 9. 1).
[3] The name is spelled both in plural and in singular.
[4] Seleucus Nicator, King of Syria 312–280 B.C.

7. Ἡ πολλὴ μὲν οὖν ὑψηλή ἐστι καὶ ψυχρά, τοιαῦτα δὲ καὶ τὰ ὑπερκείμενα τῶν Ἐκβατάνων ὄρη καὶ τὰ περὶ τὰς Ῥάγας καὶ τὰς Κασπίους πύλας καὶ καθόλου τὰ προσάρκτια μέρη τὰ ἐντεῦθεν μέχρι πρὸς τὴν Ματιανὴν[1] καὶ τὴν Ἀρμενίαν, ἡ δ' ὑπὸ ταῖς Κασπίοις πύλαις ἐν ταπεινοῖς ἐδάφεσι καὶ κοίλοις οὖσα εὐδαίμων σφόδρα ἐστὶ καὶ πάμφορος πλὴν ἐλαίας· εἰ δὲ καὶ φύεταί που, ἀλιπής τέ ἐστι καὶ ξηρά· ἱππόβοτος δὲ καὶ αὕτη ἐστὶ διαφερόντως καὶ ἡ Ἀρμενία, καλεῖται δέ τις καὶ λειμὼν Ἱππόβοτος, ὃν καὶ διεξίασιν οἱ ἐκ τῆς Περσίδος καὶ Βαβυλῶνος εἰς Κασπίους πύλας ὁδεύοντες, ἐν ᾗ πέντε[2] μυριάδας ἵππων θηλείων νέμεσθαί φασιν ἐπὶ τῶν Περσῶν, εἶναι δὲ τὰς ἀγέλας ταύτας βασιλικάς. τοὺς δὲ Νησαίους[3] ἵππους, οἷς ἐχρῶντο οἱ βασιλεῖς ἀρίστοις οὖσι καὶ μεγίστοις, οἱ μὲν ἐνθένδε λέγουσι τὸ γένος, οἱ δ' ἐξ Ἀρμενίας· ἰδιόμορφοι δέ εἰσιν, ὥσπερ καὶ οἱ Παρθικοὶ λεγόμενοι νῦν παρὰ τοὺς Ἑλλαδικοὺς καὶ τοὺς ἄλλους τοὺς παρ' ἡμῖν. καὶ τὴν βοτάνην δὲ τὴν μάλιστα τρέφουσαν τοὺς ἵππους ἀπὸ τοῦ πλεονάζειν ἐνταῦθα ἰδίως Μηδικὴν καλοῦμεν. φέρει δὲ καὶ σίλφιον ἡ χώρα, ἀφ' οὗ ὁ Μηδικὸς καλούμενος ὀπός, ἐπὶ τὸ[4] πολὺ λειπόμενος τοῦ Κυρηναϊκοῦ, ἔστι δ' ὅτε καὶ διαφέρων ἐκείνου, εἴτε παρὰ τὰς τῶν τόπων διαφοράς, εἴτε τοῦ φυτοῦ κατ' εἶδος ἐξαλλάττοντος, εἴτε καὶ παρὰ τοὺς

[1] *rw* have Μαντιανήν.

[2] For πέντε, Wesseling (note on Diodorus 17. 110), comparing Arrian 7. 13, conj. πεντεκαίδεκα.

[3] E has Νισαίους.

[4] ἐπὶ τό, Jones inserts before πολύ; Stephanus Byz. (*s.v.* Μηδία) reads οὐ πολύ.

7. Now most of the country is high and cold; and such, also, are the mountains which lie above Ecbatana and those in the neighbourhood of Rhagae and the Caspian Gates, and in general the northerly regions extending thence to Matianê and Armenia; but the region below the Caspian Gates, consisting of low-lying lands and hollows, is very fertile and productive of everything but the olive; and even if the olive is produced anywhere, it is dry and yields no oil. This, as well as Armenia, is an exceptionally good "horse-pasturing"[1] country; and a certain meadow there is called "Horse-pasturing," and those who travel from Persis and Babylon to Caspian Gates pass through it; and in the time of the Persians it is said that fifty thousand mares were pastured in it and that these herds belonged to the kings. As for the Nesaean horses, which the kings used because they were the best and the largest, some writers say that the breed came from here, while others say from Armenia. They are characteristically different in form, as are also the Parthian horses, as they are now called, as compared with the Helladic and the other horses in our country. Further, we call the grass that makes the best food for horses by the special name "Medic," from the fact that it abounds there. The country also produces silphium; whence the "Medic" juice, as it is called, which in general is inferior to the "Cyrenaic" juice, but sometimes is even superior to it, either owing to regional differences, or because of a variation in the species of the plant, or even owing to the people who extract and prepare

[1] "Hippobotos," a Homeric epithet of Argos (*e.g. Od.* 4. 99).

ὀπίζοντας καὶ σκευάζοντας, ὥστε συμμένειν πρὸς τὴν ἀπόθεσιν καὶ τὴν χρείαν.

8. Τοιαύτη μέν τις ἡ χώρα· τὸ δὲ μέγεθος πάρισός πώς ἐστιν εἰς πλάτος καὶ μῆκος· δοκεῖ δὲ μέγιστον εἶναι πλάτος [1] τῆς Μηδίας τὸ ἀπὸ τῆς τοῦ Ζάγρου ὑπερθέσεως, ἥπερ καλεῖται Μηδικὴ πύλη, εἰς Κασπίους πύλας διὰ τῆς Σιγριανῆς σταδίων τετρακισχιλίων ἑκατόν. τῷ δὲ μεγέθει καὶ τῇ δυνάμει τῆς χώρας ὁμολογεῖ καὶ ἡ περὶ τῶν φόρων ἱστορία· τῆς γὰρ Καππαδοκίας παρεχούσης τοῖς Πέρσαις κατ' ἐνιαυτὸν πρὸς τῷ ἀργυρικῷ τέλει ἵππους χιλίους καὶ πεντακοσίους, ἡμιόνους δὲ δισχιλίους, προβάτων δὲ πέντε μυριάδας, διπλάσια σχεδόν τι τούτων ἐτέλουν οἱ Μῆδοι.

9. Ἔθη [2] δὲ τὰ πολλὰ μὲν τὰ αὐτὰ τούτοις τε καὶ τοῖς Ἀρμενίοις διὰ τὸ καὶ τὴν χώραν παραπλησίαν εἶναι. τοὺς μέντοι Μήδους ἀρχηγέτας εἶναί φασι καὶ τούτοις καὶ ἔτι πρότερον Πέρσαις τοῖς ἔχουσιν αὐτοὺς καὶ διαδεξαμένοις τὴν τῆς Ἀσίας ἐξουσίαν. ἡ γὰρ νῦν λεγομένη Περσικὴ στολὴ καὶ ὁ τῆς τοξικῆς καὶ ἱππικῆς ζῆλος καὶ ἡ περὶ τοὺς βασιλέας θεραπεία καὶ κόσμος καὶ σεβασμὸς θεοπρεπὴς παρὰ τῶν ἀρχομένων εἰς τοὺς Πέρσας παρὰ Μήδων ἀφῖκται. καὶ ὅτι τοῦτ' ἀληθές, ἐκ τῆς ἐσθῆτος μάλιστα δῆλον· τιάρα γάρ τις καὶ κίταρις καὶ πῖλος καὶ χειριδωτοὶ

C 526

[1] πλάτος, Meineke emends to μῆκος, presumably in view of Strabo's general use of the two terms (see 2. 1. 32).
[2] ἔθη oz, ἔθηκε other MSS.

---

[1] i.e. robe (cf. Lat. "stola").

the juice in such a way as to conserve its strength for storage and for use.

8. Such is the nature of the country. As for its size, its length and breadth are approximately equal. The greatest breadth of Media seems to be that from the pass that leads over the Zagrus, which is called Medic Gate, to the Caspian Gates through Sigrianê, four thousand one hundred stadia. The reports on the tributes paid agree with the size and the power of the country; for Cappadocia paid the Persians yearly, in addition to the silver tax, fifteen hundred horses, two thousand mules, and fifty thousand sheep, whereas Media paid almost twice as much as this.

9. As for customs, most of theirs and of those of the Armenians are the same, because their countries are similar. The Medes, however, are said to have been the originators of customs for the Armenians, and also, still earlier, for the Persians, who were their masters and their successors in the supreme authority over Asia. For example, their "Persian" stolê,[1] as it is now called, and their zeal for archery and horsemanship, and the court they pay to their kings, and their ornaments, and the divine reverence paid by subjects to kings, came to the Persians from the Medes. And that this is true is particularly clear from their dress; for tiara,[2] citaris,[3] pilus,[4] tunics with sleeves reaching to the hands, and

[2] The royal tiara was high and erect and encircled with a diadem, while that of the people was soft and fell over on one side.

[3] A kind of Persian head-dress. Aristophanes (*Birds* 497) compares a cock's comb to it.

[4] A felt skull-cap, like a fez.

χιτῶνες καὶ ἀναξυρίδες ἐν μὲν τοῖς ψυχροῖς τόποις καὶ προσβόροις, ἐπιτήδειά ἐστι φορήματα, οἷοί εἰσιν οἱ Μηδικοί· ἐν δὲ τοῖς νοτίοις ἥκιστα· οἱ δὲ Πέρσαι τὴν πλείστην οἴκησιν ἐπὶ τῇ Ἐρυθρᾷ θαλάττῃ κέκτηνται, μεσημβρινώτεροι καὶ Βαβυλωνίων ὄντες καὶ Σουσίων· μετὰ δὲ τὴν κατάλυσιν τὴν τῶν Μήδων προσεκτήσαντό τινα καὶ τῶν προσαπτομένων Μηδίᾳ. ἀλλ' οὕτως ἐφάνη σεμνὰ καὶ τοῦ βασιλικοῦ προσχήματος οἰκεῖα τὰ ἔθη τοῖς νικήσασι καὶ[1] τὰ τῶν νικηθέντων, ὥστ' ἀντὶ γυμνητῶν καὶ ψιλῶν θηλυστολεῖν ὑπέμειναν, καὶ κατηρεφεῖς εἶναι τοῖς σκεπάσμασι.

10. Τινὲς δὲ Μήδειαν καταδεῖξαι τὴν ἐσθῆτα ταύτην φασί, δυναστεύσασαν ἐν τοῖς τόποις, καθάπερ καὶ Ἰάσονα, καὶ ἐπικρυπτομένην τὴν ὄψιν, ὅτε ἀντὶ τοῦ βασιλέως ἐξίοι· τοῦ μὲν[2] Ἰάσονος ὑπομνήματα εἶναι τὰ Ἰασόνια ἡρῷα, τιμώμενα σφόδρα ὑπὸ τῶν βαρβάρων (ἔστι δὲ καὶ ὄρος μέγα ὑπὲρ τῶν Κασπίων πυλῶν ἐν ἀριστερᾷ, καλούμενον Ἰασόνιον), τῆς δὲ Μηδείας τὴν ἐσθῆτα καὶ τοὔνομα τῆς χώρας. λέγεται δὲ καὶ Μῆδος, υἱὸς αὐτῆς, διαδέξασθαι τὴν ἀρχὴν καὶ τὴν χώραν ἐπώνυμον αὐτοῦ καταλιπεῖν. ὁμολογεῖ δὲ τούτοις καὶ τὰ κατὰ τὴν Ἀρμενίαν Ἰασόνια καὶ τὸ τῆς χώρας ὄνομα καὶ ἄλλα πλείω, περὶ ὧν ἐροῦμεν.

11. Καὶ τοῦτο δὲ Μηδικόν, τὸ βασιλέα αἱρεῖσθαι τὸν ἀνδρειότατον, ἀλλ' οὐ πᾶσιν, ἀλλὰ τοῖς ὀρείοις· μᾶλλον δὲ τὸ τοῖς βασιλεῦσι πολλὰς

[1] καί, before τά, *oz* and Meineke omit.

trousers, are indeed suitable things to wear in cold and northerly regions, such as the Medes wear, but by no means in southerly regions; and most of the settlements possessed by the Persians were on the Red Sea, farther south than the country of the Babylonians and the Susians. But after the overthrow of the Medes the Persians acquired in addition certain parts of the country that reached to Media. However, the customs even of the conquered looked to the conquerors so august and appropriate to royal pomp that they submitted to wear feminine robes instead of going naked or lightly clad, and to cover their bodies all over with clothes.

10. Some say that Medeia introduced this kind of dress when she, along with Jason, held dominion in this region, even concealing her face whenever she went out in public in place of the king; and that the Jasonian hero-chapels, which are much revered by the barbarians, are memorials of Jason (and above the Caspian Gates on the left is a large mountain called Jasonium), whereas the dress and the name of the country are memorials of Medeia. It is said also that Medus her son succeeded to the empire and left his own name to the country. In agreement with this are the Jasonia of Armenia and the name of that country[1] and several other things which I shall discuss.

11. This, too, is a Medic custom—to choose the bravest man as king; not, however, among all Medes, but only among the mountaineers. More general is the custom for the kings to have many

[1] See 11. 4. 8.

---

[2] Meineke inserts οὖν after μέν.

εἶναι γυναῖκας. τοῖς δ' ὀρείοις τῶν Μήδων καὶ πᾶσιν ἔθος τοῦτο, ἐλάττους δὲ τῶν πέντε οὐκ ἔξεστιν· ὡς δ' αὕτως τὰς γυναῖκάς φασιν ἐν καλῷ τίθεσθαι ὅτι πλείστους νέμειν ἄνδρας,[1] τῶν πέντε δὲ ἐλάττους συμφορὰν ἡγεῖσθαι. τῆς δ' ἄλλης Μηδίας εὐδαιμονούσης τελέως, λυπρά ἐστιν ἡ προσάρκτιος ὀρεινή· σιτοῦνται γοῦν ἀπὸ ἀκροδρύων, ἔκ τε μήλων ξηρῶν κοπέντων ποιοῦνται μάζας, ἀπὸ δ' ἀμυγδάλων φωχθέντων ἄρτους, ἐκ δὲ ῥιζῶν τινῶν οἶνον ἐκθλίβουσι, κρέασι δὲ χρῶνται θηρείοις, ἥμερα δὲ οὐ τρέφουσι θρέμματα. τοσαῦτα καὶ περὶ Μήδων φαμέν· περὶ δὲ τῶν νομίμων[2] κοινῇ τῆς συμπάσης Μηδίας, ἐπειδὴ ταὐτὰ[3] τοῖς Περσικοῖς γεγένηται διὰ τὴν τῶν Περσῶν ἐπικράτειαν, ἐν τῷ περὶ ἐκείνων λόγῳ φήσομεν.[4]

XIV

1. Τῆς δ' Ἀρμενίας τὰ μὲν νότια προβέβληται τὸν Ταῦρον, διείργοντα αὐτὴν ἀφ' ὅλης τῆς μεταξὺ C 527 Εὐφράτου καὶ τοῦ Τίγριος, ἣν Μεσοποταμίαν καλοῦσι, τὰ δὲ ἑωθινὰ τῇ Μηδίᾳ συνάπτει τῇ μεγάλῃ καὶ τῇ Ἀτροπατηνῇ· προσάρκτια δέ

[1] ὅτι πλείστας νέμειν τοὺς ἄνδρας Groskurd, and so Meineke, omitting the τούς; Kramer conj. ὅτι πλείστας ἔχοντας νέμειν ἄνδρας (see Kramer's note, and C. Müller's *Ind. Var. Lect.* p. 1018).

[2] νομίμων margin of *x* and the editors, for νομαδικῶν.

[3] ταὐτά, Corais, for ταῦτα; so the later editors.

[4] φήσομεν, Casaubon, for θήσομεν; so the later editors.

wives; this is the custom of the mountaineers of the Medes, and all Medes, and they are not permitted to have less than five; likewise, the women are said to account it an honourable thing to have as many husbands as possible and to consider less than five a calamity.[1] But though the rest of Media is extremely fertile, the northerly mountainous part has poor soil; at any rate, the people live on the fruits of trees, making cakes out of apples that are sliced and dried, and bread from roasted almonds; and they squeeze out a wine from certain roots; and they use the meat of wild animals, but do not breed tame animals. Thus much I add concerning the Medes. As for the institutions in common use throughout the whole of Media, since they prove to have been the same as those of the Persians because of the conquest of the Persians, I shall discuss them in my account of the latter.

XIV

1. As for Armenia, the southern parts of it have the Taurus situated in front of them,[2] which separates it from the whole of the country between the Euphrates and the Tigris, the country called Mesopotamia; and the eastern parts border on Greater Armenia and Atropatenê; and on the north

[1] So the Greek of all MSS.; but the editors since Du Theil regard the Greek text as corrupt, assuming that the women in question did not have plural husbands. Accordingly, some emend the text to make it say, "for their husbands to have as many wives as possible and consider less than five a calamity" (see critical note).

[2] The Greek implies that Armenia is *protected* on the south by the Taurus.

ἐστι τὰ ὑπερκείμενα τῆς Κασπίας θαλάττης ὄρη τὰ τοῦ Παραχοάθρα καὶ Ἀλβανοὶ καὶ Ἴβηρες καὶ ὁ Καύκασος ἐγκυκλούμενος τὰ ἔθνη ταῦτα καὶ συνάπτων τοῖς Ἀρμενίοις, συνάπτων δὲ καὶ τοῖς Μοσχικοῖς ὄρεσι καὶ Κολχικοῖς μέχρι τῶν καλουμένων Τιβαρανῶν· ἀπὸ δὲ τῆς ἑσπέρας ταῦτα ἐστι τὰ ἔθνη καὶ ὁ Παρυάδρης[1] καὶ ὁ Σκυδίσης μέχρι τῆς μικρᾶς Ἀρμενίας καὶ τῆς τοῦ Εὐφράτου ποταμίας, ἣ διειργει τὴν Ἀρμενίαν ἀπὸ τῆς Καππαδοκίας καὶ τῆς Κομμαγηνῆς.

2. Ὁ γὰρ Εὐφράτης ἀπὸ τῆς βορείου πλευρᾶς τοῦ Ταύρου τὰς ἀρχὰς ἔχων τὸ μὲν πρῶτον ῥεῖ πρὸς δύσιν διὰ τῆς Ἀρμενίας, εἶτ᾽ ἐπιστρέφει πρὸς νότον καὶ διακόπτει τὸν Ταῦρον μεταξὺ τῶν Ἀρμενίων τε καὶ Καππαδόκων καὶ Κομμαγηνῶν, ἐκπεσὼν δ᾽ ἔξω καὶ γενόμενος κατὰ τὴν Συρίαν ἐπιστρέφει πρὸς χειμερινὰς ἀνατολὰς μέχρι Βαβυλῶνος καὶ ποιεῖ τὴν Μεσοποταμίαν πρὸς τὸν Τίγριν· ἀμφότεροι δὲ τελευτῶσιν εἰς τὸν Περσικὸν κόλπον. τὰ μὲν δὴ κύκλῳ τοιαῦτα, ὀρεινὰ σχεδόν τι πάντα καὶ τραχέα, πλὴν τῶν πρὸς τὴν Μηδίαν κεκλιμένων ὀλίγων. πάλιν δὲ τοῦ λεχθέντος Ταύρου τὴν ἀρχὴν λαμβάνοντος ἀπὸ τῆς περαίας τῶν Κομμαγηνῶν καὶ τῶν Μελιτηνῶν, ἣν ὁ Εὐφράτης ποιεῖ, Μάσιον μέν ἐστι τὸ ὑπερκείμενον ὄρος τῶν ἐν τῇ Μεσοποταμίᾳ Μυγδόνων ἐκ νότου, ἐν οἷς ἡ Νίσιβίς ἐστιν· ἐκ δὲ τῶν πρὸς ἄρκτον[2] μερῶν ἡ[3] Σωφηνὴ κεῖται μεταξὺ τοῦ τε Μασίου καὶ τοῦ Ἀντιταύρου. οὗτος δ᾽ ἀπὸ τοῦ Εὐφράτου

[1] Παρύδρης is the reading of the MSS.

[2] πρὸς ἄρκτον, Kramer, for πρὸς ἄρκτων E, προσάρκτων other MSS.

are the mountains of Parachoathras that lie above the Caspian Sea, and Albania, and Iberia, and the Caucasus, which last encircles these nations and borders on Armenia, and borders also on the Moschian and Colchian mountains as far as the Tibarani, as they are called; and on the west are these nations and the mountains Paryadres and Scydises in their extent to Lesser Armenia and the river-land of the Euphrates, which latter separates Armenia from Cappadocia and Commagenê.

2. For the Euphrates, having its beginnings on the northern side of the Taurus, flows at first towards the west through Armenia, and then bends towards the south and cuts through the Taurus between Armenia, Cappadocia, and Commagenê, and then, after falling outside the Taurus and reaching the borders of Syria, it bends towards the winter-sunrise[1] as far as Babylon, and with the Tigris forms Mesopotamia; and both rivers end in the Persian Gulf. Such, then, is our circuit of Armenia, almost all parts being mountainous and rugged, except the few which verge towards Media. But since the above-mentioned Taurus[2] takes a new beginning on the far side of the Euphrates opposite Commagenê and Melitenê, countries formed by that river, Mt. Masius is the mountain which lies above the Mygdonians of Mesopotamia on the south, in whose country is Nisibis, whereas Sophenê is situated in the northern parts, between Masius and Antitaurus. The Antitaurus takes its beginning at the Euphrates

[1] See Vol. I, p. 105, note 2.
[2] Cf. 11. 12. 4.

[3] ἡ *xz* and the editors insert.

καὶ τοῦ Ταύρου τὴν ἀρχὴν λαβὼν τελευτᾷ πρὸς τὰ ἑῷα τῆς Ἀρμενίας, ἀπολαμβάνων μέσην τὴν Σωφηνήν, ἐκ θατέρου δὲ μέρους ἔχων τὴν Ἀκιλισηνὴν μεταξὺ ἱδρυμένην τοῦ Ἀντιταύρου[1] τε καὶ τῆς τοῦ Εὐφράτου ποταμίας,[2] πρὶν ἢ κάμπτειν αὐτὴν[3] ἐπὶ νότον. βασίλειον δὲ τῆς Σωφηνῆς Καρκαθιόκερτα. τοῦ δὲ Μασίου ὑπέρκειται πρὸς ἕω πολὺ κατὰ τὴν Γορδυηνὴν[4] ὁ Νιφάτης, εἶθ' ὁ Ἄβος, ἀφ' οὗ καὶ ὁ Εὐφράτης ῥεῖ καὶ ὁ Ἀράξης, ὁ μὲν πρὸς δύσιν, ὁ δὲ πρὸς ἀνατολάς· εἶθ' ὁ Νίβαρος μέχρι τῆς Μηδίας παρατείνει.

3. Ὁ μὲν οὖν Εὐφράτης εἴρηται ὃν τρόπον ῥεῖ· ὁ δὲ Ἀράξης, πρὸς τὰς ἀνατολὰς ἐνεχθεὶς μέχρι τῆς Ἀτροπατηνῆς, κάμπτει πρὸς δύσιν καὶ πρὸς ἄρκτους καὶ παραρρεῖ τὰ[5] Ἄζαρα πρῶτον, εἶτ' Ἀρτάξατα, πόλεις Ἀρμενίων· ἔπειτα διὰ τοῦ Ἀραξηνοῦ πεδίου πρὸς τὸ Κάσπιον ἐκδίδωσι πέλαγος.

C 528 4. Ἐν αὐτῇ δὲ τῇ Ἀρμενίᾳ πολλὰ μὲν ὄρη, πολλὰ δὲ ὀροπέδια, ἐν οἷς οὐδ' ἄμπελος φύεται ῥᾳδίως, πολλοὶ δ' αὐλῶνες, οἱ μὲν μέσως, οἱ δὲ καὶ σφόδρα εὐδαίμονες, καθάπερ τὸ Ἀραξηνὸν πεδίον, δι' οὗ ὁ Ἀράξης ποταμὸς ῥέων εἰς τὰ ἄκρα τῆς Ἀλβανίας καὶ τὴν Κασπίαν ἐκπίπτει θάλασσαν. καὶ μετὰ ταῦτα ἡ Σακασηνή, καὶ αὐτὴ τῇ Ἀλβανίᾳ πρόσχωρος καὶ τῷ Κύρῳ ποταμῷ, εἶθ' ἡ Γωγαρηνή· πᾶσα γὰρ ἡ χώρα

[1] Ἀντιταύρου, Du Theil, for Ταύρου; so Casaubon and C. Müller.
[2] ποταμίας, Corais from conj. of Salmasius, for μεσοποταμίας; so the later editors.
[3] *xz*, Tzschucke, and Corais read αὐτόν.

and the Taurus and ends towards the eastern parts of Armenia, thus on one side[1] enclosing the middle of Sophenê,[2] and having on its other side Acilisenê, which is situated between the Antitaurus[3] and the river-land[4] of the Euphrates, before that river bends towards the south. The royal city of Sophenê is Carcathiocerta. Above Mt. Masius, far towards the east opposite Gordyenê, lies Mt. Niphates; and then comes Mt. Abus, whence flow both the Euphrates and the Araxes, the former towards the west and the latter towards the east; and then Mt. Nibarus, which stretches as far as Media.

3. I have already described the course of the Euphrates. As for the Araxes, it first flows towards the east as far as Atropatenê, and then bends towards the west and towards the north and flows first past Azara and then past Artaxata, Armenian cities, and then, passing through the Araxene Plain, empties into the Caspian Sea.

4. In Armenia itself there are many mountains and many plateaus, in which not even the vine can easily grow; and also many valleys, some only moderately fertile, others very fertile, for instance, the Araxene Plain, through which the Araxes River flows to the extremities of Albania and then empties into the Caspian Sea. After these comes Sacasenê, this too bordering on Albania and the Cyrus River; and then comes Gogarenê. Indeed, the whole of

[1] See critical note.

[2] *i.e.* "enclosing Sophenê in a valley between itself (the Antitaurus) and the Taurus" (11. 12. 4).

[3] See critical note. [4] See critical note.

---

[4] Γορδυηνήν, Corais, for Γορδυληνήν E, Γοργοδιλήν z, Γοργοδυληνήν other MSS. [5] τά, the editors, for τήν.

αὕτη καρποῖς τε καὶ τοῖς ἡμέροις δένδρεσι καὶ τοῖς ἀειθαλέσι πληθύει, φέρει δὲ καὶ ἐλαίαν. ἔστι δὲ καὶ ἡ Φαυηνὴ[1] τῆς Ἀρμενίας ἐπαρχία καὶ ἡ Κωμισηνὴ καὶ Ὀρχιστηνή, πλείστην ἱππείαν παρέχουσα· ἡ δὲ Χορζηνὴ καὶ Καμβυσηνὴ προσβορώταταί εἰσι καὶ νιφόβολοι μάλιστα, συνάπτουσαι τοῖς Καυκασίοις ὄρεσι καὶ τῇ Ἰβηρίᾳ καὶ τῇ Κολχίδι· ὅπου φασὶ κατὰ τὰς ὑπερβολὰς τῶν ὀρῶν πολλάκις καὶ συνοδίας ὅλας[2] ἐν τῇ χιόνι καταπίνεσθαι νιφετῶν γινομένων ἐπὶ πλέον· ἔχειν δὲ καὶ βακτηρίας πρὸς τοὺς τοιούτους κινδύνους[3] παρεξαίροντας εἰς τὴν ἐπιφάνειαν ἀναπνοῆς τε χάριν καὶ τοῦ διαμηνύειν τοῖς ἐπιοῦσιν, ὥστε βοηθείας τυγχάνειν, ἀνορύττεσθαι καὶ σώζεσθαι. ἐν δὲ τῇ χιόνι βώλους πήγνυσθαί φασι κοίλας περιεχούσας χρηστὸν ὕδωρ ὡς ἐν χιτῶνι, καὶ ζῷα δὲ ἐν αὐτῇ γεννᾶσθαι· καλεῖ δὲ σκώληκας Ἀπολλωνίδης, Θεοφάνης δὲ θρῖπας· κἀν τούτοις ἀπολαμβάνεσθαι χρηστὸν ὕδωρ, περισχισθέντων[4] δὲ τῶν χιτώνων πίνεσθαι· τὴν δὲ γένεσιν τῶν ζῴων τοιαύτην εἰκάζουσιν, οἵαν τὴν τῶν κωνώπων ἐκ τῆς ἐν τοῖς μετάλλοις φλογὸς καὶ τοῦ φεψάλου.[5]

5. Ἱστοροῦσι δὲ τὴν Ἀρμενίαν, μικρὰν πρότερον οὖσαν, αὐξηθῆναι διὰ τῶν περὶ Ἀρταξίαν καὶ Ζαρίαδριν,[6] οἳ πρότερον μὲν ἦσαν Ἀντιόχου

[1] Φαυηνή (Φανηνή *orwxz*) seems corrupt; perhaps Φαυνηνή (Tzschucke, Corais) is right (cp. Φαυνῖτις below), if not Φασιανή (see Kramer's note).

[2] The words τῶν ὀρῶν after ὅλας are omitted by *gxy* and Corais. Strabo probably wrote ἐμπόρων (conj. of Corais) or ὁδοιπόρων (conj. of Meineke).

[3] Meineke inserts ἅς after κινδύνους.

this country abounds in fruits and cultivated trees and evergreens, and even bears the olive. There is also Phauenê,[1] a province of Armenia, and Comisenê, and Orchistenê, which last furnishes the most cavalry. Chorzenê and Cambysenê are the most northerly and the most subject to snows, bordering on the Caucasian mountains and Iberia and Colchis. It is said that here, on the passes over the mountains, whole caravans are often swallowed up in the snow when unusually violent snowstorms take place, and that to meet such dangers people carry staves, which they raise to the surface of the snow in order to get air to breathe and to signify their plight to people who come along, so as to obtain assistance, be dug out, and safely escape. It is said that hollow masses of ice form in the snow which contain good water, in a coat of ice as it were; and also that living creatures breed in the snow (Apollonides[2] calls these creatures "scoleces"[3] and Theophanes[4] "thripes"[5]); and that good water is enclosed in these hollow masses which people obtain for drinking by slitting open the coats of ice; and the genesis of these creatures is supposed to be like that of the gnats which spring from the flames and sparks at mines.

5. According to report, Armenia, though a small country in earlier times, was enlarged by Artaxias and Zariadris, who formerly were generals of

[1] See critical note. [2] See Vol. III, p. 234, foot-note 2.
[3] "Worms" or "larvae." [4] See foot-note on 11. 2. 2.
[5] Wood-worms.

---

[4] περισχισθέντων E *Epit.*, περισχεθέντων other MSS.
[5] φεψάλου E *Epit.*, πετάλλου D*h*, πετάλου other MSS.
[6] Ζαρίαδριν, Tyrwhitt, for Ζαριάδην; so the later editors.

τοῦ μεγάλου στρατηγοί, βασιλεύσαντες δ' ὕστερον μετὰ τὴν ἐκείνου ἧτταν, ὁ μὲν τῆς Σωφηνῆς καὶ τῆς Ἀκισηνῆς[1] καὶ Ὀδομαντίδος καὶ ἄλλων τινῶν, ὁ δὲ τῆς περὶ Ἀρτάξατα, συνηύξησαν, ἐκ τῶν περικειμένων ἐθνῶν ἀποτεμόμενοι μέρη, ἐκ Μήδων μὲν τήν τε Κασπιανὴν καὶ Φαυνῖτιν καὶ Βασοροπέδαν, Ἰβήρων δὲ τήν τε παρώρειαν τοῦ Παρυάδρου[2] καὶ τὴν Χορζηνὴν[3] καὶ Γωγαρηνήν, πέραν οὖσαν τοῦ Κύρου, Χαλύβων δὲ καὶ Μοσυνοίκων Καρηνῖτιν[4] καὶ Ξερξηνήν, ἃ τῇ μικρᾷ Ἀρμενίᾳ ἐστὶν ὅμορα ἢ καὶ μέρη αὐτῆς ἐστί, Καταόνων δὲ Ἀκιλισηνὴν[5] καὶ τὴν περὶ τὸν Ἀντίταυρον, Σύρων δὲ Ταρωνῖτιν,[6] ὥστε πάντας ὁμογλώττους εἶναι.

6. Πόλεις δ' ἐστὶ τῆς Ἀρμενίας Ἀρτάξατά τε, ἣν καὶ Ἀρταξιάσατα καλοῦσιν, Ἀννίβα κτίσαντος Ἀρταξίᾳ τῷ βασιλεῖ, καὶ Ἄρξατα, ἀμφότεραι ἐπὶ τῷ Ἀράξῃ, ἡ μὲν Ἄρξατα πρὸς τοῖς ὅροις τῆς Ἀτροπατίας,[7] ἡ δὲ Ἀρτάξατα πρὸς τῷ Ἀραξηνῷ[8] πεδίῳ, συνῳκισμένη καλῶς καὶ βασίλειον οὖσα τῆς χώρας. κεῖται δ' ἐπὶ χερρονησιάζοντος ἀγκῶνος, τὸ τεῖχος κύκλῳ προβεβλημένον τὸν ποταμὸν πλὴν τοῦ ἰσθμοῦ, τὸν ἰσθμὸν δ' ἔχει τάφρῳ καὶ χάρακι κεκλεισμένον. οὐ πολὺ δ'

C 529

[1] Ἀκισηνῆς (Ἀκιλισηνῆς editors before Kramer) is very doubtful (see Kramer's note).

[2] Παρυάδρου, Xylander, for Παιάδρου; so the later editors.

[3] Χορζηνήν, Xylander, for Χορζονήν; so the later editors.

[4] Καρηνῖτιν, Kramer, for Καρηνίτην; so the later editors.

[5] Ἀκιλισηνήν, Tzschucke, for Ἀκλισιηνήν; so the later editors.

[6] Ταρωνῖτιν, Kramer, for Ταμωνῖτις; so the later editors.

[7] Ἀτροπατίας, the editors, for Ἀτροπάτης C, Ἀτροπάτας other MSS.

Antiochus the Great,[1] but later, after his defeat, reigned as kings (the former as king of Sophenê, Acisenê, Odomantis, and certain other countries, and the latter as king of the country round Artaxata), and jointly enlarged their kingdoms by cutting off for themselves parts of the surrounding nations,—I mean by cutting off Caspianê and Phaunitis and Basoropeda from the country of the Medes; and the country along the side of Mt. Paryadres and Chorzenê and Gogarenê, which last is on the far side of the Cyrus River, from that of the Iberians; and Carenitis and Xerxenê, which border on Lesser Armenia or else are parts of it, from that of the Chalybians and the Mosynoeci; and Acilisenê and the country round the Antitaurus from that of the Cataonians; and Taronitis from that of the Syrians; and therefore they all speak the same language, as we are told.

6. The cities of Armenia are Artaxata, also called Artaxiasata, which was founded by Hannibal[2] for Artaxias the king, and Arxata, both on the Araxes River, Arxata being near the borders of Atropatia, whereas Artaxata is near the Araxene plain, being a beautiful settlement and the royal residence of the country. It is situated on a peninsula-like elbow of land and its walls have the river as protection all round them, except at the isthmus, which is enclosed by a trench and a palisade. Not

[1] Reigned as king of Syria 223–187 B.C.
[2] The Carthaginian.

---

[8] Ἀραξηνῷ, Tzschucke, for Ἀρταξενῷ *Dh*, Ἀρταξηνῷ other MSS.; so the later editors.

ἄπωθέν ἐστι τῆς πόλεως[1] τὰ Τιγράνου καὶ Ἀρταουάσδου γαζοφυλάκια, φρούρια ἐρυμνά, Βάβυρσά τε καὶ Ὀλανή· ἦν δὲ καὶ ἄλλα ἐπὶ τῷ Εὐφράτῃ. Ἀρταγήρας[2] δὲ ἀπέστησε μὲν Ἀδὼρ[3] ὁ φρούραρχος, ἐξεῖλον δ' οἱ Καίσαρος στρατηγοί, πολιορκήσαντες πολὺν χρόνον, καὶ τὰ τείχη περιεῖλον.

7. Ποταμοὶ δὲ πλείους μέν εἰσιν ἐν τῇ χώρᾳ, γνωριμώτατοι δὲ Φᾶσις μὲν καὶ Λύκος εἰς τὴν Ποντικὴν ἐκπίπτοντες θάλατταν (Ἐρατοσθένης δ' ἀντὶ τοῦ Λύκου τίθησι Θερμώδοντα οὐκ εὖ), εἰς δὲ τὴν Κασπίαν Κῦρος καὶ Ἀράξης, εἰς δὲ τὴν Ἐρυθρὰν ὅ τε Εὐφράτης καὶ ὁ Τίγρις.

8. Εἰσὶ δὲ καὶ λίμναι κατὰ τὴν Ἀρμενίαν μεγάλαι, μία μὲν ἡ Μαντιανή, Κυανῆ[4] ἑρμηνευθεῖσα, μεγίστη, ὥς φασι, μετὰ τὴν Μαιῶτιν, ἁλμυροῦ ὕδατος, διήκουσα μέχρι τῆς Ἀτροπατίας, ἔχουσα καὶ ἁλοπήγια· ἡ δὲ Ἀρσηνή, ἣν καὶ Θωπῖτιν[5] καλοῦσιν· ἔστι δὲ νιτρῖτις, τὰς δ' ἐσθῆτας ῥύπτει[6] καὶ διαξαίνει· διὰ δὲ τοῦτο καὶ ἄποτόν ἐστι τὸ ὕδωρ. φέρεται δὲ δι' αὐτῆς

[1] ἐπί, after πόλεως, Meineke omits; the editors before Kramer emended it to καί.
[2] Meineke emends Ἀρταγήρας to Ἀρτάγειρα, perhaps rightly.
[3] Meineke emends Ἀδώρ to Ἄδων, perhaps rightly.
[4] Κυανῆ E, Κυανεανή other MSS.
[5] Θωπῖτιν, Kramer, for Θωῆτιν; so the later editors.
[6] ῥύπτει (ῥήπτει C, ῥύττει *m*), Eustathius, for ῥίττει; so Xylander (cp. 11. 13. 2).

---

[1] Father and son respectively, kings of Armenia.
[2] See critical note. [3] See critical note.
[4] Mantianê (apparently the word should be spelled "Matianê"; see 11. 8. 8 and 11. 13. 2) is the lake called

far from the city are the treasuries of Tigranes and Artavasdes,[1] the strong fortresses Babyrsa and Olanê. And there were other fortresses on the Euphrates. Of these, Artageras[2] was caused to revolt by Ador,[3] its commandant, but Caesar's generals sacked it after a long siege and destroyed its walls.

7. There are several rivers in the country, but the best known are the Phasis and the Lycus, which empty into the Pontic Sea (Eratosthenes wrongly writes "Thermodon" instead of "Lycus"), whereas the Cyrus and the Araxes empty into the Caspian Sea, and the Euphrates and the Tigris into the Red Sea.

8. There are also large lakes in Armenia; one the Mantianê, which being translated means "Blue";[4] it is the largest salt-water lake after Lake Maeotis, as they say, extending as far as Atropatia; and it also has salt-works. Another is Arsenê, also called Thopitis.[5] It contains soda,[6] and it cleanses and restores clothes;[7] but because of this ingredient the water is also unfit for drinking.

"Capauta" in 11. 13. 2, Capauta meaning "Blue" and corresponding to the old Armenian name Kapoit-azow (Blue Lake), according to Tozer (note *ad loc.*), quoting Kiepert.

[5] On the position of this lake see Tozer (note *ad loc.*).

[6] The Greek word "nitron" means "soda" (carbonate of soda, our washing soda), and should not be confused with our "nitre" (potassium nitrate), nor yet translated "potash" (potassium carbonate). Southgate (*Narrative of a Tour through Armenia, Kurdistan, etc.*, Vol. II, p. 306, Eng. ed.) says that "a chemical analysis of a specimen shows it to be alkaline salts, composed chiefly of carbonate of soda and chloride" (*chlorite* in Tozer is a typographical error) "of sodium" (salt).

[7] See 11. 13. 2.

ὁ Τίγρις ἀπὸ τῆς κατὰ τὸν Νιφάτην ὀρεινῆς ὁρμηθείς, ἄμικτον φυλάττων τὸ ῥεῦμα διὰ τὴν ὀξύτητα, ἀφ' οὗ καὶ τοὔνομα, Μήδων τίγριν καλούντων τὸ τόξευμα· καὶ οὗτος μὲν ἔχει πολυειδεῖς ἰχθῦς, οἱ δὲ λιμναῖοι ἑνὸς εἴδους εἰσί· κατὰ δὲ τὸν μυχὸν τῆς λίμνης εἰς βάραθρον ἐμπεσὼν ὁ ποταμὸς καὶ πολὺν τόπον ἐνεχθεὶς ὑπὸ γῆς ἀνατέλλει κατὰ τὴν Χαλωνῖτιν· ἐκεῖθεν δ' ἤδη πρὸς τὴν Ὦπιν καὶ τὸ τῆς Σεμιράμιδος καλούμενον διατείχισμα ἐκεῖνός τε καταφέρεται, τοὺς Γορδυαίους ἐν δεξιᾷ ἀφεὶς καὶ τὴν Μεσοποταμίαν ὅλην, καὶ ὁ Εὐφράτης τοὐναντίον ἐν ἀριστερᾷ ἔχων τὴν αὐτὴν χώραν· πλησιάσαντες δὲ ἀλλήλοις καὶ ποιήσαντες τὴν Μεσοποταμίαν, ὁ μὲν διὰ Σελευκείας φέρεται πρὸς τὸν Περσικὸν κόλπον, ὁ δὲ διὰ Βαβυλῶνος, καθάπερ εἴρηταί που ἐν τοῖς πρὸς Ἐρατοσθένην καὶ Ἵππαρχον λόγοις.

9. Μέταλλα δ' ἐν μὲν τῇ Συσπιρίτιδί[1] ἐστι χρυσοῦ κατὰ τὰ Κάβαλλα, ἐφ' ἃ Μένωνα ἔπεμψεν Ἀλέξανδρος μετὰ στρατιωτῶν, ἀνήχθη[2] δ' ὑπὸ τῶν ἐγχωρίων· καὶ ἄλλα δ' ἐστὶ μέταλλα, καὶ δὴ[3] τῆς σάνδυκος[4] καλουμένης, ἣν δὴ καὶ Ἀρμένιον καλοῦσι χρῶμα, ὅμοιον κάλχῃ. οὕτω δ' ἐστὶν ἱπποβότος σφόδρα ἡ χώρα, καὶ οὐχ

[1] Συσπιρίτιδί, Groskurd, for Ὑσπιράτιδι; so Kramer (see his note), Meineke, and C. Müller (*Ind. Var. Lect.* p. 1018).

[2] For ἀνήχθη (ἀνείχθη C), Casaubon conj. ἀνῃρέθη, Tzschucke ἀνεδείχθη or ἐδείχθη, Groskurd ἀπήχθη; Corais reads ἀνεψύχθη and Meineke ἀπήγχθη.

[3] δή, Tzschucke and Corais emend to τό.

[4] σάνδυκος, Salmasius, for ὀπάνδικος; so the later editors.

[1] There must have been a second Chalonitis, one "not far from Gordyaea" (see 16. 1. 21), as distinguished from

The Tigris flows through this lake after issuing from the mountainous country near the Niphates; and because of its swiftness it keeps its current unmixed with the lake; whence the name Tigris, since the Median word for "arrow" is "tigris." And while the river has fish of many kinds, the fish in the lake are of one kind only. Near the recess of the lake the river falls into a pit, and after flowing underground for a considerable distance rises near Chalonitis.[1] Thence the river begins to flow down towards Opis and the wall of Semiramis, as it is called, leaving the Gordiaeans and the whole of Mesopotamia on the right, while the Euphrates, on the contrary, has the same country on the left. Having approached one another and formed Mesopotamia, the former flows through Seleuceia to the Persian Gulf and the latter through Babylon, as I have already said somewhere in my arguments against Eratosthenes and Hipparchus.[2]

9. There are gold mines in Syspiritis near Caballa, to which Menon was sent by Alexander with soldiers, and he was led up[3] to them by the natives. There are also other mines, in particular those of sandyx,[4] as it is called, which is also called "Armenian" colour, like chalcê.[5] The country is so very good

that in eastern Assyria, or else there is an error in the name.

[2] 2. 1. 27.

[3] "Led up" (or "inland") seems wrong. The verb has been emended to "destroyed," "imprisoned," "hanged" (Meineke), and other such words, but the translator knows of no evidence either to support any one of these emendations or to encourage any other.

[4] An earthy ore containing arsenic, which yields a bright red colour.

[5] *i.e.* purple dye. The usual spelling is calchê.

C 530 ἧττον τῆς Μηδίας, ὥστε οἱ Νησαῖοι[1] ἵπποι καὶ ἐνταῦθα γίνονται, οἷσπερ οἱ Περσῶν βασιλεῖς ἐχρῶντο· καὶ ὁ σατράπης τῆς Ἀρμενίας τῷ Πέρσῃ κατ᾽ ἔτος δισμυρίους πώλους τοῖς Μιθρακίνοις[2] ἔπεμπεν. Ἀρταουάσδης δὲ Ἀντωνίῳ χωρὶς τῆς ἄλλης ἱππείας αὐτὴν τὴν κατάφρακτον ἑξακισχιλίαν ἵππον ἐκτάξας ἐπέδειξεν, ἡνίκα εἰς τὴν Μηδίαν ἐνέβαλε σὺν αὐτῷ. ταύτης δὲ τῆς ἱππείας οὐ Μῆδοι μόνοι καὶ Ἀρμένιοι ζηλωταὶ γεγόνασιν, ἀλλὰ καὶ Ἀλβανοί, καὶ γὰρ ἐκεῖνοι καταφράκτοις χρῶνται.

10. Τοῦ δὲ πλούτου καὶ τῆς δυνάμεως τῆς χώρας σημεῖον οὐ μικρόν, ὅτι Πομπηίου Τιγράνῃ τῷ πατρὶ τῷ Ἀρταουάσδου τάλαντα ἐπιγράψαντος ἑξακισχίλια ἀργυρίου, διένειμεν αὐτίκα ταῖς δυνάμεσι τῶν Ῥωμαίων, στρατιώτῃ μὲν κατ᾽ ἄνδρα πεντήκοντα δραχμάς,[3] ἑκατοντάρχῃ δὲ χιλίας, ἱππάρχῳ[4] δὲ καὶ χιλιάρχῳ τάλαντον.

11. Μέγεθος δὲ τῆς χώρας Θεοφάνης ἀποδίδωσιν εὖρος μὲν σχοίνων ἑκατόν, μῆκος δὲ διπλάσιον, τιθεὶς τὴν σχοῖνον τετταράκοντα σταδίων· πρὸς ὑπερβολὴν δ᾽ εἴρηκεν· ἐγγυτέρω δ᾽ ἐστὶ τῆς ἀληθείας μῆκος μὲν θέσθαι τὸ ὑπ᾽ ἐκείνου λεχθὲν εὖρος,[5] εὖρος δὲ τὸ ἥμισυ ἢ μικρῷ πλεῖον. ἡ μὲν δὴ φύσις τῆς Ἀρμενίας καὶ δύναμις τοιαύτη.

[1] E has Νισαῖοι.

[2] Μιθρακίνοις, Kramer, for Μιθρακήνοις C, Μιθρακάνοις *Elorwg*, Μιθριακοῖς Corais, Μιθραϊκοῖς Groskurd.

[3] καὶ ἑκατόν, after δραχμάς, Corais would omit; so the later editors.

[4] ἱππάρχῳ, Du Theil, for ἐπάρχῳ; so the later editors.

[5] εὖρος, Groskurd inserts; so the later editors.

for "horse-pasturing," not even inferior to Media,[1] that the Nesaean horses, which were used by the Persian kings, are also bred there. The satrap of Armenia used to send to the Persian king twenty thousand foals every year at the time of the Mithracina.[2] Artavasdes,[3] at the time when he invaded Media with Antony, showed him, apart from the rest of the cavalry, six thousand horses drawn up in battle array in full armour. Not only the Medes and the Armenians pride themselves upon this kind of cavalry, but also the Albanians, for they too use horses in full armour.

10. As for the wealth and power of the country, the following is no small sign of it, that when Pompey imposed upon Tigranes, the father of Artavasdes, a payment of six thousand talents of silver, he forthwith distributed to the Roman forces as follows: to each soldier fifty drachmas, to each centurion a thousand drachmas, and to each hipparch and chiliarch a talent.

11. The size of the country is given by Theophanes:[4] the breadth one hundred "schoeni," and the length twice as much, putting the "schoenus" at forty stadia;[5] but his estimate is too high; it is nearer the truth to put down as length what he gives as breadth, and as breadth the half, or a little more, of what he gives as breadth. Such, then, is the nature and power of Armenia.

[1] See 11. 13. 7.

[2] The annual festival in honour of the Persian Sun-god Mithras.

[3] See 11. 13. 4.

[4] See foot-note on 11. 2. 2.

[5] On the variations in the meaning of "schoenus," see 17. 1. 24.

12. Ἀρχαιολογία δέ τίς ἐστι περὶ τοῦ ἔθνους τοῦδε τοιαύτη· Ἄρμενος ἐξ Ἀρμενίου, πόλεως Θετταλικῆς, ἣ κεῖται μεταξὺ Φερῶν καὶ Λαρίσης ἐπὶ τῇ Βοίβῃ, καθάπερ εἴρηται, συνεστράτευσεν Ἰάσονι εἰς τὴν Ἀρμενίαν· τούτου φασὶν ἐπώνυμον τὴν Ἀρμενίαν οἱ περὶ Κυρσίλον τὸν Φαρσάλιον καὶ Μήδιον τὸν Λαρισαῖον, ἄνδρες συνεστρατευκότες Ἀλεξάνδρῳ, τῶν δὲ μετὰ τοῦ Ἀρμένου τοὺς μὲν τὴν Ἀκιλισηνὴν οἰκῆσαι τὴν ὑπὸ τοῖς Σωφηνοῖς πρότερον οὖσαν, τοὺς δὲ ἐν τῇ Συσπιρίτιδι ἕως τῆς Καλαχηνῆς καὶ τῆς Ἀδιαβηνῆς ἔξω τῶν Ἀρμενιακῶν ὅρων.[1] καὶ τὴν ἐσθῆτα δὲ τὴν Ἀρμενιακὴν Θετταλικήν φασιν, οἷον τοὺς βαθεῖς χιτῶνας, οὓς καλοῦσι Θετταλικοὺς[2] ἐν ταῖς τραγῳδίαις, καὶ ζωννύουσι περὶ τὰ στήθη, καὶ ἐφαπτίδας, ὡς καὶ τῶν τραγῳδῶν μιμησαμένων τοὺς Θετταλούς, ἔδει μὲν γὰρ αὐτοῖς ἐπιθέτου κόσμου τοιούτου τινός, οἱ δὲ Θετταλοὶ μάλιστα βαθυστολοῦντες, ὡς εἰκός, διὰ τὸ πάντων εἶναι Ἑλλήνων βορειοτάτους καὶ ψυχροτάτους νέμεσθαι τόπους ἐπιτηδειοτάτην παρέσχοντο μίμησιν τῇ τῶν ὑποκριτῶν διασκευῇ[3] ἐν τοῖς ἀναπλάσμασιν· καὶ τὸν τῆς ἱππικῆς C 531 ζῆλόν φασιν εἶναι Θετταλικὸν καὶ τούτοις ὁμοίως καὶ Μήδοις· τὴν δὲ Ἰάσονος στρατείαν καὶ τὰ Ἰασόνια μαρτυρεῖ, ὧν τινὰ οἱ δυνάσται κατεσκεύασαν[4] παραπλησίως ὥσπερ τὸν ἐν Ἀβδήροις νεὼν τοῦ Ἰάσονος Παρμενίων.

[1] ὅρων, Xylander, for ὀρῶν; so the later editors.
[2] Θετταλικούς, Corais from conj. of Du Theil, for Αἰτωλικούς; so the later editors.
[3] τῇ . . . διασκευῇ, Kramer, for τήν . . . διασκευήν, omitting δέ after διασκευῇ; so the later editors.

12. There is an ancient story of the Armenian race to this effect: that Armenus of Armenium, a Thessalian city, which lies between Pherae and Larisa on Lake Boebe, as I have already said,[1] accompanied Jason into Armenia; and Cyrsilus the Pharsalian and Medius the Larisaean, who accompanied Alexander, say that Armenia was named after him, and that, of the followers of Armenus, some took up their abode in Acilisenê, which in earlier times was subject to the Sopheni, whereas others took up their abode in Syspiritis, as far as Calachenê and Adiabenê, outside the Armenian mountains. They also say that the clothing of the Armenians is Thessalian, for example, the long tunics, which in tragedies are called Thessalian and are girded round the breast; and also the cloaks that are fastened on with clasps, another way in which the tragedians imitated the Thessalians, for the tragedians had to have some alien decoration of this kind; and since the Thessalians in particular wore long robes, probably because they of all the Greeks lived in the most northerly and coldest region, they were the most suitable objects of imitation for actors in their theatrical make-ups. And they say that their style of horsemanship is Thessalian, both theirs and alike that of the Medes. To this the expedition of Jason and the Jasonian monuments bear witness, some of which were built by the sovereigns of the country, just as the temple of Jason at Abdera was built by Parmenion.

[1] 11. 4. 8.

[4] κατεσκεύασαν, Casaubon, for κατέσκαψαν; so the later editors.

13. Τὸν δὲ Ἀράξην κληθῆναι νομίζουσι κατὰ τὴν ὁμοιότητα τὴν πρὸς τὸν Πηνειὸν ὑπὸ τῶν περὶ τὸν Ἄρμενον ὁμωνύμως ἐκείνῳ, καλεῖσθαι γὰρ Ἀράξην κἀκεῖνον διὰ τὸ ἀπαράξαι τὴν Ὄσσαν ἀπὸ τοῦ Ὀλύμπου, ῥήξαντα τὰ Τέμπη· καὶ τὸν ἐν Ἀρμενίᾳ δέ, ἀπὸ τῶν ὀρῶν καταβάντα, πλατύνεσθαί φασι τὸ παλαιὸν καὶ πελαγίζειν ἐν τοῖς ὑποκειμένοις πεδίοις, οὐκ ἔχοντα διέξοδον, Ἰάσονα δέ, μιμησάμενον τὰ Τέμπη, ποιῆσαι τὴν διασφάγα δι' ἧς καταράττει νυνὶ τὸ ὕδωρ εἰς τὴν Κασπίαν θάλατταν, ἐκ δὲ τούτου γυμνωθῆναι τὸ Ἀραξηνὸν πεδίον, δι' οὗ τυγχάνει[1] ῥέων ἐπὶ τὸν καταράκτην ὁ ποταμός. οὗτος μὲν οὖν ὁ λόγος περὶ τοῦ Ἀράξου ποταμοῦ λεγόμενος ἔχει τι πιθανόν, ὁ δὲ Ἡροδότειος οὐ πάνυ, φησὶ γὰρ ἐκ Ματιηνῶν αὐτὸν ῥέοντα εἰς τετταράκοντα ποταμοὺς σχίζεσθαι, μερίζειν δὲ Σκύθας καὶ Βακτριανούς· καὶ Καλλισθένης δὲ ἠκολούθησεν αὐτῷ.

14. Λέγονται δὲ καὶ τῶν Αἰνιάνων τινές, οἱ μὲν τὴν Οὐιτίαν οἰκῆσαι, οἱ δ' ὕπερθε τῶν Ἀρμενίων ὑπὲρ τὸν Ἄβον καὶ τὸν Νίβαρον.[2] μέρη δ' ἐστὶ τοῦ Ταύρου ταῦτα, ὧν ὁ Ἄβος ἐγγύς ἐστι τῆς ὁδοῦ τῆς εἰς Ἐκβάτανα φερούσης παρὰ τὸν τῆς Βάριδος[3] νεών. φασὶ δὲ καὶ Θρᾳκῶν τινάς, τοὺς προσαγορευομένους Σαραπάρας, οἷον κεφαλοτόμους, οἰκῆσαι ὑπὲρ τῆς Ἀρμενίας, πλησίον Γουρανίων

[1] τυγχάνει, Kramer, for συγχαίνει CE*hi*, and margin of D; συγχέαι D*lrwx*, συμβῇ *z*, συμβαίνει *o* and editors before Kramer.

[2] Νίβαρον, Corais, for Ἴμμαρον E, Ἴμβαρον other MSS.

[3] For Βάριδος C*x*, Tzschucke and Corais read Ἀβάριδος.

13. It is thought that the Araxes was given the same name as the Peneius by Armenus and his followers because of its similarity to that river, for that river too, they say, was called Araxes because of the fact that it "cleft"[1] Ossa from Olympus, the cleft called Tempê. And it is said that in ancient times the Araxes in Armenia, after descending from the mountains, spread out and formed a sea in the plains below, since it had no outlet, but that Jason, to make it like Tempê, made the cleft through which the water now precipitates[2] itself into the Caspian Sea, and that in consequence of this the Araxene Plain, through which the river flows to its precipitate[3] descent, was relieved of the sea. Now this account of the Araxes contains some plausibility, but that of Herodotus not at all; for he says that after flowing out of the country of the Matieni it splits into forty rivers[4] and separates the Scythians from the Bactrians. Callisthenes, also, follows Herodotus.

14. It is also said of certain of the Aenianes that some of them took up their abode in Vitia and others above the Armenians beyond the Abus and the Nibarus. These two mountains are parts of the Taurus, and of these the Abus is near the road that leads into Ecbatana past the temple of Baris. It is also said that certain of the Thracians, those called "Saraparae," that is "Decapitators," took up their abode beyond Armenia near the Guranii and the

[1] "ap-arax-ae" is the Greek verb. [2] "cat-arax-ae."

[3] Again a play on the root "arax."

[4] "The Araxes discharges through forty mouths, of which all, except one, empty into marshes and shoals. . . . The one remaining mouth flows through a clear channel into the Caspian sea" (Herod. 1. 202).

καὶ Μήδων, θηριώδεις ἀνθρώπους καὶ ἀπειθεῖς, ὀρεινούς, περισκυθιστάς[1] τε καὶ ἀποκεφαλιστάς· τοῦτο γὰρ δηλοῦσιν οἱ Σαραπάραι. εἴρηται δὲ καὶ τὰ περὶ τῆς Μηδείας ἐν τοῖς Μηδικοῖς· ὥστ' ἐκ πάντων τούτων εἰκάζουσι καὶ τοὺς Μήδους καὶ Ἀρμενίους συγγενεῖς πως τοῖς Θετταλοῖς εἶναι καὶ τοῖς ἀπὸ Ἰάσονος καὶ Μηδείας.

15. Ὁ μὲν δὴ παλαιὸς λόγος οὗτος, ὁ δὲ τούτου νεώτερος καὶ κατὰ Πέρσας εἰς τὸ ἐφεξῆς μέχρι εἰς ἡμᾶς, ὡς ἐν κεφαλαίῳ πρέποι ἂν μέχρι τοσούτου λεχθείς, ὅτι κατεῖχον τὴν Ἀρμενίαν Πέρσαι καὶ Μακεδόνες, μετὰ ταῦτα οἱ τὴν Συρίαν ἔχοντες καὶ τὴν Μηδίαν· τελευταῖος δ' ὑπῆρξεν Ὀρόντης ἀπόγονος Ὑδάρνου, τῶν ἑπτὰ Περσῶν ἑνός· εἶθ' ὑπὸ τῶν Ἀντιόχου τοῦ μεγάλου στρατηγῶν τοῦ πρὸς Ῥωμαίους πολεμήσαντος διῃρέθη δίχα, Ἀρταξίου τε καὶ Ζαριάδριος· καὶ ἦρχον οὗτοι, τοῦ βασιλέως ἐπιτρέψαντος· ἡττηθέντος δ' ἐκείνου, προσθέμενοι Ῥωμαίοις καθ' C 532 αὑτοὺς ἐτάττοντο, βασιλεῖς προσαγορευθέντες. τοῦ μὲν οὖν Ἀρταξίου Τιγράνης ἦν ἀπόγονος καὶ εἶχε τὴν ἰδίως λεγομένην Ἀρμενίαν, αὕτη δ' ἦν προσεχὴς τῇ τε Μηδίᾳ καὶ Ἀλβανοῖς καὶ Ἴβηρσι μέχρι Κολχίδος καὶ τῆς ἐπὶ τῷ Εὐξείνῳ Καππαδοκίας, τοῦ δὲ Ζαριάδριος ὁ Σωφηνὸς Ἀρτάνης[2] ἔχων τὰ νότια μέρη καὶ τούτων τὰ πρὸς δύσιν μᾶλλον. κατελύθη δ' οὗτος ὑπὸ τοῦ Τιγράνου, καὶ πάντων κατέστη κύριος ἐκεῖνος. τύχαις δ' ἐχρήσατο ποικίλαις, κατ' ἀρχὰς μὲν

[1] *oxz* read περισκελιστάς.
[2] For Ἀρτάνης Steph. Byz., *s.v.* Σωφηνή, writes Ἀρσάκης, and

Medes, a fierce and intractable people, mountaineers, scalpers, and beheaders, for this last is the meaning of "Saraparae." I have already discussed Medeia in my account of the Medes;[1] and therefore, from all this, it is supposed that both the Medes and the Armenians are in a way kinsmen to the Thessalians and the descendants of Jason and Medeia.

15. This, then, is the ancient account; but the more recent account, and that which begins with Persian times and extends continuously to our own, might appropriately be stated in brief as follows: The Persians and Macedonians were in possession of Armenia; after this, those who held Syria and Media; and the last was Orontes, the descendant of Hydarnes, one of the seven Persians;[2] and then the country was divided into two parts by Artaxias and Zariadris, the generals of Antiochus the Great, who made war against the Romans; and these generals ruled the country, since it was turned over to them by the king; but when the king was defeated, they joined the Romans and were ranked as autonomous, with the title of king. Now Tigranes was a descendant of Artaxias and held what is properly called Armenia, which lay adjacent to Media and Albania and Iberia, extending as far as Colchis and Cappadocia on the Euxine, whereas the Sophenian Artanes,[3] who held the southern parts and those that lay more to the west than these, was a descendant of Zariadris. But he was overcome by Tigranes, who established himself as lord of all. The changes of fortune experienced by

[1] 11. 13. 10. [2] See Herodotus 3. 70. [3] See critical note.

so Groskurd; Tyrwhitt emends to Ἀρμενίας, making Σωφηνός a proper name (cp. 12. 2. 1).

γὰρ ὡμήρευσε παρὰ Πάρθοις, ἔπειτα δι' ἐκείνων ἔτυχε καθόδου, λαβόντων μισθὸν ἑβδομήκοντα αὐλῶνας τῆς Ἀρμενίας· αὐξηθεὶς δὲ καὶ ταῦτα ἀπέλαβε τὰ χωρία καὶ τὴν ἐκείνων ἐπόρθησε, τήν τε περὶ Νίνον[1] καὶ τὴν περὶ Ἄρβηλα· ὑπηκόους δ' ἔσχε καὶ τὸν Ἀτροπατηνὸν καὶ τὸν Γορδυαῖον, μεθ' ὧν καὶ τὴν λοιπὴν Μεσοποταμίαν, ἔτι δὲ τὴν Συρίαν αὐτὴν καὶ Φοινίκην, διαβὰς τὸν Εὐφράτην, ἀνὰ κράτος εἷλεν. ἐπὶ τοσοῦτον δ' ἐξαρθεὶς καὶ πόλιν ἔκτισε[2] πλησίον τῆς Ἰβηρίας[3] μεταξὺ ταύτης τε καὶ τοῦ κατὰ τὸν Εὐφράτην Ζεύγματος, ἣν ὠνόμασε Τιγρανόκερτα, ἐκ δώδεκα ἐρημωθεισῶν ὑπ' αὐτοῦ πόλεων Ἑλληνίδων ἀνθρώπους συναγαγών. ἔφθη δ' ἐπελθὼν Λεύκολλος ὁ τῷ Μιθριδάτῃ πολεμήσας καὶ τοὺς μὲν οἰκήτορας εἰς τὴν οἰκείαν ἑκάστου ἀπέλυσε, τὸ δὲ κτίσμα, ἡμιτελὲς ἔτι ὄν, κατέσπασε προσβαλὼν καὶ μικρὰν κώμην κατέλιπεν, ἐξήλασε δὲ καὶ τῆς Συρίας αὐτὸν καὶ τῆς Φοινίκης. διαδεξάμενος δ' Ἀρταουάσδης ἐκεῖνον τέως μὲν ηὐτύχει, φίλος ὢν Ῥωμαίοις, Ἀντώνιον δὲ προδιδοὺς Παρθυαίοις ἐν τῷ πρὸς αὐτοὺς πολέμῳ, δίκας ἔτισεν, ἀναχθεὶς γὰρ εἰς Ἀλεξάνδρειαν ὑπ' αὐτοῦ, δέσμιος πομπευθεὶς διὰ τῆς πόλεως τέως μὲν ἐφρουρεῖτο, ἔπειτ' ἀνῃρέθη,

[1] περὶ Νίνον, Xylander, for περίνιον; so the later editors.
[2] ἔκτισε, Xylander, for τίσαι; so the later editors.
[3] Ἰβηρίας seems corrupt; for conjectures see C. Müller, *Ind. Var. Lect.* p. 1019.

---

[1] This cannot be the *country* Iberia; and, so far as is known, the region in question had no *city* of that name.

Tigranes were varied, for at first he was a hostage among the Parthians; and then through them he obtained the privilege of returning home, they receiving as reward therefor seventy valleys in Armenia; but when he had grown in power, he not only took these places back but also devastated their country, both that about Ninus and that about Arbela; and he subjugated to himself the rulers of Atropenê and Gordyaea, and along with these the rest of Mesopotamia, and also crossed the Euphrates and by main strength took Syria itself and Phoenicia; and, exalted to this height, he also founded a city near Iberia,[1] between this place and the Zeugma on the Euphrates; and, having gathered peoples thither from twelve Greek cities which he had laid waste, he named it Tigranocerta; but Leucullus, who had waged war against Mithridates, arrived before Tigranes finished his undertaking and not only dismissed the inhabitants to their several home-lands but also attacked and pulled down the city, which was still only half finished, and left it a small village;[2] and he drove Tigranes out of both Syria and Phoenicia. His successor Artavasdes[3] was indeed prosperous for a time, while he was a friend to the Romans, but when he betrayed Antony to the Parthians in his war against them he paid the penalty for it, for he was carried off prisoner to Alexandreia by Antony and was paraded in chains through the city; and for a time he was kept in prison, but was afterwards

Kramer conjectures "Nisibis" (cp. 11. 12. 4); but C. Müller, more plausibly, "Carrhae." Cp. the reference to "Carrhae" in 16. 2. 23.

[2] 69 B.C. [3] See 11. 13. 4.

συνάπτοντος τοῦ Ἀκτιακοῦ πολέμου. μετ' ἐκεῖνον δὲ πλείους ἐβασίλευσαν ὑπὸ Καίσαρι καὶ Ῥωμαίοις ὄντες· καὶ νῦν ἔτι συνέχεται τὸν αὐτὸν τρόπον.

16. Ἅπαντα μὲν οὖν τὰ τῶν Περσῶν ἱερὰ καὶ Μῆδοι καὶ Ἀρμένιοι τετιμήκασι, τὰ δὲ τῆς Ἀναΐτιδος [1] διαφερόντως Ἀρμένιοι, ἔν τε ἄλλοις ἱδρυσάμενοι τόποις, καὶ δὴ καὶ ἐν τῇ Ἀκιλισηνῇ. ἀνατιθέασι δ' ἐνταῦθα δούλους καὶ δούλας. καὶ τοῦτο μὲν οὐ θαυμαστόν, ἀλλὰ καὶ θυγατέρας οἱ ἐπιφανέστατοι τοῦ ἔθνους ἀνιεροῦσι παρθένους, αἷς νόμος ἐστὶ καταπορνευθείσαις πολὺν χρόνον παρὰ τῇ θεῷ μετὰ ταῦτα δίδοσθαι πρὸς γάμον, οὐκ ἀπαξιοῦντος τῇ τοιαύτῃ συνοικεῖν οὐδενός. C 533 τοιοῦτον δέ τι καὶ Ἡρόδοτος λέγει τὸ περὶ τὰς Λυδάς· πορνεύειν γὰρ ἁπάσας. οὕτω δὲ φιλοφρόνως χρῶνται τοῖς ἐρασταῖς, ὥστε καὶ ξενίαν παρέχουσι καὶ δῶρα ἀντιδιδόασι πλείω πολλάκις ἢ λαμβάνουσιν, ἅτ' ἐξ εὐπόρων οἴκων ἐπιχορηγούμεναι· δέχονται δὲ οὐ τοὺς τυχόντας τῶν ξένων, ἀλλὰ μάλιστα τοὺς ἀπὸ ἴσου ἀξιώματος.

[1] Ἀναΐτιδος, Xylander, following *Epit.* and Eustathius (*Dionysius* 846), for Ταναΐδος; so the later editors.

---

[1] 1. 93, 199

slain, when the Actian war broke out. After him several kings reigned, these being subject to Caesar and the Romans; and still to-day the country is governed in the same way.

16. Now the sacred rites of the Persians, one and all, are held in honour by both the Medes and the Armenians; but those of Anaïtis are held in exceptional honour by the Armenians, who have built temples in her honour in different places, and especially in Acilisenê. Here they dedicate to her service male and female slaves. This, indeed, is not a remarkable thing; but the most illustrious men of the tribe actually consecrate to her their daughters while maidens; and it is the custom for these first to be prostituted in the temple of the goddess for a long time and after this to be given in marriage; and no one disdains to live in wedlock with such a woman. Something of this kind is told also by Herodotus [1] in his account of the Lydian women, who, one and all, he says, prostitute themselves. And they are so kindly disposed to their paramours that they not only entertain them hospitably but also exchange presents with them, often giving more than they receive, inasmuch as the girls from wealthy homes are supplied with means. However, they do not admit any man that comes along, but preferably those of equal rank with themselves.

lain, when the Actian war broke out. After him several kings reigned, these being subject to Caesar and the Romans; and still today the country is governed in the same way.

16. Now the sacred rites of the Persians, one and all, are held in honour by both the Medes and the Armenians; but those of Anaïtis are held in exceptional honour by the Armenians, who have built temples in her honour in different places, and especially in Acilisene. Here they dedicate to her service male and female slaves. This, indeed, is not a remarkable thing; but the most illustrious men of the tribe actually consecrate to her their daughters while maidens; and it is the custom for these first to be prostituted in the temple of the goddess for a long time and after this to be given in marriage; and no one disdains to live in wedlock with such a woman. Something of this kind is told also by Herodotus[1] in his account of the Lydian women, who, one and all, he says, prostitute themselves. And they are so kindly disposed to their paramours that they not only entertain them hospitably but also exchange presents with them, often giving more than they receive, inasmuch as the girls from wealthy homes are supplied with means. However, they do not admit any man that comes along, but preferably those of equal rank with themselves.

BOOK XII

ΙΒʹ

I

1. Καὶ ἡ Καππαδοκία[1] ἐστὶ πολυμερής τε καὶ συχνὰς δεδεγμένη μεταβολάς. οἱ δ' οὖν ὁμόγλωττοι μάλιστά εἰσιν οἱ ἀφοριζόμενοι πρὸς νότον μὲν τῷ Κιλικίῳ λεγομένῳ Ταύρῳ, πρὸς ἕω δὲ τῇ Ἀρμενίᾳ καὶ τῇ Κολχίδι καὶ τοῖς μεταξὺ ἑτερογλώττοις ἔθνεσι, πρὸς ἄρκτον δὲ τῷ Εὐξείνῳ μέχρι τῶν ἐκβολῶν τοῦ Ἅλυος, πρὸς δύσιν δὲ τῷ τε τῶν Παφλαγόνων ἔθνει καὶ Γαλατῶν τῶν τὴν Φρυγίαν ἐποικησάντων[2] μέχρι Λυκαόνων καὶ Κιλίκων τῶν τὴν τραχεῖαν Κιλικίαν νεμομένων.

2. Καὶ αὐτῶν δὲ τῶν ὁμογλώττων οἱ παλαιοὶ τοὺς Καταόνας καθ' αὑτοὺς ἔταττον, ἀντιδιαιροῦντες τοῖς Καππάδοξιν, ὡς ἑτεροεθνέσι, καὶ ἐν τῇ διαριθμήσει τῶν ἐθνῶν μετὰ τὴν Καππαδοκίαν ἐτίθεσαν τὴν Καταονίαν, εἶτα τὸν Εὐφράτην καὶ τὰ πέραν ἔθνη, ὥστε καὶ τὴν Μελιτηνὴν ὑπὸ τῇ Καταονίᾳ τάττειν, ἣ μεταξὺ κεῖται ταύτης τε καὶ τοῦ Εὐφράτου, συνάπτουσα τῇ Κομμαγηνῇ, μέρος τε τῆς Καππαδοκίας ἐστὶ δέκατον κατὰ τὴν εἰς δέκα στρατηγίας διαίρεσιν τῆς χώρας. οὕτω γὰρ
C 534 δὴ οἱ καθ' ἡμᾶς βασιλεῖς οἱ πρὸ Ἀρχελάου

[1] Before ἐστί Corais and Meineke insert δ'.
[2] ἐποικησάντων, Corais, for μετοικησάντων; so the later editors.

BOOK XII

I

1.[1] CAPPADOCIA, also, is a country of many parts and has undergone numerous changes. However, the inhabitants who speak the same language are, generally speaking, those who are bounded on the south by the "Cilician" Taurus, as it is called, and on the east by Armenia and Colchis and by the intervening peoples who speak a different group of languages, and on the north by the Euxine as far as the outlets of the Halys River, and on the west both by the tribe of the Paphlagonians and by those Galatae who settled in Phrygia and extended as far as the Lycaonians and those Cilicians who occupy Cilicia Tracheia.[2]

2. Now as for the tribes themselves which speak the same language, the ancients set one of them, the Cataonians, by themselves, contradistinguishing them from the Cappadocians, regarding the latter as a different tribe; and in their enumeration of the tribes they placed Cataonia after Cappadocia, and then placed the Euphrates and the tribes beyond it so as to include in Cataonia Melitenê, which lies between Cataonia and the Euphrates, borders on Commagenê, and, according to the division of Cappadocia into ten prefectures, is a tenth portion of the country. Indeed, it was in this way that the kings in my time who preceded Archeläus held

[1] From Xylander to Meineke the editors agree that a portion of text at the beginning of this Book is missing.
[2] "Rugged" Cilicia.

διατεταγμένην εἶχον τὴν ἡγεμονίαν τῆς Καππαδοκίας· δέκατον δ' ἐστὶ μέρος καὶ ἡ Καταονία. καθ' ἡμᾶς δὲ εἶχε στρατηγὸν ἑκατέρα ἴδιον· οὔτε δ' ἐκ τῆς διαλέκτου διαφορᾶς τινὸς ἐν τούτοις πρὸς τοὺς ἄλλους Καππάδοκας ἐμφαινομένης, οὔτε ἐκ[1] τῶν ἄλλων ἐθῶν,[2] θαυμαστὸν πῶς ἠφάνισται τελέως τὰ σημεῖα τῆς ἀλλοεθνίας. ἦσαν δ' οὖν διωρισμένοι, προσεκτήσατο δ' αὐτοὺς Ἀριαράθης ὁ πρῶτος προσαγορευθεὶς Καππαδόκων βασιλεύς.

3. Ἔστι δ' ὥσπερ χερρονήσου μεγάλης ἰσθμὸς οὗτος, σφιγγόμενος θαλάτταις δυσί, τῇ τε τοῦ Ἰσσικοῦ κόλπου μέχρι τῆς τραχείας Κιλικίας καὶ τῇ τοῦ Εὐξείνου μεταξὺ Σινώπης τε καὶ τῆς τῶν Τιβαρηνῶν παραλίας· ἐντὸς δὲ τοῦ ἰσθμοῦ λέγομεν χερρόνησον τὴν προσεσπέριον τοῖς Καππάδοξιν ἅπασαν, ἣν Ἡρόδοτος μὲν ἐντὸς Ἅλυος καλεῖ· αὕτη γάρ ἐστιν, ἧς ἦρξεν ἁπάσης Κροῖσος, λέγει δ' αὐτὸν ἐκεῖνος τύραννον ἐθνέων τῶν ἐντὸς Ἅλυος ποταμοῦ. οἱ δὲ νῦν τὴν ἐντὸς τοῦ Ταύρου καλοῦσιν Ἀσίαν, ὁμωνύμως τῇ ὅλῃ ἠπείρῳ ταύτην Ἀσίαν προσαγορεύοντες. περιέχεται δ' ἐν αὐτῇ πρῶτα μὲν ἔθνη τὰ ἀπὸ τῆς ἀνατολῆς Παφλαγόνες τε καὶ Φρύγες καὶ Λυκάονες, ἔπειτα Βιθυνοὶ καὶ Μυσοὶ καὶ ἡ Ἐπίκτητος, ἔτι δὲ Τρωὰς καὶ Ἑλλησποντία, μετὰ δὲ τούτους ἐπὶ θαλάττῃ μὲν Ἑλλήνων οἵ τε Αἰολεῖς καὶ Ἴωνες, τῶν δ' ἄλλων Κᾶρές τε καὶ Λύκιοι, ἐν δὲ τῇ μεσογαίᾳ Λυδοί. περὶ μὲν οὖν τῶν ἄλλων ἐροῦμεν ὕστερον.

[1] *τῆς*, before *τῶν ἄλλων*, is rightly omitted by *oz*.
[2] *ἐθῶν c* instead of *ἐθνῶν*; so the editors.

their several prefectures over Cappadocia. And Cataonia, also, is a tenth portion of Cappadocia. In my time each of the two countries had its own prefect; but since, as compared with the other Cappadocians, there is no difference to be seen either in the language or in any other usages of the Cataonians, it is remarkable how utterly all signs of their being a different tribe have disappeared. At any rate, they were once a distinct tribe, but they were annexed by Ariarathes, the first man to be called king of the Cappadocians.

3. Cappadocia constitutes the isthmus, as it were, of a large peninsula bounded by two seas, by that of the Issian Gulf as far as Cilicia Tracheia and by that of the Euxine as far as Sinopê and the coast of the Tibareni. I mean by "peninsula" all the country which is west of Cappadocia this side the isthmus, which by Herodotus is called "the country this side the Halys River"; for this is the country which in its entirety was ruled by Croesus, whom Herodotus calls the tyrant of the tribes this side the Halys River.[1] However, the writers of to-day give the name of Asia to the country this side the Taurus, applying to this country the same name as to the whole continent of Asia. This Asia comprises the first nations on the east, the Paphlagonians and Phrygians and Lycaonians, and then the Bithynians and Mysians and the Epictetus,[2] and, besides these, the Troad and Hellespontia, and after these, on the sea, the Aeolians and Ionians, who are Greeks, and, among the rest, the Carians and Lycians, and, in the interior, the Lydians. As for the other tribes, I shall speak of them later.

[1] 1. 6, 28. [2] The territory later "Acquired" (2. 5. 31).

4. Τὴν δὲ Καππαδοκίαν εἰς δύο σατραπείας μερισθεῖσαν ὑπὸ τῶν Περσῶν παραλαβόντες Μακεδόνες περιεῖδον[1] τὰ μὲν ἑκόντες τὰ δ' ἄκοντες εἰς βασιλείας ἀντὶ σατραπειῶν περιστᾶσαν· ὧν τὴν μὲν ἰδίως Καππαδοκίαν ὠνόμασαν καὶ πρὸς τῷ Ταύρῳ καὶ νὴ Δία μεγάλην Καππαδοκίαν, τὴν δὲ Πόντον, οἱ δὲ τὴν πρὸς τῷ Πόντῳ Καππαδοκίαν. τῆς δὲ μεγάλης Καππαδοκίας νῦν μὲν οὐκ ἴσμεν πω τὴν[2] διάταξιν· τελευτήσαντος γὰρ τὸν βίον Ἀρχελάου τοῦ βασιλεύσαντος, ἔγνω Καῖσάρ τε καὶ ἡ σύγκλητος ἐπαρχίαν εἶναι Ῥωμαίων αὐτήν. ἐπ' ἐκείνου δὲ καὶ τῶν πρὸ αὐτοῦ βασιλέων εἰς δέκα στρατηγίας διῃρημένης τῆς χώρας, πέντε μὲν ἐξητάζοντο αἱ πρὸς τῷ Ταύρῳ, Μελιτηνή, Καταονία, Κιλικία, Τυανῖτις, Γαρσαυρῖτις· πέντε δὲ λοιπαὶ Λαουιανσηνή,[3] Σαργαραυσηνή,[4] Σαραουηνή, Χαμανηνή, Μοριμηνή.[5] προσεγένετο δ' ὕστερον παρὰ Ῥωμαίων ἐκ τῆς Κιλικίας τοῖς[6] πρὸ Ἀρχελάου καὶ ἑνδεκάτη C 535 στρατηγία, ἡ περὶ Καστάβαλά τε καὶ Κύβιστρα μέχρι τῆς Ἀντιπάτρου τοῦ λῃστοῦ Δέρβης, τῷ δὲ Ἀρχελάῳ καὶ ἡ τραχεῖα περὶ Ἐλαιοῦσσαν Κιλικία καὶ πᾶσα ἡ τὰ πειρατήρια συστησαμένη.

[1] περιεῖδον, Xylander, for περιεῖλον; so the later editors.
[2] πω τήν, Tyrwhitt, for πρώτην; so the editors.
[3] Λαουιανσηνή, Kramer, for Λαουσανσηνή *l*, Λαουινασηνή other MSS.
[4] Σαργαραυσηνή, Tzschucke, for Σαργαυσηνή.
[5] Μοριμηνή, Tzschucke, for Ῥιμνηνηνή DH*ior*, Ῥιμνηνή C*xz*, Μοραμηνή *Epit.*

4. Cappadocia was divided into two satrapies by the Persians at the time when it was taken over by the Macedonians; the Macedonians willingly allowed one part of the country, but unwillingly the other, to change to kingdoms instead of satrapies; and one of these kingdoms they named "Cappadocia Proper" and "Cappadocia near Taurus," and even "Greater Cappadocia," and the other they named "Pontus," though others named it Cappadocia Pontica. As for Greater Cappadocia, we at present do not yet know its administrative divisions,[1] for after the death of king Archelaüs Caesar[2] and the senate decreed that it was a Roman province. But when, in the reign of Archelaüs and of the kings who preceded him, the country was divided into ten prefectures, those near the Taurus were reckoned as five in number, I mean Melitenê, Cataonia, Cilicia, Tyanitis, and Garsauritis; and Laviansenê, Sargarausenê, Saravenê, Chamanenê, and Morimenê as the remaining five. The Romans later assigned to the predecessors of Archelaüs an eleventh prefecture, taken from Cilicia, I mean the country round Castabala and Cybistra, extending to Derbê, which last had belonged to Antipater the pirate; and to Archelaüs they further assigned the part of Cilicia Tracheia round Elaeussa, and also all the country that had organised the business of piracy.

[1] A.D. 17. [2] Tiberius Caesar.

---

[6] τοῖς E, τῆς other MSS.

II

1. Ἔστι δ' ἡ μὲν Μελιτηνὴ παραπλησία τῇ Κομμαγηνῇ, πᾶσα γάρ ἐστι τοῖς ἡμέροις δένδροις κατάφυτος, μόνη τῆς ἄλλης Καππαδοκίας, ὥστε καὶ ἔλαιον φέρειν καὶ τὸν Μοναρίτην οἶνον τοῖς Ἑλληνικοῖς ἐνάμιλλον· ἀντίκειται δὲ τῇ Σωφηνῇ, μέσον ἔχουσα τὸν Εὐφράτην ποταμὸν καὶ αὐτὴ καὶ[1] ἡ Κομμαγηνή, ὅμορος οὖσα. ἔστι δὲ φρούριον ἀξιόλογον τῶν Καππαδόκων ἐν τῇ περαίᾳ Τόμισα. τοῦτο δ' ἐπράθη μὲν τῷ Σωφηνῷ ταλάντων ἑκατόν, ὕστερον δὲ ἐδωρήσατο Λεύκολλος τῷ Καππάδοκι συστρατεύσαντι ἀριστεῖον κατὰ τὸν πρὸς Μιθριδάτην πόλεμον.

2. Ἡ δὲ Καταονία πλατὺ καὶ κοῖλόν ἐστι πεδίον πάμφορον πλὴν τῶν ἀειθαλῶν. περίκειται δ' ὄρη ἄλλα τε καὶ Ἀμανὸς ἐκ τοῦ πρὸς νότον μέρους, ἀπόσπασμα ὂν τοῦ Κιλικίου Ταύρου, καὶ ὁ Ἀντίταυρος, εἰς τἀναντία ἀπερρωγώς. ὁ μὲν γὰρ Ἀμανὸς ἐπὶ τὴν Κιλικίαν καὶ τὴν Συριακὴν ἐκτείνεται θάλατταν πρὸς τὴν ἑσπέραν ἀπὸ τῆς Καταονίας καὶ τὸν νότον, τῇ δὲ τοιαύτῃ διαστάσει περικλείει τὸν Ἰσσικὸν κόλπον ἅπαντα καὶ τὰ μεταξὺ τῶν Κιλίκων πεδία πρὸς τὸν Ταῦρον· ὁ δ' Ἀντίταυρος ἐπὶ τὰς ἄρκτους ἐγκέκλιται καὶ μικρὸν ἐπιλαμβάνει τῶν ἀνατολῶν, εἶτ' εἰς τὴν μεσόγαιαν τελευτᾷ.

3. Ἐν δὲ τῷ Ἀντιταύρῳ τούτῳ βαθεῖς καὶ στενοί εἰσιν αὐλῶνες, ἐν οἷς ἵδρυται τὰ Κόμανα καὶ τὸ τῆς Ἐνυοῦς ἱερόν, ἣν[2] ἐκεῖνοι Μᾶ ὀνομά-

[1] καί, Xylander inserts.
[2] ἥν, Groskurd, for ὅ; so Meineke.

II

1. MELITENÊ is similar to Commagenê, for the whole of it is planted with fruit-trees, the only country in all Cappadocia of which this is true, so that it produces, not only the olive, but also the Monarite wine, which rivals the Greek wines. It is situated opposite to Sophenê; and the Euphrates River flows between it and Commagenê, which latter borders on it. On the far side of the river is a noteworthy fortress belonging to the Cappadocians, Tomisa by name. This was sold to the ruler of Sophenê for one hundred talents, but later was presented by Leucullus as a meed of valour to the ruler of Cappadocia who took the field with him in the war against Mithridates.

2. Cataonia is a broad hollow plain, and produces everything except evergreen-trees. It is surrounded on its southern side by mountains, among others by the Amanus, which is a branch of the Cilician Taurus, and by the Antitaurus, which branches off in the opposite direction; for the Amanus extends from Cataonia to Cilicia and the Syrian Sea towards the west and south, and in this intervening space it surrounds the whole of the Gulf of Issus and the intervening plains of the Cilicians which lie towards the Taurus. But the Antitaurus inclines to the north and takes a slightly easterly direction, and then terminates in the interior of the country.

3. In this Antitaurus are deep and narrow valleys, in which are situated Comana and the temple of Enyo,[1] whom the people there call "Ma." It is

[1] Goddess of war (*Iliad* 5. 333).

ζουσι· πόλις δ' ἐστὶν ἀξιόλογος, πλεῖστον μέντοι τὸ[1] τῶν θεοφορήτων πλῆθος καὶ τὸ τῶν ἱεροδούλων ἐν αὐτῇ. Κατάονες δέ εἰσιν οἱ ἐνοικοῦντες, ἄλλως μὲν ὑπὸ τῷ βασιλεῖ τεταγμένοι, τοῦ δὲ ἱερέως ὑπακούοντες τὸ πλέον· ὁ δὲ τοῦ θ' ἱεροῦ κύριός ἐστι καὶ τῶν ἱεροδούλων, οἳ κατὰ τὴν ἡμετέραν ἐπιδημίαν πλείους ἦσαν τῶν ἑξακισχιλίων, ἄνδρες ὁμοῦ γυναιξί. πρόσκειται δὲ τῷ ἱερῷ καὶ χώρα πολλή, καρποῦται δ' ὁ ἱερεὺς τὴν πρόσοδον, καὶ ἔστιν οὗτος δεύτερος κατὰ τιμὴν ἐν[2] τῇ Καππαδοκίᾳ μετὰ τὸν βασιλέα· ὡς δ' ἐπὶ τὸ πολὺ τοῦ αὐτοῦ γένους ἦσαν οἱ ἱερεῖς τοῖς βασιλεῦσι. τὰ δὲ ἱερὰ ταῦτα δοκεῖ Ὀρέστης μετὰ τῆς ἀδελφῆς Ἰφιγενείας κομίσαι δεῦρο ἀπὸ τῆς Ταυρικῆς Σκυθίας, τὰ τῆς Ταυροπόλου Ἀρτέμιδος, ἐνταῦθα δὲ καὶ τὴν πένθιμον κόμην ἀποθέσθαι, ἀφ' ἧς καὶ τοὔνομα τῇ πόλει. διὰ
C 536 μὲν οὖν τῆς πόλεως ταύτης ὁ Σάρος ῥεῖ ποταμός, καὶ διὰ τῶν συναγκειῶν[3] τοῦ Ταύρου διεκπεραιοῦται πρὸς τὰ τῶν Κιλίκων πεδία καὶ τὸ ὑποκείμενον πέλαγος.

4. Διὰ δὲ τῆς Καταονίας ὁ Πύραμος πλωτός, ἐκ μέσου τοῦ πεδίου τὰς πηγὰς ἔχων· ἔστι δὲ βόθρος ἀξιόλογος, δι' οὗ καθορᾶν[4] ἔστι τὸ ὕδωρ ὑποφερόμενον κρυπτῶς μέχρι πολλοῦ διαστήματος ὑπὸ γῆς, εἶτ' ἀνατέλλον εἰς τὴν ἐπιφάνειαν· τῷ δὲ καθιέντι ἀκόντιον ἄνωθεν εἰς τὸν βόθρον ἡ βία τοῦ ὕδατος ἀντιπράττει τοσοῦτον, ὥστε μόλις

[1] τό, inserted by *i*. [2] ἐν, Corais inserts.
[3] συναγκειῶν, the editors, for συναγγείων *oxz*, συναγκίων other MSS.
[4] καθορᾶν, Tyrwhitt, for καθαρόν ; so the editors.

a considerable city; its inhabitants, however, consist mostly of the divinely inspired people and the temple-servants who live in it. Its inhabitants are Cataonians, who, though in a general way classed as subject to the king, are in most respects subject to the priest. The priest is master of the temple, and also of the temple-servants, who on my sojourn there were more than six thousand in number, men and women together. Also, considerable territory belongs to the temple, and the revenue is enjoyed by the priest. He is second in rank in Cappadocia after the king; and in general the priests belonged to the same family as the kings. It is thought that Orestes, with his sister Iphigeneia, brought these sacred rites here from the Tauric Scythia, the rites in honour of Artemis Tauropolus, and that here they also deposited the hair [1] of mourning; whence the city's name. Now the Sarus River flows through this city and passes out through the gorges of the Taurus to the plains of the Cilicians and to the sea that lies below them.

4. But the Pyramus, a navigable river with its sources in the middle of the plain, flows through Cataonia. There is a notable pit in the earth through which one can see the water as it runs into a long hidden passage underground and then rises to the surface. If one lets down a javelin from above into the pit,[2] the force of the water resists so strongly that the javelin can hardly be immersed in it. But

[1] In Greek, "Komê," the name of the city being "Komana," or, translated into English, "Comana."

[2] At the outlet, of course.

βαπτίζεσθαι· ἀπλέτῳ[1] δὲ βάθει καὶ πλάτει πολὺς ἐνεχθεὶς ἐπειδὰν συνάψῃ τῷ Ταύρῳ, παράδοξον λαμβάνει τὴν συναγωγήν, παράδοξος δὲ καὶ ἡ διακοπὴ τοῦ ὄρους ἐστί, δι' ἧς ἄγεται τὸ ῥεῖθρον· καθάπερ γὰρ ἐν ταῖς ῥῆγμα λαβούσαις πέτραις καὶ σχισθείσαις δίχα τὰς κατὰ τὴν ἑτέραν ἐξοχὰς ὁμολόγους εἶναι συμβαίνει ταῖς κατὰ τὴν ἑτέραν εἰσοχαῖς, ὥστε κἂν συναρμοσθῆναι δύνασθαι, οὕτως εἴδομεν καὶ τὰς ὑπερκειμένας τοῦ ποταμοῦ πέτρας ἑκατέρωθεν σχεδόν τι μέχρι τῶν ἀκρωρειῶν ἀνατεινούσας ἐν διαστάσει δυεῖν ἢ τριῶν πλέθρων, ἀντικείμενα ἐχούσας τὰ κοῖλα ταῖς ἐξοχαῖς· τὸ δὲ ἔδαφος τὸ μεταξὺ πᾶν πέτρινον, βαθύ τι καὶ στενὸν τελέως ἔχον διὰ μέσου ῥῆγμα, ὥστε καὶ κύνα καὶ λαγὼ διάλλεσθαι. τοῦτο δ' ἐστὶ τὸ ῥεῖθρον τοῦ ποταμοῦ, ἄχρι χείλους πλῆρες, ὀχέτῳ[2] πλάτει προσεοικός, διὰ δὲ τὴν σκολιότητα καὶ τὴν ἐκ τοσούτου συναγωγὴν καὶ τὸ[3] τῆς φάραγγος βάθος εὐθὺς τοῖς πόρρωθεν προσιοῦσιν ὁ ψόφος βροντῇ προσπίπτει παραπλήσιος· διεκβαίνων δὲ τὰ ὄρη τοσαύτην κατάγει χοῦν ἐπὶ θάλατταν, τὴν μὲν ἐκ τῆς Καταονίας, τὴν δὲ ἐκ τῶν Κιλίκων πεδίων, ὥστε ἐπ' αὐτῷ καὶ χρησμὸς ἐκπεπτωκὼς φέρεται τοιοῦτος·

Ἔσσεται ἐσσομένοις, ὅτε Πύραμος ἀργυροδίνης,[4]
ἠιόνα προχόων,[5] ἱερὴν ἐς Κύπρον ἵκηται.

[1] ἀπλέτῳ, corr. in C, for ἀπλώτῳ; but Corais, from conj. of Tyrwhitt, writes αὐτὸ τῷ.

[2] ὀχέτῳ, Corais, for ὀχέτου; so the later editors, though Kramer conj. οὐ after ὀχέτῳ.

[3] διά, after τό, Meineke, from conj. of Kramer, deletes; others exchange the positions of the two words.

although it flows in great volume because of its immense depth and breadth, yet, when it reaches the Taurus, it undergoes a remarkable contraction; and remarkable also is the cleft of the mountain through which the stream is carried; for, as in the case of rocks which have been broken and split into two parts, the projections on either side correspond so exactly to the cavities on the other that they could be fitted together, so it was in the case of the rocks I saw there, which, lying above the river on either side and reaching almost to the summit of the mountain at a distance of two or three plethra from each other, had cavities corresponding with the opposite projections. The whole intervening bed is rock, and it has a cleft through the middle which is deep and so extremely narrow that a dog or hare could leap across it. This cleft is the channel of the river, is full to the brim, and in breadth resembles a canal; but on account of the crookedness of its course and its great contraction in width and the depth of the gorge, a noise like thunder strikes the ears of travellers long before they reach it. In passing out through the mountains it brings down so much silt to the sea, partly from Cataonia and partly from the Cilician plains, that even an oracle is reported as having been given out in reference to it, as follows: "Men that are yet to be shall experience this at the time when the Pyramus of the silver eddies shall silt up its sacred sea-beach and come to Cyprus."[1] Indeed,

[1] Cf. quotation of the same oracle in 1. 3. 7.

[4] ἀργυροδίνης, Meineke, following *Epitome* and *Oracula Sibyll.* p. 515, for εὐρυοδίνης.
[5] προχόων, for προχέων, as read in this text in 1. 3. 8.

παραπλήσιον γάρ τι κἀκεῖ συμβαίνει καὶ ἐν Αἰγύπτῳ, τοῦ Νείλου προσεξηπειροῦντος ἀεὶ τὴν θάλατταν τῇ προσχώσει· καθὸ καὶ Ἡρόδοτος μὲν δῶρον τοῦ ποταμοῦ τὴν Αἴγυπτον εἶπεν, ὁ ποιητὴς δὲ τὴν Φάρον πελαγίαν ὑπάρξαι, πρότερον οὐχ' ὡς [1] νυνὶ πρόσγειον οὖσαν τῇ Αἰγύπτῳ.

C 537 5.[2] Τρίτη δ' ἐστὶν ἱερωσύνη Διὸς Δακιήου,[3] λειπομένη ταύτης, ἀξιόλογος δ' ὅμως. ἐνταῦθα δ' ἐστὶ λάκκος ἁλμυροῦ ὕδατος, ἀξιολόγου λίμνης ἔχων περίμετρον, ὀφρύσι κλειόμενος ὑψηλαῖς τε καὶ ὀρθίαις, ὥστ' ἔχειν κατάβασιν κλιμακώδη· τὸ δ' ὕδωρ οὔτ' αὔξεσθαί φασιν, οὔτ' ἀπόρρυσιν ἔχειν οὐδαμοῦ φανεράν.

6. Πόλιν δ' οὔτε τὸ τῶν Καταόνων ἔχει πεδίον οὔθ' ἡ Μελιτηνή, φρούρια δ' ἐρυμνὰ ἐπὶ τῶν ὀρῶν, τά τε Ἀζάμορα καὶ τὸ Δάσταρκον, ὃ περιρρεῖται τῷ Καρμάλᾳ ποταμῷ. ἔχει δὲ καὶ ἱερὸν τὸ τοῦ Κατάονος Ἀπόλλωνος, καθ' ὅλον τιμώμενον τὴν Καππαδοκίαν, ποιησαμένων ἀφιδρύματα ἀπ' αὐτοῦ. οὐδὲ αἱ ἄλλαι στρατηγίαι πόλεις ἔχουσι πλὴν δυεῖν· τῶν δὲ λοιπῶν στρατηγιῶν ἐν μὲν τῇ Σαργαραυσηνῇ [4] πολίχνιόν ἐστιν Ἧρπα καὶ ποταμὸς Καρμάλας,[5] ὃς καὶ αὐτὸς εἰς τὴν Κιλικίαν ἐκδίδωσιν· ἐν δὲ ταῖς ἄλλαις ὅ τε Ἄργος, ἔρυμα ὑψηλὸν πρὸς τῷ Ταύρῳ, καὶ τὰ Νῶρα, ὃ

[1] οὐχ' ὡς, Corais, for οὔπω; so Meineke.

[2] § 5 seems to belong after § 6, as Kramer points out. Meineke transposes it in his text.

[3] Δακιήου, Jones, from conj. of C. Müller, for Δακίη οὐ. Tyrwhitt conj. Δακιήνου. Meineke, citing Marcellinus 23. 6, and Philostratus *Vit. Apollonii*, emends to Ἀσβαμαίου.

[4] Σαργαραυσήνῃ, Tzschucke, for Σαργαραυσίνῃ.

[5] Καρμάλας, Corais, for Κάρμαλος.

something similar to this takes place also in Egypt, since the Nile is always turning the sea into dry land by throwing out silt. Accordingly, Herodotus[1] calls Egypt "the gift of the Nile," while Homer[2] speaks of Pharos as "being out in the open sea," since in earlier times it was not, as now, connected with the mainland of Egypt.[3]

5.[4] The third in rank is the priesthood of Zeus Daciëus,[5] which, though inferior to that of Enyo, is noteworthy. At this place there is a reservoir of salt water which has the circumference of a considerable lake; it is shut in by brows of hills so high and steep that people go down to it by ladder-like steps. The water, they say, neither increases nor anywhere has a visible outflow.

6. Neither the plain of the Cataonians nor the country Melitenê has a city, but they have strongholds on the mountains, I mean Azamora and Dastarcum; and round the latter flows the Carmalas River. It contains also a temple, that of the Cataonian Apollo, which is held in honour thoughout the whole of Cappadocia, the Cappadocians having made it the model of temples of their own. Neither do the other prefectures, except two, contain cities; and of the remaining prefectures, Sargarausenê contains a small town Herpa, and also the Carmalas River, this too[6] emptying into the Cilician Sea. In the other prefectures are Argos, a lofty stronghold near the Taurus, and Nora, now called Neroassus, in which

[1] 2. 5. [2] *Od.* 4. 354.
[3] *i.e.* "has become, in a sense, a peninsula" (1. 3. 17).
[4] See critical note.
[5] At Morimenes (see next paragraph).
[6] Like the Sarus (12. 2. 3).

νῦν καλεῖται Νηροασσός, ἐν ᾧ Εὐμένης πολιορκούμενος ἀντέσχε πολὺν χρόνον· καθ' ἡμᾶς δὲ Σισίνου ὑπῆρξε χρηματοφυλάκιον τοῦ ἐπιθεμένου τῇ Καππαδόκων ἀρχῇ. τούτου δ' ἦν καὶ τὰ Κάδηνα, βασίλειον καὶ πόλεως κατασκευὴν ἔχον· ἔστι δὲ καὶ ἐπὶ τῶν ὅρων[1] τῶν Λυκαονικῶν τὰ Γαρσαύιρα[2] κωμόπολις· λέγεται[3] ὑπάρξαι ποτὲ καὶ αὕτη μητρόπολις τῆς χώρας. ἐν δὲ τῇ Μοριμηνῇ τὸ ἱερὸν τοῦ ἐν Οὐηνάσοις Διός, ἱεροδούλων κατοικίαν ἔχον τρισχιλίων σχεδόν τι καὶ χώραν ἱερὰν εὔκαρπον, παρέχουσαν πρόσοδον ἐνιαύσιον ταλάντων πεντεκαίδεκα τῷ ἱερεῖ· καὶ οὗτός[4] ἐστι διὰ βίου, καθάπερ καὶ ὁ ἐν Κομάνοις, καὶ δευτερεύει κατὰ τιμὴν μετ' ἐκεῖνον.

7. Δύο δὲ ἔχουσι μόνον στρατηγίαι πόλεις, ἡ μὲν Τυανῖτις τὰ Τύανα, ὑποπεπτωκυῖαν τῷ Ταύρῳ τῷ κατὰ τὰς Κιλικίας πύλας, καθ' ἃς εὐπετέσταται καὶ κοινόταται πᾶσίν εἰσιν αἱ εἰς τὴν Κιλικίαν καὶ τὴν Συρίαν ὑπερβολαί· καλεῖται δὲ Εὐσέβεια ἡ πρὸς τῷ Ταύρῳ· ἀγαθὴ δὲ καὶ πεδιὰς ἡ πλείστη. τὰ δὲ Τύανα ἐπίκειται χώματι Σεμιράμιδος τετειχισμένῳ καλῶς. οὐ πολὺ δ' ἄπωθεν ταύτης ἐστὶ τά τε Καστάβαλα καὶ τὰ Κύβιστρα, ἔτι μᾶλλον τῷ ὄρει πλησιάζοντα πολίσματα· ὧν ἐν τοῖς Κασταβάλοις ἐστὶ τὸ τῆς Περασίας Ἀρτέμιδος ἱερόν, ὅπου φασὶ τὰς ἱερείας γυμνοῖς τοῖς ποσὶ δι' ἀνθρακιᾶς βαδίζειν ἀπαθεῖς· κἀνταῦθα δέ τινες τὴν αὐτὴν θρυλοῦσιν ἱστορίαν τὴν περὶ τοῦ Ὀρέστου καὶ τῆς Ταυροπόλου, Περασίαν κεκλῆσ-

[1] ὅρων, Corais, for ὀρῶν.
[2] CD*hilrw* read τὰ γὰρ Σαύειρα (cp. Γαρσαύιρα in 12. 2. 10).
[3] After λέγεται Meineke inserts δ'.

Eumenes held out against a siege for a long time. In my time it served as the treasury of Sisines, who made an attack upon the empire of the Cappadocians. To him belonged also Cadena, which had the royal palace and had the aspect of a city. Situated on the borders of Lycaonia is also a town called Garsauira. This too is said once to have been the metropolis of the country. In Morimenê, at Venasa, is the temple of the Venasian Zeus, which has a settlement of almost three thousand temple-servants and also a sacred territory that is very productive, affording the priest a yearly revenue of fifteen talents. He, too, is priest for life, as is the priest at Comana, and is second in rank after him.

7. Only two prefectures have cities, Tyanitis the city Tyana, which lies below the Taurus at the Cilician Gates, where for all is the easiest and most commonly used pass into Cilicia and Syria. It is called "Eusebeia near the Taurus"; and its territory is for the most part fertile and level. Tyana is situated upon a mound of Semiramis,[1] which is beautifully fortified. Not far from this city are Castabala and Cybistra, towns still nearer to the mountain. At Castabala is the temple of the Perasian Artemis, where the priestesses, it is said, walk with naked feet over hot embers without pain. And here, too, some tell us over and over the same story of Orestes and Tauropolus,[2] asserting that she was

[1] Numerous mounds were ascribed to Semiramis (see 16. 1. 3).

[2] *i.e.* Artemis Tauropolus (see 12. 2. 3).

---

[4] After οὗτος Meineke inserts δ'.

θαι φάσκοντες διὰ τὸ πέραθεν κομισθῆναι. ἐν μὲν δὴ τῇ Τυανίτιδι στρατηγίᾳ τῶν λεχθεισῶν δέκα ἐστὶ πόλις[1] τὰ Τύανα (τὰς δ' ἐπικτήτους οὐ συναριθμῶ ταύταις, τὰ Καστάβαλα καὶ τὰ Κύβιστρα καὶ τὰ ἐν τῇ τραχείᾳ Κιλικίᾳ, ἐν ᾗ τὴν Ἐλαιοῦσσαν νησίον εὔκαρπον[2] συνέκτισεν Ἀρχέλαος ἀξιολόγως, καὶ τὸ πλέον ἐνταῦθα διέτριβεν), ἐν δὲ τῇ Κιλικίᾳ καλουμένῃ τὰ Μάζακα, C 538 ἡ μητρόπολις τοῦ ἔθνους· καλεῖται δ' Εὐσέβεια καὶ αὕτη, ἐπίκλησιν ἡ πρὸς τῷ Ἀργαίῳ· κεῖται γὰρ ὑπὸ τῷ Ἀργαίῳ ὄρει πάντων ὑψηλοτάτῳ καὶ ἀνέκλειπτον χιόνι τὴν ἀκρώρειαν ἔχοντι, ἀφ' ἧς φασὶν οἱ ἀναβαίνοντες (οὗτοι δ' εἰσὶν ὀλίγοι) κατοπτεύεσθαι ταῖς αἰθρίαις ἄμφω τὰ πελάγη, τό τε Ποντικὸν καὶ τὸ Ἰσσικόν. τὰ μὲν οὖν ἄλλα ἀφυῆ πρὸς συνοικισμὸν ἔχει πόλεως, ἄνυδρός τε γάρ ἐστι καὶ ἀνώχυρος διά τε τὴν ὀλιγωρίαν τῶν ἡγεμόνων καὶ ἀτείχιστος (τάχα δὲ καὶ ἐπίτηδες, ἵνα μή, ὡς ἐρύματι πεποιθότες τῷ τείχει σφόδρα, ληστεύοιεν[3] πεδίον οἰκοῦντες λόφους ὑπερδεξίους ἔχοντες καὶ ἀνεμβαλεῖς).[4] καὶ τὰ κύκλῳ δὲ χωρία ἔχει τελέως ἄφορα καὶ ἀγεώργητα, καίπερ ὄντα πεδινά· ἀλλ' ἔστιν ἀμμώδη καὶ ὑπόπετρα. μικρὸν δ' ἔτι προϊοῦσι καὶ πυρίληπτα πεδία καὶ μεστὰ βόθρων[5] πυρὸς ἐπὶ σταδίους πολλούς, ὥστε πόρρωθεν ἡ κομιδὴ

[1] πόλις, Jones, for πόλισμα.
[2] Instead of εὔκαρπον E has εὔκαιρον.
[3] ληστεύοιεν, Xylander, for πιστεύοιεν; so the later editors.
[4] ἀνεμβαλεῖς, L. Kayser (*Neue Jahrbücher* 69, 262), for ἐμβαλεῖς. Meineke follows MSS.; Kramer suggests emending καί to οὐκ; Müller-Dübner insert οὐκ after καί. *x*, however, omits καὶ ἐμβαλεῖς.

called "Perasian" because she was brought "from the other side."[1] So then, in the prefecture Tyanitis, one of the ten above mentioned is Tyana (I am not enumerating along with these prefectures those that were acquired later, I mean Castabala and Cybistra and the places in Cilicia Tracheia,[2] where is Elaeussa, a very fertile island, which was settled in a noteworthy manner by Archeläus, who spent the greater part of his time there), whereas Mazaca, the metropolis of the tribe, is in the Cilician prefecture, as it is called. This city, too, is called "Eusebeia," with the additional words "near the Argaeus," for it is situated below the Argaeus, the highest mountain of all, whose summit never fails to have snow upon it; and those who ascend it (those are few) say that in clear weather both seas, both the Pontus and the Issian Sea, are visible from it. Now in general Mazaca is not naturally a suitable place for the founding of a city, for it is without water and unfortified by nature; and, because of the neglect of the prefects, it is also without walls (perhaps intentionally so, in order that people inhabiting a plain, with hills above it that were advantageous and beyond range of missiles, might not, through too much reliance upon the wall as a fortification, engage in plundering). Further, the districts all round are utterly barren and untilled, although they are level; but they are sandy and are rocky underneath. And, proceeding a little farther on, one comes to plains extending over many stadia that are volcanic and full of fire-pits; and therefore the necessaries of life must be

[1] "perathen." [2] Cf. 12. 1. 4.

[5] βόθρων, Xylander, for βάθρων (βάραθρα *hi*, and D *man. sec.*); so the later editors.

τῶν ἐπιτηδείων. καὶ τὸ δοκοῦν δὲ πλεονέκτημα παρακείμενον ἔχει κίνδυνον· ἀξύλου γὰρ ὑπαρχούσης σχεδόν τι τῆς συμπάσης Καππαδοκίας, ὁ Ἀργαῖος ἔχει περικείμενον δρυμόν, ὥστε ἐγγύθεν ὁ ξυλισμὸς πάρεστιν, ἀλλ' οἱ ὑποκείμενοι τῷ δρυμῷ τόποι καὶ αὐτοὶ πολλαχοῦ πυρὰ ἔχουσιν, ἅμα δὲ καὶ ὕφυδροί εἰσι ψυχρῷ ὕδατι, οὔτε τοῦ πυρὸς οὔτε τοῦ ὕδατος εἰς τὴν ἐπιφάνειαν ἐκκύπτοντος, ὥστε καὶ ποάζειν τὴν πλείστην· ἔστι δ' ὅπου καὶ ἑλῶδές ἐστι τὸ ἔδαφος, καὶ νύκτωρ ἐξάπτονται φλόγες ἀπ' αὐτοῦ. οἱ μὲν οὖν ἔμπειροι φυλαττόμενοι τὸν ξυλισμὸν ποιοῦνται, τοῖς δὲ πολλοῖς κίνδυνός ἐστι, καὶ μάλιστα τοῖς κτήνεσιν, ἐμπίπτουσιν εἰς ἀδήλους βόθρους πυρός.

8. Ἔστι δὲ καὶ ποταμὸς ἐν τῷ πεδίῳ τῷ πρὸ τῆς πόλεως, Μέλας καλούμενος, ὅσον τετταράκοντα σταδίους διέχων τῆς πόλεως, ἐν ταπεινοτέρῳ τῆς πόλεως χωρίῳ τὰς πηγὰς ἔχων. ταύτῃ μὲν οὖν ἄχρηστος αὐτοῖς ἐστίν, οὐχ ὑπερδέξιον ἔχων τὸ ῥεῦμα, εἰς ἕλη δὲ καὶ λίμνας διαχεόμενος κακοῖ τὸν ἀέρα τοῦ θέρους τὸν περὶ τὴν πόλιν, καὶ τὸ λατομεῖον δὲ ποιεῖ δύσχρηστον, καίπερ εὔχρηστον ὄν· πλαταμῶνες γάρ εἰσιν, ἀφ' ὧν τὴν λιθίαν ἔχειν ἄφθονον συμβαίνει τοῖς Μαζακηνοῖς πρὸς τὰς οἰκοδομίας, καλυπτόμεναι δ' ὑπὸ τῶν ὑδάτων αἱ πλάκες ἀντιπράττουσι. καὶ ταῦτα δ' ἐστὶ τὰ ἕλη πανταχοῦ πυρίληπτα. Ἀριαράθης δ' ὁ βασιλεύς, τοῦ Μέλανος κατά τινα στενὰ ἔχοντος τὴν εἰς τὸν Εὐφράτην[1] διέξοδον, ἐμφράξας ταῦτα λίμνην πελαγίαν ἀπέ-

[1] Εὐφράτην is an error for Ἅλυν.

brought from a distance. And further, that which seems to be an advantage is attended with peril, for although almost the whole of Cappadocia is without timber, the Argaeus has forests all round it, and therefore the working of timber is close at hand; but the region which lies below the forests also contains fires in many places and at the same time has an underground supply of cold water, although neither the fire nor the water emerges to the surface; and therefore most of the country is covered with grass. In some places, also, the ground is marshy, and at night flames rise therefrom. Now those who are acquainted with the country can work the timber, since they are on their guard, but the country is perilous for most people, and especially for cattle, since they fall into the hidden fire-pits.

8. There is also a river in the plain before the city; it is called Melas, is about forty stadia distant from the city, and has its sources in a district that is below the level of the city. For this reason, therefore, it is useless to the inhabitants, since its stream is not in a favourable position higher up, but spreads abroad into marshes and lakes, and in the summertime vitiates the air round the city, and also makes the stone-quarry hard to work, though otherwise easy to work; for there are ledges of flat stones from which the Mazaceni obtain an abundant supply of stone for their buildings, but when the slabs are concealed by the waters they are hard to obtain. And these marshes, also, are everywhere volcanic. Ariarathes the king, since the Melas had an outlet into the Euphrates[1] by a certain narrow defile, dammed this and converted the neighbouring plain

[1] "Euphrates" is obviously an error for "Halys."

δειξε τὸ πλησίον πεδίον, ἐνταῦθα δὲ νησῖδάς τινας, ὡς τὰς Κυκλάδας, ἀπολαβόμενος δια-
C 539 τριβὰς ἐν αὐταῖς ἐποιεῖτο μειρακιώδεις· ἐκραγὲν δ' ἀθρόως τὸ ἔμφραγμα, ἐξέκλυσε πάλιν τὸ ὕδωρ, πληρωθεὶς δ' ὁ Εὐφράτης[1] τῆς τε τῶν Καππαδόκων πολλὴν παρέσυρε καὶ κατοικίας καὶ φυτείας ἠφάνισε πολλάς, τῆς τε τῶν Γαλατῶν τῶν τὴν Φρυγίαν ἐχόντων οὐκ ὀλίγην ἐλυμήνατο, ἀντὶ δὲ τῆς βλάβης ἐπράξαντο ζημίαν αὐτὸν τάλαντα τριακόσια, Ῥωμαίοις ἐπιτρέψαντες τὴν κρίσιν. τὸ δ' αὐτὸ συνέβη καὶ περὶ Ἥρπα· καὶ γὰρ ἐκεῖ τὸ τοῦ Καρμάλα ῥεῦμα ἐνέφραξεν, εἶτ' ἐκραγέντος τοῦ στομίου καὶ τῶν Κιλίκων τινὰ χωρία τὰ περὶ Μαλλὸν διαφθείραντος τοῦ ὕδατος, δίκας ἔτισεν τοῖς ἀδικηθεῖσιν.

9. Ἀφυὲς δ' οὖν κατὰ πολλὰ τὸ τῶν Μαζακηνῶν χωρίον ὂν[2] πρὸς κατοικίαν μάλιστα οἱ βασιλεῖς ἑλέσθαι δοκοῦσιν, ὅτι τῆς χώρας ἁπάσης τόπος ἦν μεσαίτατος οὗτος τῶν ξύλα ἐχόντων ἅμα καὶ λίθον πρὸς τὰς οἰκοδομίας καὶ χόρτον, οὗ πλεῖστον ἐδέοντο κτηνοτροφοῦντες· τρόπον γάρ τινα στρατόπεδον ἦν αὐτοῖς ἡ πόλις. τὴν δ' ἄλλην ἀσφάλειαν τὴν αὐτῶν τε καὶ σωμάτων ἐκ τῶν ἐρυμάτων[3] εἶχον τῶν ἐν τοῖς φρουρίοις, ἃ πολλὰ ὑπάρχει, τὰ μὲν βασιλικά, τὰ δὲ τῶν φίλων. ἀφέστηκε δὲ τὰ Μάζακα τοῦ μὲν Πόντου περὶ ὀκτακοσίους σταδίους πρὸς νότον, τοῦ δ' Εὐφράτου μικρὸν ἐλάττους ἢ

[1] Εὐφράτης is an error for Ἅλυς. [2] ὄν, Corais, for ὅ.

[3] Corais emends αὐτῶν to αὑτῶν and inserts τῶν before σωμάτων; and he emends ἐκ τῶν ἐρυμάτων to καὶ τῶν χρημάτων (so Meineke). Kramer proposes merely to emend σωμάτων to χρημάτων.

into a sea-like lake, and there, shutting off certain isles —like the Cyclades—from the outside world, passed his time there in boyish diversions. But the barrier broke all at once, the water streamed out again, and the Euphrates,[1] thus filled, swept away much of the soil of Cappadocia, and obliterated numerous settlements and plantations, and also damaged no little of the country of the Galatians who held Phrygia. In return for the damage the inhabitants, who gave over the decision of the matter to the Romans, exacted of him a fine of three hundred talents. The same was the case also in regard to Herpa; for there too he dammed the stream of the Carmalas River; and then, the mouth having broken open and the water having ruined certain districts in Cilicia in the neighbourhood of Mallus, he paid damages to those who had been wronged.

9. However, although the district of the Mazaceni is in many respects not naturally suitable for habitation, the kings seem to have preferred it, because of all places in the country this was nearest to the centre of the region which contained timber and stone for buildings, and at the same time provender, of which, being cattle-breeders, they needed a very large quantity, for in a way the city was for them a camp. And as for their security in general, both that of themselves and of their slaves, they got it from the defences in their strongholds, of which there are many, some belonging to the king and others to their friends. Mazaca is distant from Pontus[2] about eight hundred stadia to the south, from the Euphrates slightly less

[1] Again an error for "Halys."
[2] *i.e.* the country, not the sea.

διπλασίους, τῶν Κιλικίων δὲ πυλῶν ὁδὸν ἡμερῶν ἓξ καὶ τοῦ Κυρίνου[1] στρατοπέδου διὰ Τυάνων· κατὰ μέσην δὲ τὴν ὁδὸν κεῖται τὰ Τύανα, διέχει δὲ Κυβίστρων τριακοσίους σταδίους. χρῶνται δὲ οἱ Μαζακηνοὶ τοῖς Χαρώνδα νόμοις, αἱρούμενοι καὶ νομῳδόν, ὅς ἐστιν αὐτοῖς ἐξηγητὴς τῶν νόμων, καθάπερ οἱ παρὰ Ῥωμαίοις νομικοί. διέθηκε δὲ φαύλως αὐτοὺς Τιγράνης ὁ Ἀρμένιος, ἡνίκα τὴν Καππαδοκίαν κατέδραμεν· ἅπαντας γὰρ ἀναστάτους ἐποίησεν εἰς τὴν Μεσοποταμίαν καὶ τὰ Τιγρανόκερτα ἐκ τούτων συνῴκισε τὸ πλέον· ὕστερον δ' ἀπανῆλθον οἱ δυνάμενοι μετὰ τὴν τῶν Τιγρανοκέρτων ἅλωσιν.

10. Μέγεθος δὲ τῆς χώρας κατὰ πλάτος μὲν τὸ ἀπὸ τοῦ Πόντου πρὸς τὸν Ταῦρον ὅσον χίλιοι καὶ ὀκτακόσιοι στάδιοι, μῆκος δὲ ἀπὸ τῆς Λυκαονίας καὶ Φρυγίας μέχρι Εὐφράτου πρὸς τὴν ἕω καὶ τὴν Ἀρμενίαν περὶ τρισχιλίους. ἀγαθὴ δὲ καὶ καρποῖς, μάλιστα δὲ σίτῳ καὶ βοσκήμασι παντοδαποῖς, νοτιωτέρα δ' οὖσα τοῦ Πόντου ψυχροτέρα ἐστίν· ἡ δὲ Βαγαδανία,[2] καίπερ πεδιὰς οὖσα καὶ νοτιωτάτη πασῶν (ὑποπέπτωκε γὰρ τῷ Ταύρῳ), μόλις τῶν καρπίμων τι φέρει δένδρων, ὀναγρόβοτος[3] δ' ἐστὶ καὶ αὕτη καὶ ἡ πολλὴ τῆς ἄλλης, καὶ μάλιστα C 540 ἡ περὶ Γαρσαύιρα[4] καὶ Λυκαονίαν καὶ Μοριμηνήν. ἐν δὲ τῇ Καππαδοκίᾳ γίνεται καὶ ἡ λεγομένη Σινωπικὴ μίλτος, ἀρίστη τῶν πασῶν· ἐνάμιλλος

[1] Κυρίνου, Meineke emends to Κύρου.

[2] Βαγαδανία, Meineke, for Γαβανία E, Γαβαδανία other MSS.; Βαγαδαονία, Tzschucke, Corais, Kramer.

than double that distance, and from the Cilician Gates and the camp of Cyrus a journey of six days by way of Tyana. Tyana is situated at the middle of the journey and is three hundred stadia distant from Cybistra. The Mazaceni use the laws of Charondas, choosing also a Nomodus,[1] who, like the jurisconsults among the Romans, is the expounder of the laws. But Tigranes, the Armenian, put the people in bad plight when he overran Cappadocia, for he forced them, one and all, to migrate into Mesopotamia; and it was mostly with these that he settled Tigranocerta.[2] But later, after the capture of Tigranocerta, those who could returned home.

10. The size of the country is as follows: In breadth, from Pontus to the Taurus, about one thousand eight hundred stadia, and in length, from Lycaonia and Phrygia to the Euphrates towards the east and Armenia, about three thousand. It is an excellent country, not only in respect to fruits, but particularly in respect to grain and all kinds of cattle. Although it lies farther south than Pontus, it is colder. Bagadania, though level and farthest south of all (for it lies at the foot of the Taurus), produces hardly any fruit-bearing trees, although it is grazed by wild asses, both it and the greater part of the rest of the country, and particularly that round Garsauira and Lycaonia and Morimenê. In Cappadocia is produced also the ruddle called "Sinopean," the best in the

[1] "Law-chanter." [2] Cf. 11. 14. 15.

[3] ὀναγρόβοτος (ὀναγροβότος, Casaubon and later editors), Jones, for ἀγρόβοτος.

[4] Γαρσαύιρα *Dhioz*. For variants see C. Müller, *Ind. Var. Lect.* p. 1020 and cp. Γαρσαύιρα in 12. 2. 6.

δ' ἐστὶν αὐτῇ καὶ ἡ Ἰβηρική· ὠνομάσθη δὲ Σινωπική, διότι κατάγειν ἐκεῖσε εἰώθεσαν[1] οἱ ἔμποροι, πρὶν ἢ τὸ τῶν Ἐφεσίων ἐμπόριον μέχρι τῶν ἐνθάδε ἀνθρώπων διϊχθαι. λέγεται δὲ καὶ κρυστάλλου πλάκας καὶ ὀνυχίτου λίθου πλησίον τῆς τῶν Γαλατῶν ὑπὸ τῶν Ἀρχελάου μεταλλευτῶν εὑρῆσθαι.[2] ἦν δέ τις τόπος καὶ λίθου λευκοῦ, τῷ ἐλέφαντι κατὰ τὴν χρόαν ἐμφεροῦς, ὥσπερ ἀκόνας τινὰς οὐ μεγάλας ἐκφέρων, ἐξ ὧν τὰ λαβία τοῖς μαχαιρίοις κατεσκεύαζον· ἄλλος[3] δ' εἰς τὰς[4] διόπτρας βώλους μεγάλας ἐκδιδούς, ὥστε καὶ ἔξω κομίζεσθαι. ὅριον δ' ἐστὶ τοῦ Πόντου καὶ τῆς Καππαδοκίας ὀρεινή τις παράλληλος τῷ Ταύρῳ, τὴν ἀρχὴν ἔχουσα ἀπὸ τῶν ἑσπερίων ἄκρων τῆς Χαμμανηνῆς, ἐφ' ἧς ἵδρυται φρούριον ἀπότομον Δασμένδα,[5] μέχρι τῶν ἑωθινῶν τῆς Λαουιανσηνῆς.[6] στρατηγίαι δ' εἰσὶ τῆς Καππαδοκίας ἥ τε Χαμμανηνὴ[7] καὶ ἡ Λαουιανσηνή.[8]

11. Συνέβη δέ, ἡνίκα πρῶτον Ῥωμαῖοι τὰ κατὰ τὴν Ἀσίαν διῴκουν, νικήσαντες Ἀντίοχον, καὶ φιλίας καὶ συμμαχίας ἐποιοῦντο πρός τε τὰ ἔθνη καὶ τοὺς βασιλέας, τοῖς μὲν ἄλλοις βασιλεῦσιν αὐτοῖς καθ' ἑαυτοὺς δοθῆναι τὴν τιμὴν ταύτην, τῷ δὲ Καππάδοκι καὶ αὐτῷ δὲ τῷ ἔθνει κοινῇ. ἐκλιπόντος δὲ τοῦ βασιλικοῦ γένους, οἱ μὲν

[1] εἰώθεσαν, Groskurd, for εἰώθασιν; so the later editors.
[2] εὑρῆσθαι, Corais, for εὑρέσθαι; so the later editors.
[3] CD*hilrw* read ἄλλως.
[4] δ' εἰς τάς, Corais, for δὲ τάς; so the later editors.
[5] For the variant spellings of this name, see C. Müller (*l.c.*).

world, although the Iberian rivals it. It was named "Sinopean"[1] because the merchants were wont to bring it down thence to Sinopê before the traffic of the Ephesians had penetrated as far as the people of Cappadocia. It is said that also slabs of crystal and of onyx stone were found by the miners of Archeläus near the country of the Galatians. There was a certain place, also, which had white stone that was like ivory in colour and yielded pieces of the size of small whetstones; and from these pieces they made handles for their small swords. And there was another place which yielded such large lumps of transparent stone[2] that they were exported. The boundary of Pontus and Cappadocia is a mountain tract parallel to the Taurus, which has its beginning at the western extremities of Chammanenê, where is situated Dasmenda, a stronghold with sheer ascent, and extends to the eastern extremities of Laviansenê. Both Chammanenê and Laviansenê are prefectures in Cappadocia.

11. It came to pass, as soon as the Romans, after conquering Antiochus, began to administer the affairs of Asia and were forming friendships and alliances both with the tribes and with the kings, that in all other cases they gave this honour to the kings individually, but gave it to the king of Cappadocia and the tribe jointly. And when the royal family died out, the Romans, in accordance

[1] See 3. 2. 6.

[2] Apparently the *lapis specularis*, or a variety of mica, or isinglass, used for making window-panes.

---

[6] For variant spellings, see C. Müller (*l.c.*).

[7] For variant spellings, see C. Müller (*l.c.*).

[8] For variant spellings, see C. Müller (*l.c.*).

Ῥωμαῖοι συνεχώρουν αὐτοῖς αὐτονομεῖσθαι κατὰ τὴν συγκειμένην φιλίαν τε καὶ συμμαχίαν πρὸς τὸ ἔθνος, οἱ δὲ πρεσβευσάμενοι τὴν μὲν ἐλευθερίαν παρῃτοῦντο (οὐ γὰρ δύνασθαι φέρειν αὐτὴν ἔφασαν), βασιλέα δ' ἠξίουν αὐτοῖς ἀποδειχθῆναι. οἱ δέ, θαυμάσαντες εἴ τινες οὕτως εἶεν ἀπειρηκότες πρὸς τὴν ἐλευθερίαν,[1] ἐπέτρεψαν δ' οὖν[2] αὐτοῖς ἐξ ἑαυτῶν ἑλέσθαι κατὰ χειροτονίαν, ὃν ἂν βούλωνται·[3] καὶ εἵλοντο Ἀριοβαρζάνην, εἰς τριγονίαν δὲ προελθόντος τοῦ γένους ἐξέλιπε· κατεστάθη δ' ὁ Ἀρχέλαος, οὐδὲν προσήκων αὐτοῖς, Ἀντωνίου καταστήσαντος. ταῦτα καὶ περὶ τῆς μεγάλης Καππαδοκίας· περὶ δὲ τῆς τραχείας Κιλικίας, τῆς προστεθείσης αὐτῇ, βέλτιόν ἐστιν ἐν τῷ περὶ τῆς ὅλης Κιλικίας λόγῳ διελθεῖν.

III

1. Τοῦ δὲ Πόντου καθίστατο μὲν Μιθριδάτης ὁ Εὐπάτωρ βασιλεύς. εἶχε δὲ τὴν ἀφοριζομένην τῷ Ἅλυϊ μέχρι Τιβαρανῶν καὶ Ἀρμενίων καὶ
C 541 τῆς ἐντὸς Ἅλυος τὰ μέχρι Ἀμάστρεως καί τινων τῆς Παφλαγονίας μερῶν. προσεκτήσατο δ' οὗτος καὶ τὴν μέχρι Ἡρακλείας παραλίαν ἐπὶ τὰ δυσμικὰ μέρη, τῆς Ἡρακλείδου τοῦ Πλατωνικοῦ πατρίδος, ἐπὶ δὲ τἀναντία μέχρι Κολχίδος καὶ τῆς μικρᾶς Ἀρμενίας, ἃ δὴ καὶ προσέθηκε τῷ Πόντῳ. καὶ δὴ καὶ Πομπήιος καταλύσας ἐκεῖνον

[1] Meineke, following conj. of Kramer, indicates a lacuna before ἐπέτρεψαν.
[2] δ' οὖν omitted by editors before Kramer.
[3] βούλωνται, restored by Kramer, instead of βούλοιντο.

with their compact of friendship and alliance with the tribe, conceded to them the right to live under their own laws; but those who came on the embassy not only begged off from the freedom (for they said that they were unable to bear it), but requested that a king be appointed for them. The Romans, amazed that any people should be so tired of freedom,[1]—at any rate, they permitted them to choose by vote from their own number whomever they wished. And they chose Ariobarzanes; but in the course of the third generation his family died out; and Archelaüs was appointed king, though not related to the people, being appointed by Antony. So much for Greater Cappadocia. As for Cilicia Tracheia, which was added to Greater Cappadocia, it is better for me to describe it in my account of the whole of Cilicia.[2]

III

1. As for Pontus, Mithridates Eupator established himself as king of it; and he held the country bounded by the Halys River as far as the Tibarani and Armenia, and held also, of the country this side the Halys, the region extending to Amastris and to certain parts of Paphlagonia. And he acquired, not only the sea-coast towards the west as far as Heracleia, the native land of Heracleides the Platonic philosopher, but also, in the opposite direction, the sea-coast extending to Colchis and Lesser Armenia; and this, as we know, he added to Pontus. And in fact this country was comprised within these

[1] Something seems to have fallen out of the text here.
[2] 14. 5. 1.

ἐν τούτοις τοῖς ὅροις οὖσαν τὴν χώραν ταύτην παρέλαβε· τὰ μὲν πρὸς Ἀρμενίαν καὶ τὰ περὶ τὴν Κολχίδα τοῖς συναγωνισαμένοις δυνάσταις κατένειμε, τὰ δὲ λοιπὰ εἰς ἕνδεκα πολιτείας διεῖλε καὶ τῇ Βιθυνίᾳ προσέθηκεν, ὥστ' ἐξ ἀμφοῖν ἐπαρχίαν γενέσθαι μίαν. μεταξύ τε τῶν Παφλαγόνων τῶν μεσογαίων τινὰς βασιλεύεσθαι παρέδωκε τοῖς ἀπὸ Πυλαιμένους, καθάπερ καὶ τοὺς Γαλάτας τοῖς ἀπὸ γένους τετράρχαις. ὕστερον δ' οἱ τῶν Ῥωμαίων ἡγεμόνες ἄλλους καὶ ἄλλους ἐποιήσαντο μερισμούς, βασιλέας τε καὶ δυνάστας καθιστάντες καὶ πόλεις τὰς μὲν ἐλευθεροῦντες, τὰς δὲ ἐγχειρίζοντες τοῖς δυνάσταις, τὰς δ' ὑπὸ τῷ δήμῳ τῷ Ῥωμαίων ἐῶντες. ἡμῖν δ' ἐπιοῦσι τὰ καθ' ἕκαστα, ὡς νῦν ἔχει, λεγέσθω, μικρὰ καὶ τῶν προτέρων ἐφαπτομένοις, ὅπου τοῦτο χρήσιμον. ἀρξόμεθα δὲ ἀπὸ Ἡρακλείας, ἥπερ δυσμικωτάτη ἐστὶ τούτων τῶν τόπων.

2. Εἰς δὴ τὸν Εὔξεινον πόντον εἰσπλέουσιν ἐκ τῆς Προποντίδος ἐν ἀριστερᾷ μὲν τὰ προσεχῆ τῷ Βυζαντίῳ κεῖται, Θρᾳκῶν δ' ἐστί, καλεῖται δὲ τὰ Ἀριστερὰ τοῦ Πόντου· ἐν δεξιᾷ δὲ τὰ προσεχῆ Χαλκηδόνι, Βιθυνῶν δ' ἐστὶ τὰ πρῶτα, εἶτα Μαριανδυνῶν (τινὲς δὲ καὶ Καυκώνων φασίν), εἶτα Παφλαγόνων μέχρι Ἅλυος, εἶτα Καππαδόκων τῶν πρὸς τῷ Πόντῳ καὶ τῶν ἑξῆς μέχρι Κολχίδος· ταῦτα δὲ πάντα καλεῖται τὰ Δεξιὰ τοῦ Εὐξείνου πόντου. ταύτης δὲ τῆς παραλίας ἁπάσης ἐπῆρξεν Εὐπάτωρ, ἀρξάμενος ἀπὸ τῆς

---

[1] Between Pontus and Bithynia.

boundaries when Pompey took it over, upon his overthrow of Mithridates. The parts towards Armenia and those round Colchis he distributed to the potentates who had fought on his side, but the remaining parts he divided into eleven states and added them to Bithynia, so that out of both there was formed a single province. And he gave over to the descendants of Pylaemenes the office of king over certain of the Paphlagonians situated in the interior between them,[1] just as he gave over the Galatians to the hereditary tetrarchs. But later the Roman prefects made different divisions from time to time, not only establishing kings and potentates, but also, in the case of cities, liberating some and putting others in the hands of potentates and leaving others subject to the Roman people. As I proceed I must speak of things in detail as they now are, but I shall touch slightly upon things as they were in earlier times whenever this is useful. I shall begin at Heracleia, which is the most westerly place in this region.

2. Now as one sails into the Euxine Sea from the Propontis, one has on his left the parts which adjoin Byzantium (these belong to the Thracians, and are called "the Left-hand Parts" of the Pontus), and on his right the parts which adjoin Chalcedon. The first of these latter belong to the Bithynians, the next to the Mariandyni (by some also called Caucones), the next to the Paphlygonians as far as the Halys River, and the next to the Pontic Cappadocians and to the people next in order after them as far as Colchis. All these are called the "Right-hand Parts" of the Pontus. Now Eupator reigned over the whole of this sea-coast, beginning at Colchis

Κολχίδος μέχρι Ἡρακλείας, τὰ δ' ἐπέκεινα τὰ μέχρι τοῦ στόματος καὶ τῆς Χαλκηδόνος τῷ Βιθυνῶν βασιλεῖ συνέμενε. καταλυθέντων δὲ τῶν βασιλέων, ἐφύλαξαν οἱ Ῥωμαῖοι τοὺς αὐτοὺς ὅρους, ὥστε τὴν Ἡράκλειαν προσκεῖσθαι τῷ Πόντῳ, τὰ δ' ἐπέκεινα Βιθυνοῖς προσχωρεῖν.

3. Οἱ μὲν οὖν Βιθυνοὶ διότι πρότερον Μυσοὶ ὄντες μετωνομάσθησαν οὕτως ἀπὸ τῶν Θρᾳκῶν τῶν ἐποικησάντων, Βιθυνῶν τε καὶ Θυνῶν, ὁμολογεῖται παρὰ τῶν πλείστων, καὶ σημεῖα τίθενται τοῦ μὲν τῶν Βιθυνῶν ἔθνους τὸ μέχρι νῦν ἐν τῇ Θρᾴκῃ λέγεσθαί τινας Βιθυνούς, τοῦ δὲ τῶν Θυνῶν τὴν Θυνιάδα ἀκτὴν τὴν πρὸς Ἀπολλωνίᾳ καὶ Σαλμυδησσῷ. καὶ οἱ Βέβρυκες δὲ οἱ τούτων προεποικήσαντες τὴν Μυσίαν Θρᾷκες, ὡς εἰκάζω C 542 ἐγώ. εἴρηται δ', ὅτι καὶ αὐτοὶ οἱ Μυσοὶ Θρᾳκῶν ἄποικοί εἰσι τῶν νῦν λεγομένων Μοισῶν. ταῦτα μὲν οὕτω λέγεται.

4. Τοὺς δὲ Μαριανδυνοὺς καὶ τοὺς Καύκωνας οὐχ ὁμοίως ἅπαντες λέγουσι· τὴν γὰρ δὴ Ἡράκλειαν ἐν τοῖς Μαριανδυνοῖς ἱδρῦσθαί φασι, Μιλησίων κτίσμα, τίνες δὲ καὶ πόθεν, οὐδὲν[1] εἴρηται, οὐδὲ διάλεκτος, οὐδ' ἄλλη διαφορὰ ἐθνικὴ περὶ τοὺς ἀνθρώπους φαίνεται, παραπλήσιοι δ' εἰσὶ τοῖς Βιθυνοῖς· ἔοικεν οὖν καὶ τοῦτο Θρᾴκιον ὑπάρξαι τὸ φῦλον. Θεόπομπος δὲ Μαριανδυνόν φησι μέρους τῆς Παφλαγονίας ἄρξαντα ὑπὸ πολλῶν δυναστευομένης, ἐπελθόντα τὴν τῶν

[1] οὐδέν, Meineke emends to οὐδενί.

---

[1] See 7. 3. 2.

and extending as far as Heracleia, but the parts farther on, extending as far as the mouth of the Pontus and Chalcedon, remained under the rule of the king of Bithynia. But when the kings had been overthrown, the Romans preserved the same boundaries, so that Heracleia was added to Pontus and the parts farther on went to the Bithynians.

3. Now as for the Bithynians, it is agreed by most writers that, though formerly Mysians, they received this new name from the Thracians—the Thracian Bithynians and Thynians—who settled the country in question, and they put down as evidences of the tribe of the Bithynians that in Thrace certain people are to this day called Bithynians, and of that of the Thynians, that the coast near Apollonia and Salmydessus is called Thynias. And the Bebryces, who took up their abode in Mysia before these people, were also Thracians, as I suppose. It is stated that even the Mysians themselves are colonists of those Thracians who are now called Moesians.[1] Such is the account given of these people.

4. But all do not give the same account of the Mariandyni and the Caucones; for Heracleia, they say, is situated in the country of the Mariandyni, and was founded by the Milesians; but nothing has been said as to who they are or whence they came, nor yet do the people appear characterised by any ethnic difference, either in dialect or otherwise, although they are similar to the Bithynians. Accordingly, it is reasonable to suppose that this tribe also was at first Thracian. Theopompus says that Mariandynus ruled over a part of Paphlagonia, which was under the rule of many potentates, and then invaded and took possession of the country of the

Βεβρύκων κατασχεῖν, ἣν δ' ἐξέλιπεν, ἐπώνυμον ἑαυτοῦ καταλιπεῖν. εἴρηται δὲ καὶ τοῦτο, ὅτι πρῶτοι τὴν Ἡράκλειαν κτίσαντες Μιλήσιοι τοὺς Μαριανδυνοὺς εἱλωτεύειν ἠνάγκασαν τοὺς προκατέχοντας τὸν τόπον, ὥστε καὶ πιπράσκεσθαι ὑπ' αὐτῶν, μὴ εἰς τὴν ὑπερορίαν δέ (συμβῆναι γὰρ ἐπὶ τούτοις), καθάπερ Κρησὶ μὲν ἐθήτευεν ἡ Μνῴα[1] καλουμένη σύνοδος, Θετταλοῖς δὲ οἱ Πενέσται.

5. Τοὺς δὲ Καύκωνας, οὓς ἱστοροῦσι τὴν ἐφεξῆς οἰκῆσαι παραλίαν τοῖς Μαριανδυνοῖς μέχρι τοῦ Παρθενίου ποταμοῦ, πόλιν ἔχοντας τὸ Τίειον,[2] οἱ μὲν Σκύθας φασίν, οἱ δὲ τῶν Μακεδόνων τινάς, οἱ δὲ τῶν Πελασγῶν· εἴρηται δέ που καὶ περὶ τούτων πρότερον. Καλλισθένης δὲ καὶ ἔγραφε τὰ ἔπη ταῦτα εἰς τὸν Διάκοσμον, μετὰ τὸ

Κρῶμνάν τ' Αἰγιαλόν τε καὶ ὑψηλοὺς Ἐρυθίνους

τιθεὶς

Καύκωνας δ' αὖτ' ἦγε Πολυκλέος υἱὸς ἀμύμων,
οἳ περὶ Παρθένιον ποταμὸν κλυτὰ δώματ' ἔναιον·

παρήκειν γὰρ ἀφ' Ἡρακλείας καὶ Μαριανδυνῶν μέχρι Λευκοσύρων, οὓς καὶ ἡμεῖς Καππάδοκας προσαγορεύομεν, τό τε τῶν Καυκώνων γένος τὸ περὶ τὸ Τίειον[3] μέχρι Παρθενίου καὶ τὸ τῶν Ἑνετῶν τὸ συνεχὲς μετὰ τὸν Παρθένιον τῶν ἐχόντων τὸ Κύτωρον, καὶ νῦν δ' ἔτι Καυκωνίτας εἶναί τινας περὶ τὸν Παρθένιον·

[1] Μνῴα, the editors, for Μινῷα and Μινώα.
[2] Τίειον, the editors, for Τήιον.
[3] Τίειον, the editors, for Τήιον.

[1] Literally, "synod." [2] 8. 3. 17.

Bebryces, but left the country which he had abandoned named after himself. This, too, has been said, that the Milesians who were first to found Heracleia forced the Mariandyni, who held the place before them, to serve as Helots, so that they sold them, but not beyond the boundaries of their country (for the two peoples came to an agreement on this), just as the Mnoan class,[1] as it is called, were serfs of the Cretans and the Penestae of the Thessalians.

5. As for the Cauconians, who, according to report, took up their abode on the sea-coast next to the Mariandyni and extended as far as the Parthenius River, with Tieium as their city, some say that they were Scythians, others that they were a certain people of the Macedonians, and others that they were a certain people of the Pelasgians. But I have already spoken of these people in another place.[2] Callisthenes in his treatise on *The Marshalling of the Ships* was for inserting[3] after the words "Cromna, Aegialus, and lofty Erythini"[4] the words "the Cauconians were led by the noble son of Polycles—they who lived in glorious dwellings in the neighbourhood of the Parthenius River," for, he adds, the Cauconians extended from Heracleia and the Mariandyni to the White Syrians, whom we call Cappadocians, and the tribe of the Cauconians round Tieium extended to the Parthenius River, whereas that of the Heneti, who held Cytorum, were situated next to them after the Parthenius River, and still to-day certain "Cauconitae"[5] live in the neighbourhood of the Parthenius River.

[3] *i.e.* in the Homeric text.

[4] *Iliad* 2. 855. On the site of the Erythini ("reddish cliffs"), see Leaf, *Troy*, p. 282.

[5] Called "Cauconiatae" in 8. 3. 17.

6. Ἡ μὲν οὖν Ἡράκλεια πόλις ἐστὶν εὐλίμενος καὶ ἄλλως ἀξιόλογος, ἥ γε καὶ ἀποικίας ἔστελλεν· ἐκείνης γὰρ ἥ τε Χερρόνησος ἄποικος καὶ ἡ Κάλλατις· ἦν τε αὐτόνομος, εἶτ᾽ ἐτυραννήθη χρόνους τινάς, εἶτ᾽ ἠλευθέρωσεν ἑαυτὴν πάλιν· ὕστερον δ᾽ ἐβασιλεύθη, γενομένη ὑπὸ τοῖς Ῥωμαίοις· ἐδέξατο δ᾽ ἀποικίαν Ῥωμαίων ἐπὶ μέρει τῆς πόλεως καὶ τῆς χώρας. λαβὼν δὲ παρ᾽ Ἀντωνίου C 543 τὸ μέρος τοῦτο τῆς πόλεως Ἀδιατόριξ ὁ Δομνεκλείου, τετράρχου Γαλατῶν, υἱός, ὃ κατεῖχον οἱ Ἡρακλειῶται, μικρὸν πρὸ τῶν Ἀκτιακῶν ἐπέθετο νύκτωρ τοῖς Ῥωμαίοις καὶ ἀπέσφαξεν αὐτούς, ἐπιτρέψαντος, ὡς ἔφασκεν ἐκεῖνος, Ἀντωνίου· θριαμβευθεὶς δὲ μετὰ τὴν ἐν Ἀκτίῳ νίκην, ἐσφάγη μεθ᾽ υἱοῦ. ἡ δὲ πόλις ἐστὶ τῆς Ποντικῆς ἐπαρχίας τῆς συντεταγμένης τῇ Βιθυνίᾳ.

7. Μεταξὺ δὲ Χαλκηδόνος καὶ Ἡρακλείας ῥέουσι ποταμοὶ πλείους, ὧν εἰσὶν ὅ τε Ψίλλις καὶ ὁ Κάλπας καὶ ὁ Σαγγάριος, οὗ μέμνηται καὶ ὁ ποιητής. ἔχει δὲ τὰς πηγὰς κατὰ Σαγγίαν κώμην ἀφ᾽ ἑκατὸν καὶ πεντήκοντά που σταδίων οὗτος Πεσσινοῦντος·[1] διέξεισι δὲ τῆς ἐπικτήτου Φρυγίας τὴν πλείω, μέρος δέ τι καὶ τῆς Βιθυνίας, ὥστε καὶ τῆς Νικομηδείας ἀπέχειν[2] μικρὸν πλείους ἢ τριακοσίους σταδίους, καθ᾽ ὃ συμβάλλει ποταμὸς αὐτῷ Γάλλος, ἐκ Μόδρων τὰς ἀρχὰς ἔχων τῆς ἐφ᾽ Ἑλλησπόντῳ Φρυγίας. αὕτη δ᾽ ἐστὶν ἡ αὐτὴ τῇ ἐπικτήτῳ, καὶ εἶχον αὐτὴν οἱ Βιθυνοὶ πρότερον. αὐξηθεὶς δὲ καὶ γενόμενος πλωτός,

[1] CE*hoxz* read Πισινοῦντος.
[2] ἀπέχειν, Corais, for ἀποσχεῖν; so the later editors.

6. Now Heracleia is a city that has good harbours and is otherwise worthy of note, since, among other things, it has also sent forth colonies; for both Chersonesus[1] and Callatis are colonies from it. It was at first an autonomous city, and then for some time was ruled by tyrants, and then recovered its freedom, but later was ruled by kings, when it became subject to the Romans. The people received a colony of Romans, sharing with them a part of their city and territory. But Adiatorix, the son of Domnecleius, tetrarch of the Galatians, received from Antony that part of the city which was occupied by the Heracleiotae; and a little before the Battle of Actium he attacked the Romans by night and slaughtered them, by permission of Antony, as he alleged. But after the victory at Actium he was led in triumph and slain together with his son. The city belongs to the Pontic Province which was united with Bithynia.

7. Between Chalcedon and Heracleia flow several rivers, among which are the Psillis and the Calpas and the Sangarius, which last is mentioned by the poet.[2] The Sangarius has its sources near the village Sangia, about one hundred and fifty stadia from Pessinus. It flows through the greater part of Phrygia Epictetus, and also through a part of Bithynia, so that it is distant from Nicomedeia a little more than three hundred stadia, reckoning from the place where it is joined by the Gallus River, which has its beginnings at Modra in Phrygia on the Hellespont. This is the same country as Phrygia Epictetus, and it was formerly occupied by the Bithynians. Thus increased, and now having

[1] See 7. 4. 2. [2] *Iliad* 3. 187, 16. 719.

καίπερ πάλαι ἄπλωτος ὤν, τὴν Βιθυνίαν ὁρίζει πρὸς ταῖς ἐκβολαῖς. πρόκειται δὲ τῆς παραλίας ταύτης καὶ ἡ Θυνία νῆσος. ἐν δὲ τῇ Ἡρακλειώτιδι γίνεται τὸ ἀκόνιτον· διέχει δὲ ἡ πόλις αὕτη τοῦ ἱεροῦ τοῦ Χαλκηδονίου σταδίους χιλίους που καὶ πεντακοσίους, τοῦ δὲ Σαγγαρίου πεντακοσίους.

8. Τὸ δὲ Τίειόν ἐστι πολίχνιον οὐδὲν ἔχον μνήμης ἄξιον, πλὴν ὅτι Φιλέταιρος ἐντεῦθεν ἦν, ὁ ἀρχηγέτης τοῦ τῶν Ἀτταλικῶν βασιλέων γένους· εἶθ' ὁ Παρθένιος ποταμὸς διὰ χωρίων ἀνθηρῶν φερόμενος καὶ διὰ τοῦτο τοῦ ὀνόματος τούτου τετυχηκώς, ἐν αὐτῇ τῇ Παφλαγονίᾳ τὰς πηγὰς ἔχων· ἔπειτα ἡ Παφλαγονία καὶ οἱ Ἐνετοί. ζητοῦσι δέ, τίνας λέγει τοὺς Ἐνετοὺς ὁ ποιητής, ὅταν φῇ·

Παφλαγόνων δ' ἡγεῖτο Πυλαιμένεος λάσιον κῆρ
ἐξ Ἐνετῶν, ὅθεν ἡμιόνων γένος ἀγροτεράων.

οὐ γὰρ δείκνυσθαί φασι νῦν Ἐνετοὺς ἐν τῇ Παφλαγονίᾳ· οἱ δὲ κώμην ἐν τῷ Αἰγιαλῷ φασὶ δέκα σχοίνους ἀπὸ Ἀμάστρεως διέχουσαν. Ζηνόδοτος δὲ ἐξ Ἐνετῆς γράφει, καί φησι δηλοῦσθαι τὴν νῦν Ἀμισόν· ἄλλοι δὲ φῦλόν τι τοῖς Καππάδοξιν ὅμορον στρατεῦσαι μετὰ Κιμμερίων, εἶτ' ἐκπεσεῖν εἰς τὸν Ἀδρίαν. τὸ δὲ μάλισθ' ὁμολογούμενόν ἐστιν, ὅτι ἀξιολογώτατον ἦν τῶν Παφλαγόνων φῦλον οἱ Ἐνετοί, ἐξ οὗ ὁ Πυλαιμένης ἦν·

---

[1] "parthenius" (lit. "maidenly") was the name of a flower used in making garlands.

[2] *Iliad* 2. 851. [3] *Sc.* "called Eneti," or Enetê.

become navigable, though of old not navigable, the river forms a boundary of Bithynia at its outlets. Off this coast lies also the island Thynia. The plant called aconite grows in the territory of Heracleia. This city is about one thousand five hundred stadia from the Chalcedonian temple and five hundred from the Sangarius River.

8. Tieium is a town that has nothing worthy of mention except that Philetaerus, the founder of the family of Attalic Kings, was from there. Then comes the Parthenius River, which flows through flowery districts and on this account came by its name;[1] it has its sources in Paphlagonia itself. And then comes Paphlagonia and the Eneti. Writers question whom the poet means by "the Eneti," when he says, "And the rugged heart of Pylaemenes led the Paphlagonians, from the land of the Eneti, whence the breed of wild mules";[2] for at the present time, they say, there are no Eneti to be seen in Paphlagonia, though some say that there is a village[3] on the Aegialus[4] ten schoeni[5] distant from Amastris. But Zenodotus writes "from Enetê,"[6] and says that Homer clearly indicates the Amisus of to-day. And others say that a tribe called Eneti, bordering on the Cappadocians, made an expedition with the Cimmerians and then were driven out to the Adriatic Sea.[7] But the thing upon which there is general agreement is, that the Eneti, to whom Pylaemenes belonged, were the most notable tribe of the Paphlagonians, and that,

[4] *i.e.* Shore. [5] A variable measure (see 17. 1. 24).

[6] *i.e.* instead of "from the Eneti" (cf. 12. 3. 25).

[7] For a discussion of the Eneti, see Leaf, *Troy*, pp. 285 ff. (cf. 1. 3. 21, 3. 2. 13, and 12. 3. 25).

καὶ δὴ καὶ συνεστράτευσαν οὗτοι αὐτῷ πλεῖστοι, ἀποβαλόντες δὲ τὸν ἡγεμόνα διέβησαν εἰς τὴν Θρᾴκην μετὰ τὴν Τροίας ἅλωσιν, πλανώμενοι δ' εἰς τὴν νῦν Ἐνετικὴν ἀφίκοντο. τινὲς δὲ καὶ C 544 Ἀντήνορα καὶ τοὺς παῖδας αὐτοῦ κοινωνῆσαι τοῦ στόλου τούτου φασὶ καὶ ἱδρυθῆναι κατὰ τὸν μυχὸν τοῦ Ἀδρίου, καθάπερ ἐμνήσθημεν ἐν τοῖς Ἰταλικοῖς. τοὺς μὲν οὖν Ἐνετοὺς διὰ τοῦτ' ἐκλιπεῖν εἰκὸς καὶ μὴ δείκνυσθαι ἐν τῇ Παφλαγονίᾳ.

9. Τοὺς δὲ Παφλαγόνας πρὸς ἕω μὲν ὁρίζει ὁ Ἅλυς ποταμός, ὃς[1] ῥέων ἀπὸ μεσημβρίας μεταξὺ Σύρων τε καὶ Παφλαγόνων[2] ἐξίησι[3] κατὰ τὸν Ἡρόδοτον εἰς τὸν Εὔξεινον καλεόμενον πόντον, Σύρους λέγοντα τοὺς Καππάδοκας· καὶ γὰρ ἔτι καὶ νῦν Λευκόσυροι καλοῦνται, Σύρων καὶ τῶν ἔξω τοῦ Ταύρου λεγομένων· κατὰ δὲ τὴν πρὸς τοὺς ἐντὸς τοῦ Ταύρου σύγκρισιν, ἐκείνων ἐπικεκαυμένων τὴν χρόαν, τούτων δὲ μή, τοιαύτην τὴν ἐπωνυμίαν γενέσθαι συνέβη· καὶ Πίνδαρός φησιν, ὅτι αἱ Ἀμαζόνες Σύριον εὐρυαίχμαν δίεπον[4] στρατόν, τὴν ἐν τῇ Θεμισκύρᾳ κατοικίαν οὕτω δηλῶν. ἡ δὲ Θεμίσκυρά ἐστιν τῶν Ἀμισηνῶν, αὕτη δὲ Λευκοσύρων τῶν μετὰ τὸν Ἅλυν. πρὸς ἕω μὲν τοίνυν ὁ Ἅλυς ὅριον τῶν Παφλαγόνων, πρὸς νότον δὲ Φρύγες καὶ οἱ ἐποικήσαντες Γαλάται, πρὸς δύσιν δὲ Βιθυνοὶ καὶ Μαριανδυνοί (τὸ γὰρ τῶν Καυκώνων γένος ἐξέφθαρται τελέως πάντοθεν),

[1] ὅς, Corais inserts (see Herod. 1. 6); so the later editors.
[2] καί, before ἐξίησι, Meineke ejects.
[3] But Herodotus reads ἐξίει.
[4] δίεπον *oxz* and Meineke, for διϊπον C, δίηπον *lw*, διεῖπον other MSS. and editors.

furthermore, these made the expedition with him in very great numbers, but, losing their leader, crossed over to Thrace after the capture of Troy, and on their wanderings went to the Enetian country,[1] as it is now called. According to some writers, Antenor and his children took part in this expedition and settled at the recess of the Adriatic, as mentioned by me in my account of Italy.[2] It is therefore reasonable to suppose that it was on this account that the Eneti disappeared and are not to be seen in Paphlagonia.

9. As for the Paphlagonians, they are bounded on the east by the Halys River, "which," according to Herodotus,[3] "flows from the south between the Syrians and the Paphlagonians and empties into the Euxine Sea, as it is called"; by "Syrians," however, he means the "Cappadocians," and in fact they are still to-day called "White Syrians," while those outside the Taurus are called "Syrians." As compared with those this side the Taurus, those outside have a tanned complexion, while those this side do not, and for this reason received the appellation "white." And Pindar says that the Amazons "swayed a 'Syrian' army that reached afar with their spears," thus clearly indicating that their abode was in Themiscyra. Themiscyra is in the territory of the Amiseni; and this territory belongs to the White Syrians, who live in the country next after the Halys River. On the east, then, the Paphlagonians are bounded by the Halys River; on the south by Phrygians and the Galatians who settled among them; on the west by the Bithynians and the Mariandyni (for the race of the Cauconians has

[1] See 3. 2. 13 and 5. 1. 4. [2] 5. 1. 4. [3] 1. 6.

πρὸς ἄρκτον δὲ ὁ Εὔξεινός ἐστι. τῆς δὲ χώρας ταύτης διῃρημένης εἴς τε τὴν μεσόγαιαν καὶ τὴν ἐπὶ θαλάττῃ, διατείνουσαν ἀπὸ τοῦ Ἅλυος μέχρι Βιθυνίας ἑκατέραν, τὴν μὲν παραλίαν ἕως τῆς Ἡρακλείας εἶχεν ὁ Εὐπάτωρ, τῆς δὲ μεσογαίας τὴν μὲν ἐγγυτάτω ἔσχεν, ἧς τινὰ καὶ πέραν τοῦ Ἅλυος διέτεινε· καὶ μέχρι δεῦρο τοῖς Ῥωμαίοις ἡ Ποντικὴ ἐπαρχία ἀφώρισται· τὰ λοιπὰ δ' ἦν ὑπὸ δυνάσταις καὶ μετὰ τὴν Μιθριδάτου κατάλυσιν. περὶ μὲν δὴ τῶν ἐν τῇ μεσογαίᾳ Παφλαγόνων ἐροῦμεν ὕστερον τῶν μὴ ὑπὸ τῷ Μιθριδάτῃ, νῦν δὲ πρόκειται τὴν ὑπ' ἐκείνῳ χώραν, κληθεῖσαν δὲ Πόντον, διελθεῖν.

10. Μετὰ δὴ τὸν Παρθένιον ποταμόν ἐστιν Ἄμαστρις, ὁμώνυμος τῆς συνῳκικυίας πόλις· ἵδρυται δ' ἐπὶ χερρονήσου λιμένας ἔχουσα τοῦ ἰσθμοῦ ἑκατέρωθεν· ἦν δ' ἡ Ἄμαστρις γυνὴ μὲν Διονυσίου, τοῦ Ἡρακλείας τυράννου, θυγάτηρ δὲ Ὀξυάθρου, τοῦ Δαρείου ἀδελφοῦ τοῦ κατὰ Ἀλέξανδρον· ἐκείνη μὲν οὖν ἐκ τεττάρων κατοικιῶν συνῴκισε[1] τὴν πόλιν, ἔκ τε Σησάμου καὶ Κυτώρου καὶ Κρώμνης (ὧν καὶ Ὅμηρος μέμνηται ἐν τῷ Παφλαγονικῷ διακόσμῳ), τετάρτης δὲ τῆς Τιείου·[2] ἀλλ' αὕτη μὲν ταχὺ ἀπέστη τῆς κοινωνίας, αἱ δὲ ἄλλαι συνέμειναν, ὧν ἡ Σήσαμος ἀκρόπολις τῆς Ἀμάστρεως λέγεται. τὸ δὲ Κύτωρον ἐμπόριον ἦν ποτὲ Σινωπέων, ὠνόμασται δ' ἀπὸ Κυ-

[1] E reads συνέστησε.

[2] Τιείου, Tzschucke, Corais, and Müller-Dübner, for Τηίου; the *Epitome*, Kramer, and Meineke read Τίου.

---

[1] *i.e.* interior of Paphlagonia.

everywhere been destroyed), and on the north by the Euxine. Now this country was divided into two parts, the interior and the part on the sea, each stretching from the Halys River to Bithynia; and Eupator not only held the coast as far as Heracleia, but also took the nearest part of the interior,[1] certain portions of which extended across the Halys (and the boundary of the Pontic Province has been marked off by the Romans as far as this).[2] The remaining parts of the interior, however, were subject to potentates, even after the overthrow of Mithridates. Now as for the Paphlagonians in the interior, I mean those not subject to Mithridates, I shall discuss them later,[3] but at present I propose to describe the country which was subject to him, called the Pontus.

10. After the Parthenius River, then, one comes to Amastris, a city bearing the same name as the woman who founded it. It is situated on a peninsula and has harbours on either side of the isthmus. Amastris was the wife of Dionysius the tyrant of Heracleia and the daughter of Oxyathres, the brother of the Dareius whom Alexander fought. Now she formed the city out of four settlements, Sesamus and Cytorum and Cromna (which Homer mentions in his marshalling of the Paphlagonian ships)[4] and, fourth, Tieium. This last, however, soon revolted from the united city, but the other three remained together; and, of these three, Sesamus is called the acropolis of Amastris. Cytorum was once the emporium of the Sinopeans; it was named after

[2] Cp. J. G. C. Anderson in *Anatolian Studies presented to Sir William Mitchell Ramsay*, p. 6.
[3] 12. 3 41–42. [4] 2. 853–885.

τώρου, τοῦ Φρίξου παιδός, ὡς Ἔφορός φησι. C 545 πλείστη δὲ καὶ ἀρίστη πύξος φύεται κατὰ τὴν Ἀμαστριανήν, καὶ μάλιστα περὶ τὸ Κύτωρον. ὁ δὲ Αἰγιαλός ἐστι μὲν ἠιὼν μακρὰ πλειόνων[1] ἢ ἑκατὸν σταδίων· ἔχει δὲ καὶ κώμην ὁμώνυμον, ἧς μέμνηται ὁ ποιητής, ὅταν φῇ,

> Κρῶμνάν τ' Αἰγιαλόν τε καὶ ὑψηλοὺς Ἐρυθίνους.

γράφουσι δέ τινες,

> Κρῶμναν Κωβίαλόν τε.

Ἐρυθίνους δὲ λέγεσθαί φασι τοὺς νῦν Ἐρυθρίνους, ἀπὸ τῆς χρόας· δύο δ' εἰσὶ σκόπελοι. μετὰ δὲ Αἰγιαλὸν Κάραμβις, ἄκρα μεγάλη πρὸς τὰς ἄρκτους ἀνατεταμένη καὶ τὴν Σκυθικὴν χερρόνησον. ἐμνήσθημεν δ' αὐτῆς πολλάκις καὶ τοῦ ἀντικειμένου αὐτῇ Κριοῦ μετώπου, διθάλαττον ποιοῦντος τὸν Εὔξεινον πόντον. μετὰ δὲ Κάραμβιν Κίνωλις καὶ Ἀντικίνωλις καὶ Ἀβώνου τεῖχος, πολίχνιον, καὶ Ἀρμένη, ἐφ' ᾗ παροιμιάζονται,

> ὅστις ἔργον οὐδὲν εἶχεν Ἀρμένην ἐτείχισεν.

ἔστι δὲ κώμη τῶν Σινωπέων ἔχουσα λιμένα.

11. Εἶτ' αὐτὴ Σινώπη, σταδίους πεντήκοντα τῆς Ἀρμένης διέχουσα, ἀξιολογωτάτη τῶν ταύτῃ πόλεων. ἔκτισαν μὲν οὖν αὐτὴν Μιλήσιοι· κατασκευασαμένη δὲ ναυτικὸν ἐπῆρχε τῆς ἐντὸς Κυανέων θαλάττης, καὶ ἔξω δὲ πολλῶν ἀγώνων μετεῖχε τοῖς Ἕλλησιν· αὐτονομηθεῖσα δὲ πολὺν χρόνον οὐδὲ διὰ τέλους ἐφύλαξε τὴν ἐλευθερίαν,

Cytorus, the son of Phryxus, as Ephorus says. The most and the best box-wood grows in the territory of Amastris, and particularly round Cytorum. The Aegialus is a long shore of more than a hundred stadia, and it has also a village bearing the same name, which the poet mentions when he says, "Cromna and Aegialus and the lofty Erythini,"[1] though some write, "Cromna and Cobialus." They say that the Erythrini of to-day, from their colour,[2] used to be called Erythini; they are two lofty rocks. After Aegialus one comes to Carambis, a great cape extending towards the north and the Scythian Chersonese. I have often mentioned it, as also Criumetopon which lies opposite it, by which the Euxine Pontus is divided into two seas.[3] After Carambis one comes to Cinolis, and to Anticinolis, and to Abonuteichus,[4] a small town, and to Armenê, to which pertains the proverb, "whoever had no work to do walled Armenê." It is a village of the Sinopeans and has a harbour.

11. Then one comes to Sinopê itself, which is fifty stadia distant from Armenê; it is the most noteworthy of the cities in that part of the world. This city was founded by the Milesians; and, having built a naval station, it reigned over the sea inside the Cyaneae, and shared with the Greeks in many struggles even outside the Cyaneae; and, although it was independent for a long time, it could not eventually preserve its freedom, but was captured by

[1] *Iliad* 2. 855. [2] *i.e.* "Red."
[3] 2. 5. 22, 7. 4. 3 11. 2. 14.
[4] Literally, Wall of Abonus.

[1] μέν, before ἤ, Meineke, following the editors before Kramer, omits; *rw* read δέ.

ἀλλ' ἐκ πολιορκίας ἑάλω καὶ ἐδούλευσε Φαρνάκῃ πρῶτον, ἔπειτα τοῖς διαδεξαμένοις ἐκεῖνον μέχρι τοῦ Εὐπάτορος καὶ τῶν καταλυσάντων Ῥωμαίων ἐκεῖνον. ὁ δὲ Εὐπάτωρ καὶ ἐγεννήθη ἐκεῖ καὶ ἐτράφη· διαφερόντως δὲ ἐτίμησεν αὐτὴν μητρόπολίν τε τῆς βασιλείας ὑπέλαβεν. ἔστι δὲ καὶ φύσει καὶ[1] προνοίᾳ κατεσκευασμένη καλῶς· ἵδρυται γὰρ ἐπὶ αὐχένι χερρονήσου τινός, ἑκατέρωθεν δὲ τοῦ ἰσθμοῦ λιμένες καὶ ναύσταθμα καὶ πηλαμυδεῖα θαυμαστά, περὶ ὧν εἰρήκαμεν, ὅτι δευτέραν θήραν οἱ Σινωπεῖς ἔχουσι, τρίτην δὲ Βυζάντιοι. καὶ κύκλῳ δ' ἡ χερρόνησος προβέβληται ῥαχιώδεις ἀκτάς, ἐχούσας[2] καὶ κοιλάδας τινάς, ὡσανεὶ βόθρους πετρίνους, οὓς καλοῦσι χοινικίδας· πληροῦνται δὲ οὗτοι μετεωρισθείσης τῆς θαλάττης, ὡς καὶ διὰ τοῦτο οὐκ εὐπρόσιτον τὸ[3] χωρίον, καὶ διὰ τὸ πᾶσαν τὴν τῆς πέτρας ἐπιφάνειαν ἐχινώδη καὶ ἀνεπίβατον εἶναι γυμνῷ ποδί· ἄνωθεν μέντοι καὶ ὑπὲρ τῆς πόλεως εὔγεών C 546 ἐστι τὸ ἔδαφος καὶ ἀγροκηπίοις κεκόσμηται πυκνοῖς,[4] πολὺ δὲ μᾶλλον τὰ προάστεια. αὐτὴ δ' ἡ πόλις τετείχισται καλῶς, καὶ γυμνασίῳ δὲ καὶ ἀγορᾷ καὶ στοαῖς κεκόσμηται λαμπρῶς. τοιαύτη δὲ οὖσα δὶς ὅμως ἑάλω, πρότερον μὲν

[1] φύσει καί, Kramer, from conj. of Casaubon, for φυσικῇ.
[2] ἐχούσας, Corais, for ἔχουσα.
[3] τό, the editors insert from E.
[4] E reads πολλοῖς instead of πυκνοῖς.

---

[1] 183 B.C.
[2] Mithridates the Great.
[3] 7. 6. 2 and 12. 3. 19.
[4] "Crossing the town to the north I passed through a sally-port, and descended to the beach, where the wall was

siege, and was first enslaved by Pharnaces[1] and afterwards by his successors down to Eupator[2] and to the Romans who overthrew Eupator. Eupator was both born and reared at Sinopê; and he accorded it especial honour and treated it as the metropolis of his kingdom. Sinopê is beautifully equipped both by nature and by human foresight, for it is situated on the neck of a peninsula, and has on either side of the isthmus harbours and roadsteads and wonderful pelamydes-fisheries, of which I have already made mention, saying that the Sinopeans get the second catch and the Byzantians the third.[3] Furthermore, the peninsula is protected all round by ridgy shores, which have hollowed-out places in them, rock-cavities, as it were, which the people call "choenicides";[4] these are filled with water when the sea rises, and therefore the place is hard to approach, not only because of this, but also because the whole surface of the rock is prickly and impassable for bare feet. Higher up, however, and above the city, the ground is fertile and adorned with diversified market-gardens; and especially the suburbs of the city. The city itself is beautifully walled, and is also splendidly adorned with gymnasium and market-place and colonnades. But although it was such a city, still it was twice captured, first by Pharnaces, who

built upon a sharp decomposing shelly limestone which I was surprised to find full of small circular holes, apparently resembling those described by Strabo, under the name of 'choenicides'; but those which I saw were not above nine inches in diameter, and from one to two feet deep. There can, however, be no doubt that such cavities would, if larger, render it almost impossible for a body of men to wade on shore." (Hamilton's *Researches in Asia Minor*, 1. p. 310, quoted by Tozer.)

τοῦ Φαρνάκου παρὰ δόξαν αἰφνιδίως ἐπιπεσόντος, ὕστερον δὲ ὑπὸ Λευκόλλου καὶ τοῦ ἐγκαθημένου τυράννου, καὶ ἐντὸς ἅμα καὶ ἐκτὸς πολιορκουμένη· ὁ γὰρ ἐγκατασταθεὶς ὑπὸ τοῦ βασιλέως φρούραρχος Βακχίδης, ὑπονοῶν ἀεί τινα προδοσίαν ἐκ τῶν ἔνδοθεν, καὶ πολλὰς αἰκίας καὶ σφαγὰς ποιῶν, ἀπαγορεῦσαι τοὺς ἀνθρώπους ἐποίησε πρὸς ἄμφω, μήτ' ἀμύνασθαι δυναμένους γενναίως μήτε προσθέσθαι κατὰ συμβάσεις. ἑάλωσαν δ' οὖν· καὶ τὸν μὲν ἄλλον κόσμον τῆς πόλεως διεφύλαξεν ὁ Λεύκολλος, τὴν δὲ τοῦ Βιλλάρου σφαῖραν ἦρε καὶ τὸν Αὐτόλυκον,[1] Σθένιδος ἔργον, ὃν ἐκεῖνοι οἰκιστὴν ἐνόμιζον καὶ ἐτίμων ὡς θεόν· ἦν δὲ καὶ μαντεῖον αὐτοῦ· δοκεῖ δὲ τῶν Ἰάσονι συμπλευσάντων εἶναι καὶ κατασχεῖν τοῦτον τὸν τόπον. εἶθ' ὕστερον Μιλήσιοι τὴν εὐφυΐαν ἰδόντες καὶ τὴν ἀσθένειαν τῶν ἐνοικούντων ἐξιδιάσαντο καὶ ἐποίκους ἔστειλαν· νυνὶ δὲ καὶ Ῥωμαίων ἀποικίαν δέδεκται καὶ μέρος τῆς πόλεως καὶ τῆς χώρας ἐκείνων ἐστί. διέχει δὲ τοῦ μὲν Ἱεροῦ τρισχιλίους καὶ πεντακοσίους, ἀφ' Ἡρακλείας δὲ δισχιλίους, Καράμβεως δὲ ἑπτακοσίους σταδίους. ἄνδρας δὲ ἐξήνεγκεν ἀγαθούς, τῶν μὲν φιλοσόφων Διογένη τὸν Κυνικὸν καὶ Τιμόθεον τὸν Πατρίωνα, τῶν δὲ ποιητῶν Δίφιλον τὸν κωμικόν, τῶν δὲ συγγραφέων Βάτωνα τὸν πραγματευθέντα τὰ Περσικά.

12. Ἐντεῦθεν δ' ἐφεξῆς ἡ τοῦ Ἅλυος ἐκβολὴ

[1] Αὐτόλυκον, Xylander, for Αὐτόλυτον.

---

[1] See Plutarch, *Lucullus*, 23.

unexpectedly attacked it all of a sudden, and later by Leucullus and by the tyrant who was garrisoned within it, being besieged both inside and outside at the same time; for, since Bacchides, who had been set up by the king as commander of the garrison, was always suspecting treason from the people inside, and was causing many outrages and murders, he made the people, who were unable either nobly to defend themselves or to submit by compromise, lose all heart for either course. At any rate, the city was captured; and though Leucullus kept intact the rest of the city's adornments, he took away the globe of Billarus and the work of Sthenis, the statue of Autolycus,[1] whom they regarded as founder of their city and honoured as god. The city had also an oracle of Autolycus. He is thought to have been one of those who went on the voyage with Jason and to have taken possession of this place. Then later the Milesians, seeing the natural advantages of the place and the weakness of its inhabitants, appropriated it to themselves and sent forth colonists to it. But at present it has received also a colony of Romans; and a part of the city and the territory belong to these. It is three thousand five hundred stadia distant from the Hieron,[2] two thousand from Heracleia, and seven hundred from Carambis. It has produced excellent men: among the philosophers, Diogenes the Cynic and Timotheus Patrion; among the poets, Diphilus the comic poet; and, among the historians, Baton, who wrote the work entitled *The Persica*.

12. Thence, next, one comes to the outlet of the

[2] *i.e.* the [Chalcedonian] "Temple" on the "Sacred Cape" (see 12. 4. 2) in Chalcedonia, now called Cape Khelidini.

ποταμοῦ· ὠνόμασται δ' ἀπὸ τῶν ἁλῶν, ἃς παραρρεῖ· ἔχει δὲ τὰς πηγὰς ἐν τῇ μεγάλῃ Καππαδοκίᾳ τῆς Ποντικῆς πλησίον κατὰ τὴν Καμισηνήν, ἐνεχθεὶς δ' ἐπὶ δύσιν πολύς, εἶτ' ἐπιστρέψας πρὸς τὴν ἄρκτον διά τε Γαλατῶν καὶ Παφλαγόνων ὁρίζει τούτους τε καὶ τοὺς Λευκοσύρους. ἔχει δὲ καὶ ἡ Σινωπῖτις καὶ πᾶσα ἡ μέχρι Βιθυνίας ὀρεινὴ ὑπερκειμένη τῆς λεχθείσης παραλίας ναυπηγήσιμον ὕλην ἀγαθὴν καὶ εὐκατακόμιστον. ἡ δὲ Σινωπῖτις καὶ σφένδαμνον φύει καὶ ὀροκάρυον, ἐξ ὧν τὰς τραπέζας τέμνουσιν· ἅπασα δὲ καὶ ἐλαιόφυτός ἐστιν ἡ μικρὸν ὑπὲρ τῆς θαλάττης γεωργουμένη.

13. Μετὰ δὲ τὴν ἐκβολὴν τοῦ Ἅλυος ἡ Γαζηλωνῖτίς[1] ἐστι μέχρι τῆς Σαραμηνῆς,[2] εὐδαίμων χώρα καὶ πεδιὰς πᾶσα καὶ πάμφορος· ἔχει δὲ καὶ προβατείαν ὑποδιφθέρου καὶ μαλακῆς ἐρέας, ἧς καθ' ὅλην τὴν Καππαδοκίαν καὶ τὸν Πόντον σφόδρα πολλὴ σπάνις ἐστί· γίνονται δὲ C 547 καὶ ζόρκες, ὧν ἀλλαχοῦ σπάνις ἐστί. ταύτης δὲ τῆς χώρας τὴν μὲν ἔχουσιν Ἀμισηνοί, τὴν δ' ἔδωκε Δηιοτάρῳ Πομπήιος, καθάπερ καὶ τὰ περὶ Φαρνακίαν καὶ τὴν Τραπεζουσίαν μέχρι Κολχίδος καὶ τῆς μικρᾶς Ἀρμενίας· καὶ τούτων ἀπέδειξεν αὐτὸν βασιλέα, ἔχοντα καὶ τὴν πατρῴαν τετραρχίαν τῶν Γαλατῶν, τοὺς Τολιστοβωγίους. ἀποθανόντος δ' ἐκείνου, πολλαὶ διαδοχαὶ τῶν ἐκείνου γεγόνασι.

[1] Γαζηλωνῖτις, Meineke for Γαδιλωνῖτις; for other spellings see C. Müller (*l.c.*) and Kramer.
[2] CD*hilxz* read Ἀραμηνῆς.

[1] "salt-works." [2] *i.e.* "Pontus" (see 12. 1. 4).

Halys River. It was named from the "halae,"[1] past which it flows. It has its sources in Greater Cappadocia in Camisenê near the Pontic country;[2] and, flowing in great volume towards the west, and then turning towards the north through Galatia and Paphlagonia, it forms the boundary between these two countries and the country of the White Syrians.[3] Both Sinopitis and all the mountainous country extending as far as Bithynia and lying above the aforesaid seaboard have shipbuilding timber that is excellent and easy to transport. Sinopitis produces also the maple and the mountain-nut, the trees from which they cut the wood used for tables. And the whole of the tilled country situated a little above the sea is planted with olive trees.

13. After the outlet of the Halys comes Gazelonitis, which extends to Saramenê; it is a fertile country and is everywhere level and productive of everything. It has also a sheep-industry, that of raising flocks clothed in skins and yielding soft wool,[4] of which there is a very great scarcity throughout the whole of Cappadocia and Pontus. The country also produces gazelles, of which there is a scarcity elsewhere. One part of this country is occupied by the Amiseni, but the other was given to Deïotarus by Pompey, as also the regions of Pharnacia and Trapezusia as far as Colchis and Lesser Armenia. Pompey appointed him king of all these, when he was already in possession of his ancestral Galatian tetrarchy,[5] the country of the Tolistobogii. But since his death there have been many successors to his territories.

[3] *i.e.* Cappadocians (see 12. 3. 9).
[4] See Vol. II, p. 241, and foot-note 13. [5] See 12. 5. 1.

14. Μετὰ δὲ τὴν Γαζηλῶνα[1] ἡ Σαραμηνὴ καὶ Ἀμισός, πόλις ἀξιόλογος, διέχουσα τῆς Σινώπης περὶ ἐννακοσίους σταδίους. φησὶ δ' αὐτὴν Θεόπομπος πρώτους Μιλησίους κτίσαι,[2] . . . Καππαδόκων ἄρχοντα, τρίτον δ' ὑπ' Ἀθηνοκλέους καὶ Ἀθηναίων ἐποικισθεῖσαν, Πειραιᾶ μετονομασθῆναι. καὶ ταύτην δὲ κατέσχον οἱ βασιλεῖς, ὁ δ' Εὐπάτωρ ἐκόσμησεν ἱεροῖς καὶ προσέκτισε μέρος. Λεύκολλος δὲ καὶ ταύτην ἐπολιόρκησεν, εἶθ' ὕστερον Φαρνάκης, ἐκ Βοσπόρου διαβάς· ἐλευθερωθεῖσαν δ' ὑπὸ Καίσαρος τοῦ Θεοῦ παρέδωκεν Ἀντώνιος βασιλεῦσιν· εἶθ' ὁ τύραννος Στράτων κακῶς αὐτὴν διέθηκεν· εἶτ' ἠλευθερώθη πάλιν μετὰ τὰ Ἀκτιακὰ ὑπὸ Καίσαρος τοῦ Σεβαστοῦ, καὶ νῦν εὖ συνέστηκεν. ἔχει δὲ τήν τε ἄλλην χώραν καλὴν καὶ τὴν Θεμίσκυραν, τὸ τῶν Ἀμαζόνων οἰκητήριον, καὶ τὴν Σιδηνήν.

15. Ἔστι δὲ ἡ Θεμίσκυρα πεδίον, τῇ μὲν ὑπὸ τοῦ πελάγους κλυζόμενον, ὅσον ἑξήκοντα σταδίους τῆς πόλεως διέχον, τῇ δ' ὑπὸ τῆς ὀρεινῆς εὐδένδρου καὶ διαρρύτου ποταμοῖς, αὐτόθεν τὰς πηγὰς ἐχουσιν. ἐκ μὲν οὖν τούτων πληρούμενος ἁπάντων εἷς ποταμὸς διέξεισι τὸ πεδίον, Θερμώδων καλούμενος· ἄλλος δὲ τούτῳ πάρισος, ῥέων ἐκ τῆς καλουμένης Φαναροίας, τὸ αὐτὸ διέξεισι πεδίον, καλεῖται δὲ Ἶρις. ἔχει δὲ τὰς πηγὰς ἐν αὐτῷ τῷ Πόντῳ, ῥυεὶς δὲ διὰ πόλεως μέσης Κομάνων

[1] Γαζηλῶνα, Meineke, for Γαδιλῶνα (Γαλιδῶνα D).

[2] Certainly one or more words have fallen out here. *i* inserts καί, and *oz* καὶ εἶτα.

14. After Gazelon one comes to Saramenê, and to a notable city, Amisus, which is about nine hundred stadia from Sinopê. Theopompus says that it was first founded by the Milesians, . . .[1] by a leader of the Cappadocians, and thirdly was colonised by Athenocles and Athenians and changed its name to Peiraeus. The kings also took possession of this city; and Eupator adorned it with temples and founded an addition to it. This city too was besieged by Leucullus, and then by Pharnaces, when he crossed over from the Bosporus. After it had been set free by the deified Caesar,[2] it was given over to kings by Antony. Then Straton the tyrant put it in bad plight. And then, after the Battle of Actium,[3] it was again set free by Caesar Augustus; and at the present time it is well organised. Besides the rest of its beautiful country, it possesses also Themiscyra, the abode of the Amazons, and Sidenê.

15. Themiscyra is a plain; on one side it is washed by the sea and is about sixty stadia distant from the city, and on the other side it lies at the foot of the mountainous country, which is well-wooded and coursed by streams that have their sources therein. So one river, called the Thermodon, being supplied by all these streams, flows out through the plain; and another river similar to this, which flows out of Phanaroea, as it is called, flows out through the same plain, and is called the Iris. It has its sources in Pontus itself, and, after flowing through the middle of the city Comana in

[1] See critical note.

[2] It was in reference to his battle with Pharnaces near Zela that Julius Caesar informed the Senate of his victory by the words, "I came, I saw, I conquered."

[3] 31 B.C.

τῶν Ποντικῶν καὶ διὰ τῆς Δαζιμωνίτιδος, εὐδαίμονος πεδίου, πρὸς δύσιν, εἶτ' ἐπιστρέφει πρὸς τὰς ἄρκτους παρ' αὐτὰ τὰ Γαζίουρα, παλαιὸν βασίλειον, νῦν δ' ἔρημον, εἶτα ἀνακάμπτει πάλιν πρὸς ἕω, παραλαβὼν τόν τε Σκύλακα καὶ ἄλλους ποταμούς, καὶ παρ' αὐτὸ τὸ τῆς Ἀμασείας ἐνεχθεὶς τεῖχος, τῆς ἡμετέρας πατρίδος, πόλεως ἐρυμνοτάτης, εἰς τὴν Φανάροιαν πρόεισιν· ἐνταῦθα δὲ συμβαλὼν ὁ Λύκος αὐτῷ, τὰς ἀρχὰς ἐξ Ἀρμενίας ἔχων, γίνεται καὶ αὐτὸς Ἶρις· εἶθ' ἡ Θεμίσκυρα ὑποδέχεται τὸ ῥεῦμα καὶ τὸ Ποντικὸν πέλαγος. διὰ δὲ τοῦτο ἔνδροσόν ἐστι καὶ πόαζον ἀεὶ τὸ πεδίον τοῦτο τρέφειν ἀγέλας βοῶν τε ὁμοίως καὶ ἵππων δυνάμενον, σπόρον δὲ πλεῖστον δέχεται τὸν ἐκ τῆς ἐλύμου καὶ κέγχρου, μᾶλλον δὲ ἀνέκλειπτον· C 548 αὐχμοῦ γάρ ἐστι κρείττων ἡ εὐυδρία παντός, ὥστ' οὐδὲ λιμὸς καθικνεῖται τῶν ἀνθρώπων τούτων οὐδ' ἅπαξ· τοσαύτην δ' ὀπώραν ἐκδίδωσιν ἡ παρόρειος τὴν αὐτοφυῆ καὶ ἀγρίαν σταφυλῆς τε καὶ ὄχνης καὶ μήλου καὶ τῶν καρυωδῶν, ὥστε κατὰ πᾶσαν τοῦ ἔτους ὥραν ἀφθόνως εὐπορεῖν τοὺς ἐξιόντας ἐπὶ τὴν ὕλην· τοτὲ μὲν ἔτι κρεμαμένων τῶν καρπῶν ἐν τοῖς δένδρεσι, τοτὲ δ' ἐν τῇ πεπτωκυίᾳ φυλλάδι καὶ ὑπ' αὐτῇ κειμένων βαθείᾳ καὶ πολλῇ κεχυμένῃ. συχναὶ δὲ καὶ θῆραι παντοίων ἀγρευμάτων διὰ τὴν εὐφορίαν[1] τῆς τροφῆς.

16. Μετὰ δὲ τὴν Θεμίσκυράν ἐστιν ἡ Σιδηνή, πεδίον εὔδαιμον, οὐχ ὁμοίως δὲ καὶ κατάρρυτον, ἔχον χωρία ἐρυμνὰ ἐπὶ τῇ παραλίᾳ, τήν τε Σίδην, ἀφ' ἧς ὠνομάσθη Σιδηνή, καὶ Χάβακα

Pontus and through Dazimonitis, a fertile plain, towards the west, then turns towards the north past Gaziura itself, an ancient royal residence, though now deserted, and then bends back again towards the east, after receiving the waters of the Scylax and other rivers, and after flowing past the very wall of Amaseia, my fatherland, a very strongly fortified city, flows on into Phanaroea. Here the Lycus River, which has its beginnings in Armenia, joins it, and itself also becomes the Iris. Then the stream is received by Themiscyra and by the Pontic Sea. On this account the plain in question is always moist and covered with grass and can support herds of cattle and horses alike and admits of the sowing of millet-seeds and sorghum-seeds in very great, or rather unlimited, quantities. Indeed, their plenty of water offsets any drought, so that no famine comes down on these people, never once; and the country along the mountain yields so much fruit, self-grown and wild, I mean grapes and pears and apples and nuts, that those who go out to the forest at any time in the year get an abundant supply—the fruits at one time still hanging on the trees and at another lying on the fallen leaves or beneath them, which are shed deep and in great quantities. And numerous, also, are the catches of all kinds of wild animals, because of the good yield of food.

16. After Themiscyra one comes to Sidenê, which is a fertile plain, though it is not well-watered like Themiscyra. It has strongholds on the seaboard: Sidê, after which Sidenê was named, and Chabaca

---

[1] εὐφορίαν, Corais emends to εὐπορίαν, Meineke following.

καὶ Φάβδα· μέχρι μὲν δὴ δεῦρο Ἀμισηνή. ἄνδρες δὲ γεγόνασιν ἄξιοι μνήμης κατὰ παιδείαν ἐνταῦθα, μαθηματικοὶ μὲν Δημήτριος ὁ τοῦ Ῥαθηνοῦ καὶ Διονυσόδωρος,[1] ὁμώνυμος τῷ Μηλίῳ[2] γεωμέτρῃ, γραμματικὸς δὲ Τυραννίων, οὗ ἡμεῖς ἠκροασάμεθα.

17. Μετὰ δὲ τὴν Σιδηνὴν ἡ Φαρνακία ἐστίν, ἐρυμνὸν πόλισμα, καὶ μετὰ ταῦτα ἡ Τραπεζοῦς, πόλις Ἑλληνίς, εἰς ἣν ἀπὸ τῆς Ἀμισοῦ περὶ δισχιλίους καὶ διακοσίους σταδίους ἐστὶν ὁ πλοῦς· εἶτ' ἔνθεν εἰς Φᾶσιν χίλιοί που καὶ τετρακόσιοι, ὥστε οἱ σύμπαντες ἀπὸ τοῦ Ἱεροῦ μέχρι Φάσιδος περὶ ὀκτακισχιλίους σταδίους εἰσὶν ἢ μικρῷ πλείους ἢ ἐλάττους. ἐν δὲ τῇ παραλίᾳ ταύτῃ ἀπὸ Ἀμισοῦ πλέουσιν ἡ Ἡράκλειος ἄκρα πρῶτόν ἐστιν, εἶτ' ἄλλη ἄκρα Ἰασόνιον καὶ ὁ Γενήτης,[3] εἶτα Κύτωρος[4] πολίχνη, ἐξ ἧς συνῳκίσθη ἡ Φαρνακία, εἶτ' Ἰσχόπολις κατερηριμμένη, εἶτα κόλπος, ἐν ᾧ Κερασοῦς τε καὶ Ἑρμώνασσα, κατοικίαι μέτριαι, εἶτα τῆς Ἑρμωνάσσης πλησίον ἡ Τραπεζοῦς, εἶθ' ἡ Κολχίς· ἐνταῦθα δέ που ἐστὶ καὶ Ζυγόπολίς τις λεγομένη κατοικία. περὶ μὲν οὖν τῆς Κολχίδος εἴρηται καὶ τῆς ὑπερκειμένης παραλίας.

18. Τῆς δὲ Τραπεζοῦντος ὑπέρκεινται καὶ τῆς Φαρνακίας Τιβαρανοί τε καὶ Χαλδαῖοι καὶ Σάννοι, οὓς πρότερον ἐκάλουν Μάκρωνας, καὶ

[1] Διονυσόδωρος, the editors, for Διονυσιόδωρος.
[2] Μηλίῳ, Tyrwhitt, for Ἴκενι; so Meineke.
[3] Γενήτης, Casaubon, for γενέτης; so the later editors.
[4] Κύτωρος, an error for Κοτύωρα, Κοτύωρον, or Κοτύωρος (see C. Müller, *l.c.*).

and Phabda. Now the territory of Amisus extends to this point; and the city has produced men noteworthy for their learning, Demetrius, the son of Rhathenus, and Dionysodorus, the mathematicians, the latter bearing the same name as the Melian geometer, and Tyrranion the grammarian, of whom I was a pupil.

17. After Sidenê one comes to Pharnacia, a fortified town; and afterwards to Trapezus, a Greek city, to which the voyage from Amisus is about two thousand two hundred stadia. Then from here the voyage to Phasis is approximately one thousand four hundred stadia, so that the distance from Hieron[1] to Phasis is, all told, about eight thousand stadia, or slightly more or less. As one sails along this seaboard from Amisus, one comes first to the Heracleian Cape, and then to another cape called Jasonium, and to Genetes, and then to a town called Cytorus,[2] from the inhabitants of which Pharnacia was settled, and then to Ischopolis, now in ruins, and then to a gulf, on which are both Cerasus and Hermonassa, moderate-sized settlements, and then, near Hermonassa, to Trapezus, and then to Colchis. Somewhere in this neighbourhood is also a settlement called Zygopolis. Now I have already described[3] Colchis and the coast which lies above it.

18. Above Trapezus and Pharnacia are situated the Tibarani and Chaldaei and Sanni, in earlier times called Macrones, and Lesser Armenia; and the

[1] See 12. 3. 11.

[2] Apparently an error for "Cotyora" or "Cotyorum" or "Cotyorus."

[3] 11. 2. 15.

ἡ μικρὰ Ἀρμενία, καὶ οἱ Ἀππαῖται δε πως πλησιάζουσι τοῖς χωρίοις τούτοις, οἱ πρότερον Κερκῖται. διήκει δὲ διὰ τούτων ὅ τε Σκυδίσης, ὄρος τραχύτατον, συνάπτον τοῖς Μοσχικοῖς ὄρεσι τοῖς ὑπὲρ τῆς Κολχίδος, οὗ τὰ ἄκρα κατέχουσιν οἱ Ἑπτακωμῆται, καὶ ὁ Παρυάδρης ὁ μέχρι τῆς μικρᾶς Ἀρμενίας ἀπὸ τῶν κατὰ Σιδηνὴν
C 549 καὶ Θεμίσκυραν τόπων διατείνων καὶ ποιῶν τὸ ἑωθινὸν τοῦ Πόντου πλευρόν. εἰσὶ δ' ἅπαντες μὲν οἱ ὄρειοι τούτων ἄγριοι τελέως, ὑπερβέβληνται δὲ τοὺς ἄλλους οἱ Ἑπτακωμῆται· τινὲς δὲ καὶ ἐπὶ δένδρεσιν ἢ πυργίοις οἰκοῦσι, διὸ καὶ Μοσυνοίκους ἐκάλουν οἱ παλαιοί, τῶν πύργων μοσύνων λεγομένων. ζῶσι δ' ἀπὸ θηρείων σαρκῶν καὶ τῶν ἀκροδρύων, ἐπιτίθενται δὲ καὶ τοῖς ὁδοιποροῦσι, καταπηδήσαντες ἀπὸ τῶν ἰκρίων. οἱ δὲ Ἑπτακωμῆται τρεῖς Πομπηίου σπείρας κατέκοψαν διεξιούσας τὴν ὀρεινήν, κεράσαντες κρατῆρας ἐν ταῖς ὁδοῖς τοῦ μαινομένου μέλιτος, ὃ φέρουσιν οἱ ἀκρεμόνες τῶν δένδρων· πιοῦσι γὰρ καὶ παρακόψασιν ἐπιθέμενοι ῥᾳδίως διεχειρίσαντο τοὺς ἀνθρώπους. ἐκαλοῦντο δὲ τούτων τινὲς τῶν βαρβάρων καὶ Βύζηρες.

19. Οἱ δὲ νῦν Χαλδαῖοι Χάλυβες τὸ παλαιὸν ὠνομάζοντο, καθ' οὓς μάλιστα ἡ Φαρνακία ἵδρυται, κατὰ θάλατταν μὲν ἔχουσα εὐφυΐαν τὴν ἐκ τῆς πηλαμυδείας (πρώτιστα γὰρ ἁλίσκεται ἐνταῦθα τὸ ὄψον τοῦτο), ἐκ δὲ τῆς γῆς τὰ μέταλλα, νῦν μὲν σιδήρου, πρότερον δὲ καὶ ἀργύ-

---

[1] *i.e.* six hundred, unless the Greek word should be translated "cohort," to which it is sometimes equivalent.

Appaïtae, in earlier times called the Cercitae, are fairly close to these regions. Two mountains cross the country of these people, not only the Scydises, a very rugged mountain, which joins the Moschian Mountains above Colchis (its heights are occupied by the Heptacometae), but also the Paryadres, which extends from the region of Sidenê and Themiscyra to Lesser Armenia and forms the eastern side of Pontus. Now all these peoples who live in the mountains are utterly savage, but the Heptacometae are worse than the rest. Some also live in trees or turrets; and it was on this account that the ancients called them "Mosynoeci," the turrets being called "mosyni." They live on the flesh of wild animals and on nuts; and they also attack wayfarers, leaping down upon them from their scaffolds. The Heptacometae cut down three maniples[1] of Pompey's army when they were passing through the mountainous country; for they mixed bowls of the crazing honey which is yielded by the tree-twigs, and placed them in the roads, and then, when the soldiers drank the mixture and lost their senses, they attacked them and easily disposed of them. Some of these barbarians were also called Byzeres.

19. The Chaldaei of to-day were in ancient times named Chalybes; and it is just opposite their territory that Pharnacia is situated, which, on the sea, has the natural advantages of *pelamydes*-fishing (for it is here that this fish is first caught)[2] and, on the land, has the mines, only iron-mines at the present time, though in earlier times it also had silver-mines.[3]

[2] See 7. 6. 2 and 12. 3. 11.
[3] On these mines see Leaf, *Troy*, p. 290.

ρου. ὅλως δὲ κατὰ τοὺς τόπους τούτους ἡ παραλία στενὴ τελέως ἐστίν, ὑπέρκειται γὰρ εὐθὺς τὰ ὄρη μετάλλων πλήρη καὶ δρυμῶν, γεωργεῖται[1] δ' οὐ πολλά· λείπεται δὲ τοῖς μὲν μεταλλευταῖς ἐκ τῶν μετάλλων ὁ βίος, τοῖς δὲ θαλαττουργοῖς ἐκ τῆς ἁλιείας, καὶ μάλιστα τῶν πηλαμύδων καὶ τῶν δελφίνων· ἐπακολουθοῦντες γὰρ ταῖς ἀγέλαις τῶν ἰχθύων, κορδύλης τε καὶ θύννης καὶ αὐτῆς τῆς πηλαμύδος, πιαίνονταί τε καὶ εὐάλωτοι γίνονται διὰ τὸ πλησιάζειν τῇ γῇ προαλέστερον· δελεαζομένους μόνοι οὗτοι κατακόπτουσι τοὺς δελφῖνας καὶ τῷ στέατι πολλῷ χρῶνται πρὸς ἅπαντα.

20. Τούτους οὖν οἶμαι λέγειν τὸν ποιητὴν Ἁλιζώνους ἐν τῷ μετὰ τοὺς Παφλαγόνας καταλόγῳ·

> αὐτὰρ Ἁλιζώνων Ὀδίος καὶ Ἐπίστροφος ἦρχον
> τηλόθεν ἐξ Ἀλύβης, ὅθεν ἀργύρου ἐστὶ γενέθλη·

ἤτοι τῆς γραφῆς μετατεθείσης ἀπὸ τοῦ τηλόθεν ἐκ Χαλύβης, ἢ τῶν ἀνθρώπων πρότερον Ἀλύβων λεγομένων ἀντὶ Χαλύβων· οὐ γὰρ νῦν μὲν δυνατὸν γέγονεν ἐκ Χαλύβων Χαλδαίους λεχθῆναι, πρότερον δ' οὐκ ἐνῆν ἀντὶ Ἀλύβων Χάλυβας, καὶ ταῦτα τῶν ὀνομάτων μεταπτώσεις πολλὰς δεχομένων, καὶ μάλιστα ἐν τοῖς βαρβάροις· Σίντιες γὰρ ἐκαλοῦντό τινες τῶν Θρᾳκῶν, εἶτα Σιντοί, εἶτα Σάϊοι, παρ' οἷς φησὶν Ἀρχίλοχος τὴν ἀσπίδα ῥῖψαι·

Upon the whole, the seaboard in this region is extremely narrow, for the mountains, full of mines and forests, are situated directly above it, and not much of it is tilled. But there remains for the miners their livelihood from the mines, and for those who busy themselves on the sea their livelihood from their fishing, and especially from their catches of *pelamydes* and dolphins; for the dolphins pursue the schools of fish—the *cordylê* and the tunny-fish and the *pelamydes* themselves;[1] and they not only grow fat on them, but also become easy to catch because they are rather eager to approach the land. These are the only people who cut up the dolphins, which are caught with bait, and use their abundance of fat for all purposes.

20. So it is these people, I think, that the poet calls Halizoni, mentioning them next the after Paphlagonians in his *Catalogue.* "But the Halizones were led by Odius and Epistrophus, from Alybê far away, where is the birth-place of silver," since the text has been changed from "Chalybê far away" or else the people were in earlier times called "Alybes" instead of "Chalybes"; for at the present time it proves impossible that they should have been called "Chaldaei," deriving their name from "Chalybê," if in earlier times they could not have been called "Chalybes" instead of "Alybes," and that too when names undergo many changes, particularly among the barbarians; for instance, certain of the Thracians were called Sinties, then Sinti and then Saïi, in whose country Archilochus says he flung away his

[1] All three are species of tunny-fish.

---

[1] γεωργεῖται, Casaubon, for γεωργεῖ; so the later editors.

ἀσπίδα μὲν Σαΐων τις ἀνείλετο,[1] τὴν παρὰ[2] θάμνῳ
ἔντος ἀμώμητον κάλλιπον οὐκ ἐθέλων·

C 550 οἱ δ' αὐτοὶ οὗτοι Σαπαῖοι[3] νῦν ὀνομάζονται· πάντες γὰρ οὗτοι περὶ Ἄβδηρα τὴν οἴκησιν εἶχον καὶ τὰς περὶ Λῆμνον νήσους· ὁμοίως δὲ καὶ Βρύγοι καὶ Βρύγες[4] καὶ Φρύγες οἱ αὐτοί, καὶ Μυσοὶ[5] καὶ Μαίονες καὶ Μῄονες· οὐ χρεία δὲ πλεονάζειν. ὑπονοεῖ δὲ καὶ ὁ Σκήψιος τὴν τοῦ ὀνόματος μετάπτωσιν ἐξ Ἀλύβων εἰς Χάλυβας, τὰ δ' ἑξῆς καὶ τὰ συνῳδὰ οὐ νοῶν, καὶ μάλιστα ἐκ τίνος Ἁλιζώνους εἴρηκε τοὺς Χάλυβας, ἀποδοκιμάζει τὴν δόξαν· ἡμεῖς δ' ἀντιπαραθέντες τῇ ἡμετέρᾳ τὴν ἐκείνου καὶ τὰς τῶν ἄλλων ὑπολήψεις σκοπῶμεν.

21. Οἱ μὲν μεταγράφουσιν Ἀλαζώνων,[6] οἱ δ' Ἀμαζώνων ποιοῦντες, τὸ δ' ἐξ Ἀλύβης ἐξ Ἀλόπης ἢ[7] ἐξ Ἀλόβης,[8] τοὺς μὲν[9] Σκύθας Ἀλαζῶνας[10] φάσκοντες ὑπὲρ τὸν Βορυσθένη καὶ Καλλιπίδας καὶ ἄλλα ὀνόματα, ἅπερ Ἑλλάνικός τε καὶ Ἡρόδοτος καὶ Εὔδοξος κατεφλυάρησαν ἡμῶν, τὰς[11] δ' Ἀμαζῶνας[12] μεταξὺ Μυσίας καὶ Καρίας καὶ Λυδίας, καθάπερ Ἔφορος νομίζει, πλησίον Κυμης τῆς πατρίδος αὐτοῦ· καὶ τοῦτο μὲν ἔχεταί

[1] ἀνείλετο, omitted by MSS. except E. ἀγάλλεται, editors before Kramer (cp. 10. 2. 17 where same passage is quoted).
[2] παρά, Corais for περί; so the later editors.
[3] Σαπαῖοι, Groskurd, for Σάπαι; so the later editors.
[4] Βρύγες, *Epit.*, Βρέγες MSS.
[5] καὶ Μέρονες, before καὶ Μαίονες, Corais and later editors eject.
[6] Ἀλαζώνων, Tzschucke, for Ἀλαζίνων; so the later editors.

shield: "One of the Saïi robbed me of my shield, which, a blameless weapon, I left behind me beside a bush, against my will."[1] These same people are now named Sapaei; for all these have their abode round Abdera and the islands round Lemnos. Likewise the Brygi and Bryges and Phryges are the same people; and the Mysi and Maeones and Meïones are the same; but there is no use of enlarging on the subject. The Scepsian[2] doubts the alteration of the name from "Alybes" to "Chalybes"; and, failing to note what follows and what accords with it, and especially why the poet calls the Chalybians Halizoni, he rejects this opinion. As for me, let me place his assumption and those of the other critics side by side with my own and consider them.

21. Some change the text and make it read "Alazones," others "Amazones," and for the words "from Alybê" they read "from Alopê," or "from Alobê," calling the Scythians beyond the Borysthenes River "Alazones," and also "Callipidae" and other names—names which Hellanicus and Herodotus and Eudoxus have foisted on us—and placing the Amazons between Mysia and Caria and Lydia near Cymê, which is the opinion also of Ephorus, who was a native of Cymê. And this opinion might perhaps

[1] *Frag.* 6 (51), Bergk. Same fragment quoted in 10. 2. 17.
[2] Demetrius of Scepsis.

---

[7] ἤ, Corais inserts; so the later editors.
[8] Ἀλόβης, Tzschucke, for Ἀόλης; so the later editors.
[9] μέν, Corais, for δέ; so the later editors.
[10] Ἀλαζῶνας, Tzschucke, for Ἀλιζῶνας; so the later editors.
[11] τάς, Jones restores, instead of τούς CDw and the editors.
[12] Ἀμαζῶνας C, Ἀμαζόνας other MSS.

τινος λόγου τυχὸν ἴσως· εἴη γὰρ ἂν λέγων τὴν ὑπὸ τῶν Αἰολέων καὶ Ἰώνων οἰκισθεῖσαν ὕστερον, πρότερον δ' ὑπὸ Ἀμαζόνων· καὶ ἐπωνύμους πόλεις τινὰς εἶναί φασι, καὶ γὰρ Ἔφεσον καὶ Σμύρναν καὶ Κύμην καὶ Μύριναν. ἡ δὲ Ἀλύβη ἤ, ὥς τινες, Ἀλόπη ἢ Ἀλόβη πῶς ἂν ἐν τοῖς τόποις τούτοις ἐξητάζετο; πῶς δὲ τηλόθεν; πῶς δ' ἡ τοῦ ἀργύρου γενέθλη;

22. Ταῦτα μὲν ἀπολύεται τῇ μεταγραφῇ· γράφει γὰρ οὕτως·

> αὐτὰρ Ἀμαζώνων[1] Ὀδίος καὶ Ἐπίστροφος ἦρχον,
> ἐλθόντ' ἐξ Ἀλόπης, ὅθ' Ἀμαζονίδων γένος ἐστί.

ταῦτα δ' ἀπολυσάμενος εἰς ἄλλο ἐμπέπτωκε πλάσμα· οὐδαμοῦ γὰρ ἐνθάδε εὑρίσκεται Ἀλόπη, καὶ ἡ μεταγραφὴ δὲ παρὰ τὴν τῶν ἀντιγράφων τῶν ἀρχαίων πίστιν καινοτομουμένη ἐπὶ τοσοῦτον σχεδιασμῷ ἔοικεν. ὁ δὲ Σκήψιος οὔτε[2] τὴν τούτου δόξαν ἔοικεν ἀποδεξάμενος οὔτε τῶν περὶ τὴν Παλλήνην τοὺς Ἀλιζώνους ὑπολαβόντων, ὧν ἐμνήσθημεν ἐν τοῖς Μακεδονικοῖς· ὁμοίως διαπορεῖ καὶ πῶς ἐκ τῶν ὑπὲρ τὸν Βορυσθένην νομάδων ἀφῖχθαι συμμαχίαν τοῖς Τρωσί τις νομίσειεν· ἐπαινεῖ δὲ μάλιστα τὴν Ἑκαταίου τοῦ Μιλησίου καὶ Μενεκράτους τοῦ Ἐλαΐτου, τῶν Ξενοκράτους γνωρίμων ἀνδρός, δόξαν καὶ τὴν Παλαιφάτου, ὧν ὁ μὲν ἐν γῆς περιόδῳ φησίν· "ἐπὶ δ' Ἀλαζίᾳ πόλι[3] ποταμὸς Ὀδρύσσης[4] ῥέων διὰ Μυγδονίης[5] πεδίου

[1] *Dhiloriw* read Ἀμαζόνων.
[2] οὔτε, Corais, for οὐδέ; so the later editors.

not be unreasonable, for he may mean the country which was later settled by the Aeolians and the Ionians, but earlier by the Amazons. And there are certain cities, it is said, which got their names from the Amazons, I mean Ephesus, Smyrna, Cymê, and Myrina.[1] But how could Alybê, or, as some call it, "Alopê" or "Alobê," be found in this region, and how about "far away," and how about "the birth-place of silver"?

22. These objections Ephorus solves by his change of the text, for he writes thus: "But the Amazons were led by Odius and Epistrophus, from Alopê far away, where is the race of Amazons." But in solving these objections he has fallen into another fiction; for Alopê is nowhere to be found in this region; and, further, his change of the text, with innovations so contrary to the evidence of the early manuscripts, looks like rashness. But the Scepsian apparently accepts neither the opinion of Ephorus nor of those who suppose them to be the Halizoni near Pallenê, whom I have mentioned in my description of Macedonia.[2] He is also at loss to understand how anyone could think that an allied force came to help the Trojans from the nomads beyond the Borysthenes River; and he especially approves of the opinions of Hecataeus of Miletus, and of Menecrates of Elaea, one of the disciples of Xenocrates, and also of that of Palaephatus. The first of these says in his *Circuit of the Earth*: "Near the city Alazia is the River Odrysses, which flows out of

[1] Cf. 11. 5. 4. [2] Vol. III, p. 351, *Frag.* 27*a*.

---

[3] C reads πόλει.
[4] Ὀδρύσσης, Tzschucke, for ὁ ῥύμος D*hilorw*, ὀδρύσιος *x*.
[5] Μυγδονίης, Corais, for Μυγδόνος *xz*, Μυγδόνης other MSS.

C 551 ἀπὸ δύσιος ἐκ τῆς λίμνης τῆς Δασκυλίτιδος ἐς Ῥύνδακον ἐσβάλλει·" ἔρημον δὲ εἶναι νῦν τὴν Ἀλαζίαν λέγει, κώμας δὲ πολλὰς τῶν Ἀλαζώνων[1] οἰκεῖσθαι, δι' ὧν Ὀδρύσσης ῥεῖ, ἐν δὲ ταύταις τὸν Ἀπόλλωνα τιμᾶσθαι διαφερόντως, καὶ μάλιστα κατὰ τὴν ἐφορίαν τῶν Κυζικηνῶν. ὁ δὲ Μενεκράτης ἐν τῇ Ἑλλησποντιακῇ περιόδῳ ὑπερκεῖσθαι λέγει τῶν περὶ[2] τὴν Μύρλειαν[3] τόπων ὀρεινὴν συνεχῆ, ἣν κατῴκει τὸ τῶν Ἁλιζώνων ἔθνος· δεῖ δέ, φησί, γράφειν ἐν τοῖς δύο λάβδα, τὸν δὲ ποιητὴν ἐν τῷ ἑνὶ γράφειν διὰ τὸ μέτρον. ὁ δὲ Παλαίφατός φησιν, ἐξ Ἀμαζόνων τῶν ἐν τῇ Ἀλόπῃ οἰκούντων, νῦν δ' ἐν Ζελείᾳ,[4] τὸν Ὀδίον καὶ τὸν Ἐπίστροφον στρατεῦσαι. τί οὖν ἄξιον ἐπαινεῖν τὰς τούτων δόξας; χωρὶς γὰρ τοῦ τὴν ἀρχαίαν γραφὴν καὶ τούτους κινεῖν οὔτε τὰ ἀργυρεῖα δεικνύουσιν, οὔτε ποῦ[5] τῆς Μυρλεάτιδος Ἀλόπη ἐστίν, οὔτε πῶς οἱ ἐνθένδε ἀφιγμένοι εἰς Ἴλιον τηλόθεν ἦσαν, εἰ καὶ δοθείη Ἀλόπην[6] τινὰ γεγονέναι ἢ Ἀλαζίαν· πολὺ γὰρ δὴ ταῦτα ἐγγυτέρω ἐστὶ τῇ Τρῳάδι ἢ τὰ περὶ Ἔφεσον. ἀλλ' ὅμως τοὺς περὶ Πύγελα λέγοντας τοὺς Ἀμαζῶνας[7] μεταξὺ Ἐφέσου καὶ Μαγνησίας καὶ Πριήνης φλυαρεῖν φησὶν ὁ Δημήτριος· τὸ γὰρ τηλόθεν οὐκ ἐφαρμόττειν τῷ τόπῳ. ὁπόσῳ οὖν μᾶλλον οὐκ ἐφαρμόττει τῷ περὶ Μυσίαν καὶ Τευθρανίαν;

23. Νὴ Δία, ἀλλά φησι δεῖν ἔνια καὶ ἀκύρως προστιθέμενα δέχεσθαι, ὡς καί·

[1] *x* reads Ἀλαζόνων, other MSS. Ἀμαζόνων.
[2] περί, Corais (from Eustathius), for ὑπέρ; so the later editors.
[3] Μυρλείαν, Xylander (from Eustathius), for Μυρλίαν.
[4] Meineke emends δ' ἐν Ζελείᾳ to δὲ Ζηλείᾳ (cp. Ζέλειαν § 23).
[5] οὔτε ποῦ, Kramer, for ὅπου; so the later editors.

Lake Dascylitis from the west through the plain of Mygdonia and empties into the Rhyndacus." But he goes on to say that Alazia is now deserted, and that many villages of the Alazones, through whose country the Odrysses flows, are inhabited, and that in these villages Apollo is accorded exceptional honour, and particularly on the confines of the Cyziceni. Menecrates in his work entitled *The Circuit of the Hellespont* says that above the region of Myrleia there is an adjacent mountainous tract which is occupied by the tribe of the Halizones. One should spell the name with two *l*'s, he says, but on account of the metre the poet spells it with only one. But Palaephatus says that it was from the Amazons who then lived in Alopê, but now in Zeleia, that Odius and Epistrophus made their expedition. How, then, can the opinions of these men deserve approval? For, apart from the fact that these men also disturb the early text, they neither show us the silver-mines, nor where in the territory of Myrleia Alopê is, nor how those who went from there to Ilium were "from far away," even if one should grant that there actually was an Alopê or Alazia; for these, of course, are much nearer the Troad than the places round Ephesus. But still those who speak of the Amazons as living in the neighbourhood of Pygela between Ephesus and Magnesia and Prienê talk nonsense, Demetrius says, for, he adds, "far away" cannot apply to that region. How much more inapplicable, then, is it to the region of Mysia and Teuthrania?

23. Yes, by Zeus, but he goes on to say that some things are arbitrarily inserted in the text, for

[6] Ἀλόπην, Groskurd, for λίμνη; so later editors.
[7] Ἀμαζῶνας, Kramer, for Ἀμαζόνας; so later editors.

τῆλ' ἐξ Ἀσκανίης·

καί

Ἀρναῖος δ' ὄνομ' ἔσκε, τὸ γὰρ θέτο πότνια μήτηρ·

καί

εἵλετο δὲ κληῖδ' εὐκαμπέα χειρὶ παχείῃ Πηνελόπη.

δεδόσθω δὴ καὶ τοῦτο· ἀλλ' ἐκεῖνα οὐ δοτέα, οἷς προσέχων ὁ Δημήτριος οὐδὲ τοῖς ὑπολαβοῦσι δεῖν ἀκούειν τηλόθεν ἐκ Χαλύβης πιθανῶς ἀντείρηκε. συγχωρήσας γάρ, ὅτι, εἰ καὶ μὴ ἔστι νῦν ἐν τοῖς Χάλυψι τὰ ἀργυρεῖα, ὑπάρξαι γε ἐνεδέχετο, ἐκεῖνό γε οὐ συγχωρεῖ, ὅτι καὶ ἔνδοξα ἦν καὶ ἄξια μνήμης, καθάπερ τὰ σιδηρεῖα. τί δὲ κωλύει, φαίη τις ἄν, καὶ ἔνδοξα εἶναι, καθάπερ καὶ τὰ σιδηρεῖα; ἢ σιδήρου μὲν εὐπορία τόπον ἐπιφανῆ δύναται ποιεῖν, ἀργύρου δ' οὔ; τί δ' εἰ μὴ[1] κατὰ τοὺς ἥρωας, ἀλλὰ καθ' Ὅμηρον εἰς δόξαν ἀφῖκτο τὰ ἀργυρεῖα, ἆρα μέμψαιτό τις ἂν τὴν ἀπόφασιν τοῦ ποιητοῦ; πῶς οὖν εἰς τὸν ποιητὴν ἡ δόξα ἀφίκετο; πῶς δ' ἡ τοῦ ἐν τῇ Τεμέσῃ χαλκοῦ τῇ Ἰταλιώτιδι; πῶς δ' ἡ τοῦ Θηβαϊκοῦ πλούτου τοῦ κατ' Αἴγυπτον; καίτοι διπλάσιον σχεδόν τι διέχοντα τῶν Αἰγυπτίων Θηβῶν ἢ τῶν Χαλδαίων. C 552 ἀλλ' οὐδ'[2] οἷς συνηγορεῖ, τούτοις ὁμολογεῖ· τὰ γὰρ περὶ τὴν Σκῆψιν τοποθετῶν,[3] τὴν ἑαυτοῦ πατρίδα, πλησίον τῆς Σκήψεως καὶ τοῦ Αἰσήπου Νέαν[4] κώμην καὶ Ἀργυρίαν λέγει καὶ Ἀλαζονίαν.

[1] τί δ' εἰ μή, Corais, for οὔτι εἰ μή; so the later editors.
[2] οὐδ', Corais, for οὔτ'; so Meineke.

example, "from Ascania far away,"[1] and "Arnaeus was his name, for his revered mother had given him this name at his birth,"[2] and "Penelope took the bent key in her strong hand."[3] Now let this be granted, but those other things are not to be granted to which Demetrius assents without even making a plausible reply to those who have assumed that we ought to read "from Chalybê far away"; for although he concedes that, even if the silver-mines are not now in the country of the Chalybians, they could have been there in earlier times, he does not concede that other point, that they were both famous and worthy of note, like the iron-mines. But, one might ask, what is there to prevent them from being famous like the iron-mines? Or can an abundance of iron make a place famous but an abundance of silver not do so? And if the silver-mines had reached fame, not in the time of the heroes, but in the time of Homer, could any person find fault with the assertion of the poet? How, pray, could their fame have reached the poet? How, pray, could the fame of the copper-mine at Temesa in Italy have reached him? How the fame of the wealth of Thebes in Egypt,[4] although he was about twice as far from Thebes as from the Chaldaeans? But Demetrius is not even in agreement with those for whose opinions he pleads; for in fixing the sites round Scepsis, his birth-place, he speaks of Nea, a village, and of Argyria and Alazonia as near Scepsis

[1] *Iliad* 2. 863. [2] *Odyssey* 18. 5.
[3] *Odyssey* 21. 6. [4] *Iliad* 9. 381.

---

[3] τοποθετῶν, Casaubon, for νομοθετῶν; so the later editors.
[4] Νέαν, Meineke, for Ἐνέαν.

ταῦτα μὲν οὖν εἰ καὶ ἔστι, πρὸς ταῖς πηγαῖς ἂν εἴη τοῦ Αἰσήπου. ὁ δὲ Ἑκαταῖος λέγει ἐπέκεινα τῶν ἐκβολῶν αὐτοῦ, ὅ τε Παλαίφατος πρότερον μὲν Ἀλόπην οἰκεῖν φήσας, νῦν δὲ Ζέλειαν, οὐδὲν ὅμοιον λέγει τούτοις. εἰ δ' ἄρα ὁ Μενεκράτης, καὶ οὐδ' οὗτος τὴν Ἀλόπην ἢ Ἀλόβην ἢ ὅπως ποτὲ βούλονται γράφειν φράζει, ἥτις ἐστίν, οὐδ'[1] αὐτὸς ὁ Δημήτριος.

24. Πρὸς Ἀπολλόδωρον δὲ περὶ τῶν αὐτῶν ἐν τῷ Τρωικῷ διακόσμῳ διαλεγόμενον πολλὰ μὲν εἴρηται πρότερον, καὶ νῦν δὲ λεκτέον. οὐ γὰρ οἴεται δεῖν δέχεσθαι τοὺς Ἁλιζώνους ἐκτὸς τοῦ Ἅλυος· μηδεμίαν γὰρ συμμαχίαν ἀφῖχθαι τοῖς Τρωσὶν ἐκ τῆς περαίας τοῦ Ἅλυος. πρῶτον τοίνυν ἀπαιτήσομεν αὐτόν, τίνες εἰσὶν οἱ[2] ἐντὸς τοῦ Ἅλυος Ἁλίζωνοι, οἱ καὶ

τηλόθεν ἐξ Ἀλύβης, ὅθεν ἀργύρου ἐστὶ γενέθλη·

οὐ γὰρ ἕξει λέγειν· ἔπειτα τὴν αἰτίαν, δι' ἣν οὐ συγχωρεῖ καὶ ἐκ τῆς περαίας ἀφῖχθαί τινα συμμαχίαν· καὶ γὰρ εἰ τὰς ἄλλας ἐντὸς εἶναι τοῦ ποταμοῦ πάσας συμβαίνει πλὴν τῶν Θρᾳκῶν, μίαν γε ταύτην οὐδὲν ἐκώλυε πέραθεν ἀφῖχθαι ἐκ τῆς ἐπέκεινα τῶν Λευκοσύρων. ἢ πολεμήσαντας[3] μὲν ἦν δυνατὸν διαβαίνειν ἐκ τῶν τόπων τούτων καὶ τῶν ἐπέκεινα, καθάπερ τὰς Ἀμαζόνας καὶ Τρῆρας καὶ Κιμμερίους φασί, συμμαχήσαντας[4]

[1] οὐδ', Jones, for οὔτ'.
[2] οἱ, Corais inserts; so the later editors.
[3] πολεμήσαντας, Corais and Meineke, following z, emend to πολεμήσοντας; "idque sane arridet," says Kramer.
[4] συμμαχήσαντας, Corais and Meineke, following z, emend to συμμαχήσοντας.

and the Aesepus River. These places, then, if they really exist, would be near the sources of the Aesepus; but Hecataeus speaks of them as beyond the outlets of it; and Palaephatus, although he says that they[1] formerly lived in Alopê, but now in Zeleia, says nothing like what these men say. But if Menecrates does so, not even he tells us what kind of a place "Alopê" is or "Alobê," or however they wish to write the name, and neither does Demetrius himself.

24. As regards Apollodorus, who discusses the same subject in his *Marshalling of the Trojan Forces*, I have already said much in answer to him,[2] but I must now speak again; for he does not think that we should take the Halizoni as living outside the Halys River; for, he says, no allied force came to the Trojans from beyond the Halys. First, therefore, we shall ask of him who are the Halizoni this side the Halys and "from Alybê far away, where is the birthplace of silver." For he will be unable to tell us. And we shall next ask him the reason why he does not concede that an allied force came also from the country on the far side of the river; for, if it is the case that all the rest of the allied forces except the Thracians lived this side the river, there was nothing to prevent this one allied force from coming from the far side of the Halys, from the country beyond the White Syrians.[3] Or was it possible for peoples who fought the Trojans to cross over from these regions and from the regions beyond, as they say the Amazons and Treres and Cimmerians did, and yet impossible for people who fought as allies with them

[1] The Amazons (12. 3. 22).
[2] *e.g.* 7. 3. 6.
[3] *i.e.* Cappadocians.

δ' ἀδύνατον; αἱ μὲν οὖν Ἀμαζόνες οὐ συνεμάχουν, διὰ τὸ τὸν Πρίαμον πολεμῆσαι πρὸς αὐτὰς συμμαχοῦντα τοῖς Φρυξίν,[1]

οἵ ῥα τότ' ἦλθον Ἀμαζόνες ἀντιάνειραι

(φησὶν ὁ Πρίαμος),

καὶ γὰρ ἐγὼν ἐπίκουρος ἐὼν μετὰ τοῖσιν ἐλέγμην.

οἱ δ' ὁμοροῦντες αὐταῖς, οὐδ' οὕτως ἄπωθεν ὄντες, ὥστε χαλεπὴν εἶναι τὴν ἐκεῖθεν μετάπεμψιν, οὐδ' ἔχθρας ὑποκειμένης, οὐδὲν ἐκωλύοντο, οἶμαι, συμμαχεῖν.

25. Ἀλλ' οὐδὲ δόξαν ἔχει τοιαύτην τῶν παλαιῶν εἰπεῖν, ὡς συμφωνούντων ἁπάντων, μηδένας ἐκ τῆς περαίας τοῦ Ἅλυος κοινωνῆσαι τοῦ Τρωικοῦ πολέμου. πρὸς τοὐναντίον δὲ μᾶλλον εὕροι τις ἂν μαρτυρίας· Μαιάνδριος γοῦν ἐκ τῶν Λευκοσύρων φησὶ τοὺς Ἐνετοὺς ὁρμηθέντας συμμαχῆσαι τοῖς Τρωσίν, ἐκεῖθεν δὲ μετὰ τῶν Θρᾳκῶν ἀπᾶραι καὶ οἰκῆσαι περὶ τὸν τοῦ Ἀδρίου μυχόν, τοὺς δὲ μὴ μετασχόντας τῆς στρατείας Ἐνετοὺς C 553 Καππάδοκας γενέσθαι. συνηγορεῖν δ' ἂν δόξειε τῷ λόγῳ τούτῳ, διότι πᾶσα ἡ πλησίον τοῦ Ἅλυος Καππαδοκία, ὅση παρατείνει τῇ Παφλαγονίᾳ, ταῖς δυσὶ χρῆται διαλέκτοις καὶ τοῖς ὀνόμασι πλεονάζει τοῖς Παφλαγονικοῖς, Βάγας καὶ Βιάσας καὶ Αἰνιάτης καὶ Ῥατώτης καὶ Ζαρδώκης καὶ Τίβιος καὶ Γάσυς καὶ Ὀλίγασυς καὶ Μάνης· ταῦτα γὰρ ἔν τε τῇ Βαμωνίτιδι[2] καὶ τῇ Πι-

[1] Φρυξίν, Kramer (see *Iliad* 3. 184), for Ἴωσιν *oz*, Τρωσίν other MSS.; so the later editors.

[2] Βαμωνίτιδι MSS.; Φαζημωνίτιδι Meineke.

to do so? Now the Amazons would not fight on Priam's side because of the fact that he had fought against them as an ally of the Phrygians, against the "Amazons, peers of men, who came at that time,"[1] as Priam says, "for I too, being their ally, was numbered among them"; but since the peoples whose countries bordered on that of the Amazons were not even far enough away to make difficult the Trojan summons for help from their countries, and since, too, there was no underlying cause for hatred, there was nothing to prevent them, I think, from being allies of the Trojans.

25. Neither can Apollodorus impute such an opinion to the early writers, as though they, one and all, voiced the opinion that no peoples from the far side of the Halys River took part in the Trojan war. One might rather find evidence to the contrary; at any rate, Maeandrius says that the Eneti first set forth from the country of the White Syrians and allied themselves with the Trojans, and that they sailed away from Troy with the Thracians and took up their abode round the recess of the Adrias,[2] but that the Eneti who did not have a part in the expedition had become Cappadocians. The following might seem to agree with this account, I mean the fact that the whole of that part of Cappadocia near the Halys River which extends along Paphlagonia uses two languages which abound in Paphlagonian names, as "Bagas," "Biasas," "Aeniates," "Rhatotes," "Zardoces," "Tibius," "Gasys," "Oligasys," and "Manes," for these names are prevalent in

[1] *Iliad* 3. 189; but the text of Homer reads "on that day when the Amazons came, the peers of men."

[2] *i.e.* the Adriatic Gulf.

μολίτιδι[1] καὶ τῇ Γαζηλωνίτιδι[2] καὶ Γαζακηνῇ καὶ ἄλλαις πλείσταις χώραις ἐπιπολάζει τὰ ὀνόματα. αὐτὸς δὲ ὁ Ἀπολλόδωρος παρατίθησι τὸ τοῦ Ζηνοδότου, ὅτι γράφει·

ἐξ Ἐνετῆς, ὅθεν ἡμιόνων γένος ἀγροτεράων.

ταύτην δέ φησιν Ἑκαταῖον τὸν Μιλήσιον δέχεσθαι τὴν Ἀμισόν· ἡ δ' Ἀμισὸς εἴρηται, διότι τῶν Λευκοσύρων ἐστὶ καὶ ἐκτὸς τοῦ Ἅλυος.

26. Εἴρηται δ' αὐτῷ που, καὶ διότι ὁ ποιητὴς ἱστορίαν εἶχε τῶν Παφλαγόνων τῶν ἐν τῇ μεσογαίᾳ παρὰ τῶν πεζῇ διελθόντων τὴν χώραν, τὴν παραλίαν δ' ἠγνόει, καθάπερ[3] καὶ τὴν ἄλλην τὴν Ποντικήν· ὠνόμαζε γὰρ ἂν[4] αὐτήν. τοὐναντίον δ' ἔστιν ἀναστρέψαντα εἰπεῖν, ἐκ τῆς περιοδείας ὁρμηθέντα τῆς ἀποδοθείσης νυνί, ὡς τὴν μὲν παραλίαν πᾶσαν ἐπελήλυθε καὶ οὐδὲν τῶν ὄντων τότε ἀξίων[5] μνήμης παραλέλοιπεν, εἰ δ' Ἡράκλειαν καὶ Ἄμαστριν καὶ Σινώπην οὐ λέγει, τὰς μήπω συνῳκισμένας, οὐδὲν θαυμαστόν, τῆς δὲ μεσογαίας[6] οὐδὲν ἄτοπον εἰ μὴ εἴρηκε. καὶ τὸ μὴ ὀνομάζειν δὲ πολλὰ τῶν γνωρίμων οὐκ ἀγνοίας ἐστὶ σημεῖον, ὅπερ καὶ ἐν τοῖς ἔμπροσθεν ἐπεσημηνάμεθα· ἀγνοεῖν γὰρ αὐτὸν πολλὰ τῶν ἐνδόξων

[1] Πιμολίτιδι MSS., except DC*orxy*, which read Πημολίτιδι, the ι being changed to η in D; Meineke emends to Πημολισίτιδι (see C. Müller, *l.c.* p. 1021).

[2] Γαζηλωνίτιδι, Meineke, following conj. of Groskurd, for Ζαγλουθίτιδι *oz*, Γαζαλονίτιδι *w*, Γαζαλουίτιδι other MSS.

[3] καθάπερ, Xylander, for καίπερ; so the later editors, except Kramer, who strangely proposes ὥσπερ.

[4] ἄν, the editors insert.

[5] ἀξίων *h*, ἄξιον other MSS.

[6] τῆς δὲ μεσογαίας, Jones restores, for τὴν δὲ μεσόγαιαν (Kramer and later editors).

Bamonitis,[1] Pimolitis,[2] Gazelonitis, Gazacenê and most of the other districts. Apollodorus himself quotes the Homeric verse as written by Zenodotus, stating that he writes it as follows: "from Enetê,[3] whence the breed of the wild mules";[4] and he says that Hecataeus of Miletus takes Enetê to be Amisus. But, as I have already stated,[5] Amisus belongs to the White Syrians and is outside the Halys River.

26. Apollodorus somewhere states, also, that the poet got an account of those Paphlagonians who lived in the interior from men who had passed through the country on foot, but that he was ignorant of the Paphlagonian coast, just as he was ignorant of the rest of the Pontic coast; for otherwise he would have named them. On the contrary, one can retort and say, on the basis of the description which I have now given, that Homer traverses the whole of the coast and omits nothing of the things that were then worth recording, and that it is not at all remarkable if he does not mention Heracleia and Amastris and Sinopê, cities which had not yet been founded, and that it is not at all strange if he has mentioned no part of the interior. And further, the fact that Homer does not name many of the known places is no sign of ignorance, as I have already demonstrated in the foregoing part of my work;[6] for he says that Homer

[1] "Bamonitis" is doubtful; Meineke emends to "Phazemonitis."

[2] "Pimolitis" is doubtful; Meineke emends to "Pimolisitis."

[3] *i.e.* "Enetê" instead of "Heneti," or "Eneti" (the reading accepted by Strabo and modern scholars). See Vol. II, p. 298, foot-note 4, and also pp. 308 and 309.

[4] *Iliad* 2. 852.

[5] 12. 3. 9.

[6] 1. 2. 14, 19; 7. 3. 6–7; and 8. 3. 8.

ἔφη περὶ τὸν Πόντον, οἷον ποταμοὺς καὶ ἔθνη· ὀνομάσαι γὰρ ἄν. τοῦτο δ' ἐπὶ μέν τινων σφόδρα σημειωδῶν δοίη τις ἄν, οἷον Σκύθας καὶ Μαιῶτιν καὶ Ἴστρον. οὐ γὰρ ἂν[1] διὰ σημείων μὲν τοὺς νομάδας εἴρηκε Γαλακτοφάγους Ἀβίους τε δικαιοτάτους τ' ἀνθρώπους, καὶ ἔτι ἀγαυοὺς Ἱππημολγούς, Σκύθας δὲ οὐκ ἂν εἶπεν ἢ Σαυρομάτας ἢ Σαρμάτας, εἰ δὴ οὕτως ὠνομάζοντο ὑπὸ τῶν Ἑλλήνων, οὐδ' ἂν Θρᾳκῶν τε καὶ Μυσῶν μνησθεὶς τῶν πρὸς τῷ Ἴστρῳ αὐτὸν παρεσίγησε, μέγιστον τῶν ποταμῶν ὄντα, καὶ ἄλλως ἐπιφόρως ἔχων πρὸς τὸ τοῖς ποταμοῖς ἀφορίζεσθαι τοὺς τόπους, οὐδ' ἂν Κιμμερίους λέγων παρῆκε τὸν Βόσπορον ἢ τὴν Μαιῶτιν.

27. Ἐπὶ δὲ τῶν μὴ οὕτω σημειωδῶν ἢ μὴ τότε ἢ μὴ πρὸς τὴν ὑπόθεσιν, τί ἄν τις μέμφοιτο; οἷον τὸν Τάναϊν, δι' οὐδὲν ἄλλο γνωριζόμενον ἢ διότι C 554 τῆς Ἀσίας καὶ τῆς Εὐρώπης ὅριόν ἐστιν· ἀλλ' οὔτε τὴν Ἀσίαν οὔτε τὴν Εὐρώπην ὠνόμαζόν πω οἱ τότε, οὐδὲ διῄρητο οὕτως εἰς τρεῖς ἠπείρους ἡ οἰκουμένη· ὠνόμασε γὰρ ἄν που διὰ τὸ λίαν σημειῶδες, ὡς καὶ τὴν Λιβύην καὶ τὸν Λίβα τὸν ἀπὸ τῶν ἑσπερίων τῆς Λιβύης πνέοντα· τῶν δ' ἠπείρων μήπω διωρισμένων, οὐδὲ τοῦ Ταναΐδος ἔδει καὶ τῆς μνήμης αὐτοῦ. πολλὰ δὲ καὶ ἀξιομνημόνευτα μέν, οὐχ ὑπέδραμε δέ· πολὺ γὰρ δὴ

[1] ἄν, before διά, Groskurd inserts; so Kramer and Müller-Dübner.

---

[1] See 7. 3. 6–7.

was ignorant of many of the famous things round the Pontus, for example, rivers and tribes, for otherwise, he says, Homer would have named them. This one might grant in the case of certain very significant things, for example, the Scythians and Lake Maeotis and the Ister River, for otherwise Homer would not have described the nomads by significant characteristics as "Galactophagi" and "Abii" and as "men most just," and also as "proud Hippemolgi,"[1] and yet fail to call the Scythians either Sauromatae or Sarmatae, if indeed they were so named by the Greeks, nor yet, when he mentions the Thracians and Mysians near the Ister, pass by the Ister in silence, greatest of the rivers, and especially when he is inclined to mark the boundaries of places by rivers, nor yet, when he mentions the Cimmerians, omit any mention of the Bosporus or Lake Maeotis.

27. But in the case of things not so significant, either not at that time or for the purposes of his work, how could anyone find fault with Homer for omitting them? For example, for omitting the Tanaïs River, which is well known for no other reason than that it is the boundary between Asia and Europe. But the people of that time were not yet using either the name "Asia" or "Europe," nor yet had the inhabited world been divided into three continents as now, for otherwise he would have named them somewhere because of their very great significance, just as he mentions Libya and also the Lips, the wind that blows from the western parts of Libya. But since the continents had not yet been distinguished, there was no need of mentioning the Tanaïs either. Many things were indeed worthy of mention, but they did not occur to him; for of course

καὶ τὸ ἐπελευστικὸν εἶδος ἔν τε τοῖς λόγοις καὶ ἐν ταῖς πράξεσίν ἐστιν. ἐκ πάντων δὲ[1] τῶν τοιούτων δῆλόν ἐστιν, ὅτι μοχθηρῷ σημείῳ χρῆται πᾶς ὁ ἐκ τοῦ μὴ λέγεσθαί τι ὑπὸ τοῦ ποιητοῦ τὸ ἀγνοεῖσθαι ἐκεῖνο ὑπ' αὐτοῦ τεκμαιρόμενος. καὶ δεῖ διὰ πλειόνων παραδειγμάτων ἐξελέγχειν αὐτὸ μοχθηρὸν ὄν, πολλῷ γὰρ αὐτῷ κέχρηνται πολλοί. ἀνακρουστέον οὖν αὐτοὺς προφέροντας τὰ τοιαῦτα, εἰ καὶ ταυτολογήσομεν τὸν λόγον·[2] οἷον ἐπὶ τῶν ποταμῶν εἴ τις λέγοι, τῷ μὴ ὠνομάσθαι ἀγνοεῖσθαι, εὐήθη φήσομεν τὸν λόγον· ὅπου γε οὐδὲ Μέλητα τὸν παρὰ τὴν Σμύρναν ῥέοντα ὠνόμακε ποταμόν, τὴν ὑπὸ τῶν πλείστων λεγομένην αὐτοῦ πατρίδα, Ἕρμον ποταμὸν καὶ Ὕλλον ὀνομάζων, οὐδὲ Πακτωλὸν τὸν εἰς ταὐτὸ τούτοις ῥεῖθρον ἐμβάλλοντα, τὴν δ' ἀρχὴν ἀπὸ τοῦ Τμώλου ἔχοντα, οὗ[3] μέμνηται· οὐδ' αὐτὴν Σμύρναν λέγει, οὐδὲ τὰς ἄλλας τῶν Ἰώνων πόλεις καὶ τῶν Αἰολέων τὰς πλείστας, Μίλητον λέγων καὶ Σάμον[4] καὶ Λέσβον καὶ Τένεδον, οὐδὲ Ληθαῖον τὸν παρὰ Μαγνησίαν ῥέοντα, οὐδὲ δὴ Μαρσύαν, τοὺς εἰς τὸν Μαίανδρον ἐκδιδόντας, ἐκεῖνον ὀνομάζων καὶ πρὸς τούτοις

Ῥῆσόν θ' Ἑπτάπορόν τε Κάρησόν τε Ῥοδίον τε,

καὶ τοὺς ἄλλους, ὧν οἱ πλείους ὀχετῶν οὐκ εἰσι μείζους. πολλάς τε χώρας ὀνομάζων καὶ πόλεις

[1] Before *τῶν τοιούτων* Meineke inserts *τούτων καί*!
[2] *τὸν λόγον* seems to be an interpolation; Meineke ejects.
[3] *οὗ*, the editors, for *οὐ*.
[4] *καὶ Σάμον*, ejected by Corais and later editors on the

adventitiousness is much in evidence both in one's discourse and in one's actions. From all these facts it is clear that every man who judges from the poet's failure to mention anything that he is ignorant of that thing uses faulty evidence. And it is necessary to set forth several examples to prove that it is faulty, for many use such evidence to a great extent. We must therefore rebuke them when they bring forward such evidences, even though in so doing I shall be repeating previous argument.[1] For example, in the case of rivers, if anyone should say that the poet is ignorant of some river because he does not name it, I shall say that his argument is silly, because the poet does not even name the Meles River, which flows past Smyrna, the city which by most writers is called his birth-place, although he names the Hermus and Hyllus Rivers; neither does he name the Pactolus River, which flows into the same channel as these two rivers and rises in Tmolus, a mountain which he mentions;[2] neither does he mention Smyrna itself, nor the rest of the Ionian cities; nor the most of the Aeolian cities, though he mentions Miletus and Samos and Lesbos and Tenedos; nor yet the Lethaeus River, which flows past Magnesia, nor the Marsyas River, which rivers empty into the Maeander, which last he mentions by name, as also "the Rhesus and Heptaporus and Caresus and Rhodius,"[3] and the rest, most of which are no more than small streams. And when he names both many

[1] 12. 3. 26. [2] *Iliad* 2. 866 and 21. 835.
[3] *Iliad* 12. 20

ground that the Ionian Samos is nowhere specifically mentioned by Homer (see 10. 2. 17).

τοτὲ μὲν καὶ τοὺς ποταμοὺς καὶ ὄρη συγκαταλέγει, τοτὲ δ' οὔ· τοὺς γοῦν κατὰ τὴν Αἰτωλίαν καὶ τὴν Ἀττικὴν οὐ λέγει, οὐδ' ἄλλους πλείους· ἔτι[1] καὶ τῶν πόρρω μεμνημένος τῶν ἐγγὺς σφόδρα οὐ μέμνηται, οὐ δήπου ἀγνοῶν αὐτούς, γνωρίμους τοῖς ἄλλοις ὄντας· οὐδὲ δὴ τοὺς ἐγγὺς ἐπίσης, ὧν τοὺς μὲν ὀνομάζει, τοὺς δὲ οὔ, οἷον Λυκίους μὲν καὶ Σολύμους, Μιλύας δ' οὔ, οὐδὲ Παμφύλους οὐδὲ Πισίδας· καὶ Παφλαγόνας μὲν καὶ Φρύγας καὶ Μυσούς, Μαριανδυνοὺς δ' οὔ, οὐδὲ Θυνοὺς οὐδὲ Βιθυνοὺς οὐδὲ Βέβρυκας· Ἀμαζόνων τε μέμνηται, Λευκοσύρων δ' οὔ, οὐδὲ Σύρων οὐδὲ Καππαδόκων οὐδὲ Λυκαόνων, Φοίνικας καὶ Αἰγυπτίους καὶ Αἰθίοπας θρυλῶν· καὶ Ἀλήιον μὲν πεδίον λέγει καὶ Ἀρίμους, τὸ δὲ ἔθνος, ἐν ᾧ ταῦτα, σιγᾷ. ὁ μὲν δὴ τοιοῦτος ἔλεγχος ψευδής ἐστιν, ὁ δ' ἀληθής, ὅταν δείκνυται ψεῦδος λεγόμενόν τι. ἀλλ' οὐδ' ἐν τῷ τοιούτῳ κατορθῶν ἐδείχθη, ὅτε[2] γε ἐθάρρησε πλάσματα λέγειν τοὺς ἀγαυοὺς Ἱππημολγοὺς καὶ[3] Γαλακτοφάγους. τοσαῦτα καὶ πρὸς Ἀπολλόδωρον· ἐπάνειμι δὲ ἐπὶ τὴν ἑξῆς περιήγησιν.

C 555

28. Ὑπὲρ μὲν δὴ τῶν περὶ Φαρνακίαν καὶ Τραπεζοῦντα τόπων οἱ Τιβαρηνοὶ καὶ Χαλδαῖοι μέχρι τῆς μικρᾶς Ἀρμενίας εἰσίν. αὕτη δ' ἐστὶν εὐδαίμων ἱκανῶς χώρα· δυνάσται δ' αὐτὴν κατεῖχον ἀεί, καθάπερ τὴν Σωφηνήν, τοτὲ μὲν φίλοι

[1] ἔτι, the later editors, for ἐπεί MSS., except *lm*, which omit the word.
[2] ὅτε, Groskurd, for οὔτε; so the later editors.
[3] καί, added by *i*; so the editors.

[1] *Iliad* 2. 783.

countries and cities, he sometimes names with them the rivers and mountains, but sometimes he does not. At any rate, he does not mention the rivers in Aetolia or Attica, nor in several other countries. Besides, if he mentions rivers far away and yet does not mention those that are very near, it is surely not because he was ignorant of them, since they were known to all others. Nor yet, surely, was he ignorant of peoples that were equally near, some of which he names and some not; for example he names the Lycians and the Solymi, but not the Milyae; nor yet the Pamphylians or Pisidians; and though he names the Paphlagonians, Phrygians, and Mysians, he does not name Mariandynians or Thynians or Bithynians or Bebryces; and he mentions the Amazons, but not the White Syrians or Syrians, or Cappadocians, or Lycaonians, though he repeatedly mentions the Phoenicians and the Egyptians and the Ethiopians. And although he mentions the Aleian plain and the Arimi,[1] he is silent as to the tribe to which both belong. Such a test of the poet, therefore, is false; but the test is true only when it is shown that some false statement is made by him. But Apollodorus has not been proved correct in this case either, I mean when he was bold enough to say that the "proud Hippemolgi" and "Galactophagi" were fabrications of the poet. So much for Apollodorus. I now return to the part of my description that comes next in order.

28. Above the region of Pharnacia and Trapezus are the Tibareni and the Chaldaei, whose country extends to Lesser Armenia. This country is fairly fertile. Lesser Armenia, like Sophenê, was always in the possession of potentates, who at times were

τοῖς ἄλλοις Ἀρμενίοις ὄντες, τοτὲ δὲ ἰδιοπραγοῦντες· ὑπηκόους δ' εἶχον καὶ τοὺς Χαλδαίους καὶ Τιβαρηνούς, ὥστε μέχρι Τραπεζοῦντος καὶ Φαρνακίας διατείνειν τὴν ἀρχὴν αὐτῶν. αὐξηθεὶς δὲ Μιθριδάτης ὁ Εὐπάτωρ καὶ τῆς Κολχίδος κατέστη κύριος καὶ τούτων ἁπάντων, Ἀντιπάτρου τοῦ Σίσιδος παραχωρήσαντος αὐτῷ. ἐπεμελήθη δὲ οὕτω τῶν τόπων τούτων, ὥστε πέντε καὶ ἑβδομήκοντα φρούρια ἐν αὐτοῖς κατεσκευάσατο, οἷσπερ τὴν πλείστην γάζαν ἐνεχείρισε. τούτων δ' ἦν ἀξιολογώτατα ταῦτα· Ὕδαρα καὶ Βασγοιδάριζα καὶ Σινορία, ἐπιπεφυκὸς τοῖς ὁρίοις τῆς μεγάλης Ἀρμενίας χωρίον, διόπερ Θεοφάνης Συνορίαν παρωνόμασεν. ἡ γὰρ τοῦ Παρυάδρου πᾶσα ὀρεινὴ τοιαύτας ἐπιτηδειότητας ἔχει πολλάς, εὔυδρός τε οὖσα καὶ ὑλώδης καὶ ἀποτόμοις φάραγξι καὶ κρημνοῖς διειλημμένη πολλαχόθεν· ἐτετείχιστο γοῦν ἐνταῦθα τὰ πλεῖστα τῶν γαζοφυλακίων, καὶ δὴ καὶ τὸ τελευταῖον εἰς ταύτας κατέφυγε τὰς ἐσχατιὰς τῆς Ποντικῆς βασιλείας ὁ Μιθριδάτης, ἐπιόντος Πομπηίου, καὶ τῆς Ἀκιλισηνῆς[1] κατὰ Δάστειρα εὔυδρον ὄρος καταλαβόμενος (πλησίον δ' ἦν καὶ ὁ Εὐφράτης ὁ διορίζων τὴν Ἀκιλισηνὴν ἀπὸ τῆς μικρᾶς Ἀρμενίας) διέτριψε[2] τέως, ἕως πολιορκούμενος ἠναγκάσθη φυγεῖν διὰ τῶν ὀρῶν εἰς Κολχίδα, κἀκεῖθεν εἰς Βόσπορον. Πομπήιος δὲ περὶ τὸν τόπον τοῦτον πόλιν ἔκτισεν ἐν τῇ μικρᾷ Ἀρμενίᾳ Νικόπολιν, ἢ[3] καὶ νῦν συμμένει καὶ οἰκεῖται καλῶς.

[1] Ἀκιλισηνῆς *xz*, Ἀγγολισηνῆς other MSS.
[2] τε, before τέως, omitted by *x*; so Corais and Meineke.

friendly to the other Armenians and at times minded their own affairs. They held as subjects the Chaldaei and the Tibareni, and therefore their empire extended to Trapezus and Pharnacia. But when Mithridates Eupator had increased in power, he established himself as master, not only of Colchis, but also of all these places, these having been ceded to him by Antipater, the son of Sisis. And he cared so much for these places that he built seventy-five strongholds in them and therein deposited most of his treasures. The most notable of these strongholds were these: Hydara and Basgoedariza and Sinoria; Sinoria was close to the borders of Greater Armenia, and this is why Theophanes changed its spelling to Synoria.[1] For as a whole the mountainous range of the Paryadres has numerous suitable places for such strongholds, since it is well-watered and woody, and is in many places marked by sheer ravines and cliffs; at any rate, it was here that most of his fortified treasuries were built; and at last, in fact, Mithridates fled for refuge into these farthermost parts of the kingdom of Pontus, when Pompey invaded the country, and having seized a well-watered mountain near Dasteira in Acilisenê (near by, also, was the Euphrates, which separates Acilisenê from Lesser Armenia), he stayed there until he was besieged and forced to flee across the mountains into Colchis and from there to the Bosporus. Near this place, in Lesser Armenia, Pompey built a city, Nicopolis,[2] which endures even to this day and is well peopled.

[1] "Synoria" means "border-land."
[2] "Victory-city."

[3] ἢ, Kramer inserts; so the later editors.

29. Τὴν μὲν οὖν μικρὰν Ἀρμενίαν ἄλλοτ' ἄλλων ἐχόντων, ὡς ἐβούλοντο Ῥωμαῖοι, τὸ τελευταῖον εἶχεν ὁ Ἀρχέλαος. τοὺς δὲ Τιβαρηνοὺς καὶ Χαλδαίους μέχρι Κολχίδος καὶ Φαρνακίας καὶ Τραπεζοῦντος ἔχει Πυθοδωρίς, γυνὴ σώφρων καὶ δυνατὴ προΐστασθαι πραγμάτων. ἔστι δὲ θυγάτηρ Πυθοδώρου τοῦ Τραλλιανοῦ, γυνὴ δ' ἐγένετο Πολέμωνος καὶ συνεβασίλευσεν ἐκείνῳ χρόνον τινά, εἶτα διεδέξατο τὴν ἀρχήν, τελευτήσαντος ἐν τοῖς Ἀσπουργιανοῖς[1] καλουμένοις τῶν περὶ τὴν Σινδικὴν βαρβάρων· δυεῖν δ' ἐκ τοῦ Πολέμωνος ὄντων υἱῶν καὶ θυγατρός, ἡ μὲν ἐδόθη Κότυϊ τῷ Σαπαίῳ, δολοφονηθέντος δὲ ἐχήρευσε, παῖδας ἔχουσα ἐξ αὐτοῦ· δυναστεύει δ' ὁ πρεσβύτατος αὐτῶν· τῶν δὲ τῆς Πυθοδωρίδος υἱῶν ὁ μὲν ἰδιώτης συνδιῴκει τῇ μητρὶ τὴν ἀρχήν, ὁ δὲ νεωστὶ καθέσταται[2] τῆς μεγάλης Ἀρμενίας βασιλεύς. αὐτὴ δὲ συνῴκησεν Ἀρχελάῳ καὶ συνέμεινεν ἐκείνῳ μέχρι τέλους, νῦν δὲ χηρεύει, τά τε λεχθέντα ἔχουσα χωρία καὶ ἄλλα ἐκείνων χαριέστερα, περὶ ὧν ἐφεξῆς ἐροῦμεν.

C 556

30. Τῇ γὰρ Φαρνακίᾳ συνεχής ἐστιν ἡ Σιδηνὴ καὶ ἡ Θεμίσκυρα. τούτων δ' ἡ Φανάροια ὑπέρκειται, μέρος ἔχουσα τοῦ Πόντου τὸ κράτιστον· καὶ γὰρ ἐλαιόφυτός ἐστι καὶ εὔοινος καὶ τὰς ἄλλας ἔχει πάσας ἀρετάς. ἐκ μὲν τῶν ἑῴων μερῶν

[1] Ἀσπουργιανοῖς, Xylander, for Ἀπουργιανοῖς; so the later editors.

[2] καθέσταται, Corais, for καθίσταται; so the later editors.

---

[1] Cf. 14. 1. 42. [2] King of Odrysae (Book VII, Frag. 47).

[3] In A.D. 19 by his uncle, Rhescuporis, king of the Bosporus.

29. Now as for Lesser Armenia, it was ruled by different persons at different times, according to the will of the Romans, and finally by Archelaüs. But the Tibareni and Chaldaei, extending as far as Colchis, and Pharnacia and Trapezus are ruled by Pythodoris, a woman who is wise and qualified to preside over affairs of state. She is the daughter of Pythodorus of Tralles. She became the wife of Polemon and reigned along with him for a time, and then, when he died[1] in the country of the Aspurgiani, as they are called, one of the barbarian tribes round Sindicê, she succeeded to the rulership. She had two sons and a daughter by Polemon. Her daughter was married to Cotys the Sapaean,[2] but he was treacherously slain,[3] and she lived in widowhood, because she had children by him; and the eldest of these is now in power.[4] As for the sons of Pythodoris, one of them[5] as a private citizen is assisting his mother in the administration of her empire, whereas the other[6] has recently been established as king of Greater Armenia. She herself married Archelaüs and remained with him to the end;[7] but she is living in widowhood now, and is in possession not only of the places above mentioned, but also of others still more charming, which I shall describe next.

30. Sidenê and Themiscyra are contiguous to Pharnacia. And above these lies Phanaroea, which has the best portion of Pontus, for it is planted with olive trees, abounds in wine, and has all the other goodly attributes a country can have. On its eastern

[4] The king of Thrace. [5] Polemon II.
[6] Zenon. [7] He died in A.D. 17.

προβεβλημένη τὸν Παρυάδρην, παράλληλον αὐτῇ κατὰ μῆκος, ἐκ δὲ τῶν πρὸς δύσιν τὸν Λίθρον καὶ τὸν Ὄφλιμον. ἔστι δ' αὐλὼν καὶ μῆκος ἔχων ἀξιόλογον καὶ πλάτος, διαρρεῖ δ' αὐτὴν ἐκ μὲν τῆς Ἀρμενίας ὁ Λύκος, ἐκ δὲ τῶν περὶ Ἀμάσειαν στενῶν ο Ἶρις· συμβάλλουσι δ' ἀμφότεροι κατὰ μέσον που τὸν αὐλῶνα, ἐπὶ τῇ συμβολῇ δ' ἵδρυται πόλις, ἣν ὁ μὲν πρῶτος ὑποβεβλημένος Εὐπατορίαν ἀφ' αὑτοῦ προσηγόρευσε, Πομπήιος δ' ἡμιτελῆ καταλαβών, προσθεὶς χώραν, καὶ οἰκήτορας, Μαγνόπολιν προσεῖπεν. αὕτη μὲν οὖν ἐν μέσῳ κεῖται τῷ πεδίῳ, πρὸς αὐτῇ δὲ τῇ παρωρείᾳ τοῦ Παρυάδρου Κάβειρα ἵδρυται, σταδίοις ἑκατὸν[1] καὶ πεντήκοντά που νοτιωτέρα τῆς Μαγνοπόλεως, ὅσον καὶ Ἀμάσεια δυσμικωτέρα αὐτῆς ἐστίν· ἐν δὲ τοῖς Καβείροις τὰ βασίλεια Μιθριδάτου κατεσκεύαστο καὶ ὁ ὑδραλέτης, καὶ τὰ ζωγρεῖα καὶ αἱ πλησίον θῆραι καὶ τὰ μέταλλα.

31. Ἐνταῦθα δὲ καὶ τὸ Καινὸν χωρίον προσαγορευθέν, ἐρυμνὴ καὶ ἀπότομος πέτρα, διέχουσα τῶν Καβείρων ἔλαττον ἢ διακοσίους σταδίους· ἔχει δ' ἐπὶ τῇ κορυφῇ πηγὴν ἀναβάλλουσαν πολὺ ὕδωρ, περὶ[2] τε τῇ ῥίζῃ ποταμὸν καὶ φάραγγα βαθεῖαν. τὸ δ' ὕψος ἐξαίσιον τῆς πέτρας ἐστὶ ἄνω[3] τοῦ αὐχένος, ὥστ' ἀπολιόρκητός ἐστι, τετείχισται δὲ θαυμαστῶς, πλὴν ὅσον οἱ Ῥωμαῖοι κατέσπασαν· οὕτω δ' ἐστὶν ἅπασα ἡ κύκλῳ

[1] For ἑκατόν (ρ'), C. Müller (*Ind. Var. Lect.*, p. 1021) conj. σ' (200).
[2] περί, Meineke emends to πρός.
[3] ἄνω, Jones inserts, from proposals of Groskurd.

side it is protected by the Paryadres Mountain, in its length lying parallel to that mountain; and on its western side by the Lithrus and Ophlimus Mountains. It forms a valley of considerable breadth as well as length; and it is traversed by the Lycus River, which flows from Armenia, and by the Iris, which flows from the narrow passes near Amaseia. The two rivers meet at about the middle of the valley; and at their junction is situated a city which the first man who subjugated it[1] called Eupatoria after his own name, but Pompey found it only half-finished and added to it territory and settlers, and called it Magnopolis. Now this city is situated in the middle of the plain, but Cabeira is situated close to the very foothills of the Paryadres Mountains about one hundred and fifty stadia farther south than Magnopolis, the same distance that Amaseia is farther west than Magnopolis. It was at Cabeira that the palace of Mithridates was built, and also the water-mill; and here were the zoological gardens, and, near by, the hunting grounds, and the mines.

31. Here, also, is Kainon Chorion,[2] as it is called, a rock that is sheer and fortified by nature, being less than two hundred stadia distant from Cabeira. It has on its summit a spring that sends forth much water, and at its foot a river and a deep ravine. The height of the rock above the neck[3] is immense, so that it is impregnable; and it is enclosed by remarkable walls, except the part where they have been pulled down by the Romans. And the whole country around is so overgrown with forests, and so

[1] *i.e.* Mithridates Eupator. [2] "New Place."

[3] *i.e.* the "neck," or ridge, which forms the approach to rock (cp. the use of the word in § 39 following).

κατάδρυμος καὶ ὀρεινὴ καὶ ἄνυδρος, ὥστ᾽ ἐντὸς ἑκατὸν καὶ εἴκοσι σταδίων μὴ εἶναι δυνατὸν στρατοπεδεύσασθαι. ἐνταῦθα μὲν ἦν τῷ Μιθριδάτῃ τὰ τιμιώτατα τῶν κειμηλίων, ἃ νῦν ἐν τῷ Καπιτωλίῳ κεῖται, Πομπηίου ἀναθέντος. ταύτην δὴ τὴν χώραν ἔχει πᾶσαν ἡ Πυθοδωρίς, προσεχῆ οὖσαν τῇ βαρβάρῳ τῇ ὑπ᾽ αὐτῆς κατεχομένῃ, καὶ τὴν Ζηλῖτιν καὶ Μεγαλοπολῖτιν. τὰ δὲ Κάβειρα, Πομπηίου σκευάσαντος εἰς πόλιν καὶ καλέσαντος Διόσπολιν,[1] ἐκείνη προσκατεσκεύασε καὶ Σεβαστὴν μετωνόμασε, βασιλείῳ τε τῇ πόλει χρῆται. ἔχει δὲ καὶ τὸ ἱερὸν Μηνὸς Φαρνάκου καλούμενον,[2] τὴν Ἀμερίαν κωμόπολιν πολλοὺς ἱεροδούλους ἔχουσαν καὶ χώραν ἱεράν, ἣν ὁ ἱερώμενος ἀεὶ καρποῦται. ἐτίμησαν δ᾽ οἱ βασιλεῖς τὸ ἱερὸν τοῦτο οὕτως εἰς ὑπερβολήν, ὥστε τὸν βασιλικὸν καλούμενον ὅρκον τοῦτον[3] ἀπέφηναν Τύχην βασιλέως καὶ Μῆνα Φαρνάκου· ἔστι δὲ καὶ τοῦτο τῆς Σελήνης τὸ ἱερόν, καθάπερ τὸ ἐν Ἀλβανοῖς καὶ τὰ ἐν Φρυγίᾳ, τό τε τοῦ Μηνὸς ἐν τῷ ὁμωνύμῳ τόπῳ καὶ τὸ τοῦ Ἀσκαίου τὸ

[1] Διόσπολιν *i*, Διόπολιν other MSS.
[2] *ix* and Corais insert καί before τὴν Ἀμερίαν.
[3] C and Corais read τοῖτο instead of τοῦτον.

---

[1] "City of Zeus." [2] In Latin, "Augusta."
[3] *i.e.* established by Pharnaces.
[4] Professor David M. Robinson says (in a private communication): "I think that Μὴν Φαρνάκου equals Τύχη βασιλέως, since Μήν equals Τύχη on coins of Antioch."
[5] Goddess of the "Moon." [6] See 11. 4. 7 and 12. 8. 20.
[7] Sir William Ramsay (*Journal of Hellenic Studies* 1918,

mountainous and waterless, that it is impossible for an enemy to encamp within one hundred and twenty stadia. Here it was that the most precious of the treasures of Mithridates were kept, which are now stored in the Capitolium, where they were dedicated by Pompey. Pythodoris possesses the whole of this country, which is adjacent to the barbarian country occupied by her, and also Zelitis and Megalopolitis. As for Cabeira, which by Pompey had been built into a city and called Diospolis,[1] Pythodoris further adorned it and changed its name to Sebastê;[2] and she uses the city as a royal residence. It has also the temple of Mên of Pharnaces,[3] as it is called,—the village-city Ameria, which has many temple-servants, and also a sacred territory, the fruit of which is always reaped by the ordained priest. And the kings revered this temple so exceedingly that they proclaimed the "royal" oath as follows: "By the Fortune of the king and by Mên of Pharnaces."[4] And this is also the temple of Selenê,[5] like that among the Albanians and those in Phrygia,[6] I mean that of Mên in the place of the same name and that of Mên[7] Ascaeus[8] near the Antiocheia that is near

vol. 38, pp. 148 ff.) argues that "Mên" is a grecized form for the Anatolian "Manes," the native god of the land of Ouramma; and "Manes Ourammoas was Hellenized as Zeus Ouruda-menos or Euruda-menos." See also M. Rostovtzeff, *Social and Economic History of the Roman Empire*, p. 238, and Daremberg et Saglio, *Dict. Antiq.*, *s.v.* "Lunus."

[8] "Ascaënus" (Ἀσκαηνός) is the regular spelling of the word, the spelling found in hundreds of inscriptions, whereas Ascaeus (Ἀσκαῖος) has been found in only two inscriptions, according to Professor David M. Robinson. On this temple, see Sir W. M. Ramsay's "Excavations at Pisidian Antioch in 1912," *The Athenaeum*, London, March 8, Aug. 31, and Sept. 7, 1913.

πρὸς Ἀντιοχείᾳ τῇ πρὸς Πισιδίᾳ[1] καὶ τὸ ἐν τῇ χώρᾳ τῶν Ἀντιοχέων.

32. Ὑπὲρ δὲ τῆς Φαναροίας ἐστὶ τὰ[2] Κόμανα τὰ ἐν τῷ Πόντῳ, ὁμώνυμα τοῖς ἐν τῇ μεγάλῃ Καππαδοκίᾳ καὶ τῇ αὐτῇ θεῷ καθιερωμένα, ἀφιδρυθέντα ἐκεῖθεν, σχεδὸν δέ τι καὶ τῇ ἀγωγῇ παραπλησίᾳ κεχρημένα τῶν τε ἱερουργιῶν καὶ τῶν θεοφοριῶν καὶ τῆς περὶ τοὺς ἱερέας τιμῆς, καὶ μάλιστα ἐπὶ τῶν πρὸ τοῦ βασιλέων· ἡνίκα δὶς τοῦ ἔτους κατὰ τὰς ἐξόδους λεγομένας τῆς θεοῦ διάδημα φορῶν ἐτύγχανεν ὁ ἱερεύς, καὶ ἦν δεύτερος κατὰ τιμὴν μετὰ τὸν βασιλέα.

33. Ἐμνήσθημεν δὲ πρότερον Δορυλάου τε τοῦ τακτικοῦ, ὃς ἦν πρόπαππος τῆς μητρὸς ἡμῶν, καὶ ἄλλου Δορυλάου, ὃς ἦν ἐκείνου ἀδελφιδοῦς, υἱὸς δὲ Φιλεταίρου, καὶ διότι ἐκεῖνος τῶν ἄλλων τιμῶν παρὰ τοῦ Εὐπάτορος τῶν μεγίστων τυχὼν καὶ δὴ καὶ τῆς ἐν Κομάνοις ἱερωσύνης ἐφωράθη τὴν βασιλείαν ἀφιστὰς Ῥωμαίοις· καταλυθέντος δ' ἐκείνου, συνδιεβλήθη καὶ τὸ γένος. ὀψὲ δὲ Μοαφέρνης, ὁ θεῖος τῆς μητρὸς ἡμῶν, εἰς ἐπιφάνειαν ἦλθεν ἤδη πρὸς καταλύσει τῆς βασι-

[1] Πισιδίᾳ (as in 12. 8. 14) *i*, instead of Πισιδίαν; so Corais and Meineke.

[2] τε after τά, omitted by *x* and later editors.

---

[1] Note that Strabo, both here and in 12. 8. 14, refers to this Antioch as "the Antioch near Pisidia," not as "Pisidian Antioch," the appellation now in common use. Neither does Artemidorus (lived about 100 B.C.), as quoted by Strabo (12. 7. 2), name Antioch in his list of Pisidian cities.

[2] *i.e.* in the territory of which Antiocheia was capital. At this "remote old Anatolian Sanctuary" (not to be con-

Pisidia[1] and that of Mên in the country of the Antiocheians.[2]

32. Above Phanaroea is the Pontic Comana, which bears the same name as the city in Greater Cappadocia, having been consecrated to the same goddess and copied after that city; and I might almost say that the courses which they have followed in their sacrifices, in their divine obsessions, and in their reverence for their priests, are about the same, and particularly in the times of the kings who reigned before this, I mean in the times when twice a year, during the "exoduses"[3] of the goddess, as they are called, the priest wore a diadem[4] and ranked second in honour after the king.

33. Heretofore[5] I have mentioned Dorylaüs the tactician, who was my mother's great grandfather, and also a second Dorylaüs, who was the nephew of the former and the son of Philetaerus, saying that, although he had received all the greatest honours from Eupator and in particular the priesthood of Comana, he was caught trying to cause the kingdom to revolt to the Romans; and when he was overthrown, the family was cast into disrepute along with him. But long afterwards Moaphernes, my mother's uncle, came into distinction just before

fused with that of Mên Ascaeus near Antiocheia), "Strabo does not say what epithet Mên bore" (Ramsay in first article above cited). That of Mên Ascaeus on Mt. Kara Kuyu has been excavated by Ramsay and Calder (*J.H.S.* 1912, pp. 111–150, *British School Annual* 1911–12, XVIII, 37 ff., *J.R.S.* 1918, pp. 107–145). The other, not yet found, "may have been," according to Professor Robinson, "at Saghir."

[3] *i.e.* "solemn processions."

[4] As a symbol of regal dignity.

[5] 10. 4. 10.

λείας, καὶ πάλιν τῷ βασιλεῖ συνητύχησαν καὶ αὐτὸς καὶ οἱ ἐκείνου φίλοι, πλὴν εἴ τινες ἔφθησαν προαποστάντες αὐτοῦ, καθάπερ ὁ πάππος ἡμῶν ὁ πρὸς[1] αὐτῆς, ὃς ἰδὼν τὰ τοῦ βασιλέως κακῶς φερόμενα ἐν τῷ πρὸς Λεύκολλον πολέμῳ, καὶ ἅμα ἠλλοτριωμένος αὐτοῦ δι' ὀργήν, ὅτι ἀνεψιὸν αὐτοῦ Τίβιον καὶ υἱὸν ἐκείνου Θεόφιλον ἐτύγχανεν ἀπεκτονὼς νεωστί, ὥρμησε τιμωρεῖν ἐκείνοις τε καὶ ἑαυτῷ, καὶ λαβὼν παρὰ τοῦ Λευκόλλου C 558 πίστεις ἀφίστησιν αὐτῷ πεντεκαίδεκα φρούρια, καὶ ἐπαγγελίαι μὲν ἐγένοντο ἀντὶ τούτων μεγάλαι, ἐπελθὼν δὲ Πομπήιος ὁ διαδεξάμενος τὸν πόλεμον πάντας τοὺς ἐκείνῳ τι χαρισαμένους ἐχθροὺς ὑπέλαβε διὰ τὴν γενομένην αὐτῷ πρὸς ἐκεῖνον ἀπέχθειαν, διαπολεμήσας δὲ καὶ ἐπανελθὼν οἴκαδε ἐξενίκησεν, ὥστε τὰς τιμάς, ἃς ὑπέσχετο ὁ Λεύκολλος τῶν Ποντικῶν τισί, μὴ κυρῶσαι τὴν σύγκλητον· ἄδικον γὰρ εἶναι, κατορθώσαντος ἑτέρου τὸν πόλεμον, τὰ βραβεῖα ἐπ' ἄλλῳ γενέσθαι καὶ τὴν τῶν ἀριστείων διανομήν.

34. Ἐπὶ μὲν οὖν τῶν βασιλέων[2] οὕτω τὰ Κόμανα διῳκεῖτο, ὡς εἴρηται, παραλαβὼν δὲ Πομπήιος τὴν ἐξουσίαν Ἀρχέλαον ἐπέστησεν ἱερέα καὶ προσώρισεν αὐτῷ χώραν δίσχοινον κύκλῳ (τοῦτο δ' ἐστὶν ἑξήκοντα στάδιοι) πρὸς τῇ ἱερᾷ, προστάξας τοῖς ἐνοικοῦσι πειθαρχεῖν αὐτῷ· τούτων μὲν οὖν ἡγεμὼν ἦν καὶ τῶν τὴν πόλιν οἰκούντων ἱεροδούλων κύριος πλὴν τοῦ πιπράσκειν· ἦσαν δὲ

[1] πατρός, after πρός, omitted by editors.
[2] βασιλέων, Casaubon, for βασιλειῶν; so the later editors.

the dissolution of the kingdom, and again they were unfortunate along with the king, both Moaphernes and his relatives, except some who revolted from the king beforehand, as did my maternal grandfather, who, seeing that the cause of the king was going badly in the war with Leucullus, and at the same time being alienated from him out of wrath at his recently having put to death his cousin Tibius and Tibius' son Theophilus, set out to avenge both them and himself; and, taking pledges from Leucullus, he caused fifteen garrisons to revolt to him; and although great promises were made in return for these services, yet, when Pompey, who succeeded Leucullus in the conduct of the war, went over, he took for enemies all who had in any way favoured Leucullus, because of the hatred which had arisen between himself and Leucullus; and when he finished the war and returned home, he won so completely that the Senate would not ratify those honours which Leucullus had promised to certain of the people of Pontus, for, he said, it was unjust, when one man had brought the war to a successful issue, that the prizes and the distribution of the rewards should be placed in the hands of another man.

34. Now in the times of the kings the affairs of Comana were administered in the manner already described, but when Pompey took over the authority, he appointed Archelaüs priest and included within his boundaries, in addition to the sacred land, a territory of two schoeni (that is, sixty stadia) in circuit and ordered the inhabitants to obey his rule. Now he was governor of these, and also master of the temple-servants who lived in the city, except that he was not empowered to sell them. And even

οὐκ ἐλάττους οὐδ᾽ ἐνταῦθα τῶν ἑξακισχιλίων. ἦν δ᾽ οὗτος Ἀρχέλαος υἱὸς μὲν τοῦ ὑπὸ Σύλλα καὶ τῆς συγκλήτου τιμηθέντος, φίλος δὲ Γαβινίου τῶν ὑπατικῶν τινός. ἐκείνου δὲ πεμφθέντος εἰς Συρίαν ἧκε καὶ αὐτὸς ἐπ᾽ ἐλπίδι τοῦ κοινωνήσειν αὐτῷ παρασκευαζομένῳ πρὸς τὸν Παρθικὸν πόλεμον, οὐκ ἐπιτρεπούσης δὲ τῆς συγκλήτου, ταύτην ἀφεὶς τὴν ἐλπίδα, ἄλλην εὕρετο[1] μείζω. ἐτύγχανε γὰρ Πτολεμαῖος ὁ τῆς Κλεοπάτρας πατὴρ ὑπὸ τῶν Αἰγυπτίων ἐκβεβλημένος, θυγάτηρ δ᾽ αὐτοῦ κατεῖχε τὴν βασιλείαν, ἀδελφὴ πρεσβυτέρα τῆς Κλεοπάτρας· ταύτῃ ζητουμένου ἀνδρὸς βασιλικοῦ γένους, ἐνεχείρισεν ἑαυτὸν τοῖς συμπράττουσι, προσποιησάμενος Μιθριδάτου τοῦ Εὐπάτορος υἱὸς εἶναι,[2] καὶ παραδεχθεὶς ἐβασίλευσεν ἓξ μῆνας. τοῦτον μὲν οὖν ὁ Γαβίνιος ἀνεῖλεν ἐν παρατάξει, κατάγων τὸν Πτολεμαῖον.

35. Υἱὸς δ᾽ αὐτοῦ τὴν ἱερωσύνην παρέλαβεν· εἶθ᾽ ὕστερον Λυκομήδης, ᾧ καὶ τετράσχοινος ἄλλη προσετέθη· καταλυθέντος δὲ καὶ τούτου, νῦν ἔχει Δύτευτος, υἱὸς Ἀδιατόριγος, ὃς δοκεῖ ταύτης τυγχάνειν τῆς τιμῆς παρὰ Καίσαρος τοῦ Σεβαστοῦ δι᾽ ἀρετήν. ὁ μὲν γὰρ Καῖσαρ, θριαμβεύσας τὸν Ἀδιατόριγα μετὰ παίδων καὶ γυναικός, ἔγνω ἀναιρεῖν μετὰ τοῦ πρεσβυτάτου τῶν παίδων (ἦν δὲ πρεσβύτατος οὗτος), τοῦ δὲ δευτέρου τῶν ἀδελφῶν αὐτοῦ φήσαντος εἶναι πρεσβυτάτου πρὸς τοὺς ἀπάγοντας[3] στρατιώτας, ἔρις ἦν ἀμφοτέροις

[1] C and Corais read εὕρετο instead of εὕρατο.
[2] εἶναι, after υἱός, Tzschucke inserts; so the later editors.
[3] ἀπάγοντας, Corais, for ἀνάγοντας; so the later editors.

[1] As well as in the Cappadocian Comana (12. 2. 3).

here[1] the temple-servants were no fewer in number than six thousand. This Archelaüs was the son of the Archelaüs who was honoured by Sulla and the Senate, and was also a friend of Gabinius,[2] a man of consular rank. When Gabinius was sent into Syria, Archelaüs himself also went there in the hope of sharing with him in his preparations for the Parthian War, but since the Senate would not permit him, he dismissed that hope and found another of greater importance. For it happened at that time that Ptolemaeus, the father of Cleopatra, had been banished by the Egyptians, and his daughter, elder sister of Cleopatra, was in possession of the kingdom; and since a husband of royal family was being sought for her, Archelaüs proffered himself to her agents, pretending that he was the son of Mithridates Eupator; and he was accepted, but he reigned only six months. Now this Archelaüs was slain by Gabinius in a pitched battle, when the latter was restoring Ptolemaeus to his kingdom.

35. But his son succeeded to the priesthood; and then later, Lycomedes, to whom was assigned an additional territory[3] of four hundred schoeni; but now that he has been deposed, the office is held by Dyteutus, son of Adiatorix, who is thought to have obtained the honour from Caesar Augustus because of his excellent qualities; for Caesar, after leading Adiatorix in triumph together with his wife and children, resolved to put him to death together with the eldest of his sons (for Dyteutus was the eldest), but when the second of the brothers told the soldiers who were leading them away to execution that he was the eldest, there was a contest between the two

[2] Consul 58 B.C.; in 57 B.C. went to Syria as proconsul.
[3] See § 34.

C 559 πολὺν χρόνον, ἕως οἱ γονεῖς ἔπεισαν τὸν Δύτευτον παραχωρῆσαι τῷ νεωτέρῳ τῆς νίκης· αὐτὸν γὰρ ἐν ἡλικίᾳ μᾶλλον ὄντα ἐπιτηδειότερον κηδεμόνα τῇ μητρὶ ἔσεσθαι καὶ τῷ λειπομένῳ ἀδελφῷ· οὕτω δὲ τὸν μὲν συναποθανεῖν τῷ πατρί, τοῦτον δὲ σωθῆναι καὶ τυχεῖν τῆς τιμῆς ταύτης. αἰσθόμενος γάρ, ὡς ἔοικε, Καῖσαρ ἤδη τῶν ἀνθρώπων ἀνῃρημένων ἠχθέσθη, καὶ τούς γε[1] σωζομένους εὐεργεσίας καὶ ἐπιμελείας ἀξίους ὑπέλαβε, δοὺς αὐτοῖς ταύτην τὴν τιμήν.

36. Τὰ μὲν οὖν Κόμανα εὐανδρεῖ καὶ ἔστιν ἐμπόριον τοῖς ἀπὸ τῆς Ἀρμενίας ἀξιόλογον, συνέρχονται δὲ κατὰ τὰς ἐξόδους τῆς θεοῦ πανταχόθεν ἔκ τε τῶν πόλεων καὶ τῆς χώρας ἄνδρες ὁμοῦ γυναιξὶν ἐπὶ τὴν ἑορτήν· καὶ ἄλλοι δὲ κατ' εὐχὴν ἀεί τινες ἐπιδημοῦσι, θυσίας ἐπιτελοῦντες τῇ θεῷ. καί εἰσιν ἁβροδίαιτοι οἱ ἐνοικοῦντες, καὶ οἰνόφυτα τὰ κτήματα αὐτῶν ἐστὶ πάντα, καὶ πλῆθος γυναικῶν τῶν ἐργαζομένων ἀπὸ τοῦ σώματος, ὧν αἱ πλείους εἰσὶν ἱεραί. τρόπον γὰρ δή τινα μικρὰ Κόρινθός ἐστιν ἡ πόλις· καὶ γὰρ ἐκεῖ διὰ τὸ πλῆθος τῶν ἑταιρῶν,[2] αἳ τῆς Ἀφροδίτης ἦσαν ἱεραί, πολὺς ἦν ὁ ἐπιδημῶν καὶ ἐνεορτάζων τῷ τόπῳ· οἱ δ' ἐμπορικοὶ καὶ στρατιωτικοὶ τελέως ἐξανηλίσκοντο, ὥστ' ἐπ' αὐτῶν καὶ παροιμίαν ἐκπεσεῖν τοιαύτην·

οὐ παντὸς ἀνδρὸς εἰς Κόρινθόν ἐσθ' ὁ πλοῦς.

τὰ μὲν δὴ Κόμανα τοιαῦτα.

[1] γε, Corais, for δέ; so the later editors.
[2] *oz* read ἑταιρίδων instead of ἑταιρῶν; so Tzschucke and Corais.

for a long time, until the parents persuaded Dyteutus to yield the victory to the younger, for he, they said, being more advanced in age, would be a more suitable guardian for his mother and for the remaining brother. And thus, they say, the younger was put to death with his father, whereas the elder was saved and obtained the honour of the priesthood. For learning about this, as it seems, after the men had already been put to death, Caesar was grieved, and he regarded the survivors as worthy of his favour and care, giving them the honour in question.

36. Now Comana is a populous city and is a notable emporium for the people from Armenia; and at the times of the "exoduses"[1] of the goddess people assemble there from everywhere, from both the cities and the country, men together with women, to attend the festival. And there are certain others, also, who in accordance with a vow are always residing there, performing sacrifices in honour of the goddess. And the inhabitants live in luxury, and all their property is planted with vines; and there is a multitude of women who make gain from their persons, most of whom are dedicated to the goddess, for in a way the city is a lesser Corinth,[2] for there too, on account of the multitude of courtesans, who were sacred to Aphroditê, outsiders resorted in great numbers and kept holiday. And the merchants and soldiers who went there squandered all their money,[3] so that the following proverb arose in reference to them: "Not for every man is the voyage to Corinth." Such, then, is my account of Comana.

[1] See § 32 above, and the foot-note.
[2] See 8. 6. 20. [3] See 8. 6. 20.

37. Τὴν δὲ κύκλῳ πᾶσαν ἔχει Πυθοδωρίς, ἧς ἥ τε Φανάροιά ἐστι καὶ ἡ Ζηλῖτις καὶ ἡ Μεγαλοπολῖτις. περὶ μὲν Φαναροίας εἴρηται· ἡ δὲ Ζηλῖτις ἔχει πόλιν Ζῆλα ἐπὶ χώματι Σεμιράμιδος τετειχισμένην, ἔχουσαν τὸ ἱερὸν τῆς Ἀναΐτιδος, ἥνπερ καὶ οἱ Ἀρμένιοι σέβονται. αἱ μὲν οὖν ἱεροποιίαι μετὰ μείζονος ἁγιστείας ἐνταῦθα συντελοῦνται, καὶ τοὺς ὅρκους περὶ τῶν μεγίστων ἐνταῦθα Ποντικοὶ [1] πάντες ποιοῦνται· τὸ δὲ πλῆθος τῶν ἱεροδούλων καὶ αἱ τῶν ἱερέων τιμαὶ παρὰ μὲν τοῖς βασιλεῦσι τὸν αὐτὸν εἶχον τύπον, ὅνπερ προείπομεν, νυνὶ δὲ ἐπὶ τῇ Πυθοδωρίδι πάντ᾽ ἐστίν. ἐκάκωσαν δὲ πολλοὶ καὶ ἐμείωσαν τό τε πλῆθος τῶν ἱεροδούλων καὶ τὴν ἄλλην εὐπορίαν. ἐμειώθη δὲ καὶ ἡ παρακειμένη χώρα μερισθεῖσα εἰς πλείους δυναστείας, ἡ λεγομένη Ζηλῖτις (ἣ ἔχει πόλιν Ζῆλα ἐπὶ χώματι). τὸ παλαιὸν μὲν γὰρ οἱ βασιλεῖς οὐχ ὡς πόλιν, ἀλλ᾽ ὡς ἱερὸν διῴκουν τῶν Περσικῶν θεῶν τὰ Ζῆλα, καὶ ἦν ὁ ἱερεὺς κύριος τῶν πάντων· ᾠκεῖτο δ᾽ ὑπὸ τοῦ πλήθους τῶν ἱεροδούλων καὶ τοῦ ἱερέως, ὄντος ἐν περιουσίᾳ μεγάλῃ, καὶ τοῖς περὶ αὐτὸν οὐκ ὀλίγοις χώρα τε ὑπέκειτο ἱερὰ καὶ ἡ [2] τοῦ ἱερέως. Πομ- C 560 πήιος δὲ πολλὰς ἐπαρχίας προσώρισε τῷ τόπῳ καὶ πόλιν ὠνόμασε καὶ ταύτην καὶ τὴν Μεγαλόπολιν· συνθεὶς ταύτην τε εἰς ἓν τήν τε Κουλουπηνὴν καὶ τὴν Καμισηνήν, ὁμόρους οὔσας τῇ τε μικρᾷ Ἀρμενίᾳ καὶ τῇ Λαουιανσηνῇ, ἐχούσας ὀρυκτοὺς ἅλας καὶ ἔρυμα ἀρχαῖον τὰ Κάμισα, νῦν κατεσπασμένον· οἱ δὲ μετὰ ταῦτα ἡγεμόνες τῶν

[1] Ποντικοί, Corais, for Πολιτικοί; so the later editors.
[2] ἡ, Corais and Meineke emend to ἦν.

37. The whole of the country around is held by Pythodoris, to whom belong, not only Phanaroea, but also Zelitis and Megalopolitis. Concerning Phanaroea I have already spoken. As for Zelitis, it has a city Zela, fortified on a mound of Semiramis, with the temple of Anaïtis, who is also revered by the Armenians.[1] Now the sacred rites performed here are characterised by greater sanctity; and it is here that all the people of Pontus make their oaths concerning their matters of greatest importance. The large number of temple-servants and the honours of the priests were, in the time of the kings, of the same type as I have stated before, but at the present time everything is in the power of Pythodoris. Many persons had abused and reduced both the multitude of temple-servants and the rest of the resources of the temple. The adjacent territory, also, was reduced, having been divided into several domains—I mean Zelitis, as it is called (which has the city Zela on a mound); for in early times the kings governed Zela, not as a city, but as a sacred precinct of the Persian gods, and the priest was the master of the whole thing. It was inhabited by the multitude of temple-servants, and by the priest, who had an abundance of resources; and the sacred territory as well as that of the priest was subject to him and his numerous attendants.[2] Pompey added many provinces to the boundaries of Zelitis, and named Zela, as he did Megalopolis, a city, and he united the latter and Culupenê and Camisenê into one state; the latter two border on both Lesser Armenia and Laviansenê, and they contain rock-salt, and also an ancient fortress called Camisa, now in

[1] Cf. 11. 14. 16. [2] Cf. 12. 3. 31.

Ῥωμαίων τῶν δυεῖν πολιτευμάτων τούτων τὰ μὲν τοῖς Κομάνων ἱερεῦσι προσένειμαν, τὰ δὲ τῷ Ζήλων ἱερεῖ, τὰ δ' Ἀτεπόριγι,[1] δυνάστῃ τινὶ τοῦ τετραρχικοῦ γένους τῶν Γαλατῶν ἀνδρί· τελευτήσαντος δ' ἐκείνου, ταύτην μὲν τὴν μερίδα, οὐ πολλὴν οὖσαν, ὑπὸ Ῥωμαίοις εἶναι συμβαίνει καλουμένην ἐπαρχίαν (καὶ ἔστι σύστημα καθ' αὑτὸ τὸ πολίχνιον συνοικισάντων τὰ Κάρανα, ἀφ' οὗ καὶ ἡ χώρα Καρανῖτις λέγεται), τὰ δὲ λοιπὰ ἔχει Πυθοδωρὶς καὶ ὁ Δύτευτος.

38. Λείπεται δὲ τοῦ Πόντου τὰ[2] μεταξὺ ταύτης τε τῆς χώρας καὶ τῆς Ἀμισηνῶν καὶ Σινωπέων, πρός τε τὴν Καππαδοκίαν συντείνοντα καὶ Γαλάτας καὶ Παφλαγόνας. μετὰ μὲν οὖν τὴν Ἀμισηνῶν μέχρι τοῦ Ἅλυος ἡ Φαζημωνῖτίς ἐστιν, ἣν Πομπήιος Νεαπολῖτιν ὠνόμασε, κατὰ Φαζημῶνα κώμην πόλιν[3] ἀποδείξας τὴν κατοικίαν καὶ προσαγορεύσας Νεάπολιν. ταύτης δὲ τῆς χώρας τὸ μὲν προσάρκτιον πλευρὸν ἡ Γαζηλωνῖτις[4] συγκλείει καὶ ἡ τῶν Ἀμισηνῶν, τὸ δὲ ἑσπέριον ὁ Ἅλυς, τὸ δ' ἑῷον ἡ Φανάροια, τὸ δὲ λοιπὸν ἡ ἡμετέρα χώρα ἡ τῶν Ἀμασέων, πολὺ πασῶν πλείστη καὶ ἀρίστη. τὸ μὲν οὖν πρὸς τῇ Φαναροίᾳ μέρος τῆς Φαζημωνίτιδος λίμνη κατέχει πελαγία τὸ μέγεθος, ἡ Στιφάνη καλουμένη, πολύοψος καὶ κύκλῳ νομὰς ἀφθόνους ἔχουσα καὶ παντοδαπάς· ἐπίκειται δ' αὐτῇ φρούριον ἐρυμνόν,

[1] δ' Ἀτεπόριγι, Tzschucke, for δὲ τέποργι; so the later editors.

[2] τά, before μεταξύ, Casaubon inserts; so the later editors.

[3] πόλιν, Groskurd inserts; so Meineke.

[4] Γαζηλωνῖτις (as in 12. 3. 13), Groskurd, for Γαζιλωτός

ruins. The later Roman prefects assigned a portion of these two governments to the priests of Comana, a portion to the priest of Zela, and a portion to Ateporix, a dynast of the family of tetrarchs of Galatia; but now that Ateporix has died, this portion, which is not large, is subject to the Romans, being called a province (and this little state is is a political organisation of itself, the people having incorporated Carana into it, from which fact its country is called Caranitis), whereas the rest is held by Pythodoris and Dyteutus.

38. There remain to be described the parts of the Pontus which lie between this country and the countries of the Amisenians and Sinopeans, which latter extend towards Cappadocia and Galatia and Paphlagonia. Now after the territory of the Amisenians, and extending to the Halys River, is Phazemonitis, which Pompey named Neapolitis, proclaiming the settlement at the village Phazemon a city and calling it Neapolis.[1] The northern side of this country is bounded by Gazelonitis and the country of the Amisenians; the western by the Halys River; the eastern by Phanaroea; and the remaining side by my country, that of the Amaseians, which is by far the largest and best of all. Now the part of Phazemonitis towards Phanaroea is covered by a lake which is like a sea in size, is called Stephanê, abounds in fish, and has all round it abundant pastures of all kinds. On its shores lies a strong

[1] "New City."

Dhowz, Γαδιλωτός E, Ζηλῆτις x, Γαζηλωτός other MSS.; so Meineke.

ἔρημον νῦν, Ἰκίζαρι,[1] καὶ πλησίον βασίλειον κατεσκαμμένον·[2] ἡ δὲ λοιπὴ ψιλὴ τὸ πλέον καὶ σιτοφόρος χώρα. ὑπέρκειται δὲ τῆς τῶν Ἀμασέων τά τε θερμὰ ὕδατα τῶν Φαζημωνιτῶν, ὑγιεινὰ σφόδρα, καὶ τὸ Σαγύλιον ἐπὶ ὄρους ὀρθίου καὶ ὑψηλοῦ πρὸς ὀξεῖαν ἀνατείνοντος ἄκραν, ἔρυμα ἱδρυμένον ἔχον καὶ ὑδρεῖον δαψιλές, ὃ νῦν ὠλιγώρηται, τοῖς δὲ βασιλεῦσιν ἦν χρήσιμον εἰς πολλά. ἐνταῦθα δὲ ἑάλω καὶ διεφθάρη εἷς[3] τῶν Φαρνάκου τοῦ βασιλέως παίδων Ἀρσάκης, δυναστεύων καὶ νεωτερίζων, ἐπιτρέψαντος οὐδενὸς τῶν ἡγεμόνων· ἑάλω δὲ οὐ βίᾳ, τοῦ ἐρύματος ληφθέντος ὑπὸ Πολέμωνος καὶ Λυκομήδους, βασιλέων ἀμφοῖν, ἀλλὰ λιμῷ, ἀνέφυγε γὰρ εἰς τὸ ὄρος παρασκευῆς χωρίς, εἰργόμενος τῶν πεδίων, εὗρε δὲ καὶ τὰ ὑδρεῖα ἐμπεφραγμένα πέτραις C 561 ἠλιβάτοις· οὕτω γὰρ διετέτακτο Πομπήιος, κατασπᾶν κελεύσας τὰ φρούρια καὶ μὴ ἐᾶν χρήσιμα τοῖς ἀναφεύγειν εἰς αὐτὰ βουλομένοις ληστηρίων χάριν. ἐκεῖνος μὲν οὖν οὕτω διέταξε τὴν Φαζημωνῖτιν, οἱ δ' ὕστερον βασιλεῦσι καὶ ταύτην ἔνειμαν.

39. Ἡ δ' ἡμετέρα πόλις κεῖται μὲν ἐν φάραγγι βαθείᾳ καὶ μεγάλῃ, δι' ἧς ὁ Ἶρις φέρεται ποταμός, κατεσκεύασται[4] δὲ θαυμαστῶς προνοίᾳ τε καὶ

[1] Ἰκίζαρι is doubtful. For the variant spellings see Kramer or C. Müller.

[2] κατεσκαμμένον, Corais, for κατεσκευασμένον; so the later editors.

[3] ὑπό, Jones deletes, following J. A. R. Munro (*Hermathena*, 1900), and Sir W. M. Ramsay (*Classical Review*, 1901, p. 54), the latter likewise conjecturing εἷς for ὑπό.

[4] κατεσκεύασται D, κατεσκεύαστο other MSS.

fortress, Icizari, now deserted; and, near by, a royal palace, now in ruins. The remainder of the country is in general bare of trees and productive of grain. Above the country of the Amaseians are situated the hot springs of the Phazemonitae, which are extremely good for the health, and also Sagylium, with a stronghold situated on a high steep mountain that runs up into a sharp peak. Sagylium also has an abundant reservoir of water, which is now in neglect, although it was useful to the kings for many purposes. Here Arsaces, one of the sons of King Pharnaces, who was playing the dynast and attempting a revolution without permission from any of the prefects, was captured and slain.[1] He was captured, however, not by force, although the stronghold was taken by Polemon and Lycomedes, both of them kings, but by starvation, for he fled up into the mountain without provisions, being shut out from the plains, and he also found the wells of the reservoir choked up by huge rocks; for this had been done by order of Pompey, who ordered that the garrisons be pulled down and not be left useful to those who wished to flee up to them for the sake of robberies. Now it was in this way that Pompey arranged Phazemonitis for administrative purposes, but the later rulers distributed also[2] this country among kings.

39. My city[3] is situated in a large deep valley, through which flows the Iris River. Both by human foresight and by nature it is an admirably

[1] The translation conforms with a slight emendation of the Greek text. The MSS. make Strabo say that "Arsaces . . . was captured and slain by the sons of Pharnaces" (see critical note).

[2] *i.e.* as well as Zela and Megalopolis.

[3] Amaseia.

φύσει, πόλεως τε ἅμα[1] καὶ φρουρίου παρέχεσθαι χρείαν δυναμένη· πέτρα γὰρ ὑψηλὴ καὶ περίκρημνος, κατερρωγυῖα ἐπὶ τὸν ποταμόν, τῇ μὲν ἔχουσα τὸ τεῖχος ἐπὶ τῷ χείλει τοῦ ποταμοῦ, καθ' ὃ ἡ πόλις συνῴκισται, τῇ δ' ἀνατρέχον ἑκατέρωθεν ἐπὶ τὰς κορυφάς· δύο δ' εἰσὶ συμφυεῖς ἀλλήλαις, πεπυργωμέναι παγκάλως· ἐν δὲ τῷ περιβόλῳ τούτῳ βασίλειά τ' ἐστὶ καὶ μνήματα βασιλέων· αἱ κορυφαὶ δ' ἔχουσιν αὐχένα παντάπασι στενόν, πέντε ἢ ἓξ σταδίων ἑκατέρωθεν τὸ ὕψος, ἀπὸ τῆς ποταμίας ἀναβαίνοντι καὶ τῶν προαστείων· ἀπὸ δὲ τοῦ αὐχένος ἐπὶ τὰς κορυφὰς ἄλλη σταδιαία λείπεται πρόσβασις ὀξεῖα καὶ πάσης βίας κρείττων· ἔχει[2] δὲ καὶ ὑδρεῖα ἐντὸς ἀναφαίρετα, συρίγγων τετμημένων δυεῖν, τῆς μὲν ἐπὶ τὸν ποταμόν, τῆς δ' ἐπὶ τὸν αὐχένα· ἐπέζευκται δὲ γέφυρα τῷ ποταμῷ μία μὲν ἀπὸ τῆς πόλεως ἐπὶ τὸ προάστειον, ἄλλη δ' ἀπὸ τοῦ προαστείου πρὸς τὴν ἔξω χώραν· κατὰ γὰρ τὴν γέφυραν ταύτην ἀπολήγει τὸ ὄρος τὸ τῆς πέτρας ὑπερκείμενον. αὐλὼν δ' ἐστὶν ἀπὸ τοῦ ποταμοῦ διήκων, οὐ πλατὺς τὸ πρῶτον τελέως, ἔπειτα πλατύνεται καὶ ποιεῖ τὸ Χιλιόκωμον καλούμενον πεδίον· εἶθ' ἡ Διακοπηνὴ καὶ ἡ Πιμωλισηνὴ χώρα πᾶσα εὐδαίμων μέχρι τοῦ Ἅλυος. ταῦτα μὲν τὰ ἀρκτικὰ μέρη τῆς τῶν Ἀμασέων χώρας, μῆκος ὅσον πεντακοσίων σταδίων· ἔπειθ' ἑξῆς ἡ λοιπὴ

[1] τε ἅμα, Meineke, for ἅμα τε.
[2] D*hixz* have ἐκεῖ instead of ἔχει.

---

[1] This appears to mean that the two peaks ran up into two towers, and not that they had towers built upon them.

devised city, since it can at the same time afford the advantage of both a city and a fortress; for it is a high and precipitous rock, which descends abruptly to the river, and has on one side the wall on the edge of the river where the city is settled and on the other the wall that runs up on either side to the peaks. These peaks are two in number, are united with one another by nature, and are magnificently towered.[1] Within this circuit are both the palaces and monuments of the kings. The peaks are connected by a neck[2] which is altogether narrow, and is five or six stadia in height on either side as one goes up from the river-banks and the suburbs; and from the neck to the peaks there remains another ascent of one stadium, which is sharp and superior to any kind of force. The rock also has reservoirs of water inside it, a water-supply of which the city cannot be deprived, since two tube-like channels have been hewn out, one towards the river and the other towards the neck. And two bridges have been built over the river, one from the city to the suburbs and the other from the suburbs to the outside territory; for it is at this bridge that the mountain which lies above the rock terminates. And there is a valley extending from the river which at first is not altogether wide, but it later widens out and forms the plain called Chiliocomum;[3] and then comes the Diacopene and Pimolisene country, all of which is fertile, extending to the Halys River. These are the northern parts of the country of the Amaseians, and are about five hundred stadia in length. Then in order comes the

[2] *i.e.* isthmus-like ridge.
[3] *i.e.* "Plain of the thousand villages."

πολὺ ταύτης ἐπιμηκεστέρα μέχρι τοῦ Βαβανόμου καὶ τῆς Ξιμηνῆς, ἥπερ καὶ αὐτὴ καθήκει μέχρι πρὸς τὸν Ἅλυν· τοῦτο μὲν δὴ τὸ μῆκος, πλάτος δὲ τὸ ἀπὸ τῶν ἄρκτων πρὸς νότον ἐπί τε τὴν Ζηλῖτιν καὶ τὴν μεγάλην Καππαδοκίαν μέχρι τῶν Τρόκμων. εἰσὶ δ' ἐν τῇ Ξιμηνῇ ἅλαι ὀρυκτῶν ἁλῶν, ἀφ' ὧν εἰκάζουσιν εἰρῆσθαι Ἅλυν τὸν ποταμόν. ἔστι δὲ καὶ ἐρύματα πλείω κατεσκαμμένα ἐν τῇ ἡμετέρᾳ χώρᾳ καὶ ἔρημος γῆ πολλὴ διὰ τὸν Μιθριδατικὸν πόλεμον. ἔστι μέντοι πᾶσα μὲν εὔδενδρος, ἡ δ' ἱππόβοτος καὶ τοῖς ἄλλοις θρέμμασι πρόσφορος· ἅπασα δ' οἰκήσιμος καλῶς. ἐδόθη δὲ καὶ ἡ Ἀμάσεια βασιλεῦσι, νῦν δ' ἐπαρχία ἐστί.

40. Λοιπὴ δ' ἐστὶν ἡ ἐκτὸς Ἅλυος χώρα τῆς C 562 Ποντικῆς ἐπαρχίας, ἡ περὶ τὸν Ὄλγασσυν, συναφὴς τῇ Σινωπίδι. ἔστι δ' ὁ Ὄλγασσυς ὄρος σφόδρα ὑψηλὸν καὶ δύσβατον· καὶ ἱερὰ τοῦ ὄρους τούτου πανταχοῦ καθιδρυμένα ἔχουσιν οἱ Παφλαγόνες· περίκειται δ' ἱκανῶς χώρα ἀγαθή, ἥ τε Βλαηνὴ καὶ ἡ Δομανῖτις, δι' ἧς Ἀμνίας ῥεῖ ποταμός. ἐνταῦθα Μιθριδάτης ὁ Εὐπάτωρ τὰς Νικομήδους τοῦ Βιθυνοῦ δυνάμεις ἄρδην ἠφάνισεν, οὐδ'[1] αὐτὸς παρατυχών, ἀλλὰ διὰ τῶν στρατηγῶν· καὶ ὁ μὲν φεύγων μετ' ὀλίγων εἰς τὴν οἰκείαν ἐσώθη, κἀκεῖθεν εἰς Ἰταλίαν ἔπλευσεν, ὁ δ' ἠκολούθησε καὶ τήν τε Βιθυνίαν εἷλεν ἐξ ἐφόδου

[1] οὐδ', Corais and Meineke emend to οὐκ.

---

[1] *i.e.* "salt-works."

[2] Literally, salt obtained by digging or mining. On the salt-mines of northern India, see 5. 2. 6 and 15. 1. 30.

remainder of their country, which is much longer than this, extending to Babanomus and Ximenê, which latter itself extends as far as the Halys River. This, then, is the length of their country, whereas the breadth from the north to the south extends, not only to Zelitis, but also to Greater Cappadocia, as far as the Trocmi. In Ximenê there are "halae"[1] of rock-salt,[2] after which the river is supposed to have been called "Halys." There are several demolished strongholds in my country, and also much deserted land, because of the Mithridatic War. However, it is all well supplied with trees; a part of it affords pasturage for horses and is adapted to the raising of the other animals; and the whole of it is beautifully adapted to habitation. Amaseia was also given to kings, though it is now a province.[3]

40. There remains that part of the Pontic province which lies outside the Halys River, I mean the country round Mt. Olgassys, contiguous to Sinopis. Mt. Olgassys is extremely high and hard to travel. And temples that have been established everywhere on this mountain are held by the Paphlagonians. And round it lies fairly good territory, both Blaënê and Domanitis, through which latter flows the Amnias River. Here Mithridates Eupator utterly wiped out the forces of Nicomedes the Bithynian—not in person, however, since it happened that he was not even present, but through his generals. And while Nicomedes, fleeing with a few others, safely escaped to his home-land and from there sailed to Italy, Mithridates followed him and not only took Bithynia at the first assault but

[3] *Roman* province, of course.

καὶ τὴν Ἀσίαν κατέσχε μέχρι Καρίας καὶ Λυκίας. κἀνταῦθα δ' ἀπειδείχθη πόλις ἡ Πομπηιούπολις· ἐν δὲ τῇ πόλει ταύτῃ τὸ Σανδαρακούργιον οὐ πολὺ ἄπωθεν Πιμωλίσων, φρουρίου βασιλικοῦ κατεσκαμμένου, ἀφ' οὗ ἡ χώρα ἡ ἑκατέρωθεν τοῦ ποταμοῦ καλεῖται Πιμωλισηνή. τὸ δὲ Σανδαρακούργιον ὄρος κοῖλόν ἐστιν ἐκ τῆς μεταλλείας, ὑπεληλυθότων αὐτὸ τῶν ἐργαζομένων διώρυξι μεγάλαις· εἰργάζοντο δὲ δημοσιῶναι,[1] μεταλλευταῖς χρώμενοι τοῖς ἀπὸ κακουργίας ἀγοραζομένοις ἀνδραπόδοις· πρὸς γὰρ τῷ ἐπιπόνῳ τοῦ ἔργου καὶ θανάσιμον καὶ δύσοιστον εἶναι τὸν ἀέρα φασὶ τὸν ἐν τοῖς μετάλλοις διὰ τὴν βαρύτητα τῆς τῶν βώλων ὀδμῆς, ὥστε ὠκύμορα εἶναι τὰ σώματα. καὶ δὴ καὶ ἐκλείπεσθαι[2] συμβαίνει πολλάκις τὴν μεταλλείαν διὰ τὸ ἀλυσιτελές, πλειόνων μὲν ἢ διακοσίων ὄντων τῶν ἐργαζομένων, συνεχῶς δὲ νόσοις καὶ φθοραῖς δαπανωμένων. τοσαῦτα καὶ περὶ τοῦ Πόντου εἰρήσθω.

41. Μετὰ δὲ τὴν Πομπηιούπολιν ἡ λοιπὴ τῆς Παφλαγονίας ἐστὶ τῆς μεσογαίας μέχρι Βιθυνίας ἰοῦσι πρὸς δύσιν. ταύτης δέ, καίπερ ὀλίγης οὔσης, μικρὸν μὲν πρὸ ἡμῶν ἦρχον πλείους, νῦν δ' ἔχουσι Ῥωμαῖοι, τοῦ γένους τῶν βασιλέων ἐκλιπόντος. ὀνομάζουσι δ' οὖν τὴν ὅμορον τῇ Βιθυνίᾳ Τιμωνῖτιν καὶ τὴν Γεζατόριγος καὶ

[1] δημοσιῶναι, Corais, for δημοσίων ἀεί CD*hilrw*, δημοσίως ἀεί *xz*; so the later editors.
[2] ἐκλείπεσθαι, Corais, for ἐκλιπέσθαι; so the later editors.

---

[1] "Pompey's city." On the history of this city, see J. G. C. Anderson in *Anatolian Studies presented to Sir*

also took possession of Asia as far as Caria and Lycia. And here, too, a place was proclaimed a city, I mean Pompeiupolis;[1] and in this city is Mt. Sandaracurgium,[2] not far away from Pimolisa, a royal fortress now in ruins, after which the country on either side of the river is called Pimolisenê. Mt. Sandaracurgium is hollowed out in consequence of the mining done there, since the workmen have excavated great cavities beneath it. The mine used to be worked by publicans, who used as miners the slaves sold in the market because of their crimes; for, in addition to the painfulness of the work, they say that the air in the mines is both deadly and hard to endure on account of the grievous odour of the ore, so that the workmen are doomed to a quick death. What is more, the mine is often left idle because of the unprofitableness of it, since the workmen are not only more than two hundred in number, but are continually spent by disease and death.[3] So much be said concerning Pontus.

41. After Pompeiupolis comes the remainder of the interior of Paphlagonia, extending westwards as far as Bithynia. This country, small though it is, was governed by several rulers a little before my time, but, the family of kings having died out, it is now in possession of the Romans. At any rate, they give to the country that borders on Bithynia[4] the names "Timonitis," "the country of Gezatorix,"

William Mitchell Ramsay, p. 6. Anderson's article is of great importance in the study of the time of the composition of Strabo's *Geography*.

[2] Mt. "Realgar (red sulphuret of arsenic) mine."

[3] Hence the continual necessity of purchasing other slaves to replace them.

[4] *i.e.* as being divided up into several domains.

Μαρμωλῖτίν τε καὶ Σανισηνὴν καὶ Ποταμίαν· ἦν δέ τις καὶ Κιμιατηνή,[1] ἐν ᾗ τὰ Κιμίατα, φρούριον ἐρυμνόν, ὑποκείμενον τῇ τοῦ Ὀλγάσσυος ὀρεινῇ· ᾧ χρησάμενος ὁρμητηρίῳ Μιθριδάτης, ὁ Κτίστης προσαγορευθείς, κατέστη τοῦ Πόντου κύριος, καὶ οἱ ἀπ' αὐτοῦ τὴν διαδοχὴν ἐφύλαξαν μέχρι τοῦ Εὐπάτορος. ὕστατος δὲ τῆς Παφλαγονίας ἦρξε Δηιόταρος, Κάστορος[2] υἱός, ὁ προσαγορευθεὶς Φιλάδελφος, τὸ Μορζέου[3] βασίλειον ἔχων τὰ Γάγγρα, πολισμάτιον ἅμα καὶ φρούριον.

42. Εὔδοξος δ' ὀρυκτοὺς ἰχθῦς ἐν Παφλαγονίᾳ C 563 λέγων ἐν ξηροῖς τόποις οὐ διορίζει τὸν τόπον, ἐν ὑγροῖς δὲ περὶ τὴν Ἀσκανίαν λίμνην φησὶ τὴν ὑπὸ Κίῳ, λέγων οὐδὲν σαφές. ἐπεὶ δὲ καὶ τὴν ὅμορον τῷ Πόντῳ Παφλαγονίαν ἐκτιθέμεθα, τοῖς δὲ Παφλαγόσιν ὁμοροῦσιν οἱ Βιθυνοὶ πρὸς δύσιν, πειρασόμεθα καὶ τὰ τούτων ἐπελθεῖν· ἔπειτα λαβόντες ἀρχὴν ἄλλην ἔκ τε τούτων καὶ τῶν Παφλαγόνων τὰ ἑξῆς τούτων τὰ πρὸς νότον μέχρι τοῦ Ταύρου συνυφανοῦμεν, τὰ παράλληλα τῷ Πόντῳ καὶ τῇ Καππαδοκίᾳ· τοιαύτην γάρ τινα ὑπογράφει τάξιν καὶ μερισμὸν ἡ τῶν τόπων φύσις.

[1] Κιμιατηνή, Corais, for Κινιστηνή; so the later editors.
[2] Κάστορος, Casaubon, for Καστόρους CDhl, Καστόρου iorxz.
[3] Μορζέου, Corais, Kramer, and Meineke, for Μορζέους.

and also "Marmolitis," "Sanisenê," and "Potamia." There was also a Cimiatenê, in which was Cimiata, a strong fortress situated at the foot of the mountainous country of the Olgassys. This was used by Mithridates, surnamed Ctistes,[1] as a base of operations when he established himself as lord of Pontus; and his descendants preserved the succession down to Eupator. The last to reign over Paphlagonia was Deïotarus, the son of Castor, surnamed Philadelphus, who possessed Gangra, the royal residence of Morzeüs, which was at the same time a small town and a fortress.

42. Eudoxus mentions fish that are "dug up" in Paphlagonia "in dry places," but he does not distinguish the place; and he says that they are dug up "in moist places round the Ascanian Lake below Cius," without saying anything clear on the subject.[2] Since I am describing the part of Paphlagonia which borders on Pontus and since the Bithynians border on the Paphlagonians towards the west, I shall try to go over this region also; and then, taking a new beginning from the countries of these people and the Paphlagonians, I shall interweave my description of their regions with that of the regions which follow these in order towards the south as far as the Taurus—the regions that run parallel to Pontus and Paphlagonia; for some such order and division is suggested by the nature of the regions.

[1] *i.e.* "Founder" of Pontus as an independent kingdom; reigned 337–302 B.C.

[2] Cf. the "dug mullets" in Celtica, 4. 1. 6.

IV

1. Τὴν δὲ Βιθυνίαν ἀπὸ μὲν τῆς ἀνατολῆς ὁρίζουσι Παφλαγόνες τε καὶ Μαριανδυνοὶ καὶ τῶν Ἐπικτήτων τινές, ἀπὸ δὲ τῶν ἄρκτων ἡ Ποντικὴ θάλασσα ἡ ἀπὸ τῶν ἐκβολῶν τοῦ Σαγγαρίου μέχρι τοῦ στόματος τοῦ κατὰ Βυζάντιον καὶ Χαλκηδόνα, ἀπὸ δὲ δύσεως ἡ Προποντίς, πρὸς νότον δ' ἥ τε Μυσία καὶ ἡ Ἐπίκτητος καλουμένη Φρυγία, ἡ δ' αὐτὴ καὶ Ἑλλησποντιακὴ Φρυγία καλουμένη.

2. Ταύτης δ' ἐπὶ μὲν τῷ στόματι τοῦ Πόντου Χαλκηδὼν ἵδρυται, Μεγαρέων κτίσμα, καὶ κώμη Χρυσόπολις καὶ τὸ ἱερὸν τὸ Χαλκηδόνιον, ἔχει δ' ἡ χώρα μικρὸν[1] ὑπὲρ τῆς θαλάττης κρήνην Ἀζαριτίαν, τρέφουσαν κροκοδείλους μικρούς· ἔπειτ' ἐκδέχεται τὴν τῶν Χαλκηδονίων ἠιόνα ὁ Ἀστακηνὸς καλούμενος κόλπος, μέρος ὢν τῆς Προποντίδος, ἐν ᾧ ἡ Νικομήδεια ἔκτισται ἐπώνυμος ἑνὸς τῶν Βιθυνικῶν βασιλέων, τοῦ κτίσαντος αὐτήν· πολλοὶ δ' ὁμωνύμως ὠνομάσθησαν, καθάπερ Πτολεμαῖοι, διὰ τὴν τοῦ πρώτου δόξαν. ἦν δ' ἐν αὐτῷ τῷ κόλπῳ καὶ Ἀστακὸς πόλις, Μεγαρέων κτίσμα καὶ Ἀθηναίων καὶ μετὰ ταῦτα Δοιδαλσοῦ, ἀφ' ἧς καὶ ὁ κόλπος ὠνομάσθη. κατεσκάφη δ' ὑπὸ Λυσιμάχου· τοὺς δ' οἰκήτορας μετήγαγεν εἰς Νικομήδειαν ὁ κτίσας αὐτήν.

3. Τῷ δ' Ἀστακηνῷ κόλπος ἄλλος συνεχής ἐστιν, εἰσέχων μᾶλλον πρὸς ἀνίσχοντα ἥλιον, ἐν ᾧ Προυσιάς ἐστιν, ἡ Κίος πρότερον ὀνομασθεῖσα·

[1] μικρόν *oxz* and the editors, instead of μικράν.

IV

1. BITHYNIA is bounded on the east by the Paphlagonians and Mariandyni and some of the Epicteti; on the north by the Pontic Sea, from the outlets of the Sangarius River to the mouth of the sea at Byzantium and Chalcedon; on the west by the Propontis; and towards the south by Mysia and by Phrygia "Epictetus," as it is called, though the same is also called "Hellespontiac" Phrygia.

2. In this last country, at the mouth of the Pontus, are situated Chalcedon, founded by the Megarians, and Chrysopolis, a village, and the Chalcedonian temple; and slightly above the sea the country has a spring called Azaritia, which breeds little crocodiles. Then the Chalcedonian shore is followed by the Astacene Gulf, as it is called, a part of the Propontis; and it was on this gulf that Nicomedeia was founded, being named after one of the Bithynian kings, who founded it.[1] But many kings, for example the Ptolemies, were, on account of the fame of the first, given the same name. And on the gulf itself there was also a city Astacus, founded by the Megarians and Athenians and afterwards by Doedalsus; and it was after the city Astacus that the gulf was named. It was rased to the ground by Lysimachus, and its inhabitants were transferred to Nicomedeia by the founder of the latter.

3. Continuous with the Astacene Gulf is another gulf, which runs more nearly towards the rising sun than the former does; and on this gulf is Prusias, formerly called Cius. Cius was rased to the ground

[1] Nicomedes I, in 264 B.C.

κατέσκαψε δὲ τὴν Κίον Φίλιππος, ὁ Δημητρίου μὲν υἱός, Περσέως δὲ πατήρ, ἔδωκε δὲ Προυσίᾳ τῷ Ζήλα, συγκατασκάψαντι καὶ ταύτην καὶ Μύρλειαν ἀστυγείτονα πόλιν, πλησίον δὲ καὶ Προύσης οὖσαν· ἀναλαβὼν δ' ἐκεῖνος ἐκ τῶν
C 564 ἐρειπίων αὐτὰς ἐπωνόμασεν ἀφ' ἑαυτοῦ μὲν Προυσιάδα πόλιν τὴν Κίον, τὴν δὲ Μύρλειαν Ἀπάμειαν ἀπὸ τῆς γυναικός. οὗτος δ' ἐστὶν ὁ Προυσίας ὁ καὶ Ἀννίβαν δεξάμενος, ἀναχωρήσαντα δεῦρο μετὰ τὴν Ἀντιόχου ἧτταν, καὶ τῆς ἐφ' Ἑλλησπόντῳ Φρυγίας ἀναστὰς κατὰ συμβάσεις τοῖς Ἀτταλικοῖς, ἣν οἱ μὲν πρότερον ἐκάλουν μικρὰν Φρυγίαν, ἐκεῖνοι δ' Ἐπίκτητον ὠνόμασαν. ὑπέρκειται δὲ τῆς Προυσιάδος ὄρος, ὃ καλοῦσιν Ἀργανθώνιον. ἐνταῦθα δὲ μυθεύουσι τὸν Ὕλαν, ἕνα τῶν Ἡρακλέους ἑταίρων συμπλεύσαντα ἐπὶ τῆς Ἀργοῦς αὐτῷ, ἐξιόντα δὲ ἐπὶ ὑδρείαν ὑπὸ νυμφῶν ἁρπαγῆναι· Κίον δέ, καὶ τοῦτον Ἡρακλέους ἑταῖρον καὶ σύμπλουν, ἐπανελθόντα ἐκ Κόλχων αὐτόθι καταμεῖναι καὶ κτίσαι τὴν πόλιν ἐπώνυμον αὑτοῦ. καὶ νῦν δ' ἔτι ἑορτή τις ἄγεται παρὰ τοῖς Προυσιεῦσιν καὶ ὀρειβασία, θιασευόντων καὶ καλούντων Ὕλαν, ὡς ἂν κατὰ ζήτησιν τὴν ἐκείνου πεποιημένων τὴν ἐπὶ τὰς ὕλας ἔξοδον. πολιτευσάμενοι δὲ πρὸς Ῥωμαίους οἱ Προυσιεῖς εὐνοϊκῶς ἐλευθερίας ἔτυχον. οἱ δ' Ἀπαμεῖς[1] ἀποικίαν ἐδέξαντο Ῥωμαίων. Προῦσα δὲ ἐπὶ τῷ Ὀλύμπῳ ἵδρυται τῷ Μυσίῳ, πόλις εὐνομουμένη, τοῖς τε Φρυξὶν ὅμορος καὶ τοῖς Μυσοῖς, κτίσμα Προυσίου τοῦ πρὸς Κροῖσον[2] πολεμήσαντος.

[1] Ἀπαμεῖς, Corais, for Ἀπαμιεῖς; so the later editors.
[2] Κροῖσον is probably an error for Κῦρον (see Stephanus s.v. Προῦσα).

by Philip, the son of Demetrius and father of Perseus, and given by him to Prusias the son of Zelas, who had helped him rase both this city and Myrleia, which latter is a neighbouring city and also is near Prusa. And Prusias restored them from their ruins and named the city Cius "Prusias" after himself and Myrleia "Apameia" after his wife. This is the Prusias who welcomed Hannibal, when the latter withdrew thither after the defeat of Antiochus, and who retired from Phrygia on the Hellespont in accordance with an agreement made with the Attalici.[1] This country was in earlier times called Lesser Phrygia, but the Attalici called it Phrygia Epictetus.[2] Above Prusias lies a mountain called Arganthonium. And here is the scene of the myth of Hylas, one of the companions of Heracles who sailed with him on the Argo, and who, when he was going out to get water, was carried off by the nymphs. And when Cius, who was also a companion of Heracles and with him on the voyage, returned from Colchis, he stayed here and founded the city which was named after him. And still to this day a kind of festival is celebrated among the Prusians, a mountain-ranging festival, in which they march in procession and call Hylas, as though making their exodus to the forests in quest of him. And having shown a friendly disposition towards the Romans in the conduct of their government, the Prusians obtained freedom. Prusa is situated on the Mysian Olympus; it is a well-governed city, borders on the Phrygians and the Mysians, and was founded by the Prusias who made war against Croesus.[3]

[1] Kings of Pergamum.
[2] *i.e.* "Newly acquired," or "annexed," territory.
[3] See critical note.

4. Διορίσαι δὲ τοὺς ὅρους χαλεπὸν τούς τε Βιθυνῶν καὶ Φρυγῶν καὶ Μυσῶν καὶ ἔτι Δολιόνων τῶν περὶ Κύζικον καὶ Μυγδόνων καὶ Τρώων· καὶ διότι μὲν εἶναι δεῖ ἕκαστον[1] φῦλον χωρίς, ὁμολογεῖται. καὶ ἐπί γε τῶν Φρυγῶν καὶ τῶν Μυσῶν καὶ παροιμιάζονται·

χωρὶς τὰ Μυσῶν καὶ Φρυγῶν ὁρίσματα·

διορίσαι[2] δὲ χαλεπόν. αἴτιον δὲ τὸ τοὺς ἐπήλυδας βαρβάρους καὶ στρατιώτας ὄντας μὴ βεβαίως κατέχειν τὴν κρατηθεῖσαν, ἀλλὰ πλανήτας εἶναι τὸ πλέον, ἐκβάλλοντας καὶ ἐκβαλλομένους. ἅπαντα δὲ τὰ ἔθνη ταῦτα Θρᾴκιά τις εἰκάζοι ἄν, διὰ τὸ τὴν περαίαν νέμεσθαι τούτους, καὶ διὰ τὸ μὴ πολὺ ἐξαλλάττειν ἀλλήλων ἑκατέρους.

5. Ὅμως δ' ἐφ' ὅσον εἰκάζειν οἷόν τε, τῆς μὲν Βιθυνίας μέσην ἄν τις θείη καὶ τῆς ἐκβολῆς τοῦ Αἰσήπου τὴν Μυσίαν, ἁπτομένην τῆς θαλάττης καὶ διήκουσαν μέχρι τοῦ Ὀλύμπου σχεδὸν[3] παντός· κύκλῳ δὲ τὴν Ἐπίκτητον κειμένην ἐν τῇ μεσογαίᾳ, θαλάττης οὐδαμοῦ ἁπτομένην, διατείνουσαν δὲ μέχρι τῶν ἑῴων μερῶν τῆς Ἀσκανίας λίμνης τε καὶ χώρας, ὁμωνύμως γὰρ τῇ λίμνῃ καὶ ἡ χώρα ἐλέγετο. καὶ ἦν αὐτῆς τὸ μὲν Φρύγιον, τὸ δὲ Μύσιον, ἀπωτέρω δὲ τῆς Τροίας τὸ Φρύγιον. καὶ δὴ καὶ οὕτω δεκτέον τὸ παρὰ τῷ ποιητῇ, ὅταν φῇ·

Φόρκυς δ' αὖ Φρύγας ἦγε καὶ Ἀσκάνιος θεοειδής,
τῆλ' ἐξ Ἀσκανίης,

[1] τό, before φῦλον, E omits; so Meineke.
[2] διορίσαι E, διορισάμενοι CDhilrw, διορίσασθαι oxz.

4. It is difficult to mark the boundaries between the Bithynians and the Phrygians and the Mysians, or even those between the Doliones round Cyzicus and the Mygdonians and the Trojans. And it is agreed that each tribe is "apart" from the others (in the case of the Phrygians and Mysians, at least, there is a proverb, "Apart are the boundaries of the Mysians and Phrygians"), but that it is difficult to mark the boundaries between them. The cause of this is that the foreigners who went there, being barbarians and soldiers, did not hold the conquered country firmly, but for the most part were wanderers, driving people out and being driven out. One might conjecture that all these tribes were Thracian because the Thracians occupy the other side[1] and because the people on either side do not differ much from one another.

5. But still, as far as one is able to conjecture, one might put down Mysia as situated between Bithynia and the outlet of the Aesepus River, as touching upon the sea, and as extending as far as Olympus, along almost the whole of it; and Epictetus as lying in the interior round Mysia, but nowhere touching upon the sea, and as extending to the eastern parts of the Ascanian Lake and territory; for the territory was called by the same name as the lake. And a part of this territory was Phrygian and a part Mysian, but the Phrygian part was farther away from Troy. And in fact one should thus interpret the words of the poet when he says, "And Phorcys and godlike Ascanius led the Phrygians from afar, from Ascania."[2]

[1] *i.e.* the European side. [2] *Iliad* 2. 862.

[3] All MSS. except E read δέ after σχεδόν.

τῆς Φρυγιακῆς, ὡς οὔσης ἐγγυτέρω ἄλλης Ἀσκανίας Μυσιακῆς τῆς πρὸς τῇ νῦν Νικαίᾳ, ἧς μέμνηται, ὅταν φῇ·

C 565 Πάλμυν τ' Ἀσκάνιόν τε Μόρυν θ', υἷ' Ἱπποτίωνος,
Μυσῶν ἀγχεμάχων ἡγήτορα,
οἵ ῥ' ἐξ Ἀσκανίης ἐριβώλακος ἦλθον ἀμοιβοί.

οὐ θαυμαστὸν δ', εἰ τῶν Φρυγῶν εἰπών τινα ἡγεμόνα Ἀσκάνιον καὶ ἐξ Ἀσκανίας ἥκοντα, καὶ Μυσῶν τινὰ λέγει ἡγεμόνα Ἀσκάνιον καὶ ἐξ Ἀσκανίας ἥκοντα· πολλὴ γὰρ ἡ ὁμωνυμία παρ' αὐτῷ, καὶ ἡ ἀπὸ τῶν ποταμῶν καὶ λιμνῶν καὶ χωρίων ἐπίκλησις.

6. Καὶ τὸν Αἴσηπον δὲ τῶν Μυσῶν ὅριον παραδίδωσιν αὐτὸς ὁ ποιητής· τὴν γὰρ ὑπὲρ τοῦ Ἰλίου παρώρειαν τῆς Τροίας καταλέξας τὴν ὑπ' Αἰνείᾳ, ἣν Δαρδανίαν ἐκάλεσε, τίθησιν ἐφεξῆς πρὸς ἄρκτον καὶ[1] τὴν Λυκίαν, τὴν ὑπὸ Πανδάρῳ, ἐν ᾗ ἡ Ζέλεια· καί φησιν·

οἳ δὲ Ζέλειαν ἔναιον ὑπαὶ πόδα νείατον Ἴδης,
ἀφνειοὶ πίνοντες ὕδωρ μέλαν Αἰσήποιο
Τρῶες.

τῇ δὲ Ζελείᾳ ὑποπέπτωκε πρὸς θαλάττῃ ἐπιτάδε[2] τοῦ Αἰσήπου τὸ τῆς Ἀδραστείας πεδίον καὶ Τήρεια καὶ ἡ Πιτύα καὶ καθόλου ἡ νῦν Κυζικηνὴ ἡ πρὸς Πριάπῳ, ἣν ἐφεξῆς καταλέγει· εἶτα ἀνακάμπτει πάλιν ἐπὶ τὰ πρὸς ἕω μέρη καὶ τὰ ἐπέκεινα, ὥστε ἐμφαίνει τὴν μέχρι Αἰσήπου πέρας ἡγούμενος τῆς Τρῳάδος τὸ ἀρκτικὸν καὶ

[1] εἰς, before τὴν Λυκίαν, omitted by *oxz* and the editors.

that is, the Phrygian Ascania,[1] since his words imply that another Ascania, the Mysian, near the present Nicaea, is nearer Troy, that is, the Ascania to which the poet refers when he says, "and Palmys, and Ascanius, and Morys, son of Hippotion (Morys being leader of the Mysians, hand-to-hand fighters), who had come from deep-soiled Ascania to relieve their fellows."[2] And it is not remarkable if he speaks of one Ascanius as a leader of the Phrygians and as having come from Ascania and also of another Ascanius as a leader of the Mysians and as having come from Ascania, for in Homer identity of names is of frequent occurrence, as also the surnaming of people after rivers and lakes and places.

6. And the poet himself gives the Aesepus as a boundary of the Mysians, for after naming the foot-hills of Troy above Ilium that were subject to Aeneas, which he calls Dardania, he puts down Lycia as next towards the north, the country that was subject to Pandarus, in which Zeleia was situated; and he says, "and they that dwelt in Zeleia 'neath the nethermost foot of Mt. Ida, wealthy men, Trojans, who drink the dark water of the Aesepus."[3] Below Zeleia, near the sea, and on this side of the Aesepus, are the plain of Adrasteia, Mt. Tereia, and Pitya (that is, speaking generally, the present Cyzicenê near Priapus), which the poet names next after Zeleia;[4] and then he returns to the parts towards the east and those on the far side of the Aesepus, by which he indicates that he regards the country as far as the Aesepus as the northerly and easterly limit of the

[1] See Leaf, *Troy*, p. 301.
[2] *Iliad* 13. 792.
[3] *Iliad* 2. 824.
[4] *Iliad* 2. 828.

[2] ἐπιτάδε, Meineke from conj. of Kramer, for ἐπὶ δὲ τῷ E, ἐπὶ δὲ τό other MSS., ὑπὲρ δέ *oz*, ἀπὸ δέ *x*.

ἑῷον. ἀλλὰ μὴν μετά γε τὴν Τρωάδα ἡ Μυσία ἐστὶ καὶ ὁ Ὄλυμπος. ἡ μὲν οὖν παλαιὰ μνήμη τοιαύτην τινὰ ὑπαγορεύει τὴν τῶν ἐθνῶν θέσιν. αἱ δὲ νῦν μεταβολαὶ τὰ πολλὰ ἐξήλλαξαν, ἄλλοτ' ἄλλων ἐπικρατούντων, καὶ τὰ μὲν συγχεόντων, τὰ δὲ διασπώντων. καὶ γὰρ Φρύγες ἐπεκράτησαν καὶ Μυσοὶ μετὰ τὴν Τροίας ἅλωσιν, εἶθ' ὕστερον Λυδοὶ καὶ μετ' ἐκείνων[1] Αἰολεῖς καὶ Ἴωνες, ἔπειτα Πέρσαι καὶ Μακεδόνες, τελευταῖοι δὲ Ῥωμαῖοι, ἐφ' ὧν ἤδη καὶ τὰς διαλέκτους καὶ τὰ ὀνόματα ἀποβεβλήκασιν οἱ πλεῖστοι, γεγονότος ἑτέρου τινὸς μερισμοῦ τῆς χώρας, οὗ μᾶλλον φροντίσαι δεῖ τὰ νῦν οἷα ἔστι[2] λέγοντας, τῇ δὲ ἀρχαιολογίᾳ μετρίως προσέχοντας.

7. Ἐν δὲ τῇ μεσογαίᾳ τῆς Βιθυνίας τό τε Βιθύνιόν ἐστιν, ὑπερκείμενον τοῦ Τιείου καὶ ἔχον τὴν περὶ Σάλωνα χώραν ἀρίστην βουβοσίοις, ὅθεν ἐστὶν ὁ Σαλωνίτης τυρός, καὶ Νίκαια, ἡ μητρόπολις τῆς Βιθυνίας ἐπὶ τῇ Ἀσκανίᾳ λίμνῃ, περίκειται δὲ κύκλῳ πεδίον μέγα καὶ σφόδρα εὔδαιμον, οὐ πάνυ δὲ ὑγιεινὸν τοῦ θέρους, κτίσμα Ἀντιγόνου μὲν πρῶτον τοῦ Φιλίππου, ὃς αὐτὴν Ἀντιγονίαν προσεῖπεν, εἶτα Λυσιμάχου, ὃς ἀπὸ τῆς γυναικὸς μετωνόμασε Νίκαιαν· ἦν δ' αὕτη θυγάτηρ Ἀντιπάτρου. ἔστι δὲ τῆς πόλεως

[1] *Chioxz* have ἐκείνους.

[2] οἷα ἔστι (οἷ' ἔστι Meineke), Jones, for οἷ εται (sic) C, ὡς οἷόν τε *x*, οἴεται other MSS.; but the ὄντα of Corais is tempting.

---

[1] 12. 8. 7.

Troad. Assuredly, however, Mysia and Olympus come after the Troad. Now ancient tradition suggests some such position of the tribes as this, but the present differences are the result of numerous changes, since different rulers have been in control at different times, and have confounded together some tribes and sundered others. For both the Phrygians and the Mysians had the mastery after the capture of Troy; and then later the Lydians; and with them the Aeolians and the Ionians; and then the Persians and the Macedonians; and lastly the Romans, under whose reign most of the peoples have already lost both their dialects and their names, since a different partition of the country has been made. But it is better for me to consider this matter when I describe the conditions as they now are,[1] at the same time giving proper attention to conditions as they were in antiquity.

7. In the interior of Bithynia are, not only Bithynium, which is situated above Tieium and holds the territory round Salon, where is the best pasturage for cattle and whence comes the Salonian cheese, but also Nicaea, the metropolis of Bithynia, situated on the Ascanian Lake, which is surrounded by a plain that is large and very fertile but not at all healthful in summer. Nicaea was first founded by Antigonus[2] the son of Philip, who called it Antigonia, and then by Lysimachus, who changed its name to that of Nicaea his wife. She was the daughter of Antipater.[3] The city is sixteen stadia in

[2] King of Asia; defeated by Lysimachus at the battle of Ipsus in Phrygia (301 B.C.), and fell in that battle in his 81st year (Diodorus Siculus 20. 46–86).

[3] Appointed regent of Macedonia by Alexander in 334 B.C.

ἑκκαιδεκαστάδιος ὁ περίβολος ἐν τετραγώνῳ σχήματι· ἔστι δὲ καὶ τετράπυλος ἐν πεδίῳ C 566 κείμενος ἐρρυμοτομημένος πρὸς ὀρθὰς γωνίας, ὥστ' ἀφ' ἑνὸς λίθου κατὰ μέσον ἱδρυμένου τὸ γυμνάσιον τὰς τέτταρας ὁρᾶσθαι πύλας. μικρὸν δ' ὑπὲρ τῆς Ἀσκανίας λίμνης Ὀτροία πολίχνη, πρὸς τοῖς ὅροις ἤδη τῆς Βιθυνίας τοῖς πρὸς ἕω· εἰκάζουσι δ' ἀπὸ Ὀτρέως Ὀτροίαν καλεῖσθαι.[1]

8. Ὅτι δ' ἦν κατοικία Μυσῶν ἡ Βιθυνία, πρῶτον μαρτυρήσει Σκύλαξ ὁ Καρυανδεύς,[2] φήσας περιοικεῖν τὴν Ἀσκανίαν λίμνην Φρύγας καὶ Μυσούς, ἔπειτα Διονύσιος ὁ τὰς κτίσεις συγγράψας, ὃς τὰ[3] κατὰ Χαλκηδόνα καὶ Βυζάντιον στενά, ἃ νῦν Θρᾴκιος Βόσπορος καλεῖται, πρότερόν φησι Μύσιον Βόσπορον προσαγορεύεσθαι· τοῦτο δ' ἄν τις καὶ τοῦ Θρᾷκας εἶναι τοὺς Μυσοὺς μαρτύριον θείη· ὅ τε Εὐφορίων,

> Μυσοῖο παρ' ὕδασιν Ἀσκανίοιο

λέγων, καὶ ὁ Αἰτωλὸς Ἀλέξανδρος,

> οἳ καὶ ἐπ' Ἀσκανίων δώματ' ἔχουσι ῥοῶν
> λίμνης Ἀσκανίης ἐπὶ χείλεσιν, ἔνθα Δολίων
> υἱὸς Σιληνοῦ νάσσατο καὶ Μελίης,

τὸ αὐτὸ ἐκμαρτυροῦσιν, οὐδαμοῦ τῆς Ἀσκανίας λίμνης εὑρισκομένης ἀλλ' ἐνταῦθα μόνον.

9. Ἄνδρες δ' ἀξιόλογοι κατὰ παιδείαν γεγόνασιν ἐν τῇ Βιθυνίᾳ Ξενοκράτης τε ὁ φιλόσοφος

[1] πρότερον, after καλεῖσθαι, is omitted by *xz*.
[2] Καρυανδεύς, Casaubon, for Καρυανδρεύς; so the later editors.

circuit and is quadrangular in shape; it is situated in a plain, and has four gates; and its streets are cut at right angles, so that the four gates can be seen from one stone which is set up in the middle of the gymnasium. Slightly above the Ascanian Lake is the town Otroea, situated just on the borders of Bithynia towards the east. It is surmised that Otroea was so named after Otreus.

8. That Bithynia was a settlement of the Mysians will first be testified by Scylax the Caryandian,[1] who says that Phrygians and Mysians lived round the Ascanian Lake; and next by the Dionysius[2] who wrote on "The Foundings" of cities, who says that the strait at Chalcedon and Byzantium, now called the Thracian Bosporus, was in earlier times called the Mysian Bosporus. And this might also be set down as an evidence that the Mysians were Thracians. Further, when Euphorion[3] says, "beside the waters of the Mysian Ascanius," and when Alexander the Aetolian says, "who have their homes on the Ascanian streams, on the lips of the Ascanian Lake, where dwelt Dolion the son of Silenus and Melia,"[4] they bear witness to the same thing, since the Ascanian Lake is nowhere to be found but here alone.

9. Bithynia has produced men notable for their learning: Xenocrates the philosopher, Dionysius the

[1] This Scylax was sent by Darius Hystaspis on a voyage of exploration down the Indus, and did not return for two and a half years (Herodotus 4. 44).

[2] Dionysius of Chalcis in Euboea.

[3] See *Dictionary* in Vol. IV.

[4] Passage again cited in 14. 5. 29.

---

[3] ὅς τά, Corais, for ὅτι *CDhilo*, ἔτι *rw*, ὅτι τά *xz*; so the later editors.

καὶ Διονύσιος ὁ διαλεκτικὸς καὶ Ἵππαρχος *καὶ* Θεοδόσιος *καὶ οἱ παῖδες αὐτοῦ μαθηματικοὶ* Κλεοχάρης[1] *τε ῥήτωρ,*[2] *ὁ* Μυρλεανός, Ἀσκληπιάδης *τε ἰατρός, ὁ* Προυσιεύς.

10. Πρὸς *νότον δ' εἰσὶ τοῖς* Βιθυνοῖς *οἱ περὶ τὸν* Ὄλυμπον Μυσοί *(οὓς* Ὀλυμπηνοὺς *καλοῦσί τινες, οἱ δ'* Ἑλλησποντίους*) καὶ ἡ ἐφ'* Ἑλλησπόντῳ Φρυγία, *τοῖς δὲ* Παφλαγόσι Γαλάται, *ἀμφοτέρων τε τούτων ἔτι πρὸς νότον ἡ μεγάλη* Φρυγία *καὶ* Λυκαονία *μέχρι τοῦ* Ταύρου *τοῦ* Κιλικίου *καὶ τοῦ* Πισιδικοῦ. *ἐπεὶ δὲ τὰ τῇ* Παφλαγονίᾳ *συνεχῆ παράκειται τῷ* Πόντῳ *καὶ τῇ* Καππαδοκίᾳ *καὶ τοῖς ἤδη περιωδευμένοις ἔθνεσιν, οἰκεῖον ἂν εἴη τὰ τούτοις γειτονοῦντα μέρη προσαποδοῦναι πρῶτον, ἔπειτα τοὺς ἑξῆς τόπους παραδεῖξαι.*

V

1. Πρὸς *νότον τοίνυν εἰσὶ τοῖς* Παφλαγόσι Γαλάται· *τούτων δ' ἐστὶν ἔθνη τρία, δύο μὲν τῶν ἡγεμόνων ἐπώνυμα,* Τρόκμοι[3] *καὶ* Τολιστοβώγιοι,[4] *τὸ τρίτον δ' ἀπὸ τοῦ ἐν* Κελτικῇ *ἔθνους* Τεκτοσάγες. *κατέσχον δὲ τὴν χώραν ταύτην οἱ* Γαλάται *πλανηθέντες πολὺν χρόνον καὶ καταδραμόντες τὴν ὑπὸ τοῖς* Ἀτταλικοῖς *βασιλεῦσι χώραν καὶ*

[1] Κλεοχάρης, Meineke, for Κλεοφάνης.

[2] After ῥήτωρ Meineke wrongly emends the text to read ὅ [τε] Μυρλεανὸς Ἀσκληπιάδης [γραμματικὸς] ἰατρὸς [τε] ὁ Προυσιεύς. See Pauly-Wissowa, *s. vv.*

[3] CD*hilo* read Τρόγμοι, E Τρόγκοι.

[4] Τολιστοβώγιοι, Kramer, for Τολιστοβῶγοι; so the later editors.

dialectician, Hipparchus,[1] Theodosius and his sons the mathematicians, and also Cleochares the rhetorician of Myrleia, and Asclepiades [2] the physician of Prusa.

10. To the south of the Bithynians are the Mysians round Olympus (who by some are called the Olympeni and by others the Hellespontii) and the Hellespontian Phrygia; and to the south of the Paphlagonians are the Galatae; and still to the south of these two is Greater Phrygia, as also Lycaonia, extending as far as the Cilician and the Pisidian Taurus. But since the region continuous with Paphlagonia is adjacent to Pontus and Cappadocia and the tribes which I have already described, it might be appropriate for me first to give an account of the parts in the neighbourhood of these and then set forth a description of the places that come next thereafter.

V

1. The Galatians, then, are to the south of the Paphlagonians. And of these there are three tribes; two of them, the Trocmi and the Tolistobogii, are named after their leaders, whereas the third, the Tectosages, is named after the tribe in Celtica.[3] This country was occupied by the Galatae after they had wandered about for a long time, and after they had overrun the country that was subject to the Attalic and the Bithynian kings, until by volun-

[1] See *Dictionary* in Vol. I.

[2] The friend of Crassus; lived at the beginning of the first century B.C.

[3] See 4. 1. 13.

τοῖς Βιθυνοῖς, ἕως παρ' ἑκόντων ἔλαβον τὴν νῦν Γαλατίαν καὶ Γαλλογραικίαν λεγομένην. ἀρχηγὸς δὲ δοκεῖ μάλιστα τῆς περαιώσεως τῆς εἰς τὴν C 567 Ἀσίαν γενέσθαι Λεοννόριος. τριῶν δὲ ὄντων ἐθνῶν ὁμογλώττων καὶ κατ' ἄλλο οὐδὲν ἐξηλλαγμένων, ἕκαστον διελόντες εἰς τέτταρας μερίδας τετραρχίαν ἐκάλεσαν, τετράρχην ἔχουσαν ἴδιον καὶ δικαστὴν ἕνα καὶ στρατοφύλακα ἕνα, ὑπὸ τῷ τετράρχῃ τεταγμένους, ὑποστρατοφύλακας δὲ δύο. ἡ δὲ τῶν δώδεκα τετραρχῶν βουλὴ ἄνδρες ἦσαν τριακόσιοι, συνήγοντο δὲ εἰς τὸν καλούμενον Δρυνέμετον. τὰ μὲν οὖν φονικὰ ἡ βουλὴ ἔκρινε, τὰ δὲ ἄλλα οἱ τετράρχαι καὶ οἱ δικασταί. πάλαι μὲν οὖν ἦν τοιαύτη τις ἡ διάταξις, καθ' ἡμᾶς δὲ εἰς τρεῖς, εἶτ' εἰς δύο ἡγεμόνας, εἶτα εἰς ἕνα ἧκεν ἡ δυναστεία, εἰς Δηιόταρον, εἶτα ἐκεῖνον διεδέξατο Ἀμύντας· νῦν δ' ἔχουσι Ῥωμαῖοι καὶ ταύτην καὶ τὴν ὑπὸ τῷ Ἀμύντᾳ γενομένην πᾶσαν εἰς μίαν συναγαγόντες ἐπαρχίαν.

2. Ἔχουσι δὲ οἱ μὲν Τρόκμοι[1] τὰ πρὸς τῷ Πόντῳ καὶ τῇ Καππαδοκίᾳ· ταῦτα δ' ἐστὶ τὰ κράτιστα ὧν νέμονται Γαλάται· φρούρια δ' αὐτοῖς τετείχισται τρία, Ταούιον, ἐμπόριον τῶν ταύτῃ, ὅπου ὁ τοῦ Διὸς κολοσσὸς χαλκοῦς καὶ τέμενος αὐτοῦ ἄσυλον, καὶ Μιθριδάτιον, ὃ ἔδωκε Πομπήιος Βογοδιατάρῳ,[2] τῆς Ποντικῆς βασιλείας ἀφορίσας· τρίτον δέ πως Δανάλα,[3] ὅπου τὸν

[1] Τρόκμοι, *man. sec.* in E, Τρόγμοι other MSS.

[2] Βογοδιατάρῳ is doubtful. For various conjectures see notes of Groskurd, Kramer, and C. Müller.

[3] C reads πω instead of πως. Meineke (*Vind. Strab.*) conjectures Πωδάναλα.

tary cession they received the present Galatia, or Gallo-Graecia, as it is called. Leonnorius is generally reputed to have been the chief leader of their expedition across to Asia. The three tribes spoke the same language and differed from each other in no respect; and each was divided into four portions which were called tetrarchies, each tetrarchy having its own tetrarch, and also one judge and one military commander, both subject to the tetrarch, and two subordinate commanders. The Council of the twelve tetrarchs consisted of three hundred men, who assembled at Drynemetum, as it was called. Now the Council passed judgment upon murder cases, but the tetrarchs and the judges upon all others. Such, then, was the organisation of Galatia long ago, but in my time the power has passed to three rulers, then to two, and then to one, Deïotarus, and then to Amyntas, who succeeded him. But at the present time the Romans possess both this country and the whole of the country that became subject to Amyntas, having united them into one province.[1]

2. The Trocmi possess the parts near Pontus and Cappadocia. These are the most powerful of the parts occupied by the Galatians. They have three walled garrisons: Tavium, the emporium of the people in that part of the country, where are the colossal statue of Zeus in bronze and his sacred precinct, a place of refuge; and Mithridatium, which Pompey gave to Bogodiatarus, having separated it from the kingdom of Pontus; and third, Danala,[2]

[1] 25 B.C.

[2] See critical note.

σύλλογον ἐποιήσαντο Πομπήιός τε καὶ Λεύκολλος, ὁ μὲν ἥκων ἐπὶ τὴν τοῦ πολέμου διαδοχήν, ὁ δὲ παραδιδοὺς τὴν ἐξουσίαν καὶ ἀπαίρων ἐπὶ τὸν θρίαμβον. Τρόκμοι[1] μὲν δὴ ταῦτ' ἔχουσι τὰ μέρη, Τεκτοσάγες δὲ τὰ πρὸς τῇ μεγάλῃ Φρυγίᾳ τῇ κατὰ Πεσσινοῦντα καὶ Ὀρκαόρκους· τούτων δ' ἦν φρούριον Ἄγκυρα, ὁμώνυμος τῇ πρὸς Λυδίαν περὶ Βλαῦδον[2] πολίχνῃ Φρυγιακῇ. Τολιστοβώγιοι δὲ ὅμοροι Βιθυνοῖς εἰσὶ καὶ τῇ Ἐπικτήτῳ καλουμένῃ Φρυγίᾳ. φρούρια δ' αὐτῶν ἐστὶ τό τε Βλούκιον[3] καὶ τὸ Πήιον, ὧν τὸ μὲν ἦν βασίλειον Δηιοτάρου, τὸ δὲ γαζοφυλάκιον.

3. Πεσσινοῦς δ' ἐστὶν ἐμπόριον τῶν ταύτῃ μέγιστον, ἱερὸν ἔχον τῆς Μητρὸς τῶν θεῶν σεβασμοῦ μεγάλου τύγχανον· καλοῦσι δ' αὐτὴν Ἄγδιστιν. οἱ δ' ἱερεῖς τὸ παλαιὸν μὲν δυνάσται τινὲς ἦσαν, ἱερωσύνην καρπούμενοι μεγάλην, νυνὶ δὲ τούτων μὲν αἱ τιμαὶ πολὺ μεμείωνται, τὸ δὲ ἐμπόριον συμμένει· κατεσκεύασται δ' ὑπὸ τῶν Ἀτταλικῶν βασιλέων ἱεροπρεπῶς τὸ τέμενος ναῷ τε καὶ στοαῖς λευκολίθοις· ἐπιφανὲς δ' ἐποίησαν Ῥωμαῖοι τὸ ἱερόν, ἀφίδρυμα ἐνθένδε τῆς θεοῦ μεταπεμψάμενοι κατὰ τοὺς τῆς Σιβύλλης χρησμούς, καθάπερ καὶ τοῦ Ἀσκληπιοῦ τοῦ ἐν Ἐπιδαύρῳ. ἔστι δὲ καὶ ὄρος ὑπερκείμενον τῆς πόλεως τὸ Δίνδυμον, ἀφ' οὗ ἡ Δινδυμηνή, καθάπερ ἀπὸ τῶν Κυβέλων ἡ Κυβέλη. πλησίον

[1] CD*hilow* read Τρόγμοι instead of Τρόκμοι.

[2] Βλαῦδον, Xylander, for Βλαῦρον; so the later editors.

[3] Βλούκιον, Groskurd and Kramer would emend to Λουκήιον.

where Pompey and Leucullus had their conference, Pompey coming there as successor of Leucullus in the command of the war, and Leucullus giving over to Pompey his authority and leaving the country to celebrate his triumph. The Trocmi, then, possess these parts, but the Tectosages the parts near Greater Phrygia in the neighbourhood of Pessinus and Orcaorci. To the Tectosages belonged the fortress Ancyra, which bore the same name as the Phrygian town situated toward Lydia in the neighbourhood of Blaudus. And the Tolistobogii border on the Bithynians and Phrygia "Epictetus," as it is called. Their fortresses are Blucium and Peïum, the former of which was the royal residence of Deïotarus and the latter the place where he kept his treasures.

3. Pessinus is the greatest of the emporiums in that part of the world, containing a temple of the Mother of the gods, which is an object of great veneration. They call her Agdistis. The priests were in ancient times potentates, I might call them, who reaped the fruits of a great priesthood, but at present the prerogatives of these have been much reduced, although the emporium still endures. The sacred precinct has been built up by the Attalic kings in a manner befitting a holy place, with a sanctuary and also with porticoes of white marble. The Romans made the temple famous when, in accordance with oracles of the Sibyl, they sent for the statue of the goddess there, just as they did in the case of that of Asclepius at Epidaurus. There is also a mountain situated above the city, Dindymum, after which the country Dindymenê was named, just as Cybelê was named after Cybela.

δὲ καὶ ὁ Σαγγάριος ποταμὸς ποιεῖται τὴν ῥύσιν· C 568 ἐπὶ δὲ τούτῳ τὰ παλαιὰ τῶν Φρυγῶν οἰκητήρια Μίδου καὶ ἔτι πρότερον Γορδίου καὶ ἄλλων τινῶν, οὐδ' ἴχνη σώζοντα πόλεων, ἀλλὰ κῶμαι μικρῷ μείζους τῶν ἄλλων, οἷόν ἐστι τὸ Γόρδιον καὶ Γορβεοῦς, τὸ τοῦ Κάστορος βασίλειον τοῦ Σαωκονδαρίου, ἐν ᾧ γαμβρὸν ὄντα τοῦτον ἀπέσφαξε Δηιόταρος καὶ τὴν θυγατέρα τὴν ἑαυτοῦ· τὸ δὲ φρούριον κατέσπασε, καὶ διελυμήνατο τὸ πλεῖστον τῆς κατοικίας.

4. Μετὰ δὲ τὴν Γαλατίαν πρὸς νότον ἥ τε λίμνη ἐστὶν ἡ Τάττα, παρακειμένη τῇ μεγάλῃ Καππαδοκίᾳ τῇ κατὰ τοὺς Μοριμηνούς, μέρος δ' οὖσα τῆς μεγάλης Φρυγίας, καὶ ἡ συνεχὴς ταύτῃ μέχρι τοῦ Ταύρου, ἧς τὴν πλείστην Ἀμύντας εἶχεν. ἡ μὲν οὖν Τάττα ἁλοπήγιόν ἐστιν αὐτοφυές, οὕτω δὲ περιπήττεται ῥᾳδίως τὸ ὕδωρ παντὶ τῷ βαπτισθέντι εἰς αὐτό, ὥστε στεφάνους ἁλῶν ἀνέλκουσιν, ἐπειδὰν καθῶσι κύκλον σχοίνινον, τά τε ὄρνεα ἁλίσκεται τὰ προσαψάμενα τῷ πτερώματι τοῦ ὕδατος παραχρῆμα πίπτοντα διὰ τὴν περίπηξιν τῶν ἁλῶν.

VI

1. Τοιαύτη[1] δὴ Τάττα ἐστί. καὶ τὰ περὶ Ὀρκαόρκους καὶ Πιτνισσὸν[2] καὶ τὰ τῶν Λυκαόνων ὀροπέδια ψυχρὰ καὶ ψιλὰ καὶ ὀναγρόβοτα, ὑδάτων δὲ σπάνις πολλή· ὅπου δὲ καὶ εὑρεῖν

[1] τοιαύτη, Jones, for the corrupt ἥ τε of the MSS. For other conjectures see C. Müller (*Ind. Var. Lect.* p. 1022). Meineke inserts τοιαύτη after Τάττα.

Near by, also, flows the Sangarius River; and on this river are the ancient habitations of the Phrygians, of Midas, and of Gordius, who lived even before his time, and of certain others,—habitations which preserve not even traces of cities, but are only villages slightly larger than the others, for instance, Gordium and Gorbeus, the royal residence of Castor the son of Saocondarius, where Deïotarus, Castor's father-in-law, slew him and his own daughter. And he pulled down the fortress and ruined most of the settlement.

4. After Galatia towards the south are situated Lake Tatta, which lies alongside Greater Cappadocia near Morimenê but is a part of Greater Phrygia, and the country continuous with this lake and extending as far as the Taurus, most of which was held by Amyntas. Now Lake Tatta is a natural salt-pan; and the water so easily congeals round everything that is immersed in it, that when people let down into it rings made of rope they draw up wreaths of salt, and that, on account of the congealing of the salt, the birds which touch the water with their wings fall on the spot and are thus caught.

VI

1. Such, then, is Tatta. And the regions round Orcaorci and Pitnissus, as also the plateaus of the Lycaonians, are cold, bare of trees, and grazed by wild asses, though there is a great scarcity of water; and even where it is possible to find water, the

---

[2] Πιτνισσόν, Meineke, for Πιγνισόν.

δυνατόν, βαθύτατα φρέατα τῶν πάντων, καθάπερ ἐν Σοάτροις, ὅπου καὶ πιπράσκεται τὸ ὕδωρ (ἔστι δὲ κωμόπολις Γαρσαούρων[1] πλησίον)· ὅμως δὲ καίπερ ἄνυδρος οὖσα ἡ χώρα πρόβατα ἐκτρέφει θαυμαστῶς, τραχείας δὲ ἐρέας, καί τινες ἐξ αὐτῶν τούτων μεγίστους πλούτους ἐκτήσαντο· Ἀμύντας δ' ὑπὲρ τριακοσίας ἔσχε ποίμνας ἐν τοῖς τόποις τούτοις. εἰσὶ δὲ καὶ λίμναι, Κόραλις μὲν ἡ μείζων, ἡ δ' ἐλάττων Τρωγῖτις. ἐνταῦθα δέ που καὶ τὸ Ἰκόνιόν ἐστι, πολίχνιον εὖ συνῳκισμένον καὶ χώραν εὐτυχεστέραν ἔχον τῆς λεχθείσης ὀναγροβότου· τοῦτο δ' εἶχε Πολέμων. πλησιάζει δ' ἤδη τούτοις τοῖς τόποις ὁ Ταῦρος ὁ τὴν Καππαδοκίαν ὁρίζων καὶ τὴν Λυκαονίαν πρὸς τοὺς ὑπερκειμένους Κίλικας τοὺς Τραχειώτας. Λυκαόνων τε καὶ Καππαδόκων ὅριόν ἐστι τὸ μεταξὺ Κοροπασσοῦ, κώμης Λυκαόνων, καὶ Γαρσαούρων,[2] πολιχνίου Καππαδόκων· ἔστι δὲ τὸ μεταξὺ διάστημα τῶν φρουρίων τούτων ἑκατὸν εἴκοσί που στάδιοι.

2. Τῆς δὲ Λυκαονίας ἐστὶ καὶ ἡ Ἰσαυρικὴ πρὸς αὐτῷ τῷ Ταύρῳ ἡ τὰ Ἴσαυρα ἔχουσα κώμας δύο ὁμωνύμους, τὴν μὲν Παλαιὰν καλουμένην τὴν δὲ Νέαν[3] εὐερκῆ· ὑπήκοοι δ' ἦσαν ταύταις καὶ ἄλλαι κῶμαι συχναί, λῃστῶν δ' ἅπασαι κατοικίαι. παρέσχον δὲ καὶ Ῥωμαίοις πράγματα καὶ τῷ Ἰσαυρικῷ προσαγορευθέντι Πουβλίῳ Σερβιλίῳ, ὃν ἡμεῖς εἴδομεν, ὃς καὶ ταῦτα ὑπέταξε Ῥωμαίοις C 569 καὶ τὰ πολλὰ τῶν πειρατῶν ἐρύματα ἐξεῖλε τὰ ἐπὶ τῇ θαλάττῃ.

[1] Γαρσαούρων, Corais, for Γαρσαβόρων; so Meineke.
[2] Γαρσαούρων, Corais, for Γαρεαθύρων; so Meineke.
[3] τὴν δὲ Νέαν, Meineke inserts.

wells are the deepest in the world, just as in Soatra, where the water is actually sold (this is a village-city near Garsaüra). But still, although the country is unwatered,[1] it is remarkably productive of sheep; but the wool is coarse, and yet some persons have acquired very great wealth from this alone. Amyntas had over three hundred flocks in this region. There are also two lakes in this region, the larger being Lake Coralis and the smaller Lake Trogitis. In this neighbourhood is also Iconium, a town that is well settled and has a more prosperous territory than the above-mentioned ass-grazing country. This place was held by Polemon. Here the region in question is near the Taurus, which separates Cappadocia and Lycaonia from Cilicia Tracheia,[2] which last lies above that region. The boundary between the Lycaonians and the Cappadocians lies between Coropassus, a village of the Lycaonians, and Garsaüra, a town of the Cappadocians. The distance between these strongholds is about one hundred and twenty stadia.

2. To Lycaonia belongs also Isauricê, near the Taurus itself, which has the two Isauras, villages bearing the same name, one of which is called Old Isaura, and the other New Isaura, which is well-fortified. Numerous other villages were subject to these, and they all were settlements of robbers. They were a source of much trouble to the Romans and in particular to Publius Servilius, surnamed Isauricus, with whom I was acquainted; he subjected these places to the Romans and also destroyed most of the strongholds of the pirates that were situated on the sea.

[1] *i.e.* by streams.

[2] See 14. 5. 1.

3. Τῆς δ' Ἰσαυρικῆς ἐστὶν ἐν πλευραῖς ἡ Δέρβη, μάλιστα τῇ Καππαδοκίᾳ ἐπιπεφυκὸς τὸ τοῦ Ἀντιπάτρου τυραννεῖον τοῦ Δερβήτου· τοῦ δ' ἦν καὶ τὰ Λάρανδα· ἐφ' ἡμῶν δὲ καὶ τὰ Ἴσαυρα καὶ τὴν Δέρβην Ἀμύντας εἶχεν, ἐπιθέμενος τῷ Δερβήτῃ καὶ ἀνελὼν αὐτόν, τὰ δ' Ἴσαυρα παρὰ τῶν Ῥωμαίων λαβών· καὶ δὴ βασίλειον ἑαυτῷ κατεσκεύαζεν ἐνταῦθα, τὴν παλαιὰν Ἴσαυραν[1] ἀνατρέψας. ἐν δὲ τῷ αὐτῷ χωρίῳ καινὸν τεῖχος οἰκοδομῶν οὐκ ἔφθη συντελέσας, ἀλλὰ διέφθειραν αὐτὸν οἱ Κίλικες, ἐμβάλλοντα[2] εἰς τοὺς Ὁμοναδεῖς καὶ ἐξ ἐνέδρας ληφθέντα.

4. Τὴν γὰρ Ἀντιόχειαν ἔχων τὴν πρὸς τῇ Πισιδίᾳ μέχρι Ἀπολλωνιάδος τῆς πρὸς Ἀπαμείᾳ τῇ Κιβωτῷ καὶ τῆς παρωρείου τινὰ καὶ τὴν Λυκαονίαν ἐπειρᾶτο τοὺς ἐκ τοῦ Ταύρου κατατρέχοντας Κίλικας καὶ Πισίδας τὴν χώραν ταύτην, Φρυγῶν οὖσαν καὶ Κιλίκων,[3] ἐξαιρεῖν, καὶ πολλὰ χωρία ἐξεῖλεν ἀπόρθητα πρότερον ὄντα, ὧν καὶ Κρῆμνα· τὸ δὲ Σανδάλιον οὐδ' ἐνεχείρησε βίᾳ προσάγεσθαι, μεταξὺ κείμενον τῆς τε Κρήμνης καὶ Σαγαλασσοῦ.

5. Τὴν μὲν οὖν Κρῆμναν ἄποικοι Ῥωμαίων ἔχουσιν, ἡ Σαγαλασσὸς δ' ἐστὶν ὑπὸ τῷ αὐτῷ ἡγεμόνι τῶν Ῥωμαίων, ὑφ' ᾧ καὶ ἡ Ἀμύντου βασιλεία πᾶσα· διέχει δ' Ἀπαμείας ἡμέρας ὁδόν, κατάβασιν ἔχουσα σχεδόν τι καὶ τριάκοντα

[1] Ἴσαυραν, Meineke, for Ἰσαυρίαν.

[2] ἐμβάλλοντα, the reading of the MSS., Jones restores, for ἐμβαλόντα, the reading of Corais and later editors.

[3] καὶ Κιλίκων apparently is an error for καὶ Λυκαόνων, or else should be omitted from the text (so Meineke).

3. On the side of Isauricê lies Derbê, which lies closer to Cappadocia than to any other country and was the royal seat of the tyrant Antipater Derbetes. He also possessed Laranda. But in my time Derbê and also the two Isauras have been held by Amyntas,[1] who attacked and killed Derbetes, although he received Isaura from the Romans. And, indeed, after destroying the Old Isaura, he built for himself a royal residence there. And though he was building a new wall in the same place, he did not live to complete it, but was killed by the Cilicians, when he was invading the country of the Homonadeis and was captured by ambuscade.

4. For, being in possession of the Antiocheia near Pisidia and of the country as far as the Apollonias near Apameia Cibotus and of certain parts of the country alongside the mountain, and of Lycaonia, he was trying to exterminate the Cilicians and the Pisidians, who from the Taurus were overrunning this country, which belonged to the Phrygians and the Cilicians;[2] and he captured many places which previously had been impregnable, among which was Cremna. However, he did not even try to win Sandalium by force, which is situated between Cremna and Sagalassus.

5. Now Cremna is occupied by Roman colonists: and Sagalassus is subject to the same Roman governor to whom the whole kingdom of Amyntas was subject. It is a day's journey distant from Apameia, having a descent of about thirty stadia from the fortress. It

[1] The Galatian Amyntas who fought with Antony against Augustus at the battle of Actium (31 B.C.).

[2] See critical note.

σταδίων ἀπὸ τοῦ ἐρύματος· καλοῦσι δ' αὐτὴν καὶ Σελγησσόν· ταύτην δὲ τὴν πόλιν καὶ Ἀλέξανδρος εἷλεν. ὁ δ' οὖν Ἀμύντας τὴν μὲν Κρῆμναν εἷλεν, εἰς δὲ τοὺς Ὁμοναδέας παρελθών, οἳ ἐνομίζοντο ἀληπτότατοι, καὶ καταστὰς ἤδη κύριος τῶν πλείστων χωρίων, ἀνελὼν καὶ τὸν τύραννον αὐτῶν ἐξ ἀπάτης ἐλήφθη διὰ τῆς τοῦ τυράννου γυναικός. καὶ τοῦτον μὲν ἐκεῖνοι διέφθειραν, ἐκείνους δὲ Κυρίνιος ἐξεπόρθησε λιμῷ καὶ τετρακισχιλίους ἄνδρας ἐζώγρησε καὶ συνῴκισεν εἰς τὰς ἐγγὺς πόλεις, τὴν δὲ χώραν ἀπέλιπεν ἔρημον τῶν ἐν ἀκμῇ. ἔστι δὲ[1] ἐν ὑψηλοῖς τοῦ Ταύρου μέρεσι, κρημνοῖς ἀποτόμοις σφόδρα καὶ τὸ πλέον ἀβάτοις, ἐν μέσῳ κοῖλον καὶ εὔγεων πεδίον, εἰς αὐλῶνας πλείους διῃρημένον· τοῦτο δὲ γεωργοῦντες ᾤκουν ἐν ταῖς ὑπερκειμέναις ὀφρύσιν ἢ σπηλαίοις, τὰ πολλὰ δ' ἔνοπλοι ἦσαν καὶ κατέτρεχον τὴν ἀλλοτρίαν, ἔχοντες ὄρη τειχίζοντα τὴν χώραν αὐτῶν.

VII

1. Συναφεῖς δ' εἰσὶ τούτοις οἵ τε ἄλλοι Πισίδαι καὶ οἱ Σελγεῖς, οἵπερ εἰσὶν ἀξιολογώτατοι τῶν Πισιδῶν. τὸ μὲν οὖν πλέον αὐτῶν μέρος τὰς ἀκρωρείας τοῦ Ταύρου κατέχει, τινὲς δὲ καὶ ὑπὲρ
C 570 Σίδης καὶ Ἀσπένδου, Παμφυλικῶν πόλεων, κατέχουσι γεώλοφα χωρία, ἐλαιόφυτα πάντα, τὰ δ' ὑπὲρ τούτων, ἤδη ὀρεινά, Κατεννεῖς, ὅμοροι

[1] After δέ the MSS., except *Dhi*, add καί.

is also called Selgessus; this city was also captured by Alexander. Now Amyntas captured Cremna, and, passing into the country of the Homonadeis, who were considered too strong to capture, and having now established himself as master of most of the places, having even slain their tyrant, was caught by treachery through the artifice of the tyrant's wife. And he was put to death by those people, but Cyrinius[1] overthrew the inhabitants by starving them, and captured alive four thousand men and settled them in the neighbouring cities, leaving the country destitute of all its men who were in the prime of life. In the midst of the heights of the Taurus, which are very steep and for the most part impassable, there is a hollow and fertile plain which is divided into several valleys. But though the people tilled this plain, they lived on the overhanging brows of the mountains or in caves. They were armed for the most part and were wont to overrun the country of others, having mountains that served as walls about their country.

VII

1. Contiguous to these are the Pisidians, and in particular the Selgeis, who are the most notable of the Pisidians. Now the greater part of them occupy the summits of the Taurus, but some, situated above Sidê and Aspendus, Pamphylian cities, occupy hilly places, everywhere planted with olive-trees; and the region above this (we are now in the mountains) is occupied by the Catenneis, whose country borders

[1] Sulpicius Quirinus, governor of Syria.

Σελγεῦσι καὶ Ὁμοναδεῦσι, Σαγαλασσεῖς δ' ἐπὶ τὰ ἐντὸς τὰ πρὸς τῇ Μιλυάδι.

2. Φησὶ δ' Ἀρτεμίδωρος τῶν Πισιδῶν[1] πόλεις εἶναι Σέλγην, Σαγαλασσόν, Πετνηλισσόν, Ἄδαδα, Τυμβριάδα,[2] Κρῆμναν, Πιτυασσόν, Ἄμβλαδα, Ἀνάβουρα, Σίνδα, Ἀαρασσόν, Ταρβασσόν, Τερμησσόν· τούτων δ' οἱ μέν εἰσι τελέως ὀρεινοί, οἱ δὲ καὶ μέχρι τῶν ὑπωρειῶν καθήκοντες ἐφ' ἑκάτερα, ἐπί τε τὴν Παμφυλίαν καὶ τὴν Μιλυάδα Φρυξὶ καὶ Λυδοῖς καὶ Καρσὶν ὅμοροι, πᾶσιν εἰρηνικοῖς ἔθνεσι, καίπερ προσβόροις οὖσιν. οἱ δὲ Πάμφυλοι, πολὺ τοῦ Κιλικίου φύλου μετέχοντες, οὐ τελέως ἀφεῖνται τῶν λῃστρικῶν ἔργων, οὐδὲ τοὺς ὁμόρους ἐῶσι καθ' ἡσυχίαν ζῆν, καίπερ τὰ νότια μέρη τῆς ὑπωρείας τοῦ Ταύρου κατέχοντες. εἰσὶ δὲ τοῖς Φρυξὶν ὅμοροι καὶ τῇ Καρίᾳ Τάβαι[3] καὶ Σίνδα καὶ Ἄμβλαδα, ὅθεν καὶ ὁ Ἀμβλαδεὺς οἶνος ἐκφέρεται πρὸς διαίτας ἰατρικὰς ἐπιτήδειος.

3. Τῶν δ' οὖν ὀρεινῶν, οὓς εἶπον,[4] Πισιδῶν οἱ μὲν ἄλλοι κατὰ τυραννίδας μεμερισμένοι, καθάπερ οἱ Κίλικες, λῃστρικῶς ἤσκηνται· φασὶ δ' αὐτοῖς τῶν Λελέγων συγκαταμιχθῆναί τινας τὸ παλαιόν, πλάνητας ἀνθρώπους, καὶ συμμεῖναι διὰ τὴν ὁμοιοτροπίαν αὐτόθι. Σέλγη δὲ καὶ ἐξ ἀρχῆς μὲν ὑπὸ Λακεδαιμονίων ἐκτίσθη πόλις, καὶ ἔτι πρότερον ὑπὸ Κάλχαντος· ὕστερον δὲ καθ' αὑτὴν

[1] Πισιδῶν D, Πισιδικῶν other MSS.

[2] Ἄδαδα, Τυμβριάδα, Corais, from conj. of Wesseling, for ἀδαδάτην βριάδα; so the later editors.

[3] Τάβαι, the editors, from Stephanus (*s.v.* Ἄμβλαδα), for Ταβᾷ D, Τιαμᾷ, Τιάβαι *r*, Τιάβα other MSS.

on that of the Selgeis and the Homonadeis; but the Sagalasseis occupy the region this side the Taurus that faces Milyas.

2. Artemidorus says that the cities of the Pisidians are Selgê, Sagalassus, Petnelissus, Adada, Tymbriada, Cremna, Pityassus, Amblada, Anabura, Sinda, Aarassus, Tarbassus, and Termessus. Of these, some are entirely in the mountains, while others extend even as far as the foot-hills on either side, to both Pamphylia and Milyas, and border on the Phrygians and the Lydians and the Carians, which are all peaceable tribes, although they are situated towards the north. But the Pamphylians, who share much in the traits of the Cilician stock of people, do not wholly abstain from the business of piracy, nor yet do they allow the peoples on their borders to live in peace, although they occupy the southern parts of the foot-hills of the Taurus. And on the borders of the Phrygians and Caria are situated Tabae and Sinda, and also Amblada, whence is exported the Ambladian wine, which is suitable for use in medicinal diets.

3. Now all the rest of the above-mentioned Pisidians who live in the mountains are divided into separate tribes governed by tyrants, like the Cilicians and are trained in piracy. It is said that in ancient times certain Leleges,[1] a wandering people, intermingled with them and on account of similarity of character stayed there. Selgê was founded at first by the Lacedaemonians as a city, and still earlier by Calchas; but later it remained an independent city,

[1] See 7. 7. 2.

---

[4] *οὓς εἶπον*, Groskurd (*ἃς εἶπα* Corais), for *ὡς εἰπεῖν*; so the later editors in general.

ἔμεινεν αὐξηθεῖσα ἐκ τοῦ πολιτεύεσθαι νομίμως, ὥστε καὶ δισμυρίανδρός ποτε εἶναι. θαυμαστὴ δ' ἐστὶν ἡ φύσις τῶν τόπων· ἐν γὰρ ταῖς ἀκρωρείαις τοῦ Ταύρου χώρα μυριάδας τρέφειν δυναμένη σφόδρα εὔκαρπός ἐστιν, ὥστε καὶ ἐλαιόφυτα εἶναι πολλὰ χωρία καὶ εὐάμπελα, νομάς τε ἀφθόνους ἀνεῖσθαι παντοδαποῖς βοσκήμασι· κύκλῳ δ' ὑπέρκεινται δρυμοὶ ποικίλης ὕλης. πλεῖστος δ' ὁ στύραξ φύεται παρ' αὐτοῖς, δένδρον οὐ μέγα ὀρθηλόν,[1] ἀφ' οὗ καὶ τὰ στυράκινα ἀκοντίσματα, ἐοικότα τοῖς κρανεΐνοις·[2] ἐγγίνεται δ' ἐν τοῖς στελέχεσι ξυλοφάγου τι σκώληκος εἶδος, ὃ μέχρι τῆς ἐπιφανείας διαφαγὸν τὸ ξύλον τὸ μὲν πρῶτον πιτύροις ἢ πρίσμασιν ἐοικός τι ψῆγμα προχεῖ, καὶ σωρὸς συνίσταται πρὸς τῇ ῥίζῃ, μετὰ δὲ ταῦτα ἀπολείβεταί τις ὑγρασία δεχομένη πῆξιν ῥᾳδίαν παραπλησίαν τῇ κόμμει· ταύτης δὲ τὸ μὲν ἐπὶ τὸ ψῆγμα πρὸς τῇ ῥίζῃ κατενεχθὲν[3] ἀναμίγνυται τούτῳ τε καὶ τῇ γῇ, πλὴν ὅσον τὸ μὲν ἐν ἐπιπολῇ συστὰν διαμένει καθαρόν, τὸ δ' ἐν

C 571 τῇ ἐπιφανείᾳ τοῦ στελέχους, καθ' ἣν ῥεῖ, πήττεται, καὶ τοῦτο καθαρόν· ποιοῦσι δὲ καὶ ἐκ τοῦ μὴ καθαροῦ μῖγμα ξυλομιγές τι καὶ γεωμιγές, εὐωδέστερον τοῦ καθαροῦ, τῇ δ' ἄλλῃ δυνάμει λειπόμενον (λανθάνει δὲ τοὺς πολλούς), ᾧ πλείστῳ χρῶνται θυμιάματι οἱ δεισιδαίμονες. ἐπαινεῖται

[1] ὀρθηλόν, as Meineke suspects, might be an error for ὀρθόκαυλον ("straight-stalked").

[2] κρανεΐνοις, Tzschucke, for κραναίνοις CDE*hilorw*, κραναιίνοις *x*, κρανίνοις *z*.

[3] κατενεχθέν D, καταμιχθέν other MSS.

having waxed so powerful on account of the law-abiding manner in which its government was conducted that it once contained twenty thousand men. And the nature of the region is wonderful, for among the summits of the Taurus there is a country which can support tens of thousands of inhabitants and is so very fertile that it is planted with the olive in many places, and with fine vineyards, and produces abundant pasture for cattle of all kinds; and above this country, all round it, lie forests of various kinds of timber. But it is the styrax-tree[1] that is produced in greatest abundance there, a tree which is not large but grows straight up, the tree from which the styracine javelins are made, similar to those made of cornel-wood. And a species of wood-eating worm[2] is bred in the trunk which eats through the wood of the tree to the surface, and at first pours out raspings like bran or saw-dust, which are piled up at the root of the tree; and then a liquid substance exudes which readily hardens into a substance like gum. But a part of this liquid flows down upon the raspings at the root of the tree and mixes with both them and the soil, except so much of it as condenses on the surface of the raspings and remains pure, and except the part which hardens on the surface of the trunk down which it flows, this too being pure. And the people make a kind of substance mixed with wood and earth from that which is not pure, this being more fragrant than the pure substance but otherwise inferior in strength to it (a fact unnoticed by most people), which is used in large quantities as frankincense by the worshippers of the gods. And

[1] A species of gum-tree.
[2] Apparently some kind of wood-boring beetle.

δὲ καὶ ἡ Σελγικὴ ἶρις καὶ τὸ ἀπ' αὐτῆς ἄλειμμα. ἔχει δ' ὀλίγας προσβάσεις τὰ[1] περὶ τὴν πόλιν καὶ τὴν χώραν τὴν Σελγέων, ὀρεινὴν κρημνῶν καὶ χαραδρῶν οὖσαν πλήρη, ἃς ποιοῦσιν ἄλλοι τε ποταμοὶ καὶ ὁ Εὐρυμέδων καὶ ὁ Κέστρος, ἀπὸ τῶν Σελγικῶν ὀρῶν εἰς τὴν Παμφυλίαν ἐκπίπτοντες θάλατταν· γέφυραι δ' ἐπίκεινται ταῖς ὁδοῖς. διὰ δὲ[2] τὴν ἐρυμνότητα οὔτε πρότερον οὔθ' ὕστερον οὐδ' ἅπαξ οἱ Σελγεῖς ἐπ' ἄλλοις ἐγένοντο, ἀλλὰ τὴν μὲν ἄλλην χώραν ἀδεῶς ἐκαρποῦντο, ὑπὲρ δὲ τῆς κάτω τῆς τε ἐν τῇ Παμφυλίᾳ καὶ τῆς ἐντὸς τοῦ Ταύρου διεμάχοντο πρὸς τοὺς βασιλέας ἀεί· πρὸς δὲ τοὺς Ῥωμαίους ἐπὶ τακτοῖς τισὶ κατεῖχον τὴν χώραν· πρὸς Ἀλέξανδρον δὲ πρεσβευσάμενοι δέχεσθαι τὰ προστάγματα εἶπον κατὰ φιλίαν· νῦν δὲ ὑπήκοοι τελέως γεγόνασι, καί εἰσιν ἐν τῇ ὑπὸ Ἀμύντᾳ τεταγμένῃ πρότερον.

VIII

1. Τοῖς δὲ Βιθυνοῖς ὁμοροῦσι πρὸς νότον, ὡς ἔφην, οἱ περὶ τὸν Ὄλυμπον τὸν Μύσιον προσαγορευόμενον[3] Μυσοί τε καὶ Φρύγες ἑκάτερον δὲ τὸ ἔθνος διττόν ἐστι. Φρυγία τε γὰρ ἡ μὲν καλεῖται μεγάλη, ἧς ὁ Μίδας ἐβασίλευσε, καὶ ἧς μέρος οἱ Γαλάται κατέσχον, ἡ δὲ μικρά, ἡ ἐφ' Ἑλλησ-

[1] τά, before περί, Corais inserts; so the later editors.
[2] δέ, after διά, is omitted by all MSS. except D.
[3] προσαγορευόμενον *w*, προσαγορευόμενοι other MSS.

people praise also the Selgic iris[1] and the ointment made from it. The region round the city and the territory of the Selgians has only a few approaches, since their territory is mountainous and full of precipices and ravines, which are formed, among other rivers, by the Eurymedon and the Cestrus, which flow from the Selgic mountains and empty into the Pamphylian Sea. But they have bridges on their roads. Because of their natural fortifications, however, the Selgians have never even once, either in earlier or later times, become subject to others, but unmolested have reaped the fruit of the whole country except the part situated below them in Pamphylia and inside the Taurus, for which they were always at war with the kings; but in their relations with the Romans, they occupied the part in question on certain stipulated conditions. They sent an embassy to Alexander and offered to receive his commands as a friendly country, but at the present time they have become wholly subject to the Romans and are included in the territory that was formerly subject to Amyntas.

VIII

1. Bordering on the Bithynians towards the south, as I have said,[2] are the Mysians and Phrygians who live round the Mysian Olympus, as it is called. And each of these tribes is divided into two parts. For one part of Phrygia is called Greater Phrygia, the part over which Midas reigned, a part of which was occupied by the Galatians, whereas the other is

[1] The orris-root, used in perfumery and medicine.
[2] 12. 4. 4 f.

πόντῳ καὶ ἡ περὶ τὸν Ὄλυμπον, ἡ καὶ Ἐπίκτητος λεγομένη. Μυσία τε ὁμοίως ἥ τε Ὀλυμπηνή, συνεχὴς οὖσα τῇ Βιθυνίᾳ καὶ τῇ Ἐπικτήτῳ, ἣν ἔφη Ἀρτεμίδωρος ἀπὸ τῶν πέραν Ἴστρου Μυσῶν ἀπῳκίσθαι, καὶ ἡ περὶ τὸν Κάϊκον καὶ τὴν Περγαμηνὴν μέχρι Τευθρανίας καὶ τῶν ἐκβολῶν τοῦ ποταμοῦ.

2. Οὕτω δ' ἐνήλλακται ταῦτα ἐν ἀλλήλοις, ὡς πολλάκις λέγομεν, ὥστε καὶ τὴν περὶ τὴν Σίπυλον Φρυγίαν οἱ παλαιοὶ καλοῦσιν, ἄδηλον, εἴτε τῆς μεγάλης εἴτε τῆς μικρᾶς μέρος οὖσαν, ᾗ καὶ τὸν Τάνταλον Φρύγα καὶ τὸν Πέλοπα καὶ τὴν Νιόβην· ὁποτέρως δ' ἂν ἔχῃ, ἥ γε ἐπάλλαξις φανερά. ἡ γὰρ Περγαμηνὴ καὶ ἡ Ἐλαΐτις, καθ' ἣν ὁ Κάϊκος ἐκπίπτει, καὶ ἡ μεταξὺ τούτων Τευθρανία, ἐν ᾗ Τεύθρας καὶ ἡ τοῦ Τηλέφου ἐκτροφή, ἀνὰ μέσον ἐστὶ τοῦ τε Ἑλλησπόντου καὶ τῆς περὶ Σίπυλον καὶ Μαγνησίαν τὴν ὑπ' αὐτῷ χώρας· ὥσθ', ὅπερ ἔφην, ἔργον διορίσαι

C 572 χωρὶς τὰ Μυσῶν καὶ Φρυγῶν ὁρίσματα.

3. Καὶ οἱ Λυδοὶ καὶ οἱ Μαίονες, οὓς Ὅμηρος καλεῖ Μῄονας, ἐν συγχύσει πώς εἰσι καὶ πρὸς τούτους καὶ πρὸς ἀλλήλους· ὅτι οἱ μὲν τοὺς αὐτούς, οἱ δ' ἑτέρους φασί, πρὸς δὲ τούτους,[1] ὅτι

[1] τούτους, Kramer, for τούτοις; so the later editors.

---

[1] Cf. 12. 4. 3 and foot-note.
[2] See 7. 3. 2, 10; 12. 3. 3, and 12. 4. 8.
[3] See 12. 4. 4. [4] See 12. 4. 4.
[5] Again the Mysians and Phrygians.

called Lesser Phrygia, that on the Hellespont and round Olympus, I mean Phrygia Epictetus,[1] as it is called. Mysia is likewise divided into two parts, I mean Olympenê, which is continuous with Bithynia and Phrygia Epictetus, which, according to Artemidorus, was colonised by the Mysians who lived on the far side of the Ister,[2] and, secondly, the country in the neighbourhood of the Caïcus River and Pergamenê, extending as far as Teuthrania and the outlets of the river.

2. But the boundaries of these parts have been so confused with one another, as I have often said,[3] that it is uncertain even as to the country round Mt. Sipylus, which the ancients called Phrygia, whether it was a part of Greater Phrygia or of Lesser Phrygia, where lived, they say, the "Phrygian" Tantalus and Pelops and Niobê. But no matter which of the two opinions is correct, the confusion of the boundaries is obvious; for Pergamenê and Elaïtis, where the Caïcus empties into the sea, and Teuthrania, situated between these two countries, where Teuthras lived and where Telephus was reared, lie between the Hellespont on the one side and the country round Sipylus and Magnesia, which lies at the foot of Sipylus, on the other; and therefore, as I have said before, it is a task to determine the boundaries ("Apart are the boundaries of the Mysians and Phrygians").[4]

3. And the Lydians and the Maeonians, whom Homer calls the Mëiones, are in some way confused both with these peoples and with one another, because some say that they are the same and others that they are different; and they are confused with these people[5] because some say that the Mysians

τοὺς Μυσοὺς οἱ μὲν Θρᾷκας, οἱ δὲ Λυδοὺς εἰρήκασι, κατ' αἰτίαν παλαιὰν ἱστοροῦντες, ἣν Ξάνθος ὁ Λυδὸς γράφει καὶ Μενεκράτης ὁ Ἐλαΐτης, ἐτυμολογοῦντες καὶ τὸ ὄνομα τὸ τῶν Μυσῶν, ὅτι τὴν ὀξύην οὕτως ὀνομάζουσιν οἱ Λυδοί· πολλὴ δ' ἡ ὀξύη κατὰ τὸν Ὄλυμπον, ὅπου ἐκτεθῆναί φασι τοὺς δεκατευθέντας, ἐκείνων δὲ ἀπογόνους εἶναι τοὺς ὕστερον Μυσούς, ἀπὸ τῆς ὀξύης οὕτω προσαγορευθέντας· μαρτυρεῖν δὲ καὶ τὴν διάλεκτον· μιξολύδιον γάρ πως εἶναι καὶ μιξοφρύγιον· τέως μὲν γὰρ οἰκεῖν αὐτοὺς περὶ τὸν Ὄλυμπον, τῶν δὲ Φρυγῶν ἐκ τῆς Θρᾴκης περαιωθέντων, ἀνελόντων τε[1] τῆς Τροίας ἄρχοντα καὶ τῆς πλησίον γῆς, ἐκείνους μὲν ἐνταῦθα οἰκῆσαι, τοὺς δὲ Μυσοὺς ὑπὲρ τὰς τοῦ Καΐκου πηγὰς πλησίον Λυδῶν.

4. Συνεργεῖ δὲ πρὸς τὰς τοιαύτας μυθοποιίας ἥ τε σύγχυσις τῶν ἐνταῦθα ἐθνῶν καὶ ἡ εὐδαιμονία τῆς χώρας τῆς ἐντὸς Ἅλυος, μάλιστα δὲ τῆς παραλίας, δι' ἣν ἐπιθέσεις ἐγένοντο αὐτῇ πολλαχόθεν καὶ διὰ παντὸς ἐκ τῆς περαίας, ἢ καὶ ἐπ' ἀλλήλους ἰόντων τῶν ἐγγύς. μάλιστα μὲν οὖν κατὰ τὰ Τρωικὰ καὶ μετὰ ταῦτα τὰς ἐφόδους γενέσθαι καὶ τὰς μεταναστάσεις συνέβη, τῶν τε βαρβάρων ἅμα καὶ τῶν Ἑλλήνων ὁρμῇ τινὶ χρησαμένων πρὸς τὴν τῆς ἀλλοτρίας κατάκτησιν· ἀλλὰ καὶ πρὸ τῶν Τρωικῶν ἦν ταῦτα, τό τε γὰρ τῶν

[1] ἀνελόντων τε, Corais, for εἵλοντο τόν τε; so the later editors.

---

[1] *i.e.* the oxya-tree, a kind of beech-tree, which is called "oxya" by the Greeks, is called "mysos" by the Lydians.

[2] *i.e.* one-tenth of the people were, in accordance with some religious vow, sent out of their country to the neigh-

were Thracians but others that they were Lydians, thus concurring with an ancient explanation given by Xanthus the Lydian and Menecrates of Elaea, who explain the origin of the name of the Mysians by saying that the oxya-tree is so named by the Lydians.[1] And the oxya-tree abounds in the neighbourhood of Mt. Olympus, where they say that the decimated persons were put out[2] and that their descendants were the Mysians of later times, so named after the oxya-tree, and that their language bears witness to this; for, they add, their language is, in a way, a mixture of the Lydian and the Phrygian languages, for the reason that, although they lived round Mt. Olympus for a time, yet when the Phrygians crossed over from Thrace and slew a ruler of Troy and of the country near it, those people took up their abode there, whereas the Mysians took up their abode above the sources of the Caïcus near Lydia.

4. Contributing to the creation of myths of this kind are the confusion of the tribes there and the fertility of the country this side the Halys River, particularly that of the seaboard, on account of which attacks were made against it from numerous places and continually by peoples from the opposite mainland, or else the people near by would attack one another. Now it was particularly in the time of the Trojan War and after that time that invasions and migrations took place, since at the same time both the barbarians and the Greeks felt an impulse to acquire possession of the countries of others; but this was also the case before the Trojan War, for the

bourhood of Mt. Olympus and there dedicated to the service of some god.

Πελασγῶν ἦν φῦλον καὶ τὸ τῶν Καυκώνων καὶ Λελέγων· εἴρηται δ', ὅτι πολλαχοῦ τῆς Εὐρώπης τὸ παλαιὸν ἐτύγχανε πλανώμενα, ἅπερ ποιεῖ τοῖς Τρωσὶ συμμαχοῦντα ὁ ποιητής, οὐκ ἐκ τῆς περαίας. τά τε περὶ τῶν Φρυγῶν καὶ τῶν Μυσῶν λεγόμενα πρεσβύτερα τῶν Τρωικῶν ἐστίν· οἱ δὲ διττοὶ Λύκιοι τοῦ αὐτοῦ γένους ὑπόνοιαν παρέχουσιν, ἢ τῶν Τρωικῶν ἢ τῶν πρὸς Καρίᾳ τοὺς ἑτέρους ἀποικισάντων. τάχα δὲ καὶ ἐπὶ τῶν Κιλίκων τὸ αὐτὸ συνέβη· διττοὶ γὰρ καὶ οὗτοι· οὐ μὴν ἔχομέν γε τοιαύτην λαβεῖν μαρτυρίαν, ὅτι καὶ πρὸ τῶν Τρωικῶν ἦσαν ἤδη οἱ νῦν Κίλικες· ὅ τε Τήλεφος ἐκ τῆς Ἀρκαδίας ἀφῖχθαι νομίζοιτ' ἂν μετὰ τῆς μητρός, γάμῳ δὲ τῷ ταύτης ἐξοικειωσάμενος τὸν ὑποδεξάμενον αὐτὸν Τεύθραντα ἐνομίσθη τε ἐκείνου καὶ παρέλαβε τὴν Μυσῶν ἀρχήν.

5. Καὶ οἱ Κᾶρες δὲ νησιῶται πρότερον ὄντες καὶ C 573 Λέλεγες, ὥς φασιν, ἠπειρῶται γεγόνασι, προσλαβόντων Κρητῶν, οἳ καὶ τὴν Μίλητον ἔκτισαν, ἐκ τῆς Κρητικῆς[1] Μιλήτου Σαρπηδόνα λαβόντες κτίστην· καὶ τοὺς Τερμίλας κατῴκισαν ἐν τῇ νῦν Λυκίᾳ· τούτους δ' ἀγαγεῖν ἐκ Κρήτης ἀποίκους Σαρπηδόνα, Μίνω καὶ Ῥαδαμάνθυος ἀδελφὸν ὄντα, καὶ ὀνομάσαι Τερμίλας τοὺς πρότερον Μιλύας, ὥς φησιν Ἡρόδοτος, ἔτι δὲ πρότερον Σολύμους, ἐπελθόντα δὲ τὸν Πανδίονος

[1] Κρητικῆς oz (and the editors), Κρήτης other MSS.

---

[1] 5. 2. 4 and 7. 7. 10. [2] Cp. 12. 8. 7.
[3] Cp. 13. 1. 60. [4] 1. 173; 7. 92.

tribe of the Pelasgians was then in existence, as also that of the Cauconians and Leleges. And, as I have said before,[1] they wandered in ancient times over many regions of Europe. These tribes the poet makes the allies of the Trojans, but not as coming from the opposite mainland. The accounts both of the Phrygians and of the Mysians go back to earlier times than the Trojan War. The existence of two groups of Lycians arouses suspicion that they were of the same tribe, whether it was the Trojan Lycians or those near Caria that colonised the country of the other of the two.[2] And perhaps the same was also true in the case of the Cilicians, for these, too, were two-fold;[3] however, we are unable to get the same kind of evidence that the present tribe of Cilicians was already in existence before the Trojan War. Telephus might be thought to have come from Arcadia with his mother; and having become related to Teuthras, to whom he was a welcome guest, by the marriage of his mother to that ruler, was regarded as his son and also succeeded to the rulership of the Mysians.

5. Not only the Carians, who in earlier times were islanders, but also the Leleges, as they say, became mainlanders with the aid of the Cretans, who founded, among other places, Miletus, having taken Sarpedon from the Cretan Miletus as founder; and they settled the Termilae in the country which is now called Lycia; and they say that these settlers were brought from Crete by Sarpedon, a brother of Minos and Rhadamanthus, and that he gave the name Termilae to the people who were formerly called Milyae, as Herodotus[4] says, and were in still earlier times called Solymi, but that when Lycus the

Λύκον[1] ἀφ' ἑαυτοῦ προσαγορεῦσαι τοὺς αὐτοὺς Λυκίους. οὗτος μὲν οὖν ὁ λόγος ἀποφαίνει τοὺς αὐτοὺς Σολύμους τε καὶ Λυκίους, ὁ δὲ ποιητὴς χωρίζει· Βελλεροφόντης γοῦν, ὡρμημένος ἐκ τῆς Λυκίας,

Σολύμοισι μαχέσσατο κυδαλίμοισι.

Πείσανδρόν τε ὡσαύτως, υἱὸν αὐτοῦ, Ἄρης, ὥς φησι,

μαρνάμενον Σολύμοισι κατέκτανε·

καὶ τὸν Σαρπηδόνα δὲ ἐπιχώριόν τινα λέγει.

6. Ἀλλὰ τό γε ἆθλον προκεῖσθαι κοινὸν τὴν ἀρετὴν τῆς χώρας, ἧς λέγω, τοῖς ἰσχύουσιν ἐκ πολλῶν βεβαιοῦται[2] καὶ μετὰ τὰ Τρωικά· ὅπου καὶ Ἀμαζόνες κατεθάρρησαν αὐτῆς, ἐφ' ἃς ὅ τε Πρίαμος στρατεῦσαι λέγεται καὶ ὁ Βελλεροφόντης· πόλεις τε παλαιαὶ[3] ὁμολογοῦνται ἐπώνυμοι αὐτῶν· ἐν δὲ τῷ Ἰλιακῷ πεδίῳ κολώνη τίς ἐστιν,

ἣν ἤτοι[4] ἄνδρες Βατίειαν κικλήσκουσιν,
ἀθάνατοι δέ τε σῆμα πολυσκάρθμοιο Μυρίνης·

ἣν ἱστοροῦσι μίαν εἶναι τῶν Ἀμαζόνων, ἐκ τοῦ ἐπιθέτου τεκμαιρόμενοι· εὐσκάρθμους γὰρ ἵππους λέγεσθαι διὰ τὸ τάχος· κἀκείνην οὖν πολύσκαρθ-

[1] Λύκον E, Λύκωνα other MSS.
[2] Casaubon conj. that καὶ πρὸ τῶν Τρωικῶν has fallen out before καὶ μετά; Tzschucke conj. καὶ κατὰ τὰ Τρωικά; Corais, [ἐκ των] κατὰ τὰ Τρωικά.

son of Pandion went over there he named the people Lycians after himself. Now this account represents the Solymi and the Lycians as the same people, but the poet makes a distinction between them. At any rate, Bellerophontes set out from Lycia and "fought with the glorious Solymi."[1] And likewise his son Peisander[2] "was slain when fighting the Solymi"[3] by Ares, as he says. And he also speaks of Sarpedon as a native of Lycia.[4]

6. But the fact that the fertility of the country of which I am speaking[5] was set before the powerful as a common prize of war is confirmed by many things which have taken place even subsequent to the Trojan War,[6] since even the Amazons took courage to attack it, against whom not only Priam, but also Bellerophontes, are said to have made expeditions; and the naming of ancient cities after the Amazons attests this fact. And in the Trojan Plain there is a hill "which by men is called 'Batieia,' but by the immortals 'the tomb of the much-bounding Myrina,'"[7] who, historians say, was one of the Amazons, inferring this from the epithet "much-bounding"; for they say that horses are called "well-bounding" because of their speed, and that Myrina, therefore, was called "much-bounding"

[1] *Iliad* 6. 184.
[2] "Isander" is the spelling of the name in the Iliad.
[3] *Iliad* 6. 204.
[4] *Iliad* 6. 199.
[5] The country this side the Halys (§ 4 above).
[6] *i.e.* as well as by events during, and prior to, that war.
[7] *Iliad* 2. 813.

---

[3] τε παλαιαί *x*, τὸ πάλαι καί CD*h*, τὸ πάλαι *l*, τὸ παλαιόν *i*, παλαιαὶ καί *rw*, παλαιαί *oz*.
[4] ἤτοι, Xylander, for αἱ; so the later editors.

μον διὰ τὸ ἀπὸ τῆς ἡνιοχείας τάχος· καὶ ἡ Μύρινα οὖν ἐπώνυμος ταύτης λέγεται. καὶ αἱ ἐγγὺς δὲ νῆσοι ταῦτ' ἔπαθον διὰ τὴν ἀρετήν, ὧν Ῥόδος καὶ Κῶς ὅτι πρὸ τῶν Τρωικῶν ἤδη ὑφ' Ἑλλήνων ᾠκοῦντο, καὶ ὑφ' Ὁμήρου σαφῶς ἐκμαρτυρεῖται.

7. Μετὰ δὲ τὰ Τρωικὰ αἵ τε τῶν Ἑλλήνων ἀποικίαι καὶ αἱ Τρηρῶν καὶ αἱ Κιμμερίων ἔφοδοι καὶ Λυδῶν καὶ μετὰ ταῦτα Περσῶν καὶ Μακεδόνων, τὸ τελευταῖον Γαλατῶν, ἐτάραξαν πάντα καὶ συνέχεαν. γέγονε δὲ ἡ ἀσάφεια οὐ διὰ τὰς μεταβολὰς μόνον, ἀλλὰ καὶ διὰ τὰς τῶν συγγραφέων ἀνομολογίας, περὶ τῶν αὐτῶν οὐ τὰ αὐτὰ λεγόντων, τοὺς μὲν Τρῶας καλούντων Φρύγας, καθάπερ οἱ τραγικοί, τοὺς δὲ Λυκίους Κᾶρας, καὶ ἄλλους οὕτως. οἱ δὲ Τρῶες οὕτως ἐκ μικρῶν C 574 αὐξηθέντες, ὥστε καὶ βασιλεῖς βασιλέων εἶναι, παρέσχον καὶ τῷ ποιητῇ λόγον, τίνα χρὴ καλεῖν Τροίαν, καὶ τοῖς ἐξηγουμένοις ἐκεῖνον. λέγει μὲν γὰρ καὶ κοινῶς ἅπαντας Τρῶας τοὺς συμπολεμήσαντας αὐτοῖς, ὥσπερ καὶ Δαναοὺς καὶ Ἀχαιοὺς τοὺς ἐναντίους· ἀλλ' οὐ δήπου Τροίαν καὶ τὴν Παφλαγονίαν ἐροῦμεν, νὴ Δία, οὐδὲ τὴν Καρίαν ἢ τὴν ὅμορον αὐτῇ Λυκίαν. λέγω δ', ὅταν οὕτω φῇ,

Τρῶες μὲν κλαγγῇ τ' ἐνοπῇ τ' ἴσαν·[2]

ἐκ δὲ τῶν ἐναντίων,

οἱ δ' ἄρ' ἴσαν σιγῇ μένεα πνείοντες Ἀχαιοί.[3]

καὶ ἄλλως δὲ λέγει πολλαχῶς. ὅμως δέ, καίπερ τοιούτων ὄντων, πειρατέον διαιτᾶν ἕκαστα εἰς

[1] See 14. 2. 7. [2] *Iliad* 3. 2. [3] *Iliad* 3. 8.

because of the speed with which she drove her chariot. Myrina, therefore, is named after this Amazon. And the neighbouring islands had the same experience because of their fertility; and Homer clearly testifies that, among these, Rhodes and Cos were already inhabited by Greeks before the Trojan War.[1]

7. After the Trojan War the migrations of the Greeks and the Trerans, and the onsets of the Cimmerians and of the Lydians, and, after this, of the Persians and the Macedonians, and, at last, of the Galatians, disturbed and confused everything. But the obscurity has arisen, not on account of the changes only, but also on account of the disagreements of the historians, who do not say the same things about the same subjects, calling the Trojans Phrygians, as do the tragic poets, and the Lycians Carians; and so in the case of other peoples. But the Trojans, having waxed so strong from a small beginning that they became kings of kings, afforded both the poet and his expounders grounds for enquiring what should be called Troy; for in a general way he calls "Trojans" the peoples, one and all, who fought on the Trojan side, just as he called their opponents both "Danaans" and "Achaeans"; and yet, of course, we shall surely not speak of Paphlagonia as a part of Troy, nor yet Caria, nor the country that borders on Caria, I mean Lycia. I mean when the poet says, "the Trojans advanced with clamour and with a cry like birds,"[2] and when he says of their opponents, "but the Achaeans advanced in silence, breathing rage."[3] And in many ways he uses terms differently. But still, although such is the case, I must try to arbitrate the several details to the best

δύναμιν· ὅ τι δ' ἂν διαφύγῃ τῆς παλαιᾶς ἱστορίας, τοῦτο μὲν ἐατέον, οὐ γὰρ ἐνταῦθα τὸ τῆς γεωγραφίας ἔργον, τὰ δὲ νῦν ὄντα λεκτέον.

8. Ἔστι τοίνυν ὄρη δύο ὑπερκείμενα τῆς Προποντίδος, ὅ τε Ὄλυμπος ὁ Μύσιος καὶ ἡ Ἴδη. τῷ μὲν οὖν Ὀλύμπῳ τὰ τῶν Βιθυνῶν ὑποπέπτωκε, τῆς δὲ Ἴδης μεταξὺ καὶ τῆς θαλάττης ἡ Τροία κεῖται, συνάπτουσα τῷ ὄρει· περὶ μὲν οὖν ταύτης ἐροῦμεν ὕστερον καὶ τῶν συνεχῶν αὐτῇ πρὸς νότον, νῦν δὲ περὶ τῶν Ὀλυμπηνῶν καὶ τῶν ἐφεξῆς μέχρι τοῦ Ταύρου παραλλήλων τοῖς προεφωδευμένοις λέγωμεν. ἔστι τοίνυν ὁ Ὄλυμπος κύκλῳ μὲν εὖ[1] συνοικούμενος, ἐν δὲ τοῖς ὕψεσι δρυμοὺς ἐξαισίους ἔχων καὶ λῃστήρια δυναμένους ἐκτρέφειν τόπους εὐερκεῖς, ἐν οἷς καὶ τύραννοι συνίστανται πολλάκις, οἱ δυνάμενοι συμμεῖναι πολὺν χρόνον· καθάπερ Κλέων ὁ καθ' ἡμᾶς τῶν λῃστηρίων ἡγεμών.

9. Οὗτος δ' ἦν μὲν ἐκ Γορδίου κώμης, ἣν ὕστερον αὐξήσας ἐποίησε πόλιν καὶ προσηγόρευσεν Ἰουλιόπολιν· λῃστηρίῳ δ' ἐχρῆτο καὶ ὁρμητηρίῳ κατ' ἀρχὰς τῷ καρτερωτάτῳ τῶν χωρίων,[2] ὄνομα Καλλυδίῳ· ὑπῆρξε δ' Ἀντωνίῳ μὲν χρήσιμος, ἐπελθὼν ἐπὶ τοὺς ἀργυρολογοῦντας Λαβιήνῳ,[3] καθ' ὃν χρόνον ἐκεῖνος τὴν Ἀσίαν κατέσχε, καὶ κωλύσας τὰς παρασκευάς· ἐν δὲ τοῖς Ἀκτιακοῖς ἀποστὰς Ἀντωνίου τοῖς Καίσαρος προσέθετο

[1] εὖ, Mannert, for οὐ; so the editors.
[2] χωρίων, Corais, for χωρῶν; so the later editors.
[3] Λαβιήνῳ, Xylander, for Λαβίνῳ *Chi*, Λαβήνῳ other MSS.

[1] 13. 1. 34, 35.
[2] Quintus Labienus, son of Titus Labienus the tribune.

of my ability. However, if anything in ancient history escapes me, I must leave it unmentioned, for the task of the geographer does not lie in that field, and I must speak of things as they now are.

8. Above the Propontis, then, there are two mountains, the Mysian Olympus and Mt. Ida. Now the region of the Bithynians lies at the foot of Olympus, whereas Troy is situated between Mt. Ida and the sea and borders on the mountain. As for Troy, I shall describe it and the parts adjacent to it towards the south later on,[1] but at present let me describe the country of Mt. Olympus and the parts which come next in order thereafter, extending as far as the Taurus and lying parallel to the parts which I have previously traversed. Mt. Olympus, then, is not only well settled all round but also has on its heights immense forests and places so well-fortified by nature that they can support bands of robbers; and among these bands there often arise tyrants who are able to maintain their power for a long time; for example, Cleon, who in my time was chieftain of the bands of robbers.

9. Cleon was from the village Gordium, which he later enlarged, making it a city and calling it Juliopolis; but from the beginning he used the strongest of the strongholds, Callydium by name, as retreat and base of operations for the robbers. And he indeed proved useful to Antony, since he made an attack upon those who were levying money for Labienus[2] at the time when the latter held possession of Asia,[3] and he hindered his preparations, but in the course of the Actian War, having revolted from Antony, he joined the generals of

[3] 40–39 B.C.

στρατηγοῖς, καὶ ἐτιμήθη πλέον ἢ κατ' ἀξίαν, προσλαβὼν τοῖς παρ' Ἀντωνίου δοθεῖσι καὶ τὰ παρὰ τοῦ Καίσαρος· ὥστ' ἀντὶ λῃστοῦ δυνάστου περιέκειτο σχῆμα, ἱερεὺς μὲν ὢν τοῦ Ἀβρεττηνοῦ[1] Διός, Μυσίου θεοῦ, μέρος δ' ἔχων ὑπήκοον τῆς Μωρηνῆς (Μυσία δ' ἐστὶ καὶ αὕτη, καθάπερ ἡ Ἀβρεττηνή), λαβὼν δὲ ὕστατα καὶ τὴν ἐν τῷ Πόντῳ τῶν Κομάνων ἱερωσύνην, εἰς ἣν κατελθὼν ἐντὸς μηνιαίου χρόνου κατέστρεψε τὸν βίον· C 575 νόσος δ' ἐξήγαγεν αὐτὸν ὀξεῖα, εἴτ' ἄλλως ἐπιπεσοῦσα ἐκ τῆς ἄδην πλησμονῆς, εἴθ', ὡς ἔφασαν οἱ περὶ τὸ ἱερόν, κατὰ μῆνιν τῆς θεοῦ· ἐν γὰρ τῷ περιβόλῳ τοῦ τεμένους ἡ οἴκησίς ἐστιν ἥ τε τοῦ ἱερέως καὶ τῆς ἱερείας, τὸ δὲ τέμενος χωρὶς τῆς ἄλλης ἁγιστείας διαφανέστατα τῆς τῶν ὑείων κρεῶν βρώσεως καθαρεύει, ὅπου γε καὶ ἡ ὅλη πόλις, οὐδ' εἰσάγεται εἰς αὐτὴν ὗς· ὁ δ' ἐν τοῖς πρώτοις τὸ λῃστρικὸν ἦθος ἐπεδείξατο εὐθὺς κατὰ τὴν πρώτην εἴσοδον τῇ παραβάσει τούτου τοῦ ἔθους, ὥσπερ οὐχ ἱερεὺς εἰσεληλυθώς, ἀλλὰ διαφθορεὺς τῶν ἱερῶν.

10. Ὁ μὲν δὴ Ὄλυμπος τοιόσδε, περιοικεῖται δὲ πρὸς ἄρκτον μὲν ὑπὸ τῶν Βιθυνῶν καὶ Μυγδόνων καὶ Δολιόνων, τὸ δὲ λοιπὸν ἔχουσι Μυσοὶ καὶ Ἐπίκτητοι. Δολίονας μὲν οὖν μάλιστα καλοῦσι τοὺς περὶ Κύζικον ἀπὸ Αἰσήπου ἕως Ῥυνδάκου καὶ τῆς Δασκυλίτιδος λίμνης, Μυγδόνας δὲ τοὺς ἐφεξῆς τούτοις μέχρι τῆς Μυρλειανῶν χώρας· ὑπέρκεινται δὲ τῆς Δασκυλίτιδος

[1] Ἀβρεττηνοῦ, Xylander, for Ἀβρετατηνοῦ CD*hilrw*, Ἀβρεττανοῦ *oz*, Ἀβρυτανοῦ *ux*.

Caesar and was honoured more than he deserved, since he also received, in addition to what Antony had given him, what Caesar gave him, so that he was invested with the guise of dynast, from being a robber, that is, he was priest of Zeus Abrettenus, a Mysian god, and held subject a part of Morenê, which, like Abrettenê, is also Mysian, and received at last the priesthood of Comana in Pontus, although he died within a month's time after he went down to Comana. He was carried off by an acute disease, which either attacked him in consequence of excessive repletion or else, as the people round the temple said, was inflicted upon him because of the anger of the goddess; for the dwelling of both the priest and the priestess is within the circuit of the sacred precinct, and the sacred precinct, apart from its sanctity in other respects, is most conspicuously free from the impurity of the eating of swine's flesh; in fact, the city as a whole is free from it; and swine cannot even be brought into the city. Cleon, however, among the first things he did when he arrived, displayed the character of the robber by transgressing this custom, as though he had come, not as priest, but as corrupter of all that was sacred.

10. Such, then, is Mt. Olympus; and towards the north it is inhabited all round by the Bithynians and Mygdonians and Doliones, whereas the rest of it is occupied by Mysians and Epicteti. Now the peoples round Cyzicus, from the Aesepus River to the Rhyndacus River and Lake Dascylitis, are for the most part called Doliones, whereas the peoples who live next after these as far as the country of the Myrleians are called Mygdonians. Above Lake Dascylitis lie two other lakes, large ones, I mean

ἄλλαι δύο λίμναι μεγάλαι, ἥ τε Ἀπολλωνιᾶτις ἥ τε Μιλητοπολῖτις· πρὸς μὲν οὖν τῇ Δασκυλίτιδι Δασκύλιον πόλις, πρὸς δὲ τῇ Μιλητοπολίτιδι Μιλητούπολις, πρὸς δὲ τῇ τρίτῃ Ἀπολλωνία ἡ ἐπὶ Ῥυνδάκῳ λεγομένη· τὰ πλεῖστα δὲ τούτων ἐστὶ Κυζικηνῶν νυνί.

11. Ἔστι δὲ νῆσος ἐν τῇ Προποντίδι ἡ Κύζικος συναπτομένη γεφύραις δυσὶ πρὸς τὴν ἤπειρον, ἀρετῇ μὲν κρατίστη, μεγέθει δὲ ὅσον πεντακοσίων σταδίων τὴν περίμετρον· ἔχει δὲ ὁμώνυμον πόλιν πρὸς αὐταῖς ταῖς γεφύραις καὶ λιμένας δύο κλειστοὺς καὶ νεωσοίκους πλείους τῶν διακοσίων· τῆς δὲ πόλεως τὸ μὲν ἔστιν ἐν ἐπιπέδῳ, τὸ δὲ πρὸς ὄρει· καλεῖται δ' Ἄρκτων ὄρος·[1] ὑπέρκειται δ' ἄλλο Δίνδυμον μονοφυές, ἱερὸν ἔχον τῆς Δινδυμήνης μητρὸς θεῶν, ἵδρυμα τῶν Ἀργοναυτῶν. ἔστι δ' ἐνάμιλλος ταῖς πρώταις τῶν κατὰ τὴν Ἀσίαν ἡ πόλις μεγέθει τε καὶ κάλλει καὶ εὐνομίᾳ πρός τε εἰρήνην καὶ πόλεμον· ἔοικέ τε τῷ παραπλησίῳ τύπῳ κοσμεῖσθαι, ὥσπερ ἡ τῶν Ῥοδίων καὶ Μασσαλιωτῶν καὶ Καρχηδονίων τῶν πάλαι. τὰ μὲν οὖν πολλὰ ἐῶ, τρεῖς δ' ἀρχιτέκτονας τοὺς ἐπιμελουμένους οἰκοδομημάτων τε δημοσίων καὶ ὀργάνων, τρεῖς δὲ καὶ θησαυροὺς κέκτηται, τὸν μὲν ὅπλων, τὸν δ' ὀργάνων, τὸν δὲ σίτου· ποιεῖ δὲ τὸν σῖτον ἄσηπτον ἡ Χαλκιδικὴ γῆ[1] μιγνυμένη. ἐπεδείξαντο δὲ τὴν ἐκ τῆς παρασκευῆς ταύτης ὠφέλειαν ἐν τῷ Μιθριδατικῷ

[1] γῆ, omitted by all MSS. except F.

---

[1] *i.e.* "Mountain of the Bears."

Lake Apolloniatis and Lake Miletopolitis. Near Lake Dascylitis is the city Dascylium, and near Lake Miletopolitis Miletopolis, and near the third lake "Apollonia on Rhyndacus," as it is called. But at the present time most of these places belong to the Cyziceni.

11. Cyzicus is an island in the Propontis, being connected with the mainland by two bridges; and it is not only most excellent in the fertility of its soil, but in size has a perimeter of about five hundred stadia. It has a city of the same name near the bridges themselves, and two harbours that can be closed, and more than two hundred ship-sheds. One part of the city is on level ground and the other is near a mountain called "Arcton-oros."[1] Above this mountain lies another mountain, Dindymus; it rises into a single peak, and it has a temple of Dindymenê, mother of the gods, which was founded by the Argonauts. This city rivals the foremost of the cities of Asia in size, in beauty, and in its excellent administration of affairs both in peace and in war. And its adornment appears to be of a type similar to that of Rhodes and Massalia and ancient Carthage. Now I am omitting most details, but I may say that there are three directors who take care of the public buildings and the engines of war, and three who have charge of the treasure-houses, one of which contains arms and another engines of war and another grain. They prevent the grain from spoiling by mixing Chalcidic earth[2] with it. They showed in the Mithridatic war the advantage resulting from this preparation of theirs; for when the king unexpectedly came over

[2] Apparently a soil containing lime carbonate.

πολέμῳ. ἐπελθόντος γὰρ αὐτοῖς ἀδοκήτως τοῦ βασιλέως πεντεκαίδεκα μυριάσι καὶ ἵππῳ πολλῇ καὶ κατασχόντος τὸ ἀντικείμενον ὄρος, ὃ καλοῦσιν Ἀδραστείας, καὶ τὸ προάστειον, ἔπειτα καὶ διάραντος εἰς τὸν ὑπὲρ τῆς πόλεως αὐχένα καὶ
C 576 προσμαχομένου πεζῇ τε καὶ κατὰ θάλατταν τετρακοσίαις ναυσίν, ἀντέσχον πρὸς ἅπαντα οἱ Κυζικηνοί, ὥστε καὶ ἐγγὺς ἦλθον τοῦ ζωγρίᾳ λαβεῖν τὸν βασιλέα ἐν τῇ διώρυγι ἀντιδιορύττοντες, ἀλλ' ἔφθη φυλαξάμενος καὶ ἀναλαβὼν ἑαυτὸν ἔξω τοῦ ὀρύγματος· ὀψὲ δὲ ἴσχυσεν εἰσπέμψαι τινὰς νύκτωρ ἐπικούρους ὁ τῶν Ῥωμαίων στρατηγὸς Λεύκολλος· ὤνησε δὲ καὶ λιμὸς τῷ τοσούτῳ πλήθει τῆς στρατιᾶς ἐπιπεσών, ὃν οὐ προείδετο ὁ βασιλεύς, ὡς ἀπῆλθε πολλοὺς ἀποβαλών. Ῥωμαῖοι δ' ἐτίμησαν τὴν πόλιν, καὶ ἔστιν ἐλευθέρα μέχρι νῦν καὶ χώραν ἔχει πολλὴν τὴν μὲν ἐκ παλαιοῦ, τὴν δὲ τῶν Ῥωμαίων προσθέντων. καὶ γὰρ τῆς Τρῳάδος ἔχουσι τὰ πέραν τοῦ Αἰσήπου τὰ περὶ τὴν Ζέλειαν καὶ τὸ τῆς Ἀδραστείας πεδίον, καὶ τῆς Δασκυλίτιδος λίμνης τὰ μὲν ἔχουσιν ἐκεῖνοι, τὰ δὲ Βυζάντιοι· πρὸς δὲ τῇ Δολιονίδι καὶ τῇ Μυγδονίδι νέμονται πολλὴν μέχρι τῆς Μιλητοπολίτιδος λίμνης καὶ τῆς Ἀπολλωνιάτιδος αὐτῆς, δι' ὧν χωρίων καὶ ὁ Ῥύνδακος ῥεῖ ποταμός, τὰς ἀρχὰς ἔχων ἐκ τῆς Ἀζανίτιδος· προσλαβὼν δὲ καὶ ἐκ τῆς Ἀβρεττηνῆς Μυσίας ἄλλους τε καὶ Μάκεστον ἀπ' Ἀγκύρας τῆς Ἀβαειτίδος[1] ἐκδίδωσιν εἰς τὴν Προποντίδα κατὰ Βέσβικον νῆσον. ἐν ταύτῃ δὲ τῇ νήσῳ τῶν Κυζικηνῶν ὄρος ἐστὶν εὔδενδρον

[1] Ἀβαειτίδος, Kramer, for Ἀβασίτιδος; so the later editors.

against them with one hundred and fifty thousand men and with a large cavalry, and took possession of the mountain opposite the city, the mountain called Adrasteia, and of the suburb, and then, when he transferred his army to the neck of land above the city and was fighting them, not only on land, but also by sea with four hundred ships, the Cyziceni held out against all attacks, and, by digging a counter-tunnel, all but captured the king alive in his own tunnel; but he forestalled this by taking precautions and by withdrawing outside his tunnel. Leucullus, the Roman general, was able, though late, to send an auxiliary force to the city by night; and, too, as an aid to the Cyziceni, famine fell upon that multitudinous army, a thing which the king did not foresee, because he suffered a great loss of men before he left the island. But the Romans honoured the city; and it is free to this day, and holds a large territory, not only that which it has held from ancient times, but also other territory presented to it by the Romans; for, of the Troad, they possess the parts round Zeleia on the far side of the Aesepus, as also the plain of Adrasteia, and, of Lake Dascylitis, they possess some parts, while the Byzantians possess the others. And in addition to Dolionis and Mygdonis they occupy a considerable territory extending as far as Lake Miletopolitis and Lake Apolloniatis itself. It is through this region that the Rhyndacus River flows; this river has its sources in Azanitis, and then, receiving from Mysia Abrettenê, among other rivers, the Macestus, which flows from Ancyra in Abäeitis, empties into the Propontis opposite the island Besbicos. In this island of the Cyziceni is a well-

'Αρτάκη· καὶ νησίον ὁμώνυμον πρόκειται τούτου, καὶ πλησίον ἀκρωτήριον Μέλανος καλούμενον ἐν παράπλῳ τοῖς εἰς Πρίαπον κομιζομένοις ἐκ τῆς Κυζίκου.

12. Τῆς δ' ἐπικτήτου Φρυγίας Ἀζανοί[1] τέ εἰσι καὶ Νακολία καὶ Κοτιάειον καὶ Μιδάειον[2] καὶ Δορύλαιον πόλεις καὶ Κάδοι· τοὺς δὲ Κάδους ἔνιοι τῆς Μυσίας φασίν. ἡ δὲ Μυσία κατὰ τὴν μεσόγαιαν ἀπὸ τῆς Ὀλυμπηνῆς ἐπὶ τὴν Περγαμηνὴν καθήκει καὶ τὸ Καΐκου λεγόμενον πεδίον, ὥστε μεταξὺ κεῖσθαι τῆς τε Ἴδης καὶ τῆς Κατακεκαυμένης, ἣν οἱ μὲν Μυσίαν, οἱ δὲ Μαιονίαν φασίν.

13. Ὑπὲρ δὲ τῆς Ἐπικτήτου πρὸς νότον ἐστὶν ἡ μεγάλη Φρυγία, λείπουσα[3] ἐν ἀριστερᾷ τὴν Πεσσινοῦντα καὶ τὰ περὶ Ὀρκαόρκους καὶ Λυκαονίαν, ἐν δεξιᾷ δὲ Μαίονας καὶ Λυδοὺς καὶ Κᾶρας· ἐν ᾗ ἐστὶν ἥ τε Παρώρειος λεγομένη Φρυγία καὶ ἡ πρὸς Πισιδίαν καὶ τὰ περὶ Ἀμόριον καὶ Εὐμένειαν καὶ Σύνναδα, εἶτα Ἀπάμεια ἡ Κιβωτὸς λεγομένη καὶ Λαοδίκεια, αἵπερ εἰσὶ μέγισται τῶν κατὰ τὴν Φρυγίαν πόλεων· περίκειται δὲ ταύταις πολίσματα καί[4] . . . . Ἀφροδισιάς, Κολοσσαί, Θεμισώνιον, Σαναός, Μητρόπολις, Ἀπολλωνιάς· ἔτι δὲ ἀπωτέρω τούτων Πέλται, Τάβαι,[5] Εὐκαρπία, Λυσιάς. C 577

[1] Ἀζανοί (as in Stephanus), the editors, for Ἀζάνιοι.

[2] Μιδάειον, Tzschucke, for Μιδάιον; so the later editors.

[3] λείπουσα, Corais, for λιποῦσα; so the later editors.

[4] Corais omits καί and supplies the lacuna of about fifteen letters with ἄλλα τε καί, in reference to which Kramer says, "substantivum potius videatur excidisse, velut χωρία vel simile quid." Jones conjectures χωρία, ἄλλα τε καί (fourteen letters).

wooded mountain called Artacê; and in front of this mountain lies an isle bearing the same name; and near by is a promontory called Melanus, which one passes on a coasting-voyage from Cyzicus to Priapus.

12. To Phrygia Epictetus belong the cities Azani, Nacolia, Cotiäeium, Midäeium, and Dorylaeum, and also Cadi, which, according to some writers, belongs to Mysia. Mysia extends in the interior from Olympenê to Pergamenê, and to the plain of Caïcus, as it is called; and therefore it lies between Mt. Ida and Catacecaumenê, which latter is by some called Mysian and by others Maeonian.

13. Above Phrygia Epictetus towards the south is Greater Phrygia, which leaves on the left Pessinus and the region of Orcaorci and Lycaonia, and on the right the Maeonians and Lydians and Carians. In Epictetus are Phrygia "Paroreia,"[1] as it is called, and the part of Phrygia that lies towards Pisidia, and the parts round Amorium and Eumeneia and Synnada, and then Apameia Cibotus, as it is called, and Laodiceia, which two are the largest of the Phrygian cities. And in the neighbourhood of these are situated towns, and. . . .,[2] Aphrodisias, Colossae, Themisonium, Sanaüs, Metropolis, and Apollonias; but still farther away than these are Peltae, Tabae, Eucarpia, and Lysias.

[1] *i.e.* the part of Phrygia "along the mountain."

[2] There is a lacuna in the MSS. at this point (see critical note) which apparently should be supplied as follows: "places, among others."

---

[5] Τάβαι, Corais, for Ταβαίαι *x*, Ταμέαι *hi*, Ταβέαι other MSS.; so the later editors.

14. Ἡ μὲν οὖν Παρώρεια ὀρεινήν τινα ἔχει ῥάχιν ἀπὸ τῆς ἀνατολῆς ἐκτεινομένην ἐπὶ δύσιν· ταύτῃ δ' ἑκατέρωθεν ὑποπέπτωκέ τι πεδίον μέγα καὶ πόλεις πλησίον αὐτῆς, πρὸς ἄρκτον μὲν Φιλομήλιον, ἐκ θατέρου δὲ μέρους Ἀντιόχεια ἡ πρὸς Πισιδίᾳ καλουμένη, ἡ μὲν ἐν πεδίῳ κειμένη πᾶσα, ἡ δ' ἐπὶ λόφου, ἔχουσα ἀποικίαν Ῥωμαίων· ταύτην δ' ᾤκισαν Μάγνητες οἱ πρὸς Μαιάνδρῳ. Ῥωμαῖοι δ' ἠλευθέρωσαν τῶν βασιλέων, ἡνίκα τὴν ἄλλην Ἀσίαν Εὐμένει παρέδοσαν τὴν ἐντὸς τοῦ Ταύρου· ἦν δ' ἐνταῦθα καὶ ἱερωσύνη τις Μηνὸς Ἀρκαίου, πλῆθος ἔχουσα ἱεροδούλων καὶ χωρίων ἱερῶν· κατελύθη δὲ μετὰ τὴν Ἀμύντου τελευτὴν ὑπὸ τῶν πεμφθέντων ἐπὶ τὴν ἐκείνου κληρονομίαν. Σύνναδα δ' ἐστὶν οὐ μεγάλη πόλις· πρόκειται δ' αὐτῆς ἐλαιόφυτον πεδίον ὅσον ἑξήκοντα σταδίων· ἐπέκεινα δ' ἐστὶ Δοκιμία κώμη, καὶ τὸ λατόμιον Συνναδικοῦ λίθου (οὕτω μὲν Ῥωμαῖοι καλοῦσιν, οἱ δ' ἐπιχώριοι Δοκιμίτην καὶ Δοκιμαῖον),[1] κατ' ἀρχὰς μὲν μικρὰς βώλους ἐκδιδόντος τοῦ μετάλλου, διὰ δὲ τὴν νυνὶ πολυτέλειαν τῶν Ῥωμαίων κίονες ἐξαιροῦνται μονόλιθοι μεγάλοι, πλησιάζοντες τῷ ἀλαβαστρίτῃ λίθῳ κατὰ τὴν ποικιλίαν· ὥστε, καίπερ πολλῆς οὔσης τῆς ἐπὶ θάλατταν ἀγωγῆς τῶν τηλικούτων φορτίων, ὅμως καὶ κίονες καὶ πλάκες εἰς Ῥώμην κομίζονται θαυμασταὶ κατὰ τὸ μέγεθος καὶ κάλλος.

[1] Δοκιμαῖον, Xylander, for Δοκιμαίαν; so the later editors.

---

[1] 190 B.C. Strabo refers to Eumenes II, king of Pergamum, who reigned 197–159 B.C.

14. Now Phrygia Paroreia has a kind of mountainous ridge extending from the east towards the west; and below it on either side lies a large plain. And there are cities near it: towards the north, Philomelium, and, on the other side, the Antiocheia near Pisidia, as it is called, the former lying wholly in a plain, whereas the latter is on a hill and has a colony of Romans. The latter was settled by Magnetans who lived near the Maeander River. The Romans set them free from their kings at the time when they gave over to Eumenes[1] the rest of Asia this side the Taurus. Here there was also a priesthood of Mên Arcaeus,[2] which had a number of temple-slaves and sacred places, but the priesthood was destroyed after the death of Amyntas by those who were sent thither as his inheritors. Synnada is not a large city; but there lies in front of it a plain planted with olives, about sixty stadia in circuit.[3] And beyond it is Docimaea, a village, and also the quarry of "Synnadic" marble (so the Romans call it, though the natives call it "Docimite" or "Docimaean"). At first this quarry yielded only stones of small size, but on account of the present extravagance of the Romans great monolithic pillars are taken from it, which in their variety of colours are nearly like the alabastrite marble; so that, although the transportation of such heavy burdens to the sea is difficult, still, both pillars and slabs, remarkable for their size and beauty, are conveyed to Rome.

[2] "Arcaeus" appears to be an error for "Ascaeus" (see 12. 3. 31 and foot-note on "Mên Ascaeus").

[3] Or does Strabo mean sixty stadia in extent?

15. Ἀπάμεια δ' ἐστὶν ἐμπόριον μέγα τῆς ἰδίως λεγομένης Ἀσίας, δευτερεῦον μετὰ τὴν Ἔφεσον· αὕτη γὰρ καὶ τῶν ἀπὸ τῆς Ἰταλίας καὶ τῆς Ἑλλάδος ὑποδοχεῖον κοινόν ἐστιν. ἵδρυται δὲ ἡ Ἀπάμεια ἐπὶ ταῖς ἐκβολαῖς τοῦ Μαρσύου ποταμοῦ, καὶ ῥεῖ διὰ μέσης τῆς πόλεως ὁ ποταμός, τὰς ἀρχὰς ἀπὸ τῆς πόλεως[1] ἔχων· κατενεχθεὶς δ' ἐπὶ τὸ προάστειον σφοδρῷ καὶ κατωφερεῖ τῷ ῥεύματι συμβάλλει πρὸς τὸν Μαίανδρον, προσειληφότα καὶ ἄλλον ποταμὸν Ὀργᾶν, δι' ὁμαλοῦ φερόμενον πρᾶον καὶ μαλακόν· ἐντεῦθεν δ' ἤδη γενόμενος μέγας[2] Μαίανδρος τέως μὲν διὰ τῆς Φρυγίας φέρεται, ἔπειτα διορίζει τὴν Καρίαν καὶ τὴν Λυδίαν κατὰ τὸ Μαιάνδρου καλούμενον πεδίον, σκολιὸς ὢν εἰς ὑπερβολήν, ὥστε ἐξ ἐκείνου τὰς σκολιότητας ἁπάσας μαιάνδρους καλεῖσθαι· τελευτῶν δὲ καὶ τὴν[3] Καρίαν αὐτὴν διαρρεῖ[4] τὴν ὑπὸ τῶν Ἰώνων νῦν κατεχομένην καὶ μεταξὺ Μιλήτου καὶ Πριήνης ποιεῖται τὰς ἐκβολάς. ἄρχεται δὲ ἀπὸ Κελαινῶν, λόφου τινός, ἐν ᾧ πόλις ἦν ὁμώνυμος τῷ λόφῳ· ἐντεῦθεν δ' ἀναστήσας τοὺς ἀνθρώπους ὁ Σωτὴρ Ἀντίοχος εἰς C 578 τὴν νῦν Ἀπάμειαν τῆς μητρὸς ἐπώνυμον τὴν πόλιν ἐπέδειξεν Ἀπάμας, ἣ θυγάτηρ μὲν ἦν Ἀρταβάζου, δεδομένη δ' ἐτύγχανε πρὸς γάμον Σελεύκῳ τῷ Νικάτορι. ἐνταῦθα δὲ μυθεύεται τὰ περὶ τὸν Ὄλυμπον καὶ τὸν Μαρσύαν καὶ

[1] Instead of ἀπό C. Müller conj. οὐκ ἄπωθεν; Corais inserts παλαιᾶς between τῆς and πόλεως; Kramer conj. ἀκροπόλεως.
[2] μέγας is omitted by all MSS. except *oruz*.
[3] καὶ τήν, Corais, for κατά; so the later editors.
[4] διαρρεῖ, Casaubon, for διαιρεῖ; so the later editors.

15. Apameia is a great emporium of Asia, I mean Asia in the special sense of that term,[1] and ranks second only to Ephesus; for it is a common entrepôt for the merchandise from both Italy and Greece. Apameia is situated near the outlets of the Marsyas River, which flows through the middle of the city and has its sources in the city;[2] it flows down to the suburbs, and then with violent and precipitate current joins the Maeander. The latter receives also another river, the Orgas, and traverses a level country with an easy-going and sluggish stream; and then, having by now become a large river, the Maeander flows for a time through Phrygia and then forms the boundary between Caria and Lydia at the Plain of Maeander, as it is called, where its course is so exceedingly winding that everything winding is called "meandering." And at last it flows through Caria itself, which is now occupied by the Ionians, and then empties between Miletus and Prienê. It rises in a hill called Celaenae, on which there is a city which bears the same name as the hill; and it was from Celaenae that Antiochus Soter[3] made the inhabitants move to the present Apameia, the city which he named after his mother Apama, who was the daughter of Artabazus and was given in marriage to Seleucus Nicator. And here is laid the scene of the myth of Olympus and of

[1] *i.e.* Asia Minor.

[2] *i.e.* in the city's *territory*, unless the text is corrupt and should be emended to read, "having its sources in Celaenae" (Groskurd), or "not far away from the city" (C. Müller), or "in the old city" (Corais) of Celaenae, whence, Strabo later says, "Antiochus made the inhabitants move to the present Apameia" (see critical note).

[3] Antiochus "the Saviour."

τὴν ἔριν, ἣν ἤρισεν ὁ Μαρσύας πρὸς Ἀπόλλωνα. ὑπέρκειται δὲ καὶ λίμνη φύουσα κάλαμον τὸν εἰς τὰς γλώττας τῶν αὐλῶν ἐπιτήδειον, ἐξ ἧς ἀπολείβεσθαί[1] φασι τὰς πηγὰς ἀμφοτέρας, τήν τε τοῦ Μαρσύου καὶ τὴν τοῦ Μαιάνδρου.

16. Ἡ δὲ Λαοδίκεια, μικρὰ πρότερον οὖσα, αὔξησιν ἔλαβεν ἐφ' ἡμῶν καὶ τῶν ἡμετέρων πατέρων, καίτοι κακωθεῖσα ἐκ πολιορκίας ἐπὶ Μιθριδάτου τοῦ Εὐπάτορος· ἀλλ' ἡ τῆς χώρας ἀρετὴ καὶ τῶν πολιτῶν τινὲς εὐτυχήσαντες μεγάλην ἐποίησαν αὐτήν, Ἱέρων μὲν πρότερον, ὃς πλειόνων ἢ δισχιλίων ταλάντων κληρονομίαν κατέλιπε τῷ δήμῳ πολλοῖς τ' ἀναθήμασιν ἐκόσμησε τὴν πόλιν, Ζήνων δὲ ὁ ῥήτωρ ὕστερον καὶ ὁ υἱὸς αὐτοῦ Πολέμων, ὃς καὶ βασιλείας ἠξιώθη διὰ τὰς ἀνδραγαθίας ὑπ' Ἀντωνίου μὲν πρότερον, ὑπὸ Καίσαρος δὲ τοῦ Σεβαστοῦ μετὰ ταῦτα. φέρει δ' ὁ περὶ τὴν Λαοδίκειαν τόπος προβάτων ἀρετὰς οὐκ εἰς μαλακότητα[2] μόνον τῶν ἐρίων, ᾗ καὶ τῶν Μιλησίων διαφέρει, ἀλλὰ καὶ εἰς τὴν κοραξὴν[3] χρόαν, ὥστε καὶ προσοδεύονται λαμπρῶς ἀπ' αὐτῶν· ὥσπερ καὶ οἱ Κολοσσηνοὶ ἀπὸ τοῦ ὁμωνύμου χρώματος πλησίον οἰκοῦντες. ἐνταῦθα δὲ καὶ ὁ Κάπρος καὶ ὁ Λύκος συμβάλλει τῷ Μαιάνδρῳ ποταμῷ, ποταμὸς εὐμεγέθης, ἀφ' οὗ καὶ ἡ πρὸς τῷ Λύκῳ Λαοδίκεια λέγεται. ὑπέρκειται δὲ τῆς πόλεως ὄρος Κάδμος, ἐξ οὗ καὶ ὁ Λύκος ῥεῖ, καὶ

[1] ἀπολείβεσθαι is emended to ὑπολείβεσθαι by Tzschucke, Kramer, and Müller-Dübner.
[2] μαλακότητα, Kramer, for μαλακότητας; so the later editors.
[3] κοραξήν, the editors, for κοραξίν.

Marsyas and of the contest between Marsyas and Apollo. Above is situated a lake which produces the reed that is suitable for the mouth-pieces of pipes; and it is from this lake that pour the sources of both the Marsyas and the Maeander.

16. Laodiceia, though formerly small, grew large in our time and in that of our fathers, even though it had been damaged by siege in the time of Mithridates Eupator.[1] However, it was the fertility of its territory and the prosperity of certain of its citizens that made it great: at first Hieron, who left to the people an inheritance of more than two thousand talents and adorned the city with many dedicated offerings, and later Zeno the rhetorician and his son Polemon,[2] the latter of whom, because of his bravery and honesty, was thought worthy even of a kingdom, at first by Antony and later by Augustus. The country round Laodiceia produces sheep that are excellent, not only for the softness of their wool, in which they surpass even the Milesian wool, but also for its raven-black colour,[3] so that the Laodiceians derive splendid revenue from it, as do also the neighbouring Colosseni from the colour which bears the same name.[4] And here the Caprus River joins the Maeander, as does also the Lycus, a river of good size, after which the city is called the "Laodiceia near Lycus."[5] Above the city lies Mt. Cadmus,

[1] King of Pontus 120–63 B.C.

[2] Polemon I, king of Pontus and the Bosporus, and husband of Pythodoris.

[3] Cf. 3. 2. 6.

[4] *i.e.* the "Colossian" wool, dyed purple or madder-red (see Pliny 25. 9. 67 and 21. 9. 27).

[5] *i.e.* to distinguish it from the several other Laodiceias.

ἄλλος ὁμώνυμος τῷ ὄρει. τὸ πλέον δ' οὗτος ὑπὸ γῆς ῥυείς, εἶτ' ἀνακύψας συνέπεσεν εἰς ταὐτὸ τοῖς ἄλλοις ποταμοῖς, ἐμφαίνων ἅμα καὶ τὸ πολύτρητον τῆς χώρας καὶ τὸ εὔσειστον· εἰ γάρ τις ἄλλη, καὶ ἡ Λαοδίκεια εὔσειστος, καὶ τῆς πλησιοχώρου δὲ Κάρουρα.

17. Ὅριον δέ[1] ἐστι τῆς Φρυγίας καὶ τῆς Καρίας τὰ Κάρουρα· κώμη δ' ἐστὶν αὕτη πανδοχεῖα ἔχουσα καὶ ζεστῶν ὑδάτων ἐκβολάς, τὰς μὲν ἐν τῷ ποταμῷ Μαιάνδρῳ, τὰς δ' ὑπὲρ τοῦ χείλους. καὶ δή ποτέ φασι πορνοβοσκὸν αὐλισθέντα ἐν τοῖς πανδοχείοις σὺν πολλῷ πλήθει γυναικῶν, νύκτωρ γενομένου σεισμοῦ, συναφανισθῆναι πάσαις. σχεδὸν δέ τι καὶ πᾶσα εὔσειστός ἐστιν ἡ περὶ τὸν Μαίανδρον χώρα, καὶ ὑπόνομος πυρί τε καὶ ὕδατι μέχρι τῆς μεσογαίας. διατέτακε γὰρ ἀπὸ τῶν πεδίων ἀρξαμένη πᾶσα ἡ τοιαύτη κατασκευὴ τῆς χώρας εἰς τὰ Χαρώνια, τό τε ἐν Ἱεραπόλει καὶ τὸ ἐν Ἀχαράκοις[2] τῆς Νυσαΐδος καὶ τὸ περὶ Μαγνησίαν καὶ Μυοῦντα· εὔθρυπτός τε γάρ ἐστιν ἡ γῆ καὶ ψαθυρά, πλήρης τε ἁλμυρίδων καὶ εὐεκπύρωτός ἐστι. τάχα δὲ καὶ ὁ Μαίανδρος διὰ τοῦτο σκολιός, ὅτι πολλὰς μεταπτώσεις λαμβάνει τὸ ῥεῖθρον, καὶ πολλὴν χοῦν κατάγων

[1] δὲ Κάρουρα. Ὅριον δέ, the editors, for Κάρουρα δὲ ὅριον.

[2] Ἀχαρακοῖς, Tzschucke, for Χαρακοῖς; so the later editors.

---

[1] See 5. 4. 5, and the note on "Plutonia."

[2] *i.e.* sodium chloride (salt), and perhaps other salts found

whence the Lycus flows, as does also another river of the same name as the mountain. But the Lycus flows under ground for the most part, and then, after emerging to the surface, unites with the other rivers, thus indicating that the country is full of holes and subject to earthquakes; for if any other country is subject to earthquakes, Laodiceia is, and so is Carura in the neighbouring country.

17. Carura forms a boundary between Phrygia and Caria. It is a village; and it has inns, and also fountains of boiling-hot waters, some in the Maeander River and some above its banks. Moreover, it is said that once, when a brothel-keeper had taken lodging in the inns along with a large number of women, an earthquake took place by night, and that he, together with all the women, disappeared from sight. And I might almost say that the whole of the territory in the neighbourhood of the Maeander is subject to earthquakes and is undermined with both fire and water as far as the interior; for, beginning at the plains, all these conditions extend through that country to the Charonia,[1] I mean the Charonium at Hierapolis and that at Acharaca in Nysaïs and that near Magnesia and Myus. In fact, the soil is not only friable and crumbly but is also full of salts[2] and easy to burn out.[3] And perhaps the Maeander is winding for this reason, because the stream often changes its course and, carrying down much silt, adds the silt at different times to

in soil, as, for example, sodium carbonate and calcium sulphate—unless by the plural of the word Strabo means merely "salt-particles," as Tozer takes it.

[3] On "soil which is burnt out," see Vol. II, p. 454, footnote 1.

ἄλλοτ' ἄλλῳ μέρει τῶν αἰγιαλῶν προστίθησι· τὸ δὲ πρὸς τὸ πέλαγος βιασάμενος[1] ἐξωθεῖ. καὶ δὴ καὶ τὴν Πριήνην ἐπὶ θαλάττῃ πρότερον οὖσαν μεσόγαιαν πεποίηκε τετταράκοντα σταδίων προσχώματι.

18. Καὶ ἡ Κατακεκαυμένη δέ, ἥπερ ὑπὸ Λυδῶν καὶ Μυσῶν κατέχεται, διὰ τοιαῦτά τινα τῆς προσηγορίας τετύχηκε ταύτης· ἥ τε Φιλαδέλφεια, ἡ πρὸς αὐτῇ πόλις, οὐδὲ τοὺς τοίχους ἔχει πιστούς, ἀλλὰ καθ' ἡμέραν τρόπον τινὰ σαλεύονται καὶ διίστανται· διατελοῦσι δὲ προσέχοντες τοῖς πάθεσι τῆς γῆς καὶ ἀρχιτεκτονοῦντες πρὸς αὐτά.[2] καὶ τῶν ἄλλων δὲ πόλεων Ἀπάμεια μὲν καὶ πρὸ τῆς Μιθριδάτου στρατείας ἐσείσθη πολλάκις, καὶ ἔδωκεν ἐπελθὼν ὁ βασιλεὺς ἑκατὸν τάλαντα ε.ς ἐπανόρθωσιν, ὁρῶν ἀνατετραμμένην τὴν πόλιν. λέγεται δὲ καὶ ἐπ' Ἀλεξάνδρου παραπλήσια συμβῆναι· διόπερ εἰκός ἐστι καὶ τὸν Ποσειδῶ τιμᾶσθαι παρ' αὐτοῖς, καίπερ μεσογαίοις οὖσι, καὶ ἀπὸ Κελαινοῦ τοῦ Ποσειδῶνος ἐκ Κελαινοῦς, μιᾶς τῶν Δαναΐδων, γενομένου κεκλῆσθαι τὴν πόλιν ἐπώνυμον,[3] ἢ διὰ τὸν λίθον καὶ τὴν ἀπὸ τῶν ἐκπυρώσεων μελανίαν. καὶ τὰ περὶ Σίπυλον δὲ καὶ τὴν ἀνατροπὴν αὐτοῦ μῦθον οὐ δεῖ τίθεσθαι· καὶ γὰρ νῦν τὴν Μαγνησίαν τὴν ὑπ'

[1] βιασάμενος, Xylander, for βιασαμένους; so the later editors.
[2] αὐτά, Groskurd, for αὐτήν; so the later editors.
[3] ἐπώνυμον, the editors, for ὁμώνυμον.

---

[1] "At the present day the coastline has been advanced so far, that the island of Lade, off Miletus, has become a hill in the middle of a plain" (Tozer, *op. cit.*, p. 288).

different parts of the shore; however, it forcibly thrusts a part of the silt out to the high sea. And, in fact, by its deposits of silt, extending forty stadia, it has made Prienê, which in earlier times was on the sea, an inland city.[1]

18. Phrygia "Catacecaumenê,"[2] which is occupied by Lydians and Mysians, received its appellation for some such reason as follows: In Philadelphia, the city near it, not even the walls are safe, but in a sense are shaken and caused to crack every day. And the inhabitants are continually attentive to the disturbances in the earth and plan all structures with a view to their occurrence. And, among the other cities, Apameia was often shaken by earthquakes before the expedition of King Mithridates, who, when he went over to that country and saw that the city was in ruins, gave a hundred talents for its restoration; and it is said that the same thing took place in the time of Alexander. And this, in all probability, is why Poseidon is worshipped in their country, even though it is in the interior,[3] and why the city was called Celaenae,[4] that is, after Celaenus, the son of Poseidon by Celaeno, one of the daughters of Danaüs, or else because of the "blackness" of the stone, which resulted from the burn-outs. And the story of Mt. Sipylus and its ruin should not be put down as mythical, for in our own times Magnesia, which lies at the foot of it, was

[2] "Burnt up."

[3] Poseidon was not only the god of the sea, but also the "earth-shaker" (ἐνοσίχθων or ἐνοσίγαιος), an epithet frequently used in Homer.

[4] *i.e.* "Black."

αὐτῷ κατέβαλον σεισμοί, ἡνίκα καὶ Σάρδεις καὶ τῶν ἄλλων τὰς ἐπιφανεστάτας κατὰ πολλὰ μέρη διελυμήναντο· ἐπηνώρθωσε δ' ὁ ἡγεμών,[1] χρήματα ἐπιδούς, καθάπερ καὶ πρότερον ἐπὶ τῆς γενομένης συμφορᾶς Τραλλιανοῖς (ἡνίκα τὸ γυμνάσιον καὶ ἄλλα μέρη συνέπεσεν) ὁ πατὴρ αὐτοῦ καὶ τούτοις καὶ Λαοδικεῦσιν.

19. Ἀκούειν δ' ἔστι καὶ τῶν παλαιῶν συγγραφέων, οἷά φησιν ὁ τὰ Λύδια συγγράψας Ξάνθος, διηγούμενος, οἷαι μεταβολαὶ κατέσχον πολλάκις τὴν χώραν ταύτην, ὧν ἐμνήσθημέν που καὶ ἐν τοῖς πρόσθεν. καὶ δὴ καὶ τὰ περὶ τὸν Τυφῶνα πάθη ἐνταῦθα μυθεύουσι καὶ τοὺς Ἀρίμους καὶ τὴν Κατακεκαυμένην ταύτην εἶναί φασιν· οὐκ ὀκνοῦσι δὲ καὶ τὰ μεταξὺ Μαιάνδρου καὶ Λυδῶν ἅπανθ' ὑπονοεῖν τοιαῦτα καὶ διὰ τὸ πλῆθος τῶν λιμνῶν καὶ ποταμῶν καὶ τοὺς πολλαχοῦ κευθμῶνας τῆς γῆς. ἡ δὲ μεταξὺ Λαοδικείας καὶ Ἀπαμείας λίμνη καὶ βορβορώδη καὶ ὑπόνομον[1] τὴν ἀποφορὰν ἔχει, πελαγία οὖσα· φασὶ δὲ καὶ δίκας εἶναι τῷ Μαιάνδρῳ μεταφέροντι τὰς χώρας, ὅταν περικρουσθῶσιν οἱ ἀγκῶνες, ἁλόντι[2] δὲ τὰς ζημίας ἐκ τῶν πορθμικῶν διαλύεσθαι τελῶν.

[1] ὑπόνομον, Meineke emends to ὑπονόμου. Corais conj. ὑπόνοσον, Kramer ἐπίνοσον. T. G. Tucker (*Classical Quarterly* III, p. 101) would insert καθ' before ὑπόνομον and translate: "It has a smell after the manner of a sewer."

[2] ἁλόντι, Jones, from conj. of Capps, for ἁλόντες; others, following conj. of Xylander, emend to ἁλόντος.

---

[1] *i.e.* Tiberius (see Tacitus, *Annals* 2. 47).

laid low by earthquakes, at the time when not only Sardeis, but also the most famous of the other cities, were in many places seriously damaged. But the emperor[1] restored them by contributing money; just as his father in earlier times, when the inhabitants of Tralleis suffered their misfortune (when the gymnasium and other parts of the city collapsed), restored their city, as he also restored the city of the Laodiceians.

19. One should also hear the words of the ancient historians, as, for example, those of Xanthus, who wrote the history of Lydia, when he relates the strange changes that this country often underwent, to which I have already referred somewhere in a former part of my work.[2] And in fact they make this the setting of the mythical story of the Arimi and of the throes of Typhon, calling it the Catacecaumenê[3] country. Also, they do not hesitate to suspect that the parts of the country between the Maeander River and the Lydians are all of this nature, as well on account of the number of the lakes and rivers as on account of the numerous hollows in the earth. And the lake[4] between Laodiceia and Apameia, although like a sea,[5] emits an effluvium that is filthy and of subterranean origin. And they say that lawsuits are brought against the god Maeander for altering the boundaries of the countries on his banks, that is, when the projecting elbows of land are swept away by him; and that when he is convicted the fines are paid from the tolls collected at the ferries.

[2] 1. 3. 4.
[3] Cp. 13. 4. 11.
[4] Now called Chardak Ghieul.
[5] *i.e.* in size and depth.

20. Μεταξὺ δὲ τῆς Λαοδικείας καὶ τῶν Καρούρων ἱερόν ἐστι Μηνὸς Κάρου καλούμενον, τιμώμενον ἀξιολόγως. συνέστηκε[1] δὲ καθ' ἡμᾶς διδασκαλεῖον Ἡροφιλείων ἰατρῶν μέγα ὑπὸ Ζεύξιδος, καὶ μετὰ ταῦτα Ἀλεξάνδρου τοῦ Φιλαλήθους, καθάπερ ἐπὶ τῶν πατέρων τῶν ἡμετέρων ἐν Σμύρνῃ τὸ τῶν Ἐρασιστρατείων ὑπὸ Ἱκεσίου, νῦν δ' οὐχ ὁμοίως τι συμβαίνει.[2]

21. Λέγεται δέ τινα φῦλα Φρύγια οὐδαμοῦ δεικνύμενα, ὥσπερ οἱ Βερέκυντες· καὶ Ἀλκμὰν λέγει,

Φρύγιον ηὔλησε μέλος τὸ Κερβήσιον.

καὶ βόθυνός τις λέγεται Κερβήσιος ἔχων ὀλεθρίους ἀποφοράς· ἀλλ' οὗτός γε δείκνυται, οἱ δ' ἄνθρωποι οὐκέθ' οὕτω λέγονται. Αἰσχύλος δὲ συγχεῖ ἐν τῇ Νιόβῃ· φησὶ γὰρ ἐκείνη μνησθήσεσθαι[3] τῶν περὶ Τάνταλον,

οἷς ἐν Ἰδαίῳ πάγῳ
Διὸς πατρῴου βωμός ἐστι,

καὶ πάλιν,

Σίπυλον Ἰδαίαν ἀνὰ χθόνα·

καὶ ὁ Τάνταλος λέγει,

[1] Instead of συνέστηκε *rw*, Corais and Meineke read συνέστη.

[2] For τι συμβαίνει, Corais conj. ἔτι συμμένει; and Meineke so reads.

[3] μνησθήσεσθαι, Casaubon, for μνησθήσεται; so the later editors.

20. Between Laodiceia and Carura is a temple of Mên Carus, as it is called, which is held in remarkable veneration. In my own time a great Herophileian[1] school of medicine has been established by Zeuxis, and afterwards carried on by Alexander Philalethes,[2] just as in the time of our fathers the Erasistrateian school[3] was established by Hicesius, although at the present time the case is not at all the same as it used to be.[4]

21. Writers mention certain Phrygian tribes that are no longer to be seen; for example, the Berecyntes. And Alcman says, "On the pipe he played the Cerbesian, a Phrygian melody." And a certain pit that emits deadly effluvia is spoken of as Cerbesian. This, indeed, is to be seen, but the people are no longer called Cerbesians. Aeschylus, in his *Niobê*, confounds things that are different; for example, Niobê says that she will be mindful of the house of Tantalus, "those who have an altar of their paternal Zeus on the Idaean hill";[5] and again, "Sipylus in

[1] Herophilus was one of the greatest physicians of antiquity. He was born at Chalcedon in Bithynia, and lived at Alexandria under Ptolemy I, who reigned 323–285 B.C. His specialty was dissection; and he was the author of several works, of which only fragments remain.

[2] Alexander of Laodiceia; author of medical works of which only fragments remain.

[3] Erasistratus, the celebrated physician and anatomist, was born in the island of Ceos and flourished 300–260 B.C.

[4] The Greek for this last clause is obscure and probably corrupt. Strabo means either that schools like the two mentioned "no longer arise" or that one of the two schools mentioned (more probably the latter) "no longer flourishes the same as before." To ensure the latter thought Meineke (from conj. of Corais) emends the Greek text (see critical note).

[5] *Frag.* 162, 2 (Nauck).

σπείρω δ' ἄρουραν δώδεχ' ἡμερῶν ὁδόν,
Βερέκυντα χῶρον, ἔνθ' Ἀδραστείας ἕδος,
Ἴδη τε μυκηθμοῖσι καὶ βρυχήμασιν
βρέμουσι[1] μήλων πᾶν τ' Ἐρέχθειον[2] πέδον.

[1] βρέμουσι, Tzschucke and Corais, following Casaubon, for ἕρπουσι; Meineke conj. πρέπουσι.
[2] τ' Ἐρέχθειον, conj. of Meineke, for δ' ἐρεχθεῖ.

the Idaean land";[1] and Tantalus says, "I sow furrows that extend a ten days' journey, Berecyntian land, where is the site of Adrasteia, and where both Mt. Ida and the whole of the Erechtheian plain resound with the bleatings and bellowings of flocks."[2]

[1] *Frag.* 163 (Nauck).
[2] *Frag.* 158, 2 (Nauck).

APPENDIX

THE ITHACA-LEUCAS PROBLEM[1]

Homer (*e.g.*, *Od.* 9. 21–27) presents Odysseus as the king of a group of islands off the west coast of Greece (cf. the trip of Telemachus to Pylus), which consisted of four large islands (Ithaca, Dulichium, Samê, and Zacynthus) and of a number of smaller ones. Near the mouth of the Corinthian Gulf there is such a group of islands, the larger of which are Leucas, Ithaca (Thiaki), Cephallenia, and Zacynthus (Zante).

It is often stated, however, that Leucas is a peninsula, not an island. It is separated from the mainland by a lagoon too shallow for the passage of ships (Leaf, *Homer and History*, p. 144); and for this reason the Corinthians, in the reign of Cypselus (655–625 B.C.), "dug a canal through the isthmus of the peninsula and made Leucas an island" (Strabo 10. 2. 8). Other ancient writers agree with Strabo in speaking of Leucas as a peninsula (Scholiast on *Odyssey*, 24. 376; Scylax, *Periplus*, 34; Ovid, *Metamorphoses*, 15. 289; Plutarch, *De sera numinis vindicta*, 7. 552 E; Pliny, *Nat. Hist.* 4. 2; see also Manly, *Ithaca or Leucas?* pp. 25–29).

[1] In the preparation of this note the translator must record his indebtedness to two of his pupils, Miss Marion L. Ayer, M.A., and Whitney Tucker, B.A., each of whom wrote an able paper on the subject. A Bibliography prepared by them will be found at the end of this note.

This tradition has made it necessary to find the fourth island, as well as to identify each of the others. Scholars are agreed upon only one identification, that of the modern Zante with the Homeric Zacynthus; indeed, some have despaired of making Homer's references to the islands agree with geographical reality, on the ground that, as Strabo (*e.g.* in 1. 2. 9) insists, Homer was wont purposely to mingle false elements with true; and so, for example, Wilamowitz (*Arch. Anzeiger*, 1903, p. 43) says that Dulichium is "nowhere to be found."

Until the end of the nineteenth century the prevailing view was that Thiaki was Ithaca and that Cephallenia was Samê; while Dulichium was sought in various places (see Manly, *op. cit.*, pp. 10-12), being identified by some with the western part of Cephallenia (Pausanias 6. 15. 7; cf. Strabo 10. 2. 14), by Strabo with one of the Echinades, called Dolicha (8. 2. 2, 8. 3. 8, 10. 2. 10, 10. 2. 19; cf. Schol. on *Iliad*, 2. 625), and by Bunbury (*Hist. Ancient Geog.* I, p. 70) with Leucas. The difficulty was that Dulichium, the missing island, seems from Homer's references to it (*e.g.* in *Od.* 14. 335 and 16. 247) to have been the largest and richest of the group. Samê was supposed to be Cephallenia because of the existence there, in classical times, of a city of Samus (see Strabo 10. 2. 10).

In 1894 Draheim (*Woch. f. Kl. Philol.*, 1894, 63) wondered that no one had ever doubted the identification of Ithaca with Thiaki, and suggested that Leucas would better fit the Homeric description. In 1900 Dörpfeld announced his theory, that Ithaca was Leucas, Samê was Thiaki, and Dulichium was Cephallenia. Immediately there arose a heated

discussion, with a number of scholars taking sides or producing new variations of the theories presented. Among Dörpfeld's supporters are Cauer, Gössler, Leaf, Seymour, and von Marées; among his opponents are Allen, Bérard, Brewster, Manly, Shewan, Vollgraff, Wilamowitz, and Bürchner.

The chief arguments in support of the Ithaca-Leucas theory, as set forth by Dörpfeld, Gössler, and Leaf, are as follows: (1) In *Od.* 9. 21–28 the geographical position of Ithaca is described as "low in the sea," which they explain as "near the shore" (Dörpfeld, *Leukas*, pp. 11 f., 28–30; Gössler, *Leukas-Ithaca*, pp. 34–36); and as "farthest up towards the darkness," in contrast with the other islands, which lie "toward the dawn and the sun." The ancients confused west and north along this coast, and so "towards the darkness" means towards the north by our compasses (Dörpfeld, *op. cit.*, pp. 8–10, 26–28; Gössler, *op. cit.*, pp. 36–40). Both these expressions fit Leucas very well, but Thiaki not at all. (2) The little island of Asteris, where the suitors lay in wait for Telemachus, must be Arcudi, between Leucas and Thiaki, since this island fits the Homeric description, whereas Dascalio, the only island between Thiaki and Cephallenia, does not (Dörpfeld, *op. cit.*, pp. 14–16, 34–36; Gössler, *op. cit.*, pp. 49–52; Leaf, *op. cit.*, pp. 148, 151 f.). (3) Since Ithaca was connected with the mainland by a ferry, it must be close to the mainland, like Leucas, not far off, like Thiaki (Dörpfeld, *op. cit.*, pp. 12, 30–32; Gössler, *op. cit.*, pp. 47 f.). (4) Ithaca must lie between Thesprotia and Dulichium, in view of Odysseus's story of his trip to Ithaca (*Od.* 14. 334–359); this story would exclude Thiaki (Dörpfeld, *op. cit.*, pp.

14, 34; Gössler, *op. cit.*, pp. 45 f.; Leaf, *op. cit.*, p. 153). Many other passages in Homer are produced to reinforce the conclusion. The name of the island was transferred from Leucas to Thiaki as a result of the Dorian invasion; the Dorians drove the people of Ithaca out of their own island, whereupon they crossed over to the next island (Samê), conquered it, and changed its name to Ithaca (Dörpfeld, *op. cit.*, pp. 17 f., 25; Gössler, *op. cit.*, pp. 75–77; Leaf, *op. cit.*, pp. 154–156).

The supporters of Thiaki attack all of Dörpfeld's arguments, on various grounds; for instance, they accuse him of misinterpreting the text in connection with the "ferry" (*Od.* 20. 187 f.), and they object to his conclusions from the text in many passages, as *Od.* 9. 25, "low in the sea." Then they proceed to identify on Thiaki the topographical features of the Ithaca of Odysseus; but they do not agree in their discussion of these features, nor in the identification of the other islands. Most of them regard Cephallenia, or a part of it, as Samê; but Croiset and Brewster find Samê in Leucas. As to Dulichium there is great difference of opinion: Croiset and Brewster identify it with Cephallenia; Goekoop, Rothe, Gruhn, and Michael with the western part of Cephallenia; Bunbury, Vollgraff, Allen, Shewan, Stürmer, and Bury with Leucas; Lang, Manly, and Cserép with one of the Echinades; and Bérard (*Les Phéniciens et l'Odyssée*, II, pp. 421–446) with the small island of Meganisi, near Leucas. All these scholars, however, hold that the geographical position of Thiaki agrees with the Homeric description of Ithaca, or that the discrepancies are so slight that they can be ignored

or set down to poetic licence—as Bérard (*op. cit.*, II, pp. 409, 480–494), who, in trying to prove that Asteris is the modern Dascalio, admits that the description does not agree with reality, but argues that the topography of Asteris is in part invented by the poet and in part transferred from the near-by island of Cephallenia.

One group of scholars, including some of those already mentioned, hold that Homer lived in Asia Minor and was therefore not familiar with the home of Odysseus; and so they ascribe apparent inaccuracies to the ignorance of the poet. Wilamowitz is the most prominent of this group, and explains (*Arch. Anzeiger*, 1903, p. 44; *Homerische Untersuchungen*, pp. 26 f.) that Homer knew only a few place-names, with a little vague information about the region. Belzner (*Land und Heimat des Odysseus*), adopting this view, disregards actual geography and invents a group of islands in this neighbourhood, which, he says, would correspond to Homer's description.

Goekoop (*Ithaque la Grande*) thinks that Ithaca, Dulichium, and Samê are different parts of Cephallenia.

Through the maze of this controversy the present translator, as one of the "more Homeric," seems to see a preponderance of evidence in favour of Leucas as the Homeric Ithaca; but the problem still remains open to further investigation.[1]

[1] Two very recent works on this subject, by W. Dörpfeld and Sir Rennell Rodd (see under *Partial Bibliography*), appeared too late for consideration in the above *Appendix*. The translator has not yet seen the former, but has read, on the very day of transmitting the final page-proofs of the present volume, the modest and charming little book of the latter, who makes an able plea for the traditional Ithaca.

or set down to poetic licence—as Bérard (*op. cit.*, II, pp. 409, 430, 434), who, in trying to prove that Asteris is the modern Dascalio, admits that the description does not agree with reality, but argues that the topography of Asteris is in part invented by the poet and in part transferred from the near-by island of Cephallenia.

One group of scholars, including some of those already mentioned, hold that Homer lived in Asia Minor and was therefore not familiar with the home of Odysseus; and so they ascribe apparent inaccuracies to the ignorance of the poet. Wilamowitz is the most prominent of this group, and explains (*Arch. Anzeiger*, 1903, p. 44; *Homerische Untersuchungen*, pp. 26 f.) that Homer knew only a few place names, with a little vague information about the region. Belzner (*Land und Heimat des Odysseus*), adopting this view, disregards actual geography and invents a group of islands in this neighbourhood, which, he says, would correspond to Homer's description.

Goekoop (*Ithaque la Grande*) thinks that Ithaca, Dulichium, and Same are different parts of Cephallenia.

Through the maze of this controversy the present translator, as one of the "more Homeric," seems to see a preponderance of evidence in favour of Leucas as the Homeric Ithaca; but the problem still remains open to further investigation.[1]

[1] Two very recent works on this subject, by W. Dörpfeld and Sir Rennell Rodd (see under *Journal Bibliography*), appeared too late for consideration in the above *Appendix*. The translator has not yet seen the former but has read, on the very day of transmitting the final page-proofs of the present volume, the modest and charming little book of the latter, who makes an able plea for the traditional Ithaca.

PARTIAL BIBLIOGRAPHY

OF THE ITHACA-LEUCAS PROBLEM

Allen, T. W.—The Homeric Catalogue (*J. H. S.* 30, 1910).

Belzner, E.—*Land und Heimat des Odysseus.* Munich, 1915.

Bérard, V.—*Les Phéniciens et l'Odyssée.* Paris, 1902.

Brewster, F.—Ithaca: a Study of the Homeric Evidence (*Harvard Studies in Classical Philology,* 31, 1920). Asteris. (*Harvard Studies,* 33, 1922.) Ithaca, Dulichium, Samê, and Wooded Zacynthus (*Harvard Studies,* 36, 1925).

Bunbury, E. H.—*History of Ancient Geography.* London, 1883.

Bürchner, L.—Ithake; and Leukas, Leukadia; both (*s.vv.*) in Pauly-Wissowa.

Bury, J. B., in the *Cambridge Ancient History.*

Cauer, P.—Erfundenes und Überliefertes bei Homer, pp. 14–17 (*N. Jahrbücher,* 8, 1905). *Grundfragen der Homerkritik,* 3rd ed., Leipzig, 1923.

Croiset, M.—Observations sur la légende primitive d'Ulysse (*Académie des Inscriptions et Belles-Lettres,* 1911).

Cserép, J.—*Homeros Ithakeja.* 1908.

Dörpfeld, W.—Das Homerische Ithaka (*Mélanges Perrot,* Paris, 1902). Leukas-Ithaka (*Archäologischer Anzeiger,* 1904). [These two articles were republished together as *Leukas.* Athens, 1905.] *Die Heimkehr des Odysseus.* Munich, 1924. Zur Leukas-Ithaka Frage (*Philologus,* 1926). *Alt-Ithaka: Ein Beitrag zur Homer-Frage; Studien und Ausgrabungen auf der Insel Leukas-Ithaka* (a work in 2 vols.). Verlag Richard Uhde, München-Grätelfing, 1927.

Draheim, H.—*Die Ithaka Frage.* Berlin, 1903.

Engel, E.—*Der Wohnsitz des Odysseus.* Leipzig, 1912.

Goekoop, A. E. H.—*Ithaque la Grande.* Athens, 1908.

Gössler, P.—*Leukas-Ithaka, die Heimat des Odysseus.* Stuttgart, 1904.

BIBLIOGRAPHY

Gröschl, J.—*Dörpfelds Leukas-Ithaka-Hypothese.* Friedek, 1907.

Gruhn, A.—Ithaka (*N. Phil. Rundschau*, 1906).

Lang, G.—*Untersuchungen zur Geographie der Odyssee.* Karlsruhe, 1905.

Lang, N.—*Odysseus Hazaja.* Budapest, 1902.

Leaf, W.—*Homer and History.* London, 1915. *Strabo on the Troad.* Cambridge, 1923.

Manly, W. G.—Ithaca or Leucas? (*Univ. of Missouri Studies*, 1903).

Marées, W. von—Die Ithakalegende auf Thiaki (*Neue Jahrbücher*, 17, 1906). *Karten von Leukas.* Berlin, 1907.

Michael, H.—*Das Homerische und das Heutige Ithaka.* Jauer, 1902. *Die Heimat des Odysseus.* Jauer, 1905.

Monro, D B.—The Place and Time of Homer (*Class. Rev.*, 19, 1905).

Partsch, J.—Die Insel Leukas (*Petermanns Mittheilungen*, 1890). Das Alter der Inselnatur von Leukas (*Petermanns Mitth.*, 1907).

Pavlatos.—Ἡ Πατρὶς τοῦ Ὀδυσσέως. Athens, 1906.

Robert, C.—Ithaca (*Hermes*, 44, 1909).

Rodd, Sir Rennell.—*Homer's Ithaca: A Vindication of Tradition.* Edward Arnold and Co., London, 1927.

Rothe, C.—*Die Odyssee als Dichtung.* Paderborn, 1914.

Seymour, T. D.—*Life in the Homeric Age.* New York, 1907.

Shewan, A.—Recent Homeric Literature (*Class. Phil.*, 7, 1912). Leukas-Ithaka (*J. H. S.*, 34, 1914). Beati Possidentes Ithakistae (*Class. Phil.*, 12, 1917). Meges and Dulichium, and also Asteris and the Voyage of Telemachus (*Class. Phil.*, 19, 1924). Asteris and Dulichium (*Class. Phil.*, 21, 1926).

Stürmer, F., in *Berl. Phil. Wochenschrift*, 1913, 1660. *Rhapsodien der Odyssee.* Würzburg, 1921.

Vollgraff, W.—Dulichium-Leukas. (*Neue Jahrbücher*, 19, 1907). Fouilles d'Ithaque (*B. C. H.*, 29, 1905).

A PARTIAL DICTIONARY OF PROPER NAMES[1]

A

Abus, Mt., 321, 335
Acarnanians, the, 23, 65
Achaei, the, 191, 203, 207
Acheloüs River, the, 23, 55, 57
Achilles, temple of, 197
Acilisenê, 297, 321, 325, 333, 341
Actian War, the, 341, 497
Actium, 25, 63, 165
Acusilaüs the Argive (see foot-note on p. 115), on the Cabeiri, 115
Adiatorix, son of Domnecleius, tetrarch of the Galatians, received from Antony a part of Heracleia, 379
Admetus, king of Pherae in Thessaly, 15
Ador (Adon ?), commandant of Artageras (Artageira ?), 327
Aenianians, the, 25
Aeolians, settlers in Euboea, 13
Aeschines, reproached by Demosthenes for engaging in Phrygian rites, 109
Aeschylus the tragic poet, on the city of Euboea, 15; on the worship of Cotys and Dionysus, 105; confounds things that are different, 519
Aetolia, divided into Old Aetolia and Aetolia Epictetus, 27
Aetolians, the, 23, 65
Aetolus, son of Endymion, 77, 79
Aïclus, coloniser of Eretria, 13
Alazonius River, the, 219, 221
Albania, 187, 207
Albanians, the, 223; description of, 226
Alcmaeon, son of Amphiaraüs, 71
Alcman of Sardis (fl. about 625 B.C., founder of Doric Lyric poetry), on the Carystian wine, 11; on the Erysichaeans, 65; on the "Andreia" (public messes), 151; on the Cerbesian melody, 519
Alexander Philalethes the physician of Laodiceia, contemporary of Strabo, 519
Alexander the Aetolian poet (b. about 315 B.C.), on the Ascanian Lake, 465
Alexander the Great, consorted with Thalestria, queen of the Amazons, 237; his exploits exaggerated to glorify him, 239, 247, 255; eluded by Spitamenes and Bessus, 269; went to the Iaxartes River, 271; fewer tribes subdued by him than by the Greeks, 279; broke up Bactrian custom, 283; founder and destroyer of cities in Bactriana and Sogdiana, 283; married Rhoxana, 285; did not attempt expedition against certain Scythians, 287; captured Sagalassus (Selgessus), 479
Althaemenes the Argive, founder of cities in Crete, 143, 149
Amaltheia, the horn of, 57, 59
Amanus Mt., the, 295, 351
Amardi (Mardi), the, 249, 259, 269 305
Amaseia, 397, 429, 445
Amastris, a city named after Queen Amastris, 385
Amastris, wife of Dionysius the tyrant of Heracleia, daughter of Oxyathres, and founder of the city Amastris, 385
Amazons, the, 231, 405, 493
Ambracian Gulf, the, 25
Amisus, 211, 395, 399
Amphilochians, the, 23
Amphilochus, 73

[1] A complete index will appear in the last volume.

A PARTIAL DICTIONARY OF PROPER NAMES

Amphitryon, 47, 59
Amyntas, king of Galatia, successor of Deïotarus, 469; owned three hundred flocks, 475; slew Antipater Derbetes, 477
Anactorium, 25, 33
Anadatus, Persian deity, 263
Anaïtis, temple of, 263; worshipped by Medes and Armenians, 341
Anariacae (Parsii), the, 249, 269
Ancyra, 471
Andron (see foot-note 2 on p. 126), on Cephallenia and Dulichium, 49; on the foreigners in Crete, 127
Antenor, settled at the recess of the Adriatic, 383
Antigonia (see Nicaea), 463
Antigonus the son of Philip (see note 2 on p. 463), 463
Antiocheia near Pisidia, 477, 507
Antiochus Soter, king of Syria 280–261 B.C., founded Antiocheia, 279; removed inhabitants of Celaenae to Apameia, 509
Antiochus the Great (reigned over Syria 223–187 B.C.), 325; conquered by the Romans, 369, 457
Antipater Derbetes, the pirate, 349, and tyrant, 476
Antipater the son of Sisis, ruler of Lesser Armenia, yielded to Mithridates Eupator, 425
Antitaurus Mt., the, 295, 299, 319, 351
Antonius, Gaius, uncle of Marcus Antonius, 47
Antony, Marcus, nephew of Gaius, 47; his expedition against the Parthians, 305; betrayed by Artavasdes the king of the Armenians, 307; appointed Archelaüs king of Cappadocia, 371; gave part of Heracleia to Adiatorix, 379; gave over Amisus to kings, 395; aided by Cleon, 497; had high regard for Polemon, 511
Aorsi, the, 191, 243
Apameia Cibotus, 505, 509, 515
Aparni, the, 249, 261, 275
Apollo Selinuntius, 7; Marmarinus, 11; the Actian, 25; Leucatas, 33; leader of the Muses, 95; Aegletan, 161; born in Delos, 163; Sminthian, 169; Cataonian, 357
Apollodorus (see *Dictionary* in vol. i), on Samos, 39; on Asteria, 51; on Mts. Chalcis and Taphiassus, 63; on the Erysicheians, 65; on the Hyantes, 81; praised by Strabo, 83; on the dimensions of Crete, 123; on the Ochus River, 255; on the distance from the Caspian Gates to Rhagae and Hecatompylus, 273; on the Greeks as masters of Ariana and India, 279; on Bactriana, 281; on the distance from Hyrcania to Artemita, 291; on the Halizoni, 413, 415; on Enetê, 417; on the fabrications of Homer, 423
Apollonides (see vol. iii, p. 234, footnote 2), on Atropatian Media, 303; on certain insects in Armenia, 323
Arabians, settlers in Euboea, 13
Arachosia, 277
Aracynthus, Mt., 27
Aragus River, the, 217, 221
Aratus of Soli (fl. 270 B.C.), author of the astronomical poems *Phaenomena* and *Diosemeia* and also a work entitled *Catalepton* (see p. 167); on Dictê, 139; on Pholegandros, 161; on Gyaros, 167
Araxene Plain, the, 321, 335
Araxes River, the, 187, 225, 265, 321, 327, 335
Araxus, Cape, 57
Archardeüs River, the, 243
Archelaüs, father of Archelaüs the priest of Comana; honoured by Sulla and the Roman Senate, 437
Archelaüs, given kingdom of Cappadocia by Antony (36 B.C.), 345, 349; an eleventh prefecture assigned to his predecessors, 349; spent most of his time at Elaeussa, 361; miners of, 369; appointed king by Antony, 371; second husband of Pythodoris, 427
Archelaüs, priest of Comana, son of the Archelaüs who was honoured by Sulla and the Roman Senate, 435
Archemachus the Euboean (see footnote on p. 84), on the Curetes, 85
Archilochus the Iambic poet (fl. about 685 B.C.), born in Paros, 169; robbed of shield by one of the Saii, 403
Argaeus Mt., the, 361, 363
Aria, 277
Ariarathres (d. 220 B.C.), " first man to be called king of the Cappa-

A PARTIAL DICTIONARY OF PROPER NAMES

docians," 347; dammed the Melas and Carmalus Rivers, 363
Ariobarzanes, king of Cappadocia 93–63 B.C.; chosen by the people, 371
Aristion, tyrant of Athens (see footnote 4 on p. 167); caused revolt of Delos, 167
Aristobulus of Cassandreia, served under Alexander the Great in Asia and wrote a history of his life; on the trees of Hyrcania, the Oxus River, and on imports from India, 253; on the Polytimetus River, 285
Ariston, the peripatetic philosopher, 169
Aristotle of Chalcis (apparently flourished in fourth century B.C.); author of a work on Euboea; on the colonisation of Euboea, 5; on that of Italy and Sicily by the Chalcidians, 13
Aristotle of Stageira (384–322 B.C.), prince of ancient philosophers; died at Chalcis, 19
Arius River, the, 277
Armenia, 187, 209, 231, 301, 307, 321
Armenia, Lesser, 423, 427
Armenians, the, 185; castes among, 221; customs of, 313; ancient story of, 333; clothing of, 333; worshippers of Anaïtis, 341
Armenus, companion of Jason, called eponymous hero of Armenia, 231, 333
Arsaces a son of Pharnaces, captured and slain by Polemon I, 446
Arsaces the Scythian (or Bactrian), king of Parthia (about 250 B.C.), 275; fled from Seleucus Callinicus (king of Syria 246–226 B.C.), 269
Arsenê (Thopitis), Lake, 327
Arsinoê (Canopa), founded by Arsinoê, wife of Ptolemy II, 65
Artanes (Arsaces? or Armenias?), the Sophenian, descendant of Zariadris, 337
Artavasdes, king of the Armenians, betrayed Antony, 307; treasury of, 325; cavalry of, 331; paraded in chains, imprisoned, and slain, 339
Artaxata (Artaxiasata), 321, 325
Artaxias, general of Antiochus the Great, and king, enlarged Armenia, 323, 337
Artemidorus (see *Dictionary* in vol. ii), on Mt. Chalcis, or Chalcia, 63; on the perimeter of Crete, 123; enumerates fifteen Cyclades, 165; on the Cercetae and other peoples in Asia Minor, 207; on the cities of the Pisidians, 481
Artemis Amarynthia, 17; Perasian, 359; Tauropolus, 353
Artemita, 291
Asander (ruler of the Bosporus, by act of Augustus), 201
Asclepiades the physician of Prusa (fl. about 50 B.C.), 467
Asia, description of, 183; twofold meaning of term, 347
Aspionus, satrapy of, 281
Aspurgiani, the, attacked by Polemon, 201
Asteria (Asteris), 51
Astyages (reigned 594–559 B.C.), the last king of Media, 307
Ateporix, Galatian dynast, 443
Athena, the Nedusian, 169
Athenians, the, hospitable to things foreign, 109
Athenocles of Athens, colonised Amisus, 395
Atropates, satrap of Media under Alexander, 303

B

Babylon, 319, 329
Bacchides, commander of garrison at Sinopê, 391
Bacchylides, the poet, native of Iulis in Ceos, 169
Bactra (Zariaspa), 271, 281
Bactriana, 263, 275
Bagadania, 367
Baris, temple of, 335
Bata, village and harbour, 205
Baton (fl. second half of third century B.C.), born at Sinopê and the author of *The Persica*, 391
Bebryces, the, 375
Berecyntes, the, worshippers of Rhea, 99
Bessus, Persian who escaped from Alexander, fleeing to the Chorasmii, 269, 289
Billarus, the globe of, 391
Bion (fl. about 250 B.C.), the Borysthenite philosopher, emulated by Ariston, 169

A PARTIAL DICTIONARY OF PROPER NAMES

A PARTIAL DICTIONARY OF PROPER NAMES

A PARTIAL DICTIONARY OF PROPER NAMES

Dulichium (Dolicha), 35, 47, 55
Dyteutus, appointed priest of Comana by Augustus, 437

E

Ecbatana, 303, 307, 309, 335
Echinades Islands, the, 55
Eisadici, the, 241
Elaeussa, 361
Elixus River, the, 169
Ellops, the son of Ion, founder of Ellopia in Euboea, 7
Elymaei, the, 301, 309
Emoda, Mt., 259
Eneti, the, 381
Enyo, goddess of war, temple of, 351; priesthood of, 357
Ephesus, founded by the Amazons, 237
Ephors, the Spartan, 151
Ephorus (see *Dictionary* in vol. i), on names of cities of Acarnanians, 33; denies that they joined Trojan expedition, 71; makes Acarnania subject to Alcmaeon, 73; on the Curetes, 75; on the kinship of the Eleians and Aetolians, 79; on Minos, 131; on the good laws of Crete, 133; on the hundred cities in Crete, 143; on the Cretan constitution, 145; on the Cretan institutions, 147, 153; on the reason why Lycurgus went to Crete, 151; says Cytorum was named after Cytorus the son of Phrixus, 387; on the abode of the Amazons, 405
Epimenides the wizard (see footnote 2 on p. 141), native of Phaestus, 141
Erasistratus (fl. in first half of third century B.C.), the physician, born in Ceos, 169
Eratosthenes (see *Dictionary* in vol. i); on the distance from Cyrenaea to Criumetopon, 125; on the "Caspius" (Caucasus), 209; on the circuit of the Caspian Sea, 245; on the Oxus River, 253; says Alexander built fleet out of firwood from India, 257; on the abodes of various Asiatic peoples, and on various distances in Asia, 269; author of divisions of Asia, 301; wrongly writes "Thermodon" River instead of "Lycus," 327
Eretria, 11, 15, 17
Euboea (Macris), description of, 3; subject to earthquakes, 15
Euboeans, the, as soldiers, 21
Eucratides (king of Bactriana from about 181 to 161 B.C.), 275, 281
Eudoxus of Cnidus (see *Dictionary* in vol. i), praised by Polybius, 81; on Crete, 121; describes a "marvellous" place in Hyrcania, 257; called foister of names, 405; on certain fish in Paphlagonia, 453
Eumenes of Cardia, after death of Alexander (323 B.C.) became ruler of Cappadocia, Paphlagonia and Pontus; long held out against a siege by Antigonus, 359
Eumenes the king of Pergamum (see note on p. 506), 507
Eupatoria (Magnopolis), 429
Euphorion (see *Dictionary* in vol. iv), on the Mysian Ascanius, 465
Euphrates, the, 297, 317; course of, 319, 321, 329, 351
Euripides, on the worship of Dionysus and Rhea, 101, 113; on "things divine," 213; on a strange custom of the barbarians of the Caucasus, 291
Euripus, the, 5, 13
Euthydemus, caused revolt of Bactriana, 275
Evenus (Lycormas), the River, 29, 63

G

Gabinius (consul 58 B.C., proconsul to Syria 57 B.C.), 437
Galatia (Gallo-Grecia), 469
Galatians, the, 467, 485, 495
Gallus River, the, 379
Gargarians, the, 233
Garsauira, 359, 367
Gazelonitis, 393, 417, 443
Gelae, the, 249, 259
Gelon, tyrant of Syracuse (d. 478 B.C.), drove Chalcidians out of Sicilian Euboea, 23
Geraestus, 3, 11
Glaucus River, the, 211, 219
Gogarenê, 321, 325
Gordium (Juliopolis), 497
Gordyaean Mts., the, 299

A PARTIAL DICTIONARY OF PROPER NAMES

Gorgus, son of Cypselus the tyrant of Corinth, 33
Gortyna, 127, 137
Gyaros, the island, visited by Strabo, 165

H

Halizones (Halizoni), the, 403
Halys River, the, 189, 345, 383; origin of the name, 391
Hannibal, the Carthaginian, founder of Artaxata in Armenia, 325; welcomed by Prusias, 457
Hecataeus of Miletus (see *Dictionary* in vol. i); approved by Demetrius, 407, 413; identifies Enetê with Amisus, 417
Helius (the Sun), worshipped as god, 229, 265
Hellanicus (see *Dictionary* in vol. i), on the Aetolian cities, 29; on Cephallenia, 49; author of *Phoronis*, on the Curetes, 111; called untrustworthy, 247; foister of names, 405
Heniochi, the, 191, 203, 205, 207
Heracleia in Pontus, 273, 371, 373, 379
Heracleides the Platonic philosopher, born at Heracleia in Pontus, 371
Heracles, destroyer of Oechalia, 17; married Deïaneira, 57; made expedition to India, 239; sailed on the Argo, 457
Hermonassa, 199, 399
Herodotus, on the destruction of Old Eretria, 17; on the long hair of Leonidas' soldiers, 89; on the Cabeiri, 115; called untrustworthy, 247; on the Araxes River, 335; on prostitution of Lydian women, 341; on "the country this side the Halys River," 347; calls Egypt "the gift of the Nile," 357; by "Syrians" means "Cappadocians," 383; foister of names, 405; on the Termilae (Milyae), 491
Hesiod, on the origin of the Satyrs and Curetes, 111
Hieron, benefactor of Laodiceia, his native city, 511
Hieronymus (see foot-note 2 on p. 123), on the dimensions of Crete, 123
Hippaïtae (Cercitae), the, 401
Hippus River, the, 211, 217
Histiaeotis (Hestiaeotis), 7
Homer, 33, 35, 39, 41, 43, 47, 49, 65, 75, 127, 129, 137, 153 ("Homer, who was living in Chios"), 161 (reputed to have been buried in the isle of Ios), 357, 381, 385, 405, 411, 417, 419, 487, 495
Homonadeis, the, 479, 481
Hortensius, Quintus (consul 69 B.C.), married Marcia, wife of Cato, 273
Hydarnes (one of the Seven Persians who conspired against the Magi in 521 B.C.), 337
Hylas, companion of Heracles on the Argo and worshipped by the Prusians, 457
Hypsicrates, the historian, on the Amazons, 233
Hyrcania, 249, 261, 293
Hyrcanian (Caspian) Sea, the, 189

I

Iaxartes River, the, 269, 281, 287
Iberia, 187, 207, 217
Iberians, the; origin of the name, 215; description of, 219
Icarius, father of Penelope, settler in Acarnania, 69
Iconium, 475
Ida, Mt., in Crete, 125
Imaüs (or Imaïus), Mt., 259, 289
India, 271, 289
Indus River, the, 277
Ios, the island, where Homer was reputed to have been buried, 161
Iphigeneia, 353
Iris River, the, 395, 429
Isaura (Old and New), 475
Issus, 289
Ithaca, 39, 41

J

Jason, expedition of, 211, 231, 239, 315, 333, 335, 391

L

Labienus, in command of Asia (40–39 B.C.), 497
Laertes, father of Odysseus, 67
Lagetas, maternal ancestor of Strabo, 135

A PARTIAL DICTIONARY OF PROPER NAMES

373, 385; born and reared at Sinopê, 387; adorned Amisus, 395; master of Colchis and other places, 425; fled to Pontus, 425; kept his treasures at Kainon Chorion, 431; conquered Nicomedes, 449; besieged Laodiceia, 511; restored Apameia, 515
Mithridates of Pergamum, contemporary of Strabo, robbed oracle of Phrixus, 213
Mithridatic War, the, 449, 501
Moaphernes, uncle of Strabo's mother, governor of Colchis, 213, 433
Morimenê, 349, 359, 367
Moschian Mts., the, 209, 299, 319, 401
Mosynoeci, the, 325, 401
Muses, the, worship of, 95
Mygdonians, the, 319, 499
Myrtuntium, the salt-lake, 61
Mysia, 459, 487, 505
Mysians, the, 375, 405, 491

N

Nabiani, the, 243
Naxos, 169
Nearchus (see *Dictionary* in vol. i), on the tribes in Greater Media, 309
Neleus River, the, 21
Neroassus (Nora), 357
Nesaea, a district of Hyrcania, 253, 261
Nesaean horses, the, 311, 331
Nibarus, Mt., 321, 335
Nicaea (Antigonia). 463
Nicator, Seleucus (king of Syria 312–280 B.C.), founded Heracleia, 309
Nicomedes the Bithynian, 449
Niobê, 487, 519
Niphates, Mt., 299, 305, 321
Nisibis, 299, 319
Nisyros, 177
Nora (Neroassus), 357

O

Ochê, Mt., 7
Ochus River, the, 253, 259, 285
Odrysses River, the, 407
Odysseus, leader of the Cephallenians, 49
Oechalia, destroyed by Heracles, 17
Olynthus, 13, 29, 65
Omanus, the Persian deity, 263
Onesicritus (see *Dictionary* in vol. i), on the traits of the Bactrians, 281
Orestes, 353, 359
Oreus (formerly Histiaea), 7, 9
Orontes, descendant of Hydarnes, took Armenia, 337
Orpheus, 109
Orphic rites, beginning of the, 105 121
Oxeiae (Thoae) Islands, the, 55
Oxus River, the, 253, 269, 281, 287
Oxyartes, 283
Oxylus, son of Haemon and leader of the Heracleidae, 77

P

Palaephatus (author of a work now extant *On Incredible Things*); opinions of, approved by Demetrius, 407; on the Amazons, 409, 413
Palaerus, city in Acarnania, 61
Panticapaeum, 197; metropolis of the European Bosporians, 199
Panxani, the, 243
Paphlagonia, 381
Paphlagonians, the, 345, 383
Parachoathras, Mt., 259, 269, 299, 319
Paraetaceni, the, 301, 309
Parmenion, general under Philip and Alexander; builder of the temple of Jason at Abdera, 333
Parnassus, Mt., 25
Paropamisus, Mt., 259
Paros, birthplace of Archilochus, 169
Parthenius River, the, 377, 381
Parthia, 271, 275
Parthians, the, 185, 259
Paryadres Mountains, the, 209, 299, 319, 401, 429
Patmos, the isle, 173
Patrocles (see *Dictionary* in vol. i), on the Cadusii and the Caspian Sea, 251; on the Oxus River, 253; on the Iaxartes River, 287; on the possibility of sailing from India to Hyrcania, 289
Pelasgians, the, 125, 377, 491
Penthilus, son of Orestes, 13
Perrhaebians, the, 25
Persians, the, customs of, 313, 495
Pessinus, 471, 505
Phaedrus, the Athenian general, destroyer of Styra, 11

A PARTIAL DICTIONARY OF PROPER NAMES

Phaestus, 141
Phanagoreia (Phanagoreium), metropolis of the Asiatic Bosporians, 199
Phanaroea, 395, 427
Pharnaces, ruler of the Bosporus, 201, 243; robbed oracle of Phrixus, 213; subjugated Sinopê, 387; besieged Amisus, 395
Pharnacia, 399, 401, 427
Pharos, "out in the open sea," according to Homer, 357
Phasis River, the, 211, 219, 327
Pherecydes of Leros (see foot-note 2 on p. 171), on Dulichium, 49; on the Cyrbantes and the Cabeiri, 115
Pherecydes of Syros (see *Dictionary* in vol. i), 171
Philadelphia, 515
Philetaerus, founder of the family of Attalic kings, born at Tieium, 381
Philip, son of Demetrius and father of Perseus, rased Cius, 457
Philip II (father of Alexander the Great), outraged the cities subject to Olynthus, 13
Philistides, tyrant under Philip, 7
Phocylides the gnomic poet (b. 560 B.C.), on the Lerians, 173
Pholegandros, by Aratus called "Iron" Island, 161
Phrixus, expedition of, 211; oracle of, 213; city of (now Ideëssa), 215
Phrygia, 487
Phrygia, Greater, 485, 505
Phrygia, Lesser (see Phrygia Epictetus), 487
Phrygia Catacecaumenê, 515
Phrygia Epictetus (Lesser Phrygia), 455, 457, 459, 505
Phrygia Paroreia, 507
Pindar (see *Dictionary* in vol. iii), on the worship of Dionysus and Rhea, 99; on the Isle of Delos, 183; says that the Amazons swayed a Syrian army, 383
Pindus, Mt., 23
Pisidians, the, 185
Pissuri, the, 261
Plato, called philosophy music, 95; on the Bendideian rites, 109; on the good laws of Crete, 133
Polemon I (see foot-note on p. 193), sacked Tanaïs, 193; attacked the Aspurgiani, 201; got Colchis, 213; husband of Pythodoris, 427; son of Zeno the rhetorician and highly esteemed by Antony and Augustus, 511
Polybius (see *Dictionary* in vol. i), praises Ephorus, 81
Polycleitus of Larissa, author of a history of Alexander the Great; on the Caspian Sea, 255
Pompey the Great, friend of Poseidonius, 187; in Armenia and Iberia, 221; fought the Albanians, 227; accompanied by Theophanes, 233; enlarged Zela, 263; imposed tribute upon Tigranes, 331; took over Pontus, 373; presented territories to Deïotarus, 393; his army partly slaughtered by the Heptacomitae, 401; enlarged Eupatoria, calling it Magnopolis, 429; dedicated treasures of Mithridates in Capitolium, 431; successor of Leucullus in Asia, 435, 471; appointed Archelaus priest of Comana, 435; founded the city Neapolis in Phazemonitis, 443; his conference with Leucullus at Danala (Podanala?), 471
Pontus (Cappadocia Pontica), 349, 371, 385
Poseidonius (see *Dictionary* in vol. i), praised by Strabo, 83; on the width of the isthmus between Colchis and the mouth of the Cyrus River, of that between Lake Maeotis and the Ocean, and of that between Pelusium and the Red Sea, 187; friend of Pompey, 187; wrote history of Pompey, 189; on the earthquakes round Rhagae, 273; on the Council of the Parthians, 277
Priam, 415
Procles, founder of Sparta as metropolis, 149
Prometheus Bound, 239
Psillis River, the, 379
Ptolemy Auletes, father of Cleopatra, banished by the Egyptians, 437
Ptolemy Philadelphus, husband of his sister Arsinoê, 65
Ptolemy Philopator (reigned 222–205 B.C.), began a wall round Gortyna, 137
Publius Servilius Isauricus (contemporary of Strabo), subjugator of Isaura, 475
Pylaemenes (hero in Trojan war),

R

S

T

A PARTIAL DICTIONARY OF PROPER NAMES

PRINTED IN GREAT BRITAIN BY RICHARD CLAY AND COMPANY, LTD., BUNGAY, SUFFOLK.

THE LOEB CLASSICAL LIBRARY

VOLUMES ALREADY PUBLISHED

Latin Authors

AMMIANUS MARCELLINUS. Translated by J. C. Rolfe. 3 Vols.
APULEIUS: THE GOLDEN ASS (METAMORPHOSES). W. Adlington (1566). Revised by S. Gaselee.
ST. AUGUSTINE: CITY OF GOD. 7 Vols. Vol. I. G. H. McCracken. Vol. VI. W. C. Greene.
ST. AUGUSTINE, CONFESSIONS OF. W. Watts (1631). 2 Vols.
ST. AUGUSTINE, SELECT LETTERS. J. H. Baxter.
AUSONIUS. H. G. Evelyn White. 2 Vols.
BEDE. J. E. King. 2 Vols.
BOETHIUS: TRACTS and DE CONSOLATIONE PHILOSOPHIAE. Rev. H. F. Stewart and E. K. Rand.
CAESAR: ALEXANDRIAN, AFRICAN and SPANISH WARS. A. G. Way.
CAESAR: CIVIL WARS. A. G. Peskett.
CAESAR: GALLIC WAR. H. J. Edwards.
CATO: DE RE RUSTICA; VARRO: DE RE RUSTICA. H. B. Ash and W. D. Hooper.
CATULLUS. F. W. Cornish; TIBULLUS. J. B. Postgate; PERVIGILIUM VENERIS. J. W. Mackail.
CELSUS: DE MEDICINA. W. G. Spencer. 3 Vols.
CICERO: BRUTUS, and ORATOR. G. L. Hendrickson and H. M. Hubbell.
[CICERO]: AD HERENNIUM. H. Caplan.
CICERO: DE ORATORE, etc. 2 Vols. Vol. I. DE ORATORE, Books I. and II. E. W. Sutton and H. Rackham. Vol. II. DE ORATORE, Book III. De Fato; Paradoxa Stoicorum; De Partitione Oratoria. H. Rackham.
CICERO: DE FINIBUS. H. Rackham.
CICERO: DE INVENTIONE, etc. H. M. Hubbell.
CICERO: DE NATURA DEORUM and ACADEMICA. H. Rackham.
CICERO: DE OFFICIIS. Walter Miller.
CICERO: DE REPUBLICA and DE LEGIBUS; SOMNIUM SCIPIONIS. Clinton W. Keyes.

Cicero: De Senectute, De Amicitia, De Divinatione. W. A. Falconer.

Cicero: In Catilinam, Pro Flacco, Pro Murena, Pro Sulla. Louis E. Lord.

Cicero: Letters to Atticus. E. O. Winstedt. 3 Vols.

Cicero: Letters to His Friends. W. Glynn Williams. 3 Vols.

Cicero: Philippics. W. C. A. Ker.

Cicero: Pro Archia Post Reditum, De Domo, De Haruspicum Responsis, Pro Plancio. N. H. Watts.

Cicero: Pro Caecina, Pro Lege Manilia, Pro Cluentio, Pro Rabirio. H. Grose Hodge.

Cicero: Pro Caelio, De Provinciis Consularibus, Pro Balbo. R. Gardner.

Cicero: Pro Milone, In Pisonem, Pro Scauro, Pro Fonteio, Pro Rabirio Postumo, Pro Marcello, Pro Ligario, Pro Rege Deiotaro. N. H. Watts.

Cicero: Pro Quinctio, Pro Roscio Amerino, Pro Roscio Comoedo, Contra Rullum. J. H. Freese.

Cicero: Pro Sestio, In Vatinium. R. Gardner.

Cicero: Tusculan Disputations. J. E. King.

Cicero: Verrine Orations. L. H. G. Greenwood. 2 Vols.

Claudian. M. Platnauer. 2 Vols.

Columella: De Re Rustica. De Arboribus. H. B. Ash, E. S. Forster and E. Heffner. 3 Vols.

Curtius, Q.: History of Alexander. J. C. Rolfe. 2 Vols.

Florus. E. S. Forster; and Cornelius Nepos. J. C. Rolfe.

Frontinus: Stratagems and Aqueducts. C. E. Bennett and M. B. McElwain.

Fronto: Correspondence. C. R. Haines. 2 Vols.

Gellius, J. C. Rolfe. 3 Vols.

Horace: Odes and Epodes. C. E. Bennett.

Horace: Satires, Epistles, Ars Poetica. H. R. Fairclough.

Jerome: Selected Letters. F. A. Wright.

Juvenal and Persius. G. G. Ramsay.

Livy. B. O. Foster, F. G. Moore, Evan T. Sage, and A. C. Schlesinger and R. M. Geer (General Index). 14 Vols.

Lucan. J. D. Duff.

Lucretius. W. H. D. Rouse.

Martial. W. C. A. Ker. 2 Vols.

Minor Latin Poets: from Publilius Syrus to Rutilius Namatianus, including Grattius, Calpurnius Siculus, Nemesianus, Avianus, and others with "Aetna" and the "Phoenix." J. Wight Duff and Arnold M. Duff.

Ovid: The Art of Love and Other Poems. J. H. Mozley.

Ovid: Fasti. Sir James G. Frazer.

Ovid: Heroides and Amores. Grant Showerman.

Ovid: Metamorphoses. F. J. Miller. 2 Vols.

Ovid: Tristia and Ex Ponto. A. L. Wheeler.

Persius. Cf. Juvenal.

Petronius. M. Heseltine; Seneca: Apocolocyntosis. W. H. D. Rouse.

Plautus. Paul Nixon. 5 Vols.

Pliny: Letters. Melmoth's Translation revised by W. M. L. Hutchinson. 2 Vols.

Pliny: Natural History. H. Rackham and W. H. S. Jones. 10 Vols. Vols. I.–V. and IX. H. Rackham. Vols. VI. and VII. W. H. S. Jones.

Propertius. H. E. Butler.

Prudentius. H. J. Thomson. 2 Vols.

Quintilian. H. E. Butler. 4 Vols.

Remains of Old Latin. E. H. Warmington. 4 Vols. Vol. I. (Ennius and Caecilius.) Vol. II. (Livius, Naevius, Pacuvius, Accius.) Vol. III. (Lucilius and Laws of XII Tables.) (Archaic Inscriptions.)

Sallust. J. C. Rolfe.

Scriptores Historiae Augustae. D. Magie. 3 Vols.

Seneca: Apocolocyntosis. Cf. Petronius.

Seneca: Epistulae Morales. R. M. Gummere. 3 Vols.

Seneca: Moral Essays. J. W. Basore. 3 Vols.

Seneca: Tragedies. F. J. Miller. 2 Vols.

Sidonius: Poems and Letters. W. B. Anderson. 2 Vols.

Silius Italicus. J. D. Duff. 2 Vols.

Statius. J. H. Mozley. 2 Vols.

Suetonius. J. C. Rolfe. 2 Vols.

Tacitus: Dialogues. Sir Wm. Peterson. Agricola and Germania. Maurice Hutton.

Tacitus: Histories and Annals. C. H. Moore and J. Jackson. 4 Vols.

Terence. John Sargeaunt. 2 Vols.

Tertullian: Apologia and De Spectaculis. T. R. Glover. Minucius Felix. G. H. Rendall.

Valerius Flaccus. J. H. Mozley.

Varro: De Lingua Latina. R. G. Kent. 2 Vols.

Velleius Paterculus and Res Gestae Divi Augusti. F. W. Shipley.

Virgil. H. R. Fairclough. 2 Vols.

Vitruvius: De Architectura. F. Granger. 2 Vols.

Greek Authors

ACHILLES TATIUS. S. Gaselee.

AELIAN: ON THE NATURE OF ANIMALS. A. F. Scholfield. 3 Vols.

AENEAS TACTICUS, ASCLEPIODOTUS and ONASANDER. The Illinios Greek Club.

AESCHINES. C. D. Adams.

AESCHYLUS. H. Weir Smyth. 2 Vols.

ALCIPHRON, AELIAN, PHILOSTRATUS: LETTERS. A. R. Benner and F. H. Fobes.

ANDOCIDES, ANTIPHON, Cf. MINOR ATTIC ORATORS.

APOLLODORUS. Sir James G. Frazer. 2 Vols.

APOLLONIUS RHODIUS. R. C. Seaton.

THE APOSTOLIC FATHERS. Kirsopp Lake. 2 Vols.

APPIAN: ROMAN HISTORY. Horace White. 4 Vols.

ARATUS. Cf. CALLIMACHUS.

ARISTOPHANES. Benjamin Bickley Rogers. 3 Vols. Verse trans.

ARISTOTLE: ART OF RHETORIC. J. H. Freese.

ARISTOTLE: ATHENIAN CONSTITUTION, EUDEMIAN ETHICS, VICES AND VIRTUES. H. Rackham.

ARISTOTLE: GENERATION OF ANIMALS. A. L. Peck.

ARISTOTLE: METAPHYSICS. H. Tredennick. 2 Vols.

ARISTOTLE: METEROLOGICA. H. D. P. Lee.

ARISTOTLE: MINOR WORKS. W. S. Hett. On Colours, On Things Heard, On Physiognomies, On Plants, On Marvellous Things Heard, Mechanical Problems, On Indivisible Lines, On Situations and Names of Winds, On Melissus, Xenophanes, and Gorgias.

ARISTOTLE: NICOMACHEAN ETHICS. H. Rackham.

ARISTOTLE: OECONOMICA and MAGNA MORALIA. G. C. Armstrong; (with Metaphysics, Vol. II.).

ARISTOTLE: ON THE HEAVENS. W. K. C. Guthrie.

ARISTOTLE: ON THE SOUL. PARVA NATURALIA. ON BREATH. W. S. Hett.

ARISTOTLE: ORGANON—Categories, On Interpretation, Prior Analytics. H. P. Cooke and H. Tredennick.

ARISTOTLE: ORGANON—Posterior Analytics, Topics. H. Tredennick and E. S. Foster.

ARISTOTLE: ORGANON—On Sophistical Refutations.
On Coming to be and Passing Away, On the Cosmos. E. S. Forster and D. J. Furley.

ARISTOTLE: PARTS OF ANIMALS. A. L. Peck; MOTION AND PROGRESSION OF ANIMALS. E. S. Forster.

Aristotle: Physics. Rev. P. Wicksteed and F. M. Cornford. 2 Vols.
Aristotle: Poetics and Longinus. W. Hamilton Fyfe; Demetrius on Style. W. Rhys Roberts.
Aristotle: Politics. H. Rackham.
Aristotle: Problems. W. S. Hett. 2 Vols.
Aristotle: Rhetorica Ad Alexandrum (with Problems. Vol. II.). H. Rackham.
Arrian: History of Alexander and Indica. Rev. E. Iliffe Robson. 2 Vols.
Athenaeus: Deipnosophistae. C. B. Gulick. 7 Vols.
St. Basil: Letters. R. J. Deferrari. 4 Vols.
Callimachus: Fragments. C. A. Trypanis.
Callimachus, Hymns and Epigrams, and Lycophron. A. W. Mair; Aratus. G. R. Mair.
Clement of Alexandria. Rev. G. W. Butterworth.
Colluthus. Cf. Oppian.
Daphnis and Chloe. Thornley's Translation revised by J. M. Edmonds; and Parthenius. S. Gaselee.
Demosthenes I.: Olynthiacs, Philippics and Minor Orations. I.–XVII. and XX. J. H. Vince.
Demosthenes II.: De Corona and De Falsa Legatione. C. A. Vince and J. H. Vince.
Demosthenes III.: Meidias, Androtion, Aristocrates, Timocrates and Aristogeiton, I. and II. J. H. Vince.
Demosthenes IV.–VI.: Private Orations and In Neaeram. A. T. Murray.
Demosthenes VII.: Funeral Speech, Erotic Essay, Exordia and Letters. N. W. and N. J. DeWitt.
Dio Cassius: Roman History. E. Cary. 9 Vols.
Dio Chrysostom. J. W. Cohoon and H. Lamar Crosby. 5 Vols.
Diodorus Siculus. 12 Vols. Vols. I.–VI. C. H. Oldfather. Vol. VII. C. L. Sherman, Vols. IX. and X. R. M. Geer. Vol. XI. F. Walton.
Diogenes Laeritius. R. D. Hicks. 2 Vols.
Dionysius of Halicarnassus: Roman Antiquities. Spelman's translation revised by E. Cary. 7 Vols.
Epictetus. W. A. Oldfather. 2 Vols.
Euripides. A. S. Way. 4 Vols. Verse trans.
Eusebius: Ecclesiastical History. Kirsopp Lake and J. E. L. Oulton. 2 Vols.
Galen: On the Natural Faculties. A. J. Brock.
The Greek Anthology. W. R. Paton. 5 Vols.
Greek Elegy and Iambus with the Anacreontea. J. M. Edmonds. 2 Vols.

The Greek Bucolic Poets (Theocritus, Bion, Moschus). J. M. Edmonds.

Greek Mathematical Works. Ivor Thomas. 2 Vols.

Herodes. Cf. Theophrastus: Characters.

Herodotus. A. D. Godley. 4 Vols.

Hesiod and The Homeric Hymns. H. G. Evelyn White.

Hippocrates and the Fragments of Heracleitus. W. H. S. Jones and E. T. Withington. 4 Vols.

Homer: Iliad. A. T. Murray. 2 Vols.

Homer: Odyssey. A. T. Murray. 2 Vols.

Isaeus. E. W. Forster.

Isocrates. George Norlin and LaRue Van Hook. 3 Vols.

St. John Damascene: Barlaam and Ioasaph. Rev. G. R. Woodward and Harold Mattingly.

Josephus. H. St. J. Thackeray and Ralph Marcus. 9 Vols. Vols. I.–VII.

Julian. Wilmer Cave Wright. 3 Vols.

Lucian. 8 Vols. Vols. I.–V. A. M. Harmon. Vol. VI. K. Kilburn.

Lycophron. Cf. Callimachus.

Lyra Graeca. J. M. Edmonds. 3 Vols.

Lysias. W. R. M. Lamb.

Manetho. W. G. Waddell: Ptolemy: Tetrabiblos. F. E. Robbins.

Marcus Aurelius. C. R. Haines.

Menander. F. G. Allinson.

Minor Attic Orators (Antiphon, Andocides, Lycurgus, Demades, Dinarchus, Hypereides). K. J. Maidment and J. O. Burtt. 2 Vols.

Nonnos: Dionysiaca. W. H. D. Rouse. 3 Vols.

Oppian, Colluthus, Tryphiodorus. A. W. Mair.

Papyri. Non-Literary Selections. A. S. Hunt and C. C. Edgar. 2 Vols. Literary Selections (Poetry). D. L. Page.

Parthenius. Cf. Daphnis and Chloe.

Pausanias: Description of Greece. W. H. S. Jones. 4 Vols. and Companion Vol. arranged by R. E. Wycherley.

Philo. 10 Vols. Vols. I.–V.; F. H. Colson and Rev. G. H. Whitaker. Vols. VI.–IX.; F. H. Colson.

Philo: two supplementary Vols. (*Translation only.*) Ralph Marcus.

Philostratus: The Life of Apollonius of Tyana. F. C. Conybeare. 2 Vols.

Philostratus: Imagines; Callistratus: Descriptions. A. Fairbanks.

Philostratus and Eunapius: Lives of the Sophists. Wilmer Cave Wright.

Pindar. Sir J. E. Sandys.

Plato: Charmides, Alcibiades, Hipparchus, The Lovers, Theages, Minos and Epinomis. W. R. M. Lamb.

Plato: Cratylus, Parmenides, Greater Hippias, Lesser Hippias. H. N. Fowler.

Plato: Euthyphro, Apology, Crito, Phaedo, Phaedrus. H. N. Fowler.

Plato: Laches, Protagoras, Meno, Euthydemus. W. R. M. Lamb.

Plato: Laws. Rev. R. G. Bury. 2 Vols.

Plato: Lysis, Symposium, Gorgias. W. R. M. Lamb.

Plato: Republic. Paul Shorey. 2 Vols.

Plato: Statesman, Philebus. H. N. Fowler; Ion. W. R. M. Lamb.

Plato: Theaetetus and Sophist. H. N. Fowler.

Plato: Timaeus, Critias, Clitopho, Menexenus, Epistulae. Rev. R. G. Bury.

Plutarch: Moralia. 15 Vols. Vols. I.–V. F. C. Babbitt. Vol. VI. W. C. Helmbold. Vol. VII. P. H. De Lacy and B. Einarson. Vol. IX. E. L. Minar, Jr., F. H. Sandbach, W. C. Helmbold. Vol. X. H. N. Fowler. Vol. XII. H. Cherniss and W. C. Helmbold.

Plutarch: The Parallel Lives. B. Perrin. 11 Vols.

Polybius. W. R. Paton. 6 Vols.

Procopius: History of the Wars. H. B. Dewing. 7 Vols.

Ptolemy: Tetrabiblos. Cf. Manetho.

Quintus Smyrnaeus. A. S. Way. Verse trans.

Sextus Empiricus. Rev. R. G. Bury. 4 Vols.

Sophocles. F. Storr. 2 Vols. Verse trans.

Strabo: Geography. Horace L. Jones. 8 Vols.

Theophrastus: Characters. J. M. Edmonds. Herodes, etc. A. D. Knox.

Theophrastus: Enquiry into Plants. Sir Arthur Hort, Bart. 2 Vols.

Thucydides. C. F. Smith. 4 Vols.

Tryphiodorus. Cf. Oppian.

Xenophon: Cyropaedia. Walter Miller. 2 Vols.

Xenophon: Hellenica, Anabasis, Apology, and Symposium. C. L. Brownson and O. J. Todd. 3 Vols.

Xenophon: Memorabilia and Oeconomicus. E. C. Marchant.

Xenophon: Scripta Minora. E. C. Marchant.

IN PREPARATION

Greek Authors

Aristotle: History of Animals. A. L. Peck.

Plotinus: A. H. Armstrong.

Latin Authors

Babrius and Phaedrus. Ben E. Perry.

DESCRIPTIVE PROSPECTUS ON APPLICATION

London WILLIAM HEINEMANN LTD
Cambridge, Mass. HARVARD UNIVERSITY PRESS